Cladocera as Model Organisms in Biology

Developments in Hydrobiology 107

Series editor

H. J. Dumont

Cladocera as Model Organisms in Biology

Proceedings of the Third International Symposium on Cladocera, held in Bergen, Norway, 9–16 August 1993

Edited by

Petter Larsson and Lawrence J. Weider

Reprinted from Hydrobiologia, vol. 307 (1995)

Kluwer Academic Publishers
Dordrecht / Boston / London

Library of Congress Cataloging-in-Publication Data

Cladocera as model organisms in biology / edited by Petter Larsson &
Lawrence J. Weider.
p. cm. -- (Developments in hydrobiology ; 107)
"Papers presented at the Third International Symposium on
Cladocera, held in Bergen, Norway, 9-16 August 1993"--P.
ISBN 0-7923-3471-X (hb : alk. paper)
1. Cladocera--Congresses. I. Larsson, Petter. II. Weider,
Lawrence J. III. International Symposium on Cladocera (3rd : 1993 :
Bergen, Norway) IV. Series.
QL444.B83C57 1995
595.3'2--dc20 95-13976

ISBN 0-7923-3471-X

Published by Kluwer Academic Publishers,
P.O. Box 17, 3300 AA Dordrecht, The Netherlands.

Kluwer Academic Publishers incorporates
the publishing programmes of
D. Reidel, Martinus Nijhoff, Dr W. Junk and MTP Press.

Sold and distributed in the U.S.A. and Canada
by Kluwer Academic Publishers,
101 Philip Drive, Norwell, MA 02061, U.S.A.

In all other countries, sold and distributed
by Kluwer Academic Publishers,
P.O. Box 322, 3300 AH Dordrecht, The Netherlands.

Printed on acid-free paper

Contents

Feeding, growth, reproduction

Behaviour

Predator–prey, host–parasite interactions

Predation and competition in zooplankton communities

Preface

Cladocera as Model Organisms in Biology

This volume contains 34 papers presented at the Third International Symposium on Cladocera, held in Bergen, Norway, 9–16 August 1993. The symposium followed the tradition established by two earlier symposia on Cladocera in Budapest in 1985 and Tatranska Lomnica, Czechoslovakia in 1989.

A total of 112 participants from 30 different countries presented 76 lectures and 20 posters. In connection with the symposium, an exhibition, which presented unpublished drawings by G.O. Sars and P.E. Müller of various cladoceran species and morphotypes, was on display. A handwritten manuscript by G.O. Sars dating from 1862, which contained unpublished descriptions and illustrations of cladocerans, had been translated from Norwegian to English prior to the symposium, and was distributed to all participants.

The symposium was organized by Petter Larsson (chairman) and Ole T. Kleiven from the Institute of Zoology, University of Bergen, and Anders Hobæk from the Norwegian Institute for Water Research (NIVA). Scandinavian Travelers a/s and a number of students from the Institute of Zoology helped with the logistical arrangements. Financial support was obtained from the Norwegian Research Council for Science and Humanities, the Nordic Committee for Ecology, the International Science Foundation (USA) and the University of Bergen. Upon arrival, the County of Bergen held a reception for the participants. A mid-conference excursion was arranged to the mountains and fjords in the neighborhood of Bergen.

After an opening address from the chairman, Catherine Duigan presented a memorial talk on the scientific career of the late David G. Frey, who had been a driving force in previous cladoceran symposia. Dave was active to the very end with his pursuit of a better understanding of the taxonomy and biogeography of the chydorids. This volume contains his final direct contribution to the field, and was submitted posthumously by Sarah E. Frey. Following this talk, Marit E. Christiansen presented a lecture on the life of G.O. Sars and his significance to the field of Crustacean Systematics.

Certain organisms such as the ubiquitous laboratory rat, the drosophilid fruitflies, the nematode *C. elegans*, and the microbe *E. coli*, have served historically as appropriate model organisms for many research areas in modern biology ranging from the molecular to the population/community level. Without the knowledge gleaned from the study of such model organisms, our understanding of many biological phenomena would be lacking.

The group of small crustaceans called Cladocera has fascinated scientists since the invention of the microscope. As demonstrated in the present symposium issue, we believe that the cladocerans can serve as good model organisms in a variety of research areas. The variety of topics that can be studied using cladocerans was clearly demonstrated in the diversity of presentations that were given at the symposium. Topics included: Phylogeny and Systematics, Ecological Genetics, Zoogeography, Feeding, Growth, and Reproduction, Predatory–Prey Interactions, Population Dynamics and Life Histories, Diel-Activity, Community Ecology, Chemical Communication and Behaviour, Parasite-Host Interactions, and Paleolimnology.

In the closing session, it was decided that the next symposium on Cladocera will be held in Slovenia in 1996.

PETTER LARSSON
LARRY WEIDER

Key to the photograph from the Third International Symposium on Cladocera, 9–16 August, 1993, Bergen, Norway

1. Unknown
2. Unknown
3. J. Lüning
4. M. Boersma
5. B.J.G. Flik
6. P. Spaak
7. J. Müller
8. M.R. Miracle
9. R. Tollrian
10. Unknown
11. K. Szeroczynska
12. W. Jurasz
13. L. De Meester
14. E. Stuchlik
15. J. Green
16. G. Fryer
17. J. Ringelberg
18. N. Aladin
19. H. Ketelaars
20. N. Korovchinksy
21. J.I. Koksvik
22. K. Arbaciauskas
23. B. Faafeng
24. Z. Niewiadomski
25. Unknown
26. E.M. Lysebo
27. Unknown
28. V. Alekseev
29. T. Andrew
30. A. Herzig
31. D. Flössner
32. Unknown
33. A. Hobæk
34. B. Walseng
35. Unknown
36. H. Stibor
37. I. Trubetskova
38. M. Teschner
39. A. Sell
40. Unknown
41. W. DeMott
42. O.T. Kleiven
43. Z. Horicka
44. P. Larsson
45. A. Brancelj
46. G. Halvorsen
47. L.V. Polischuk
48. B.J. Hann
49. T. Onbe
50. W. Horn
51. Mrs Røen
52. K. Schwenk
53. J.P. Nilssen
54. H. Hansen
55. A.K.L. Schartau
56. R.L. Willey
57. I. Hudec
58. P. Ritter
59. A. Langeland
60. H. Mumm
61. D. Fuhlendorf
62. M. Walls
63. S.Repka
64. G. Dave
65. C. Lauren-Määttä
66. A. (Smith) Sartonov
67. M. Seaman
68. F. Forró
69. E.W. Roessler
70. H. Dumont
71. E. van Donk
72. S.-W. Kim
73. Mrs Onbe
74. U.I. Röen
75. O.V. Kvam
76. M. Cerny
77. D.O. Hessen
78. K. Christoffersen
79. Unknown
80. E. Bergstrand
81. K. Wong
82. W. Hollwell
83. L.J. Weider
84. T. Reede
85. Unknown
86. Unknown
87. R. Demelo
88. C. Duigan
89. C. Cotten
90. S.A. Haider
91. D. Berner
92. S.E. Frey
93. F. Margaritoria
94. L. Angatyr
95. D. Chiavelli
96. Unknown
97. J. Martinez-Jeronimo
98. M.E. Christiansen
99. M.J. Boavida

Hydrobiologia **307**: xii–xiv, 1995.
P. Larsson & L.J. Weider (eds), Cladocera as Model Organisms in Biology.

List of Participants at the 3rd International Symposium on Cladocera

Anna ABAFFY-BOTHAR, Hungarian Danube Research Station of the Hungarian Academy of Science, Javorka Sandor u. 14, 2131-Göd, Hungary

Nikolai ALADIN, Zoological Institute, University emb. 1, St. Petersburg 199034, Russia

Victor ALEKSEEV, Zoological Institute, Academy of Science, University emb. 1, St. Petersburg 199034, Russia

Tony T.E. ANDREW, Department of Biology and Biomedical Science, University of Ulster, Coleraine, CO., Derry BT52 1SA, N. Ireland, United Kingdom

Lars ANGANTYR, Miljøbiologisk Laboratorium, Beugnebjerg vej 5, DK-3050 Humlebæk, Denmark

Kestutis ARBACIAUSKAS, Institute of Ecology, Academijos Str. 2, Vilnius 2600, Lithuania

Eva BERGSTRAND, Institute of Freshwater Research, S-17011 Drottningholm, Sweden

Dorothy BERNER, Department of Biology, Temple University TU015-00, Philadelphia, PA 19122, USA

Maria José BOAVIDA, Departamento de Zoologia e Antropologia, Faculdade de Ciencias da Universidade de Lisboa, Campo Grande C2, 1700 Lisboa, Portugal

Maarten BOERSMA, Centre for Limnology, Rijksstraatweg 6, NL-3631 AC, Nieuwersluis, The Netherlands

Anton BRANCELJ, Institute of Biology, University of Lubljana, Karlovska 19, 61000 Lubljana, Slovenia

Maria Jose CARAMUJO, Departamento de Zoologia e Antropologia, Faculdade de Ciencias da Universidade de Lisboa, Campo Grande C2, 1700 Lisboa, Portugal

Martin CERNY, Department of Hydrobiology, Faculty of Science, Charles University, Vinicná 7, CS-12844 Prague, Czech Republic

Rama CHENGALATH, Canadian Museum of Nature, Ottawa K1P GP4, Canada

Deborah CHIAVELLI, Department of Biology, University of Mississippi, University, MS 38677, USA

Guttorm N. CHRISTENSEN, Norwegian College of Fishery, University of Tromsø, N-9000 Tromsø, Norway

Marit E. CHRISTIANSEN, Museum of Zoology, University of Oslo, Sarsgt. 1, N-0562 Oslo, Norway

Kirsten CHRISTOFFERSEN, University of Copenhagen, Freshwater Biology Laboratory, 51 Helsingørsg, 3400 Hillerød, Denmark

Clara COTTEN, Department of Biology, Indiana University, Jordan Hall, Bloomington, IN 47405, USA

Maria Cristina CRISPIM, Departamento de Zoologia e Antropologia, Faculdade de Ciencias da Universidade de Lisboa, Campo Grande C2, 1700 Lisboa, Portugal

Geir DAHL-HANSEN, Norwegian College of Fishery, University of Tromsø, N-9000 Tromsø, Norway

Göran DAVE, Department of Zoophysiology, University of Gothenburg, Medisinaregatan 18, S-41390 Göteborg, Sweden

Luc DE MEESTER, Laboratory of Animal Ecology, University of Ghent, K.L. Ledeganckstraat 35, B-9000 Gent, Belgium

Rita DEMELO, Department of Zoology, University of Guelph, Guelph, Ontario N1G 2W1, Canada

William R. DEMOTT, Department of Biology and Crooked Lake Biological Station, Indiana-Purdue University, Fort Wayne, IN 46805, USA

Bart DE STASIO, Biology Department, Lawrence University, Appleton, WI 54912, USA

Catherin DUIGAN, Countryside Council for Wales, Plast Penrhos, FP Bangor, Gwyn LL57 2LQ, Wales, United Kingdom

Henri DUMONT, Laboratory of Animal Ecology, University of Ghent, K.L. Ledeganckstraat 35, B-9000 Gent, Belgium

Annie DUNCAN, Department of Biology, Royal Holloway, University of London, Egham, Surrey TW20 0EX, England, United Kingdom

Fatin A.R. ELHIGZI, Institute of Zoology, University of Bergen, Allégt. 41, N-5007 Bergen, Norway

Andrzej ENGELMAYER, Department of Hydrobiology, University of Warsaw, Nowy Swiat 67, 00-046 Warsaw, Poland

Bjørn FAAFENG, Norwegian Institute for Water Research (NIVA), Box 69, Korsvoll, N-0808 Oslo, Norway

Ben J.G. FLIK, Department of Aquatic Ecology, University of Amsterdam, NL-1098 SM Amsterdam, The Netherlands

Luigi Naselli FLORES, Dipartimento di Szienzie Bot., Via Archirafi 38, Palermo, Italy

Dietrich FLÖSSNER, Institute of Ecology, University of Jena, Beuthenbergstr. 11, Jena, Germany
Laszlò FORRò, Zoological Department, Hungarian Natural History Museum, Baross u. 13, H-1088 Budapest, Hungary
Sarah Elisabeth FREY, 2625 South Smith Road, Bloomington, IN 47401, USA
Geoffrey FRYER, Elleray Cottage, Windermere, Cumbria LA23 1AW, England, United Kingdom
Dörthe FUHLENDORF, Max-Planck-Institut für Limnologie, Postfach 165, D-24302 Plön, Germany
Sabine GIESSLER, Institute of Zoology, University of Munich, Seidlstrasse 25, D-80335 München, Germany
Jim GREEN, 17 King Edwards Grove, Teddington, Middx. TW11 9LY, England, United Kingdom
Shawkat HAIDER, Institute of Zoology, University of Bergen, Allégt. 41, N-5007 Bergen, Norway
Gunnar HALVORSEN, Norwegian Institute for Nature Research (NINA), Box 1037 Blindern, N-0315 Oslo, Norway
Brenda J. HANN, Department of Zoology, University of Manitoba, Winnipeg, MB R3T 2N2, Canada
Heidi HANSEN, County Environmental Administration, N-1501 Moss, Norway
Alois HERZIG, Biological Station, A-7141 Illmitz, Austria
Dag O. HESSEN, Biological Institute, University of Oslo, Box 1027 Blindern, N-0315 Oslo, Norway
Jaana HIETALA, Department of Biology, Ecological Zoology, University of Turku, SF-20500 Turku, Finland
Anders HOBÆK, Norwegian Institute for Water Research (NIVA), Thormøhlensgt. 55, HIB, N-5008 Bergen, Norway
Werner HOLLWEDEL, Oldenburgerstrasse 16A, D-26316 Varel, Germany
Zuzanna HORICKA, Department of Hydrobiology, Faculty of Science, Charles University, Vinicná 7, CS-12844 Prague, Czech Republic
Wolfgang HORN, Sächsische Akademie der Wissenschaften zu Leipzig, Arbeitsgruppe "Limnologie von Talsperren", Neunzehnhainerstrasse 14, D-09514 Wünschendorf, Germany
Igor HUDEC, Institute for Ecology of Agricultural Landscape, Slovakian Acad. of Science, Löfflerova 10, 040 02 Kosice, Slovakia
Erik JEPPESEN, National Environmental Research Institute, Vejlsøvej 25, Box 314, DK-8600 Silkeborg, Denmark
Wojciech JURASZ, Department of Invertebrate Zoology, University of Lòdz, Banacha 12/16, 91 072 Lòdz, Poland
Henk KETELAARS, Water Storage Corporation, Brabantse Biesbosch Ltd., Petrusplaat, P.O. Box 61, NL-4250 DB Werkendam, The Netherlands
Se-Wha KIM, Department of Environment and Health, Yongin University, Yongin 449-714, Korea
Ole T. KLEIVEN, Institute of Zoology, University of Bergen, Allégt. 41, N-5007 Bergen, Norway
Michiyor KOBAYASHI, Department of Biology, Faculty of Science, Niigata University, Niigata 950-21, Japan
Sanna KOIVISTO, Department of Systems Ecology, Stockholm University, S-10691 Stockholm, Sweden
Jan Ivar KOKSVIK, The Museum, University of Trondheim, Erling Skakkesgt., N-7004 Trondheim, Norway
Nikolai KOROVCHINSKY, A.N. Severtsov Institute of Animal Evolutionary Morphology and Ecology, Russian Academy of Sciences, Leninsky Prospect 33, 117071 Moscow, Russia
Olaug Vetti KVAM, Institute of Zoology, University of Bergen, Allégt. 41, N-5007 Bergen, Norway
Arnfinn LANGELAND, Norwegian Institute for Nature Research (NINA), Tungasletta 2, N-7005 Trondheim, Norway
Petter LARSSON, Institute of Zoology, University of Bergen, Allégt. 41, N-5007 Bergen, Norway
Camilla LAURÉN-MÄÄTTÄ, Laboratory of Ecological Zoology, Department of Biology, University of Turku, SF-20500 Turku, Finland
Julia LÜNING, Department of Zoology, University of Munich, Seidlstrasse 25, D-80335 München 2, Germany
Elisabeth M. LYSEBO, Institute of Zoology, University of Bergen, Allégt. 41, N-5007 Bergen, Norway
Fiorenza MARGARITORA, Department of Animal and Human Biology, University "La Sapienza", Rome, Italy
Fernando MARTINEZ-JERONIMO, Escuela Nat. de Ciencia Biològicas, Lab. de Ecol. Apado., Postal 63-432, Mexico City, D.F. 02801, Mexico
Laura MARTINEZ TABCHE, Escuela Nat. de Ciencia Biològicas, Lab. de Ecol. Apado., Postal 63-432, Mexico City, D.F. 02801, Mexico
Dermot MCKEE, Centre for Population Biology, Imperial College at Silwood Park, Ascot, Berkshire SL5 7PY, England, United Kingdom
Maria R. MIRACLE, Ecology, Faculty of Biological Sciences, Universidad de Valencia, 46100 Burjassot, Valencia, Spain
Jakob MÜLLER, Zoological Institute, Department of Population Biology, University of Mainz, Saarstrasse 21, D-55099 Mainz, Germany
Heike MUMM, Max-Planck-Institut für Limnologie, Postfach 165, D-24302 Plön, Germany

Zdzislaw NIEWIADOMSKI, Department of Freshwater Biology and Ecology, bl. Nowa Zootechnika 315, 10-957 Olsztyn, Kortowo, Poland
Jens Petter NILSSEN, Risör, Norway
Takashi ONBE, Faculty of Applied Biological Sciences, Hiroshima University, Kagamiyama 144, Higashi-Hiroshima, Japan
Erla Björk ÖRNÓLFSDÓTTIR, Institute of Biology, State University of Iceland, Grensasvegur 12, 108 Reykjavik, Iceland
Joanna PIJANOWSKA, Department of Hydrobiology, University of Warsaw, Nowy Swiat 67, 00046 Warszawa, Poland
Leonard V. POLISHCHUK, Department of Invertebrate Zoology and White Sea Biological Station, Biology Faculty, Moscow State University, Moscow 119899, Russia
Paul PRIMICERIO, Norwegian College of Fishery, University of Tromsø, N-9000 Tromsø, Norway
Tineke REEDE, Department of Aquatic Ecology, University of Amsterdam, Kruislaan 320, NL-1098 SM Amsterdam, The Netherlands
Sari REPKA, Centre for Limnology, Rijksstraatweg 6, NL-3631 AC Nieuwersluis, The Netherlands
Joop RINGELBERG, Department of Aquatic Ecology, University of Amsterdam, Kruislaan 320, NL-1098 SM Amsterdam, The Netherlands
Pia RITTER, Institut für Hydrobiologie der TU Dresden, Mommsenstrasse 13, D-01069 Dresden, Germany
Ulrik Ib RØEN, Zoological Museum, University of Copenhagen, Univ. Parken 15, DK-2100 København Ø, Denmark
Ewald W. ROESSLER, Escuela de Postgrado, Santafe de Bogota, Colombia
Ann Kristin Lien SCHARTAU, Norwegian Institute for Nature Research (NINA), Tungasletta 2, N-7004 Trondheim, Norway
Klaus SCHWENK, Centre for Limnology, Rijksstraatweg 6, NL-3631 AC Nieuwersluis, The Netherlands
Maitland SEAMAN, Department of Zoology, University of the Orange Free State, P.O. Box 339, 9300 Bloemfontein, South Africa
Anne SELL, Institute of Hydrobiology, University of Technology, Zellescher Weg 40, D-01217 Dresden, Germany
Miroslaw SLUSARCZYK, Department of Hydrobiology, University of Warsaw, Nowy Swiat 67, 00046 Warszawa, Poland
Alison SMITH, Cold Regions Research Lab., 72 Lyme Road, Hanover, NH 03755, USA
Piet SPAAK, Centre for Limnology, Netherlands Institute of Ecology, Rijksstraatweg 6, NL-3631 AC Nieuwersluis, The Netherlands
Herwig STIBOR, Max-Planck-Institut für Limnologie, Postfach 165, D-24302 Plön, Germany
Gisela STOLPE, Max-Planck Institut für Limnologie, Postfach 165, D-24302 Plön, Germany
Raymond G. STROSS, Department of Biological Sciences, University of New York at Albany, Albany, NY 12222, USA
Evzen STUCHLIK, Department of Hydrobiology, Faculty of Science, Charles University, Vinicná 7, CS-12844 Prague, Czech Republic
Krystyna SZEROCZYNSKA, Institute of Geological Sciences, Polish Academy of Science, Al Zwirki i Wigury 93, 02-089 Warszawa, Poland
Susumu TANAKA, Department of Biology, Faculty of Education, Toyama, Japan
Martina TESCHNER, Max-Planck-Institut für Limnologie, Postfach 165, D-24302 Plön, Germany
Ralph TOLLRIAN, Max-Planck-Institut für Limnologie, Postfach 165, D-24302 Plön, Germany
Irina TRUBETSKOVA, Institute of Zoology, Belorussian Academy of Science, Minsk, Belorussia
Ellen VAN DONK, Agricultural University of Wageningen, Department of Nature Conservation, P.O. Box 8080, NL-6700 DD Wageningen, The Netherlands
Lorena VILLALOBOS, Limnologisches Institut, Univ. Konstanz, Postfach 5560, D-78464 Konstanz, Germany
Mari WALLS, University of Turku, Department of Biology, SF-20500 Turku, Finland
Bjørn WALSENG, Norwegian Institute of Nature Research (NINA), Postboks 1037, Blindern, N-0315 Oslo, Norway
Lawrence J. WEIDER, Max-Planck-Institut für Limnologie, Postfach 165, D-24302 Plön, Germany
Ruth WILLEY, Department of Biological Sciences, Univ. of Illinois at Chicago, Chicago, IL 60680, USA
Kim WONG, The Chinese University of Hong Kong, Department of Biology, Shatin NT, Hong Kong
Iris D. ZELLMER, Strasse des Friedens 3, D-06193 Wallwitz, Germany

Hydrobiologia **307**: xv–xvi, 1995.
P. Larsson & L.J. Weider (eds), *Cladocera as Model Organisms in Biology.*

An appreciation of the life and work of David G. Frey (1915–1992)

Catherine A. Duigan[1] & Clara A. Cotten[2]
[1] *Countryside Council for Wales, Ffordd Penrhos, Bangor, Gwynedd, Wales, UK*
[2] *Department of Biology, Indiana University, Bloomington, IN 47405, USA*

Introduction

The Third International Symposium on Cladocera met without one of its founding figures, Prof. David G. Frey, who passed away on 1 April 1992. Tribute should be paid to him as he was a founding member of the International Symposium on Cladocera and actively involved in all subsequent meetings. As we all appreciate, his outstanding life's work was dedicated to the study of the taxonomy and ecology of Cladocera, especially the chydorids, and he continually promoted their potential in palaeolimnological studies.

Early days

David Grover Frey was born in Hartford, Wisconsin on 10 October 1915. He attended the University of Wisconsin where he was awarded his B.A. (1936), M.A. (1938) and Ph.D. (1940). There he had an insider's view of the early development of the science of limnology in North America, now characterised as the 'Birge-Juday Era'. Chancey Juday was David Frey's nominal academic supervisor for his Ph.D. thesis. In Wisconsin, E. A. Birge carried out the earliest systematic and ecological studies of Cladocera in North America. Birge (1918) advised that 'the student of Cladocera should presume that any species is probably intercontinental'. From our viewpoint, it seems ironic that David Frey would devote several decades of research to the demolition of Birge's proposals on cladoceran biogeography and systematics. In 1940 David Frey left Wisconsin to take up an appointment as an aquatic biologist in the U.S. Fish and Wildlife Service. After service as a biologist in the US Naval Reserve he returned to the U.S. in 1947 and took up his first academic appointment at the University of North Carolina. During this time he became interested in the origin and development of the Carolina Bay Lakes which lead him into his first palaeolimnological investigations.

Indiana University

In 1950, David Frey joined the faculty of Indiana University (IU) at Bloomington as an Associate Professor. He was promoted to Professor in 1955 and remained at IU for the rest of his life, teaching courses in ornithology and limnology, supervising graduate students and carrying out an extensive research programme. As a student himself, and then as a limnology teacher, David Frey realised there was a serious lack of textbooks for students. As a descendant of German-speaking emigrants to the US, he was ideally qualified to collaborate with F. E. J. Fry on the translation of Franz Ruttner's 'Grundriss der Limnologie' (1952). During this time he developed an ambition to work with Ruttner in Europe.

Palaeolimnology

In 1953, a Guggenheim Fellowship and Fulbright Senior Fellowship gave David Frey the opportunity to teach and work with Ruttner in Austria. He analysed cores from Längsee using palynology-based techniques in an attempt to reconstruct the stratification history of the lake and the biological consequences of the onset of meromixis (Frey, 1955). Some chitinous fragments recovered from the lake sediments were not immediately identifiable. During a research period at the British Museum, J. P. Harding suggested that the remains could be cladoceran.

Subsequent palaeolimnological studies established the Cladocera as 'the diatoms of the freshwater animals' (Frey, 1961). David Frey identified this period as a turning point in his research career because he realised that the interpretation of palaeoecological data was dependent on a firm understanding of the taxonomy, modern ecology and biogeography of the Cladocera.

Non-cosmopolitanism in the chydorids

After studying the fine morphological details of European Cladocera, David Frey realised that North American chydorid species with the same names were not identical. This realisation led to the development of his hypothesis of 'non-cosmopolitanism' among chydorid species (Frey, 1986). In a series of publications he demonstrated that 'species' which were claimed to be cosmopolitan are in fact groups or complexes of morphologically similar species with each member species having a more restricted distribution than the original species or complex.

Retirement

Following retirement from teaching, David Frey embarked on an extensive tour of the southern hemisphere in search of indicators of the separation of Gondwanaland. His earlier research led him to the development of a hypothesis to account for the intercontinental distribution in the southern hemisphere of species groups with similar gross morphology but without dispersal of resting eggs. He proposed that various 'protospecies' evolved on the original land mass. With the separation of the continents, the intervening oceans became barriers to dispersal and the subsequent separate evolution of the isolates led to the development of new species that retained many of the original morphological features of the 'protospecies' (Frey, 1987a).

Prof. Frey's library and laboratory in Bloomington continued to be a focus point for Cladocera researchers. His library and Cladocera collection (over 10 000 samples) have been deposited in the National Museum of Natural History, Smithsonian Institution. His experience and authority were appreciated by the larger scientific community and as a result he held numerous positions of responsibility and received several notable awards. After thirty years of research on Cladocera, Prof. Frey (1987b) thanked the National Science Foundation for supporting 'an individual such as me, who knew in general what he wanted to do and how to do it but didn't allow sufficient time to accomplish all the research proposed'. In later days David talked about the need for a worldwide atlas of the known chydorid taxa but unfortunately David ran out of time and now we feel the loss of this scholar, mentor and friend. It is the end of the 'David G. Frey Era' of Cladocera research.

References

Birge, E. A, 1918. The water fleas (Cladocera). In: H. B. Ward & G. C. Whipple (eds), Fresh-water Biology. John Wiley & Sons, New York: 676–740.

Frey, D. G., 1955. Längsee: a history of meromixis. Mem. Ist. ital Idrobiol., Supp. 8: 141–164.

Frey, D. G., 1961. Developmental history of Schleinsee. Proc. Int. Assoc. Limnol. 14: 271–278.

Frey, D. G., 1986. The non-cosmopolitanism of the chydorid Cladocera: implications for biogeography and evolution. In: Gore, R. H. & Heck, K. L. (eds), Crustacean Biogeography, Balkema: Rotterdam: 237–256.

Frey, D. G., 1987a. North American *Chydorus faviformis* (Cladocera, Chydoridae) and the honeycombed taxa of other continents. Phil. Trans. r. Soc. Lond. B 315: 353–402.

Frey, D. G., 1987b. The taxonomy and biogeography of the Cladocera. Hydrobiologia 145 (Dev. Hydrobiol. 35): 5–17.

Hydrobiologia **307**: 1–7, 1995.
P. Larsson & L.J. Weider (eds), Cladocera as Model Organisms in Biology.

What can molecular markers tell us about the evolutionary history of *Daphnia* species complexes?

Klaus Schwenk[1], Andrea Ender[2] & Bruno Streit[2]
[1]*Netherlands Institute of Ecology, Centre for Limnology, Rijksstraatweg 6, 3631 AC Nieuwersluis, The Netherlands*
[2]*Zoologisches Institut, J. W. Goethe-Universität, Siesmayerstraße 70, 60054 Frankfurt, Germany*

Key words: Biogeography, *Daphnia longispina* complex, interspecific hybridization, phylogeny, reticulate evolution, speciation

Abstract

Despite the wealth of information on the ecology of *Daphnia* species, the systematics and phylogeny of the genus is still unresolved. The taxonomic uncertainties are based in part on the phenomenon of interspecific hybridization, which has been well documented for species of the *D. galeata/cucullata/hyalina* complexes. The occurrence of syntopic populations of up to three species and their three hybrids suggest niche differentiation, but very little is known about genetic divergence of hybridizing taxa and the potential consequences of hybridization (i.e. introgression). Since an operational species definition is necessarily based on information on the evolutionary mechanisms that result into the splitting of lineages, ecological and genetic consequences of interspecific hybridization have to be considered. In order to reveal the significance of hybridization and introgression, we combined several aspects of phylogenetic investigations within the *D. galeata/cucullata/hyalina* complexes. Furthermore we demonstrate how molecular markers contribute to an evaluation of species complexes.

Introduction

Species and speciation concepts are among the most controversial, and hence most frequently discussed, topics in evolutionary biology (for review see Otte & Endler, 1989). In contrast to the relatively high number of theoretical contributions, only a few publications address this subject empirically (e.g. Arnold *et al.*, 1991; Sturmbauer & Meyer, 1992; Gill *et al.*, 1993). However, during the last decade the amount of experimental work on speciation has increased substantially, facilitated by methods to measure genetic variation at the DNA level. In particular, sequencing techniques, amplification of DNA using the polymerase chain reaction (PCR) and DNA fingerprinting techniques have become available for population genetic research and have provided genetic information at different taxonomic levels (e.g. Arnheim *et al.*, 1990). Recent studies on DNA variation of several species have focussed on the population structure of closely related species, biogeographical patterns, and phylogenetic relationships (for review see Hewitt *et al.*, 1991). Molecular approaches in systematics have contributed significantly to our understanding of evolutionary processes (e.g. Hillis & Moritz, 1990).

Among the Cladocera, a peculiar situation is observed. Although the genus *Daphnia* is among the most intensively studied freshwater groups in terms of ecology and population biology (e.g. see Peters & de Bernardi, 1987), the phylogenetic relationships remain unresolved (Hebert *et al.*, 1989). Uncertainties are caused by high phenotypic variability of species and the occurrence of local races and hybridization between taxa. In particular, among members of the *D. longispina* complex (*D. galeata*, *D. hyalina*, *D. cucullata*, and their hybrids), ecological investigations are confronted with difficulties in identifying species (Flössner & Kraus, 1986). Since these species numerically dominate many zooplankton communities of freshwater lakes and are considered as key species in aquatic food chains, an operational definition of species seems to be of general interest. In the past few years, the *D. longispina* complex has been investigated with emphasis on interspecific hybridization and its ecological and genetic consequences (for review see Schwenk & Spaak, 1995). Mort (1991) concluded that among species of the *D. longispina* complex, interspecific hybridization is one of the most important processes influencing the genotypic and phenotypic structuring of populations. In general, hybridization among animal and plant taxa is often accompanied by introgression, the transfer of mitochondrial or nuclear genes between populations or species *via* backcrosses (see Barton & Hewitt, 1985; Harrison, 1990; Arnold, 1992). Since interspecific hybridization and subsequent introgression cause fragmentation and shuffling of genomes, they are of potential importance for processes of species formation (Harrison, 1990).

Lynch (1985) developed a speciation theory for cladocerans in order to provide an operational definition of species and a solution to the taxonomic problems. Based on existing knowledge about reproductive modes as well as ecological and genetic data among the Cladocera, he proposed a founder effect model of speciation. In general, founder effect models of speciation are based on three assumptions: (1) a dramatic reduction in population size; (2) founder individuals have high levels of genetic variability; and (3) adaptive divergence in response to a change in environmental factors (Templeton, 1980). For cladocerans, these assumptions translate into the following conditions: (a) the source populations are year-round residents in permanent lakes and sexual reproduction is rare; (b) the founder population colonizes a small, ephemeral site, or the distance between founder and source populations is high; (c) species lack an efficient transport mechanism and founders start new populations from a small number of eggs, and (d) founders face intense selection before the next bout of sexual reproduction takes place (Lynch, 1985).

Recent ecological studies among members of the *D. longispina* complex (Weider & Wolf, 1991; Weider, 1993; Spaak, 1994) and population genetic surveys (Wolf & Mort, 1986; Gießler, 1987; Wolf, 1987; Hebert *et al.*, 1989; Taylor & Hebert, 1992) give some indication about the validity of these assumptions. Populations of *D. galeata*, *D. hyalina* and *D. cucullata* exist in large permanent lakes, they reproduce mainly asexually (cyclic parthenogenesis), and dispersal is passive and probably ineffective. Thus, if a new site is colonized, the number of founders is likely to be very small. Lakes vary considerably in trophic state, presence of predators and various abiotic parameters, thus *Daphnia* species are likely to face different selective pressures, depending on the particular habitat. Furthermore, the substantial intra- and interspecific variation in certain life history traits (size at maturity, intrinsic rate of increase, *r*) is one of the necessary conditions for adaptive divergence (Weider, 1993). In addition, population genetic analyses revealed a high number of multi-locus genotypes per population (Gießler, 1987; Müller, 1993; Spaak 1994). In summary, it seems that most of the basic assumptions of the founder effect model are met. However, two questions concerning dispersal and reproductive isolation of species (what is the frequency of migration and how isolated are populations?) remain unanswered.

Unfortunately, current knowledge about the biogeography and systematics of species of the *D. longispina* complex is far too limited to rigorously test this founder speciation model. However, recent ecological and genetic investigations have broadened our knowledge about species of the *D. longispina* complex. Ecological research has focused on fitness comparisons between hybrids and parentals, selection processes, and population dynamics (Weider, 1993; Spaak, 1994). Genetic studies using molecular markers, have enabled us to discriminate among species, estimate genetic divergence and infer biogeographic patterns. This essay is intended to review and to combine findings on mitochondrial and nuclear DNA variation in order to (i) analyze consequences of interspecific hybridization for phylogenetic relationships and speciation, and (ii) to give an example of how molecular methods can contribute to an analysis of species complexes.

Interspecific hybridization

Hybrids produced by *D. galeata*, *D. hyalina* and *D. cucullata* have been reported from several sites in Central and Western Europe (Northern Germany: Wolf & Mort, 1986, Bavarian Lakes: Gießler, 1987, Bohemia: Hebert *et al.*, 1989; Western Germany: Müller, 1993; Eastern Germany: Flössner, 1993; The Netherlands: Spaak & Hoekstra, 1993). Lakes often differ in the composition of species and hybrids, harbouring e.g., all three species and their interspecific hybrids, or only one of the three species and one hybrid, or only one hybrid. Furthermore, if species and hybrid populations occur syntopically, they show different seasonal fluctuations in abundances (Wolf, 1987; Spaak & Hoekstra, 1993; Müller & Seitz, 1994). Due to differences in environmental factors among habitats, such as predation or food quality, spatial variation of hybrids and species have been observed (e.g. Weider, 1993). Investigations of species and hybrid compositions in various European lake districts indicate that hybridization is a general phenomenon within the *D. longispina* complex. On average, hybrids occur in ~30% of the lakes investigated (Schwenk, unpublished results).

Allozyme surveys of species of the *D. longispina* complex have revealed backcrossed individuals at relatively low frequencies (Wolf & Mort, 1986; Hebert *et al.*, 1989), but a few populations consisted of only backcrosses (Hebert *et al.*, 1989; Taylor & Hebert, 1992). However, the general opinion has been that backcrosses are rare among populations of the *D. longispina* complex (Wolf & Mort, 1986; Hebert *et al.*, 1989; Taylor & Hebert, 1992). These results are likely to be influenced by the use of a very limited number of diagnostic allozyme loci. Two species-specific markers, the number used in most studies, will lead to an underestimation of backcrosses of 50%, because 25% of the backcross genotypes will exhibit the same pattern as the F_1-hybrid and 25% will exhibit the same pattern as one of the parental genotypes (Nason & Ellstrand, 1993). Recent approaches have attempted to overcome limitations of morphological and allozyme analyses such as the low number of species-specific markers and the problem of distinguishing between homology and analogy for phylogenetic analyses (Ender, 1993; Ender & Schierwater, 1993; Schierwater *et al.*, 1993; Schwenk, 1993; Taylor & Hebert, 1993; Streit *et al.*, 1994).

Nuclear and mitochondrial DNA markers

Patterns of interspecific hybridization were examined by studying both nuclear and mitochondrial DNA variation. RAPD analysis (Welsh & McClelland, 1990; Williams *et al.*, 1990) was used to reveal nuclear genetic variation (Ender, 1993; Ender & Schierwater, 1993; Schierwater *et al.*, 1993), and a segment of mitochondrial DNA (mtDNA) was amplified, sequenced, and PCR restriction fragment length polymorphisms (PCR-RFLP) were detected (Schwenk, 1993).

To address questions such as introgression or phylogenetic relationships, several molecular markers from representative samples are required. The 'classical' approach, the use of allozyme electrophoresis or southern blotting techniques, require laboratory cultures of *Daphnia* clones (that are maintained *via* parthenogenetic reproduction). On the other hand, PCR-based methods allow work with a single DNA preparation from an individual, thus the effort per sample is relatively low. Furthermore, simple storage methods for preservation of tissue (i.e. in ethanol) are sufficient, and the analysis of field samples and even genetic analysis of resting eggs has become feasible. For example, to reveal the genetic pattern of a nuclear marker (i.e. one RAPD-fingerprint), only 1% of the DNA of one adult daphnid is required (personal observation).

The RAPD technology enables investigation of numerous anonymous nuclear loci. RAPD patterns were analyzed by considering the following criteria: (1) homology of bands, (2) reproducibility (different PCR thermocyclers and DNA isolations), (3) parsimony analysis (algorithm: Dollo, 0/1 matrix, 20:1; heuristic search with bootstrap, 100 replicates, confidence region 50%; addition sequence: random, 10 replicates; branch swapping algorithm: TBR), (4) in order to avoid an overestimation of the number of diagnostic markers, several individuals per species from geographically distant locations have been chosen.

In principle, the screening of F_1 hybrid genomes should reveal a strictly additive pattern of parent-specific markers (Arnold *et al.*, 1991; Crawford *et al.*, 1993). Any deviation from that 1:1 relationship enables an exclusion of F_1 hybrids and indicates backcrossed genotypes. Genetic distances between genotypes could even discriminate between hybrid classes, such as F_1-hybrids, F_2-hybrids, B_1 (backcrosses), B_n (later generation backcrosses) if the sample size is large enough and a sufficient number of markers is available.

Since our sample size was limited (N = 23) we compared the amount of species-specific markers found in hybrids simply to reveal whether hybrids are F_1-hybrids (equal proportions of parental markers) or not (backcrosses; unequal amounts of parental markers). A comparison of 306 characters of hybrid and parental clones, revealed that hybrids differ in their genetic relationships to parental species. *D. galeata* × *hyalina* hybrids exhibited more nuclear markers of *D. hyalina* (42–58%) than of *D. galeata* (25–33%), *D. cucullata* × *galeata* hybrids appeared to exhibit similar amounts of nuclear markers from *D. cucullata* and *D. galeata*. The unequal distribution of nuclear DNA towards one of the parental species, in the case of *D. galeata* × *hyalina* hybrids indicates that those individuals were probably the result of backcrossing. This result is in agreement with distance analyses between hybrids and parental clones using allozyme electrophoresis; hybrids are usually more closely related to one of the parentals (Gießler, 1987; Müller, 1993; Spaak, 1994).

Due to maternal inheritance of mitochondrial DNA among animals, patterns of mtDNA variation reveal matrilines (Harrison, 1989). Species-specific markers revealed only one maternal species for all *D. galeata* × *hyalina* hybrids: *D. galeata*. The *D. cucullata* × *galeata* and *D. cucullata* × *hyalina* hybrids always had *D. cucullata* as the maternal species (Schwenk, 1993). In contrast to our findings, Taylor and Hebert (1993) reported that bidirectional hybridization was common among neighboringly sympatric *D. rosea* and *D. galeata mendotae* populations in North America. Since the frequency of mitochondrial DNA of the dominant species was higher in hybrids, parentage seems to reflect abundances of parental species, thus directionality seems to depend on ecological rather than genetic factors. The main difference between the species complexes of *D. rosea/galeata mendotae* and *D. galeata/cucullata/hyalina* complexes is that in the latter, species and hybrids usually occur syntopically. Additionally, abundances of parental species do not depend on a prevalence for a defined habitat (within the *D. galeata mendotae/rosea* complex) but on certain food/predator regimes, which fluctuate during the year (Spaak, 1994). If the directionality of hybridization is determined by the abundance of parental species, we expect that parentage will vary over time. Future studies of several *Daphnia* populations where hybrids and parental species occur syntopically will show whether unidirectional mitochondrial gene flow among species of the *D. longispina* complex is a general phenomenon.

In order to reveal the evolutionary history of the *D. galeata/hyalina/cucullata* complex, we compared the two phylogenies, one based on nuclear DNA data (Ender 1993), and the other based on mitochondrial DNA data (Schwenk 1993). This comparison shows that the trees differ in branching order (Fig. 1). The mitochondrial tree indicates that *D. galeata* is more closely related to *D. cucullata*, whereas on the basis of nuclear DNA, *D. galeata* seems to be more closely related to *D. hyalina*. Such a pattern can be explained by either nuclear or mitochondrial DNA introgression. Patterns of differential gene flow, namely different frequencies of nuclear and mitochondrial introgression, have been observed in several other species complexes (Arnold, 1992). In most cases, cytoplasmic genes, have crossed species boundaries more readily than nuclear genes (Powell, 1983; Marchant *et al.*, 1988; Aubert & Solignac, 1990). Several hypotheses have been proposed to explain this pattern. For example, Barton & Jones (1983) and Powell (1983) suggested selection against alien nuclear, but not mitochondrial, genes. Since mitochondrial and nuclear introgression occur independently, they might lead to nonconcordant phylogenetic patterns (e.g. Solignac & Monnerot, 1986).

Allozyme investigations of *D. galeata*, *D. hyalina* and *D. cucullata* support the nuclear DNA phylogeny (e.g. Gießler, 1987; Müller, 1993). The mtDNA based topology indicates mitochondrial DNA introgression among *D. galeata* and *D. cucullata* species. A rather high sequence divergence between the species (11–15%) suggests ancient mitochondrial DNA introgression. However, these results do not exclude the possibility of recent mitochondrial gene flow, because sample sizes were too small to detect low frequencies of extant introgression.

The consequences of interspecific hybridization

Nuclear DNA data suggest that within *D. longispina* species complexes, several hybrid and species classes, such as pure species, F_1-hybrids and backcrosses can co-occur. These findings are supported by morphological studies on *Daphnia* hybrids (Flössner, 1993). Gradual transitions in characters (used for species discrimination) between hybrids and parentals indicate that various classes of recombinants are produced. Furthermore, allozyme studies of Taylor and Hebert (1992) and Hebert *et al.* (1989) have clearly shown that backcrossed genotypes can even be the dominant genotypes in a lake. Thus, *Daphnia* hybrid 'populations' should be characterized as hybrid swarms rather than as defined intermediate hybrid classes.

The observed patterns of introgression suggest reticulate evolution among species of the *D. longispina* complex. Clones formed through interspecific hybridization and backcrossing serve as a 'link' between species. Thus, a species phylogeny will be characterized by a nonconcordance of species-specific markers and by reticulate rather than hierarchical patterns of evolution (e.g. Harrison, 1990; Arnold 1992). The occurrence of interspecific hybridization *per se* provides evidence that species are separated only by semi-permeable barriers, thus reproductive isolation has not been established completely (either in sympatry

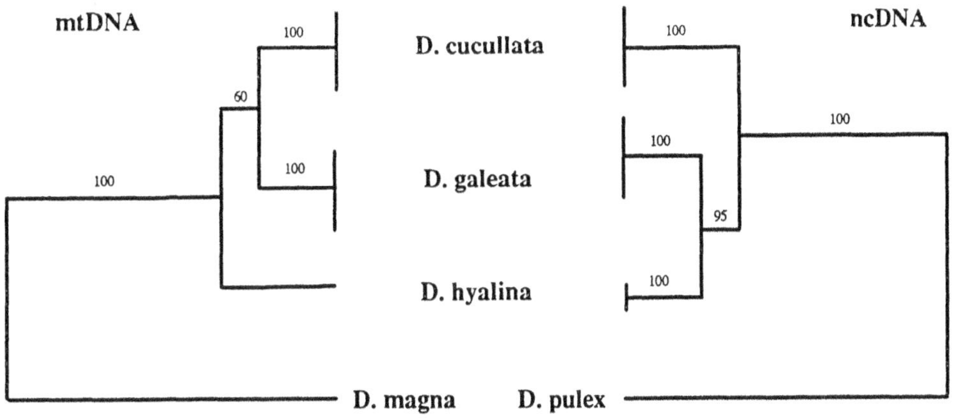

Fig. 1. Molecular phylogenies of *Daphnia* spp. based on mitochondrial and nuclear DNA data. Both trees were constructed using parsimony algorithms (mtDNA: PHYLIP, version 3.3; Felsenstein, 1985 and ncDNA: PAUP, version 3.0s; Swofford, 1990). Vertical lines at the end of a branch indicate that more than one clone was investigated. Numbers at nodes represent bootstrap values for 100 replicates. Phylogenies were constructed using sequence information of amplified mtDNA segments (Schwenk, 1993), and amplification of polymorphic nuclear DNA (Ender, 1993; Schierwater *et al.*, 1993).

or allopatry). If no pre- or postmating barriers between species exist, the extent of introgression depends on the degree of selection against hybrids or recombinants. Various levels of introgression can thus be detected. Unlimited introgression will lead to a unification of gene pools; however, if introgression is restricted (e.g. low fitness of backcrosses), only a limited number of genes will cross species boundaries.

Despite hybridization processes and substantial morphological and ecological similarities of *Daphnia* species, mtDNA analysis has revealed distinct phylogenetic separation of species. Compared with sequence comparisons among other arthropods, such as *Drosophila spp.*, *Daphnia spp.* show higher sequence divergence among species. In addition, nuclear markers reveal clear interspecific differentiation as well (Gießler, 1987; Ender, 1993). This pattern raises the question: how are species boundaries maintained despite the possibility of mitochondrial and nuclear gene flow between taxa? Investigations from laboratory crossing experiments and fitness comparisons of those recombinant genotypes and parentals are necessary to assess levels of reproductive isolation.

Genetic variability within the *D. longispina* complex can be disentangled into two aspects; 1) intraspecific variation, and 2) variation generated by interspecific hybridization. The first aspect, intraspecific variation, seems to be relatively low compared with pond-dwelling *Daphnia* species or other animal species (Mort, 1991), but hybrid populations exhibit relatively high proportions of polymorphic loci and high average heterozygosities (Müller, 1993). Since hybrid clones are able to maintain their genetic variation through asexual reproduction, the hybrids form a very heterogeneous group. Additionally, mitochondrial DNA variation and backcrossed genotypes with various combinations of nuclear and mitochondrial genes are bound to increase the level of genetic diversity (Lewontin & Birch, 1966; DePamphilis & Wyatt, 1990). Thus, interspecific hybridization is a factor potentially facilitating founder effects within the *D. longispina* complex.

With regard to speciation, patterns of interspecific hybridization have been explained by two different hypotheses (see Otte & Endler, 1989). First, species complexes have been described as an evolutionary step towards a split into two different lineages, thus hybridization was considered as a side-effect of the speciation process (sympatric speciation). Alternatively, species form hybrids as a result of secondary contact (allopatric or parapatric speciation) after a time of geographic isolation, e.g. during glaciation. Several genetic and biogeographic analyses of various hybrid taxa support the second hypothesis, but only a few cases have been explained using the first hypothesis (Harrison, 1990). As to which of these two explanations might be valid for the *D. longispina* complex, future research will need to focus on biogeography and phylogenetic relationships to provide further clarification of the mode(s) of speciation in this group. However, the application of molecular markers is bound to facilitate a systematic re-evaluation of the *D. longispina* complex. Nuclear DNA markers (e.g. RAPD analysis) could be used to identify previously indistinguishable taxa within *Daphnia* species complexes. Furthermore, taxonomic traits used for species and hybrid identification could be re-interpreted. Amplification of mitochondrial DNA and subsequent digestion using restriction endonucleases, or DNA sequencing will reveal patterns such as population structure, matrilines and phylogenetic relationships. Based on this information, processes like introgression, incipient reproductive isolation and speciation could be analyzed. DNA analysis of single animals and even single resting eggs offers the opportunity to investigate processes of hybridization and speciation in previously intractable detail.

Acknowledgements

We thank Maarten Boersma, Piet Spaak, Ramesh D. Gulati, Jos M.M. van Damme, Peter van Dijk, Peter Shaw, Onno van Tongeren, Koos Vijverberg, Thomas Städler, Lawrence J. Weider and two anonymous referees for valuable comments on a previous version of the manuscript. Research was funded by the Deutsche Forschungsgemeinschaft (Mo 492/1-2) and the Netherlands Institute of Ecology (Centre for Limnology).

References

Arnheim, N., T. White & W. E. Rainey, 1990. Application of PCR: organismal and population biology. BioScience 40: 174–182.

Arnold, M. L., 1992. Natural hybridization as an evolutionary process. Ann. Rev. Ecol. Syst. 23: 237–261.

Arnold, M. L., C. M. Buckner & J. J. Robinson, 1991. Pollen-mediated introgression and hybrid speciation in Louisiana irises. Proc. natn. Acad. Sci. USA 88: 1398–1402.

Aubert, J. & M. Solignac, 1990. Experimental evidence for mitochondrial DNA introgression between *Drosophila* species. Evolution 44: 1272–1282.

Barton, N. H. & G. M. Hewitt, 1985. Analysis of hybrid zones. Ann. Rev. Ecol. Syst. 16: 113–148.

Barton, N. H. & J. S. Jones, 1983. Mitochondrial DNA: new clues about evolution. Nature 306: 317–318.

Crawford, D. J., S. Brauner, M. B. Cosner & T. F. Stuessy, 1993. Use of RAPD markers to document the origin of the intergeneric hybrid *xMargyracaena skottsbergii* (Rosaceae) on the Juan Fernandez Islands. Am. J. Bot. 80: 89–92.

Crease, T. J., M. Lynch & K. Spitze, 1990. Hierarchical analysis of population genetic variation in mitochondrial and nuclear genes of *Daphnia pulex*. Mol. Biol. Evol. 7: 444–458.

DePamphilis, C. W. & R. Wyatt, 1990. Electrophoretic confirmation of interspecific hybridization in *Aescules* (Hippocastanaceae) and the genetic structure of a broad hybrid zone. Evolution 44: 1295–1317.

Ender, A., 1993. Identifizierung von nuklearen DNA-Markern in *Daphnia*-Hybrid-Komplexen: RAPD-Analyse. Diploma thesis. J.W. Goethe-Universität, Frankfurt, Germany.

Ender, A. & B. Schierwater, 1993. Identification of nuclear markers in *Daphnia* hybrid complexes using Random Amplified Polymorphic DNA (RAPD). Verh. dt. zool. Ges. 86.1: 45.

Felsenstein, J., 1985. Confidence limits on phylogenies: an approach using the bootstrap. Evolution 39: 783–791.

Flössner, D., 1993. Zur Kenntnis einiger *Daphnia*-Hybriden. Limnologica 23: 71–79.

Flössner, D. & K. Kraus, 1986. On the taxonomy of the *Daphnia hyalina-galeata* complex (Crustacea: Cladocera). Hydrobiologia 137: 97–115.

Gießler, S., 1987. Mikroevolution und Populationsgenetik im *Daphnia galeata/hyalina/cucullata*-Komplex (Crustacea: Cladocera). Eine Freilandanalyse. Ph.D. thesis. Universität München, Germany.

Gill, F. B., A. M. Mostrom & A. L. Mack, 1993. Speciation in North American chickadees. 1. Patterns of mtDNA genetic divergence. Evolution 47: 195–212.

Harrison, R. G., 1989. Animal mitochondrial DNA as a genetic marker in population and evolutionary biology. Trends Ecol. Evol. 4: 6–11.

Harrison, R. G., 1990. Hybrid zones: windows on evolutionary process. Oxford Surv. Evol. Biol. 7: 69–128.

Hebert, P. D. N., S. S. Schwartz & J. Hrbácek, 1989. Patterns of genotypic diversity in Czechoslovakian *Daphnia*. Heredity 62: 207–216.

Hewitt, G. M., A. W. B. Johnston & J. P. W. Young, 1991. Molecular techniques in taxonomy. NATO ASI Series H: Cell Biology 57. Springer-Verlag, Berlin.

Hillis, D. M. & C. Moritz (eds), 1990. Molecular systematics. Sinauer Associates, Sunderland, Mass.

Lewontin, R. C. & L. C. Birch, 1966. Hybridization as a source of variation for adaptation to new environments. Evolution 20: 315–336.

Lynch, M., 1985. Speciation in the Cladocera. Verh. int. Ver. Limnol. 22: 3116–3123.

Marchant, A. D., M. L. Arnold & P. Wilkinson, 1988. Gene flow across a chromosomal tension zone. I. Relicts of ancient hybridization. Heredity 61: 321–328.

Mort, M. A., 1991. Bridging the gap between ecology and genetics: the case of freshwater zooplankton. Trends Ecol. Evol. 6: 41–45.

Müller, J., 1993. Räumliche und zeitliche Variabilität der genetischen Struktur natürlicher Cladocerenpopulationen (Crustacea: Cladocera). Ph.D. thesis. J. Gutenberg-Universität, Mainz, Germany.

Müller, J. & A. Seitz, 1993. Habitat partitioning and differential vertical migration of some *Daphnia* genotypes in a lake. Arch. Hydrobiol. Beih. Ergebn. Limnol. 39: 167–174.

Müller, J. & A. Seitz, 1994. Influence of differential natality and mortality on temporal fluctuations of *Daphnia* genotypes in natural populations. In Beaumont, A. R. (ed.), Genetics and Evolution of Aquatic Organisms. Chapman & Hall, Lond.: 342–350.

Nason, J. D. & N. C. Ellstrand, 1993. Estimating the frequencies of genetically distinct classes of individuals in hybridized populations. J. Hered. 84: 1–12.

Otte, D. & J. A. Endler (eds), 1989. Speciation and its consequences. Sinauer Associates, Sunderland, Mass., 679 pp.

Peters, R. H. & R. de Bernardi (eds), 1987. *Daphnia*. Memorie Dell'Istituto Italiano di Idrobiologia. 45. Verbania Pallanza, Palanza, Italy, 502 pp.

Powell, J. R., 1983. Interspecific cytoplasmic gene flow in the absence of nuclear gene flow: evidence from *Drosophila*. Proc. natn. Acad. Sci. USA 80: 492–495.

Schierwater, B., A. Ender & B. Streit, 1993. The use of RAPD markers for molecular phylogenies. Verh. dt. zool. Ges. 86.1: 58.

Schwenk, K., 1993. Interspecific hybridization in *Daphnia*: distinction and origin of hybrid matrilines. Mol. Biol. Evol. 10: 1289–1302.

Schwenk, K. & P. Spaak, 1995. Evolutionary and ecological consequences of interspecific hybridization in cladocerans. Experientia (in press).

Sneath, P. H. A. & R. R. Sokal, 1973. Numerical Taxonomy. W.H. Freeman, San Francisco.

Solignac, M. & M. Monnerot, 1986. Race formation, speciation, and introgression within *Drosophila simulans*, *D. mauritiana*, and *D. sechellia* inferred from mitochondrial DNA analysis. Evolution 40: 531–539.

Spaak, P., 1994. Genetical ecology of a coexisting *Daphnia* hybrid species complex. Ph.D. thesis, University of Utrecht, The Netherlands.

Spaak, P. & J. R. Hoekstra, 1993. Clonal structure of the *Daphnia* population in Lake Maarsseveen: its implications for diel vertical migration. Arch. Hydrobiol. Beih. Ergebn. Limnol. 39: 157–165.

Streit, B., T. Städler, K. Schwenk, A. Ender, K. Kuhn & B. Schierwater, 1994. Natural hybridization in freshwater animals: ecological implications and molecular approaches. Naturwissenschaften 81: 65–73.

Sturmbauer, C. & A. Meyer, 1992. Genetic divergence, speciation and morphological stasis in a lineage of African cichlid fishes. Nature 358: 578–581.

Swofford, D. L., 1990. *PAUP: Phylogenetic analysis using parsimony, version 3.0.s* Illinois Natl. Hist. Surv., Champaign, IL.

Taylor, D. J. & P. D. N. Hebert, 1992. *Daphnia galeata mendotae* as a cryptic species complex with interspecific hybrids. Limnol. Oceanogr. 37: 658–665.

Taylor, D. J. & P. D. N. Hebert, 1993. Habitat-dependent hybrid parentage and differential introgression between neighboringly sympatric *Daphnia* species. Proc. natn. Acad. Sci. USA 90: 7079–7083.

Templeton, A. R., 1980. The theory of speciation *via* the founder principle. Genetics 94: 1011–1038.

Welsh, J. & M. McClelland, 1990. Fingerprinting genomes using PCR with arbitrary primers. Nucleic Acids Res. 18: 7213–7218.

Weider, L. J., 1993. Niche breadth and life history variation in a hybrid *Daphnia* complex. Ecology 74: 935–943.

Weider, L. J. & H. G. Wolf, 1991. Life-history variation in a hybrid species complex of *Daphnia*. Oecologia 87: 506–513.

Williams, J. G. K., A. R. Kubelik, K. J. Livak, J. A. Rafalski & S. V. Tingey, 1990. DNA polymorphisms amplified by arbitrary primers are useful as genetic markers. Nucleic Acids Res. 18: 6531–6535.

Wolf, H. G., 1987. Interspecific hybridization between *Daphnia hyalina*, *D. galeata*, and *D. cucullata* and seasonal abundances of these species and their hybrids. Hydrobiologia 145 (Dev. Hydrobiol. 35): 213–217.

Wolf, H. G. & M. A. Mort, 1986. Inter-specific hybridization underlies phenotypic variability in *Daphnia* populations. Oecologia 68: 507–511.

Hydrobiologia **307**: 9–14, 1995.
P. Larsson & L.J. Weider (eds), Cladocera as Model Organisms in Biology.

Genetic variation in *Simocephalus* (Anomopoda: Daphniidae) in North America: patterns and consequences

B. J. Hann
Department of Zoology, University of Manitoba, Winnipeg, Manitoba, R3T 2N2, Canada

Key words: genetic variation, Anomopoda *Simocephalus*, species complexes

Abstract

The presumption of intercontinental distributions of many species of Anomopoda has been reinforced by their considerable morphological uniformity and ease of passive dispersal via ephippial eggs. To test the validity of this dogma, genetic variation among taxa in the cladoceran genus *Simocephalus* was examined on a continental scale. Genetic variability (percentage polymorphic loci = 8.8%, individual heterozygosity = 3.4%) was comparable to that determined on a local scale in *Simocephalus* and somewhat less than for other zooplankton groups. Four species complexes were distinguished allozymically with unique allelic substitutions found for at least two loci between these taxa. Eight species in North America were differentiated within these complexes (Nei s genetic distance, $D > 0.30$), all of which were clearly separated from two European species. Both *S.* cf. *vetulus* and *S.* cf. *serrulatus* are broadly distributed in North America, and intraspecific divergence is relatively low. In contrast, *S.* cf. *congener* and *S.* cf. *exspinosus* represent species complexes, both consisting of several species, differentiated allozymically and morphologically. The concept of generalist species with moderate genetic variation throughout a broad range and specialist species, genetically depauperate, with restricted distributions, is explored.

Introduction

The cosmopolitan distribution of species was commonly accepted within most genera of Anomopoda until the last decade. The inter- continental divergence of several chydorid species was clearly demonstrated on morphological criteria (Frey, 1982, 1986, 1987a, b; Michael & Frey, 1983, 1984). The recognition of extensive genetic and morphological differentiation in *Daphnia* (Hebert & Finston, 1993; Taylor & Hebert, 1992) and *Simocephalus* (Hann & Hebert, 1982, 1986) in the family Daphniidae, *Eurycercus* (Frey, 1973, 1978; Hann, 1982, 1990) in the family Chydoridae, and *Bosmina* (Lieder, 1983) in the family Bosminidae strongly indicates narrower species distributions to be the general pattern within the Anomopoda. It suggests also the existence of many more species of anomopods than previously recognized on the basis of gross morphological assessments.

The genus *Simocephalus* is an exclusively littoral zone representative of the family Daphniidae, distributed worldwide. In North America, only three species: *S. vetulus* (O. F. Müller, 1776), *S. serrulatus* (Koch, 1841), and *S. exspinosus* (Koch, 1841) have been recognized in identification manuals of North American aquatic invertebrate fauna (Edmondson, 1959; Pennak, 1978). However, previous genetic and morphological studies have shown the existence of several as yet undescribed species in North America (Hann, 1987; Hann & Hebert, 1982, 1986).

The objective of this study was to assess allozymic variation in populations of *Simocephalus* in North America, and to make initial continental and intercontinental comparisons of populations as necessary to adequately investigate the inter-taxon and intra-taxon differentiation. Analysis of phenetic (and eventually phylogenetic) relationships among clearly defined species will provide the essential framework and foundation for further studies of biogeography and speciation within the genera and families of Anomopoda.

Table 1. Location of sampling sites and codes for populations included in the study.

Code	Locality	Geographic coordinates
ONT-1	Cedar Springs	42 °13 N, 82 ° 0′W
ONT-2	Bloomfield Rd.	42 °14′N, 82 ° 3′W
ONT-3	Rondeau Park	42 °10′N, 81 °50′W
ONT-4	Rondeau Cottage	42 °11′N, 81 °52′W
ONT-5	Ojibway Park	42 °20 N, 83 ° 0′W
ONT-6	Lukerville	42 ° 5′N, 83 ° 1′W
ONT-7	Strawberry Pond	42 ° 4′N, 83 ° 1′W
ONT-8	Holiday Beach	41 °49′N, 83 ° 0′W
ONT-9	Point Pelee	41 °55′N, 82 °30′W
MAN-1	Churchill	58 °53′N, 94 ° 2′W
MAN-2	Delta Marsh	50 ° 7′N, 99 °19′W
MAN-3	Nopiming Park	50 °35 N, 95 °40′W
NWT-1	Inuvik	69 ° 5′N, 134 °10′W
IND-1	Hackenberg Lake	41 °58′N, 87 °43′W
IND-2	Oliver Lake	41 °56′N, 87 °40′W
ILL-1	Vermilion Lake	40 ° 1′N, 87 °50′W
OHI-1	Nature Center	39 °18′N, 84 °39′W
OHI-2	Logan Pond	38 °48′N, 83 °10′W
FLA-1	St. Augustine	29 °56′N, 81 °10′W
FLA-2	Tallahassee	30 °51′N, 84 ° 6′W
FLA-3	Lowry Lake	29 °52′N, 81 °15′W
GER-1	Plön	54 ° 9′N, 10 °30′W
UK-1	Loughborough	52 °55′N, 1 °20′W
UK-2	Blackbrook Res.	52 °40′N, 2 °25′W

Materials and methods

Populations of several species of *Simocephalus* were collected from ponds and the littoral zone of lakes (24 localities) over a wide geographic range, extending from Inuvik, Northwest Territories, Canada to Florida, USA (Table 1), as well as from sites in the United Kingdom and Germany. Collections were made using a 100 μm mesh net by sweeping through the aquatic macrophytes. Individuals in each population were assigned to a species (or species group) on morphological criteria (Edmondson, 1959; Flössner, 1972; Hann, in prep.).

Electrophoresis was carried out on individual parthenogenetic females using Titan III cellulose acetate gels (Helena Laboratories, Beaumont, Texas) and staining procedures detailed in Hebert & Beaton (1989). Seven enzyme loci, known from previous studies (Hann & Hebert, 1982, 1986) to be polymorphic in *Simocephalus* species, were analyzed: phosphoglucose isomerase (PGI), phosphoglucomutase (PGM), lactate dehydrogenase (LDH), malate dehydrogenase (MDH), glucose-6-phosphate dehydrogenase (G6P), glutamate oxaloacetate transaminase (GOT), and aldehyde oxidase (AO). Alleles were numbered according to relative migration rates, with 1 being slowest and 6 fastest.

Allelic frequencies for all loci and all populations were calculated from composite genotypic data accumulated for each individual. Ward *et al.* (1991) confirmed the codominant Mendelian inheritance of *Simocephalus* electromorphs, hence genotypes were inferred from gel banding patterns. Heterozygote patterns were consistent with their known quaternary structure (Ward, 1977), and with data for several *Daphnia* species (Hebert, 1974), also in the family Daphniidae.

Hardy-Weinberg expected frequencies were calculated using Levene's (1949) correction for small sample sizes. Goodness-of-fit of the actual genotype frequencies to the Hardy-Weinberg expected proportions was tested using the standard chi-square (χ^2) test. Rare allele classes were pooled where necessary, and the chi-square value calculated with one degree of freedom. The high number of simultaneous tests required a corrected significance level to control the family-wise error rate. Using a simple Bonferroni correction, the corrected alpha is 0.05/39 = 0.00128. Individual heterozygosity values and percentage polymorphism (assuming use of 0.95 frequency criterion) within populations were determined by direct count (Swofford & Selander, 1989).

Allozyme data were analyzed phenetically for patterns of relationship and genetic differentiation. Nei's (1978) unbiased genetic distance (*D*) and genetic identity (*I*) were calculated among all pairs of samples. Populations and species were clustered using the unweighted pair-group (UPGMA) algorithm based on Nei's D values, and summary dendrograms were produced. All procedures used BIOSYS-1 (version 1.7) developed by Swofford & Selander (1989).

Results

Allozyme studies revealed abundant genetic variation within the genus, with individual heterozygosity averaging 3.4% and percentage polymorphism 8.8%. Species were identified by unique genotypic complexes, summarized for each species in Table 2. Four species or species complexes containing eight species

Table 2. Alleles present at each gene locus sampled for all species included. Individual heterozygosity, percent polymorphic, and mean sample size per locus are given for each species.

Taxon	Gene locus PGM	PGI	LDH	MDH	GOT	AO	G6P	Individual heterozygosity	Percent polymorphic	Mean sample size per locus
S. cf. vetulus	2,3	2,3,4	1,2,3	1,2	1,2,3	1,2,4	3	4.8	14.3	56
S. vetulus	6	2,3	2	3	2,3	1	3	4.8	14.3	7
S. cf. serrulatus	1,2	4,5	1,2	2,3	2,3,4	1,2	2	2.3	7.1	39
S. exspinosus congener	6	5	3	2	1,3	3	3	1.4	7.1	6
S. cf. congener 1	2	5	2,3	1,2	1,3	2,3	3	3.0	9.4	12
S. cf. congener 2	5	5	2,3	2	1	3	3	7.1	7.1	12
S. cf. exspinosus 1	2,4	6	2	1	1	4	3	0.0	0	17
S. cf. exspinosus 2	4	4	4	2	3	2	3	0.0	0	2
S. cf. exspinosus 3	3	5	3	2,3	2,3	4	2	3.4	14.3	19
S. cf. exspinosus 4	3,4	4	4	3	2	4	2	7.1	14.3	10

were distinguished in North America (D>0.30), and two European cognate species were clearly differentiated from their American counterparts (Fig. 1). Species were separated by two or more unique allele substitutions, not simply on the basis of allele frequency differences.

Most loci were in Hardy-Weinberg equilibrium; however, in seven of 39 instances, significant deviations were detected (one heterozygote excess, six heterozygote deficiencies). There was no evidence of hybridization among any species.

Intraspecific genetic diversity

S. cf. vetulus

This species is the most widely distributed in North America, with records from Florida to as far north as Inuvik, Churchill, and Iqaluit (Frobisher Bay), NWT, but not reported from Tuktoyaktuk (Hann, unpubl.) or the Canadian arctic archipelago (Hann & Hebert, 1986). Nonetheless, intraspecific differentiation was very slight (D<0.10) (Fig. 1). There was, however, a distinct dichotomy between the North American and European populations of *S. vetulus*, represented by samples from Germany and the U.K. (I = 0.483, D = 2.070). There are allelic substitutions at three loci (PGM, MDH, AO). This degree of differentiation is comparable to that found between non-sibling species of other assayed organisms (Ayala, 1983), but little morphological distinctiveness has yet been identified between the North American and European populations (Hann, unpubl.).

S. cf. serrulatus

S. cf. *serrulatus* does not occur as far north as *S.* cf. *vetulus* but appears to have a broad temperate zone distribution. Populations from Florida were the most dissimilar on the basis of allozyme data (I = 0.845, D = 1.183). No European representatives of the cognate species were available for comparative study.

S. cf. congener

There was considerable distance between the European populations of this species, known as *S. exspinosus congener*, and the North American species (I = 0.668, D = 1.497). There were three allelic substitutions (PGM, LDH, GOT) between these two groups of species. Within North America, populations were quite distinct, and the Florida and Ohio populations were the most divergent (I = 0.680, D = 1.471) from those in southern Ontario and Manitoba, with allelic replacements at two loci (PGM, GOT).

S. cf. exspinosus

Several distinct entities were masquerading under the umbrella of this species name. Besides showing a high level of allozymic differentiation and numerous allelic substitutions at all loci, there were several distinct morphotypes in North America, represented by populations from southern Ontario, Ohio, Manitoba, and Florida, as yet undescribed. In addition, on the basis of data available, each of these taxa appeared to be severely restricted in spatial distribution.

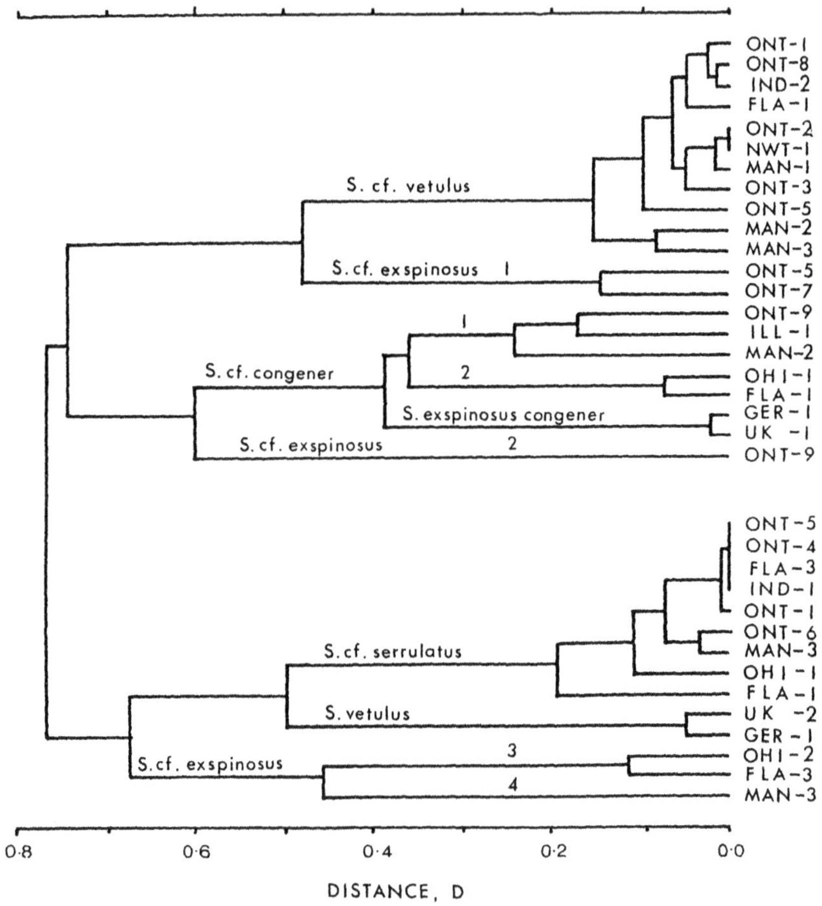

Fig. 1. UPGMA dendrogram of taxa of *Simocephalus* using allozyme data and Nei's unbiased genetic distance. Population codes are as in Table 1. Numbers associated with branches of the dendrogram for *S.* cf. *congener* and *S.* cf. *exspinosus* represent putative species designated on the basis of genetic distance evidence.

Interspecific genetic diversity

The UPGMA dendrogram (Fig. 1) indicated genetic relatedness among putative species of *Simocephalus*. There were ten clearly delimited taxa ($D>0.30$) in the tree: *S.* cf. *vetulus* (North America), *S. vetulus* (Europe), *S.* cf. *serrulatus* (North America), *S.* cf. *congener* (species 1 and 2) (North America), *S. exspinosus congener* (Europe), and *S.* cf. *exspinosus* (species 1, 2, 3 and 4) (North America). Only the *S.* cf. *exspinosus* species complex appears to be polyphyletic, suggesting that this is an artificial grouping of taxa) which further allozymic and morphological work with more populations, may clarify.

Discussion

The intra-specific genetic variation in *Simocephalus* is lower than in other zooplankton groups (Boileau & Hebert, 1988; Taylor & Hebert, 1992). All species of *Simocephalus* produce a single ephippial egg (Fryer, 1991a), in contrast with *Daphnia* species and copepods which produce more than one diapausing egg. Thus, a colonizing population may be established from one egg with one genotype (Ward *et al.*, 1991). Ephippial eggs of *Simocephalus* do not float as readily as those of *Daphnia*, further hampering dispersal capability. It is pertinent that *S.* cf. *vetulus*, the most widely distributed taxon, has the smallest, and perhaps the lightest, ephippium.

Genetically-determined allozyme polymorphisms are a rich source of taxonomic characters which can be diagnostic tools in identifying closely related species, sibling species, or even incipient species (Ayala, 1983; Thorpe, 1983; Nevo *et al.*, 1983). This was especially appropriate for *Simocephalus* in that the allozyme differences documented for each species were not clines in allelic frequency, but complete allelic substitutions, permitting clear differentiation of taxa. The amount of genetic distance that clearly indicates specific difference for many other invertebrate taxa has been designated as $D>0.20$ (Ayala, 1983; Thorpe, 1983; Richardson *et al.*, 1986). In this analysis of the extent of divergence among species of *Simocephalus*, $D>0.30$ was employed to identify putative species because this estimate of D is probably an overestimate due the small number of loci chosen *a priori* for their rapid evolution (i.e. only commonly polymorphic loci were used).

Simocephalus represents a complex of closely related, ecologically and functionally similar species (Fryer, 1991a). There are a few widespread, common species, e.g. *S.* cf. *vetulus* and *S.* cf. *serrulatus*, showing high clonal diversity (Hann & Hebert, 1986), and several other rare taxa (as currently distinguished), each consisting of several species, with restricted, local distributions, and with low clonal diversity. 'Environmental tolerance' as described by Lynch & Gabriel (1987) is comprised of two components, an optimum and a breadth of adaptation. The sensitivity of a population is a function of both parameters. It is possible that common, generalist species have large numbers of clones, each with narrow optima and tolerances and, consequently, a narrow breadth of adaptation and little plasticity. Conversely, rare, specialist species may have a smaller number of clones, each with specific optima but broader tolerances and physiological plasticity (Lynch & Gabriel, 1987).

The geographic pattern of divergence of species in *Simocephalus* in North America suggests that the greatest amount of differentiation has occurred in the southeastern USA. Several factors are potentially involved in producing this pattern, one of which is certainly the amount of time available for population differentiation. In areas of the continent that were not affected by the last glaciation, populations have been able to persist for vastly longer periods of time than the less than 10 000 years available for populations in the northern half of the continent. In addition to the divergence between species in North America, there is considerable genetic separation of the European species from those currently bearing the same names in North America. Such intercontinental differentiation merely confirms the non-cosmopolitan pattern of species distributions in Anomopoda. As *Simocephalus* has existed distinct from *Daphnia* since at least the early Cretaceous (Fryer, 1991b), there has been at least 120 million years during which to evolve these discrete taxonomic lineages.

Among the families of Anomopoda, the designation of subgenera is variable and does not follow clearly established criteria. For example, subgeneric categories have been established within *Daphnia*, *Bosmina* (Lieder, 1983), and *Eurycercus* (Frey, 1975). Sufficient interspecific genetic distance has been documented in *Simocephalus* to warrant designation of subgenera. In addition, morphological evidence corroborates the pattern of genetic differentiation (Hann, 1987, unpubl.). However, further morphological comparison of taxa, especially within *S.* cf. *exspinosus*, presently polyphyletic according to phenetic analyses, is essential prior to this decision.

Acknowledgments

This study was funded by an operating grant from the Natural Sciences and Engineering Research Council of Canada.

References

Ayala, F. J., 1983. Enzymes as taxonomic characters. In G. S. Oxford & D. Rollinson (ed.), Protein polymorphism: Adaptive and taxonomic significance. Academic Press, London: 3–26.

Boileau, M. G. & P. D. N. Hebert, 1988. Electrophoretic characterization of two closely related species of *Leptodiaptomus*. Biochem. Syst. Ecol. 16: 329–332.

Edmondson, W. T., 1959. Freshwater Biology, 2nd edn. John Wiley & Sons, Inc., New York.

Flössner, D., 1972. Krebstiere, Crustacea, Kiemen- und Blattfüsser, Branchiopoda, Fischläuse, Branchiura. Die Tierwelt Deutschlands, Tiel 60.

Frey, D. G., 1973. Comparative morphology and biology of three species of *Eurycercus* (Chydoridae, Cladocera) with a description of *Eurycercus macrocanthus* sp. nov. Int. Rev. ges. Hydrobiol. 58: 221–267.

Frey, D. G., 1975. Subgeneric differentiation within *Eurycercus* (Cladocera, Chydoridae) and a new species from Northern Sweden. Hydrobiologia 46: 263–300.

Frey, D. G., 1978. A new species of *Eurycercus* (Cladocera, Chydoridae) from the southern United States. Tulane Stud. in Zool. Bot. 20: 1–25.

Frey, D. G., 1982. Questions concerning cosmopolitanism in Cladocera. Arch. Hydrobiol. 93: 484–502.

Frey, D. G., 1986. The non-cosmopolitanism of chydorid Cladocera: implications for biogeography and evolution. In K. L. Heck & R. H. Gore (eds), Crustacean Issues, Vol. 4, Crustacean Biogeography. A. A. Balkema, Rotterdam.

Frey, D. G., 1987a. The North American *Chydorus faviformis* (Cladocera, Chydoridae) and the honeycombed taxa of other continents. Phil. Trans. r. Soc., Lond. B315: 353–402.

Frey, D. G., 1987b. The taxonomy and biogeography of the Cladocera. Hydrobiologia 145 (Dev. Hydrobiol. 35): 5–17.

Fryer, G., 1991a. Functional morphology and the adaptive radiation of the Daphniidae (Branchiopoda: Anomopoda). Phil. trans. r. Soc., Lond. B331: 1–99.

Fryer, G., 1991b. A daphniid ephippium (Branchiopoda: Anomopoda) of Cretaceous age. Zool. J. Linn. Soc. 102: 163–167.

Hann, B. J., 1982. Two new species of *Eurycercus (Bullatifrons)* from eastern North America (Chydoridae, Cladocera). Taxonomy, ontogeny, and biology. Int. Rev. ges. Hydrobiol. 67: 585–610.

Hann, B. J., 1987. Naturally occurring interspecific hybridization in *Simocephalus* (Cladocera, Daphniidae): its potential significance. Hydrobiologia 145 (Dev. Hydrobiol. 35): 219–224.

Hann, B. J., 1990. Redescription of *Eurycercus (Teretifrons) glacialis* (Cladocera, Chydoridae), and description of a new species, *E. (T.) nigracanthus*, from Newfoundland, Canada. Can. J. Zool. 68: 2146–2157.

Hann, B. J. & P. D. N. Hebert, 1982. Re-interpretation of genetic variation in *Simocephalus* (Cladocera, Daphniidae). Genetics 102: 101–107.

Hann, B. J. & P. D. N. Hebert, 1986. Genetic variation and population differentiation in species of *Simocephalus* (Cladocera, Daphniidae). Can. J. Zool. 64: 2246–2256.

Hebert, P. D. N., 1974. Enzyme variability in natural populations of *Daphnia magna*. II. Genotypic frequencies in permanent populations. Genetics 77: 323–334.

Hebert, P. D. N. & M. J. Beaton, 1989. Methodologies for allozyme analysis using cellulose acetate electrophoresis. Helena Laboratories, Beaumont, Texas, USA.

Hebert, P. D. N. & T. L. Finston, 1993. A taxonomic reevaluation of North American *Daphnia* (Crustacea: Cladocera). I. The *D. similis* complex. Can. J. Zool. 71: 209–225.

Levene, H., 1949. On a matching problem arising in genetics. Ann. Math. Stat. 20: 91–94.

Lieder, U., 1983. Revision of the genus *Bosmina* Baird, 1845 (Crustacea, Cladocera). Int. Rev. ges. Hydrobiol. 68: 121–139.

Lynch, M. & W. Gabriel, 1987. Environmental tolerance. Am. Nat. 129: 283–303.

Michael, R. G. & D. G. Frey, 1983. Assumed amphi-Atlantic distribution of *Oxyurella tenuicaudis* (Cladocera, Chydoridae) denied by a new species from North America. Hydrobiologia 106: 3–35.

Michael, R. G. & D. G. Frey, 1984. Separation of *Disparalona leei* (Chien, 1970) in North America from *D. rostrata* (Koch, 1841) in Europe (Cladocera, Chydoridae). Hydrobiologia 114: 81–108.

Nei, M., 1978. Estimation of average heterozygosity and genetic distance from a small number of individuals. Genetics 89: 583–590.

Nevo, E., A. Beileo & R. Ben-Shlomo, 1983. The evolutionary significance of genetic diversity: ecological, demographic and life history correlates. In G. S. Mani (ed.), Evolutionary dynamics of genetic diversity. Lecture Notes in Biomath. 53: 13–213.

Pennak, R. W., 1978. Freshwater Invertebrates of the United States. 2nd edn. John Wiley & Sons, Inc., New York.

Richardson, B. J., P. R. Baverstock & M. Adams, 1986. Allozyme electrophoresis. A Handbook for animal systematics and population studies. Academic Press, NY., 410 pp.

Swofford, D. L. & R. K. Selander, 1989. BIOSYS-1 (Version 1.7): a computer program for the analysis of allelic variation in population genetics and biochemical systematics. Illinois Natural History Survey, Champaign.

Taylor, D. J. & P. D. N. Hebert, 1992. *Daphnia galeata mendotae* as a cryptic species complex with interspecific hybrids. Limnol. Oceanogr. 37: 658–665.

Thorpe, J. P., 1983. Enzyme variation, genetic distance and evolutionary divergence in relation to levels of taxonomic separation. In G. S. Oxford & D. Rollinson (eds), Protein polymorphism: adaptive and taxonomic significance. Academic Press, London: 131–153.

Ward, R. D., 1977. Relationship between enzyme heterozygosity and quaternary structure. Biochem. Genet. 15: 123–135.

Ward, R. D., R. M. Tollit & M. A. Bickerton, 1991. Genetic variation and reproductive mode in populations of *Simocephalus vetulus* (Branchiopoda: Cladocera) from the East Midlands of England. Freshwat. Biol. 25: 41–50.

Hydrobiologia **307**: 15–23, 1995.
P. Larsson & L.J. Weider (eds), Cladocera as Model Organisms in Biology.

Large Hardy-Weinberg equilibrium deviations in the *Daphnia longispina* of Lake El Tobar

Charles E. King[1], Maria R. Miracle[2] & Eduardo Vicente[2]
[1]*Department of Zoology, Oregon State University, Corvallis, Oregon 97331, USA*
[2]*Department of Ecology, Faculty of Biological Sciences, University of Valencia, 46100 Burjasot (Valencia), Spain*

Key words: Daphnia, allozymes, Hardy-Weinberg equilibrium, parthenogenesis

Abstract

The population structure of *Daphnia longispina* in Lake El Tobar, Spain was studied by measuring variation at the aldehyde oxidase (AO), phosphoglucose isomerase (PGI) and phosphoglucose mutase (PGM) loci in each of 1337 individuals from four collections. In 9 of the 12 comparisons between observed allele frequencies and those expected by Hardy-Weinberg equilibrium there was an excess of heterozygotes. We found 27 of the potential number of 54 composite electromorphs ('clones') based on the three allozymes. Clone diversities were rather high in all collections. Three clones reached frequencies of over 25% and different clones were dominant in each of the four collections. Strong temporal variation was found in the genetic structure of this *Daphnia* population. This variation was driven by changes in the relative frequencies of the component clones in the lake rather than by a recruitment of novel clones into the population.

We conclude with a consideration of the role of models relating allele and genotype frequencies in populations of cyclical parthenogens. Because the breeding system of these populations infrequently involves recombination between clones, models such as the Hardy-Weinberg have limited value in providing meaningful measures of population structure.

Introduction

Daphnia populations may be viewed as dynamic assemblages of interacting clones. In comparison with most other zooplankton taxa, their genetic structure and the role of natural selection acting on *Daphnia* in different types of habitats such as intermittent ponds, permanent ponds and lakes has been extensively studied (e.g., Weider, 1985; Hebert, 1987; Lynch, 1987; Wolf, 1988). The importance of interspecific hybridization and the possibility of genetic transfers between cyclic parthenogens and obligate asexual clones are now well documented in *Daphnia* (e.g., Wolf & Mort, 1986; Hebert *et al.*, 1988, 1989; Crease *et al.*, 1989; Taylor & Hebert, 1992, 1993; Weider, 1993; Spaak, 1994). However, it is still unclear whether and under what circumstances a seasonal succession of genetically distinct clones takes place.

In this paper we demonstrate that both allele and clone frequencies of *Daphnia longispina* in Lake El Tobar (Spain) show temporal variation from one collection period to the next. The isolation of this lake precludes a high frequency of exchanges between its population and those from other localities. While the causes of these temporal changes are difficult to assess they clearly lead to departures from Hardy-Weinberg equilibrium.

Materials and methods

Field methods

El Tobar is a karstic lake in the Cuenca mountains of Spain (UTM 30TWK806888) at an altitude of 1250 m. It consists of two main basins. The large basin (57 Ha, 12 m maximum depth) is an elongated limestone sink, fed mainly by underwater springs located along its eastern shore. The small basin (10 Ha, 19.5 m maximum depth) is located in an indentation of the northern shore. Its conductivity is about 0.6 mS cm^{-1} in the mixolimnion from the surface to the chemocline which is at a depth of about 12 m. Below the mixolimnion, conductivity drastically changes to reach almost 200 mS cm^{-1} at the bottom. The small basin lies at the base of a steep cliff, is more sheltered from the wind than the main basin and, because it protrudes laterally, is more isolated from the main water flow (NE-SW). More information on the morphometry, hydrology, chemistry and planktonic populations of this lake can be found in Vicente *et al.* (1993) and Miracle *et al.* (1993).

All samples were taken in the center of the small basin using ropes attached to trees and rocks on the shore to fix the position of the collecting boat. Vertical profiles of temperature, conductivity, oxygen and light penetration were measured *in situ* with the appropriate sensors (WTW and Li-Cor instruments). Water samples for chemical analyses and phytoplankton were taken with a Ruttner bottle. Zooplankton was filtered *in situ* from water samples taken with a double Van Dorn bottle (5.41 capacity and 35 μm mesh) and a Patalas trap (25 l capacity and 100 μm mesh).

Collections were made on four occasions: August 11, September 21–23, November 19, 1991 and April 22, 1992. For the last three sampling dates, day and night 1 m-interval profiles of zooplankton were taken. Formalin preserved samples were counted under an inverted microscope (Miracle *et al.*, 1993). All *Daphnia* from these samples were measured and the numbers of eggs per female counted. In September, when vertical heterogeneity was highest, four profiles, from different times of the day, were examined.

Genetic analysis of *Daphnia* was based on live samples taken by a net tow at the surface and with Van Dorn or Patalas traps at 5 and 10 m depths at noon and midnight of the above mentioned dates, except for August from which only noon samples from surface and 11 m were analyzed. Small supplementary samples from 5 and 9 m depths were also made in August to screen for electrophoretic variation. In September four collections were made on successive days, two at noon and two at midnight. Immediately after collection, in a caravan on the shore of the lake, adult individuals from each of the three depths were isolated and placed singly in 0.5 ml eppendorf tubes with 5 μl of tris-glycine tray buffer. They were then frozen on dry ice and transported to the laboratory in Valencia where they were stored at −80 °C.

Electrophoresis

Isozymes were detected by means of cellulose acetate electrophoresis (Hebert & Payne, 1985) of whole-animal homogenates using single daphnids. Individuals from several depths (surface net tow, 5, 9 and 11 m) of the August collection were screened for 13 enzymes. No variation was found in MDH (malate dehydrogenase, E.C. 1.1.1.37), APK (arginine phosphokinase, E.C. 2.7.3.3), IDH (isocitrate dehydrogenase, E.C. 1.1.1.42), FUM (fumarate hydratase, E.C. 4.2.1.2), 6PGDH (6-phosphogluconate dehydrogenase, E.C. 1.1.1.44) or GOT (glutamate oxaloacetate transaminase, E.C. 2.6.1.2). Variation was present in LDH (lactate dehydrogenase, E.C. 1.1.1.27), G6PDH (glucose-6-phosphate dehydrogenase, E.C. 1.1.1.49), MPI (mannose phosphate isomerase, E.C. 5.3.1.8), and ME (malic enzyme, E.C. 1.1.1.40); however these enzymes were not selected for further analysis because of technical limitations.

Three additional enzymes were found to be highly variable in the El Tobar samples, had strong activity with consistently good resolution, and produced readily interpretable patterns. Each of the individuals in this study was scored for genotype at these three loci. AO (aldehyde oxidase, E.C. 1.2.3.1) and PGI (phosphoglucose isomerase, E.C. 5.3.1.9) are dimeric enzymes and each had two alleles at a single locus. PGM (phosphoglucomutase, E.C. 5.4.2.2) is a monomer and had three alleles at a single locus. These three enzymes were used to score 78 individuals from the August collections, 600 from September, 341 from November and 258 from April.

Groups of phenotypically identical individuals based on the three enzymes we assayed are most precisely designated as either 'composite electromorphs' or 'multilocus genotypes' since these terms make no presumption regarding the variation – or lack of variation – present at other loci. The scoring of additional loci might well reveal that one or more of the different composite electromorphs could be further subdivided into distinct groups. This potential exists for any electrophoretic study of field populations. With this qualification in mind and for brevity and as a matter of convenience, the different composite electromorphs will be referred to herein as different 'clones'.

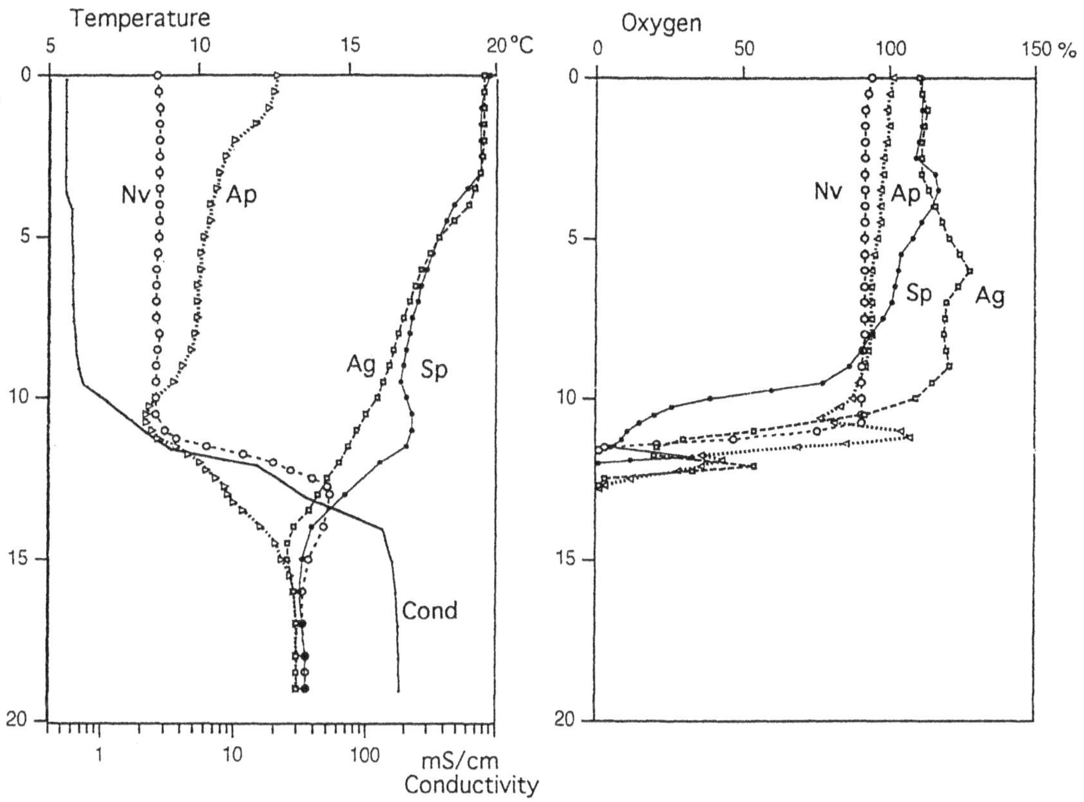

Fig. 1. Vertical profiles of temperature (°C) and oxygen (% saturation) for each of the collections (Ag: 11 August 91, Sp: 23 September 91, Nv: 19 November 91, Ap: 22 April 92. Conductivity (Cond, mS cm^{-1}) was about constant over the course of the study, only the September profile is plotted.

Calculations

Genetic similarities between collection dates were measured with Nei's Index of genetic identity (I) which quantifies the similarity of population structure in two samples. In a population of sexually reproducing individuals, the appropriate measures of population structure are allele frequencies. However, when reproduction is predominantly by parthenogenesis the allele frequencies will be functions of the rates of reproduction of the component clones (King, 1993). Therefore our measures of the genetic structure of the El Tobar *Daphnia* will be based on clonal composition and frequencies:

$$I = \frac{\sum(X_i Y_i)}{[\sum(X_i^2) \sum(Y_i^2)]^{1/2}},$$

where X_i is the relative frequency of clone *i* in population *X* and Y_i is the relative frequency of clone *i* in population *Y*. Genetic distance, *D*, is related to genetic identity by

$$D = -\log_e I.$$

In parthenogenetic organisms, clone diversity, *C*, is analogous to allelic diversity in a sexually reproducing population. Knowing p_i, the frequency of the ith clone, *C* is given by

$$C = 1 - \sum_{i=1}^{n} p_i^2.$$

C has a minimum value of 0 when the population is composed of a single clone, and a maximum value of 1 in an infinitely diverse population.

Results

Environmental variation in El Tobar

Lake El Tobar has fresh water from the surface to a depth of 10 to 12 m and from that point to the bottom

Table 1. Population parameters (means, M, and standard deviations, SD) of *Daphnia* in El Tobar: total individuals/m^2, maximum density in the vertical profile, percentage of adults and males with respect to the total, eggs per adult female, mean brood size and size at first reproduction.

	September				November		April	
	10–12 m		Total column					
	M	SD	M	SD	M	SD	M	SD
Ind/m^2	24240	(3500)	106160	(11800)	166340	(400)	20700	(3200)
Max. density (ind/l)	18		24		6		7	
Adults %	77	(8)	46	(6)	16	(0.8)	25	(1.5)
Males %	3.8	(1.7)	2.1	(0.4)	1.0	(0.5)	5.6	(4)
Eggs/adult female	0.2	(0.1)	0.3	(0.0)	1.4	(0.3)	2.2	(0.2)
Mean brood size	2.2	(0.4)	2.2	(0.4)	3.6	(1.1)	4.2	(1.6)
Size at 1st reprod.,mm	1.6		1.5		1.4		1.4	

Adults: individuals >1.23 mm (size of the smallest egg-bearing female)

has an anoxic salt water layer. Extensive changes in temperature, oxygen, light penetration and chlorophyll occur at the freshwater-salt water interface (Figs 1 and 2) and in all four collections there was a peak of stratifying algae at about the 12 m level. In general, most of the *Daphnia* occurred in the top ten meters which were dominated by the chlorophyte *Pedimonas minor* (>50% of all individuals) in September, and by the diatom *Cyclotella ocellata* (>60%) in November. *Rhodomonas minuta* was the predominant (>60%) phytoplankton species in the following April. Clearly, there is a great deal of variation in the chemical, physical and biological features of lake El Tobar.

Dynamics of the El Tobar Daphnia

The *Daphnia* species of Lake El Tobar belongs to the taxonomically difficult 'longispina' group of species *rosea- longispina-hyalina* that has been referred to as *D. longispina* in other Spanish investigations (Margalef, 1953; Armengol, 1978; Alonso, 1985; Miracle, 1978). As shown in Fig. 3, *Daphnia* occurs mainly in the upper waters of El Tobar during the mixing period and only during the summer stratification does it have a bimodal distribution with a deep abundance peak at the top of the fresh-salt-water interface (around 10–11 m). This abundance peak is associated with the oxycline and is therefore above and separated from the chlorophyll peak and the steep salinity gradient.

The dynamics of this population have not been studied in detail, but in Table 1 we summarize results that demonstrate variation in population density, fecundity

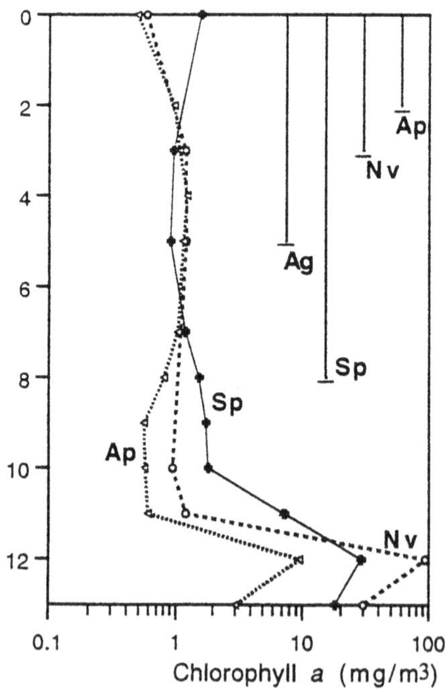

Fig. 2. Secchi depths and vertical profiles of chlorophyll (mg ml^{-1}) at the times of the September, November and April collections.

and reproductive parameters. In September, when the population has a bimodal distribution, fecundity rates were low but overall abundance and the proportion of adults were very high, particularly in the deeper strata. Also a large percentage of the deep individuals collected in September were infected with the microsporidian *Gurleya vavrai* (Green, 1974). After the autumn overturn the deep peak of *D. longispina* disappeared (Fig. 3)

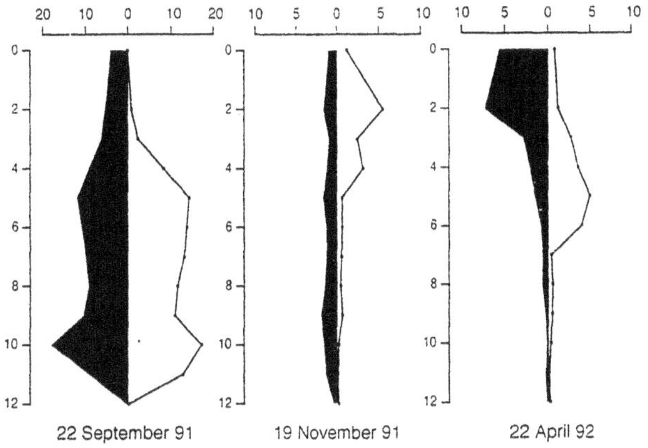

Fig. 3. Day (unshaded) and night (shaded) vertical profiles of abundance (ind. l^{-1}) in the September, November and April collections.

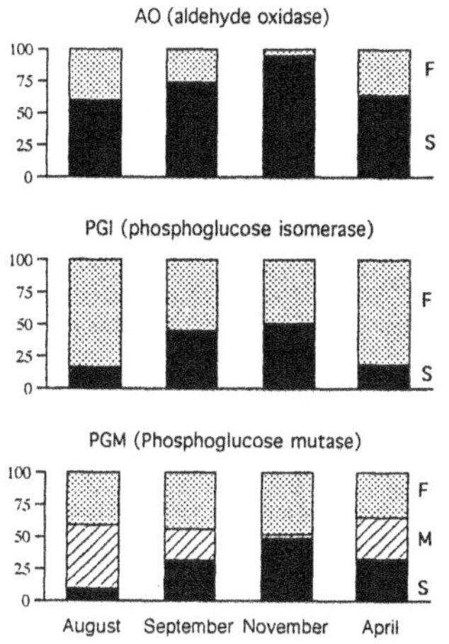

Fig. 4. Allele frequencies at the AO, PGI and PGM loci for each collection.

and the population density decreased to about 15% of its September valve.

The November population was predominantly composed of young individuals that had higher fecundity rates than were observed in September. In addition, the size at first reproduction was lower than in September and the microsporidian infection was almost nonexistent. Because of the low temperature in November, developmental times were much longer. Because there was also a decrease of phytoplankton production and shift in dominance of the algal groups between our September and November collections, it is clear that there are a number of environmental factors that could account for the observed density reduction and changes in the age structure of the *Daphnia* population.

The April *D. longispina* population had a low density, but both mean brood size and fecundity rates were quite high (Table 1). Curiously, the April population also had the highest proportion of males we observed. Ephippia were not observed in any of the four sampling dates, but were seen in previous years (8 November 1989 personal observations).

Allele and genotype frequencies

Two alleles, F and S, were present at the AO locus. Allele S had a frequency of 60% in August, 74% in September and 95% in November. In the April samples the frequency of S, 64%, was similar to that of the preceding August (Fig. 4). Similarly, two alleles, F and S, were present at the PGI locus and allele S increased from 17% in August to 45% in September and to 51% in November; by the following April the frequency of S, 18%, was again close to that of the preceding August. At the PGM locus three alleles were found, S, M and F. The frequency of allele F was relatively stable in the four collections while S increased from a frequency of 10% in August, to 32% in September, and to 49% in November. Over the same period allele M decreased from its frequency of 49% in August to 24% in September and to 3% in November. In April the three alleles had similar frequencies (33, 32 and 35%).

The observed allele frequencies can be used to generate expected Hardy-Weinberg genotype frequencies (Fig. 5). Significant heterozygote excesses, as tested with chi square statistics, were found for PGM in all collections, for AO in all except November, and for PGI in September and November. We also employed an alternative method to evaluate heterozygote frequencies, Wright's (1965) F coefficients (see also Nei, 1977). F_{IS} characterizes the average departure from Hardy-Weinberg expected frequencies of heterozygotes at each locus taken over the four collections. FIT measures the deviations from the expected frequencies of heterozygotes obtained by combining the four collections into a single data set. For both coefficients, negative values indicate heterozygote excesses while positive values indicate heterozygote deficiencies. Reinforcing the results of the chi square analysis, both coefficients yield negative values for all three loci (Table 2).

Table 2. F coefficients for each allozyme locus averaged over the individual collections (F_{IS}) and for the pooled data from the four collections (F_{IT}). Negative values indicate departures from random recombination in the direction of heterozygote excesses.

	AO	PGI	PGM
F_{IS}	−0.40	−0.07	−0.44
F_{IT}	−0.50	−0.13	−0.44

Table 3. Clone number, diversity and frequencies in the four collections. Composite electromorph identifications are presented in the order AO genotype - PGI genotype - PGM genotype.

		Frequency (%)			
Clone	Electromorph	Aug	Sep	Nov	Apr
ET1	SS - SF - SS	3.8	25.0	52.8	6.8
ET2	SF - SF - MF	23.1	30.3	0.6	1.6
ET3	SF - FF - MF	51.3	0.8	1.2	30.3
ET4	SS - FF - SM	12.8	3.2	0.9	12.7
ET5	SS - FF - SF	1.3	3.8	5.3	0.4
ET6	SS - SS - SF	0.0	3.5	8.8	2.0
ET7	SS - SF - FF	1.3	3.8	5.0	1.6
ET8	SS - SS - FF	0.0	2.3	5.9	0.4
ET9	SF - FF - SF	0.0	2.0	4.7	17.9
ET10	SF - FF - SM	0.0	8.7	2.9	11.2
All other clones:		6.4	16.6	11.9	15.2
Sample size (no. individuals):		78	600	341	258
Number of Clones:		9	21	16	22
Clone Diversity		0.66	0.83	0.70	0.84

Table 4. Comparisons of the genetic distance (D) and genetic identity (I) between the clones in the four collections. D is presented above the diagonal, I below the diagonal.

	Collection			
	Aug	Sep	Nov	Apr
Aug	–	0.96	2.35	0.26
Sep	0.37	–	0.46	1.30
Nov	0.09	0.65	–	1.37
Apr	0.77	0.27	0.27	–

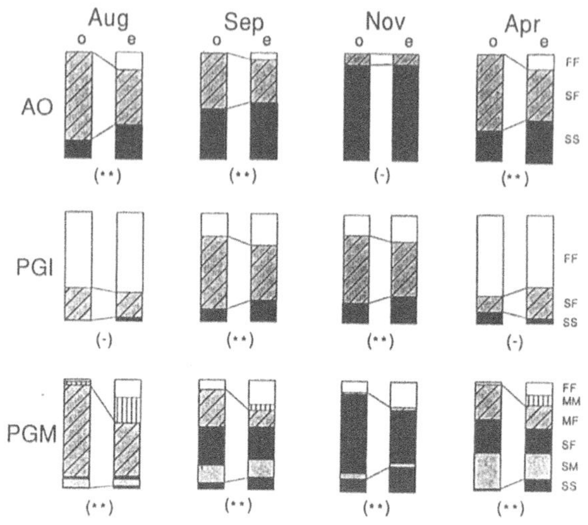

Fig. 5. Observed and expected (Hardy-Weinberg) genotype frequencies for each locus and collection. Significant ($p<0.01$) differences between pairs by a chi-square test are indicated by '(**)'; non-significant ($p>0.05$) differences are indicated by '(-)'.

Clone frequencies

Different composite electromorphs clearly represent different clones. However, as emphasized earlier, additional variation may occur at non-assayed loci so the observed number of clones reported in this paper is a conservative estimate.

Since both the AO and PGI loci had two alleles and the PGM locus had three, a maximum of 54 clones is potentially detectable using this electrophoretic protocol. We scored a total of 1337 individuals for variation at the three loci and found 27 of the 54 clones. Although SF heterozygotes of AO were common, no FF homozygotes were observed at this locus. The total absence of an expected homozygote has also been observed in *Daphnia* from other lakes. For instance, Jacobs (1990) studied the *Daphnia cuculata* in Klostersee lake for over 2.5 years and never found one of the homozygotes at the PGI locus. However, the missing genotype was common in a relatively close dystrophic lake with very different ecological conditions.

The numbers and diversities of clones in each collection are presented in Table 3. Clonal diversity was highest in September and April.

Ten of the clones we detected had a frequency of at least 5% in one or more of the collections (Table 3). Six of the ten were present in all four collections. Only three clones-ET1, ET2 and ET3-attained frequencies greater than 25% and each was numerically dominant in a different collection. ET3 was most abundant in

August and then almost disappeared until the following April when it again was the most abundant clone. ET2 was abundant in both August and September and had its highest frequencies in the deep waters where the population density reach a maximum at about 10 to 11 m. ET1 was abundant in both September and November and had its highest densities in the upper waters. More detailed information on the dynamics of ET1 and ET2 is presented in King & Miracle (1995).

The April–August and the September–November populations had much higher genetic identities than any of the other pairs (Table 4). Interestingly, the August and November populations had the greatest genetic distance.

Discussion

The first large-lake studies of allozyme frequencies of *Daphnia* (Mort & Wolf, 1986) reported that they were generally in Hardy-Weinberg equilibrium and stable through time, in contrast to those in permanent ponds which show heterozygote excess and rapid frequency changes (see Hebert 1987 for a review). However, heterozygote excess and temporal changes in other lakes have been found to be in the range reported from permanent ponds (e.g. Jacobs, 1990). In El Tobar, which is clearly classified as a lake, 9 of the 12 allele frequency measures showed significant deviations from Hardy-Weinberg equilibrium and clonal frequencies changed from one sampling date to the next. Paralleling Jacobs (1990) results, ephippial females in El Tobar are uncommon, and males occur in very low frequencies over an extended period.

One phenomenon that can cause departures from Hardy-Weinberg equilibrium is hybridization between species. Indeed, hybridization between species of *Daphnia* and recombination among their backcrosses have been clearly demonstrated by Wolf & Mort (1986) and by Hebert *et al.* (1989) in Europe and by Taylor & Hebert (1992, 1993) in North America. Because of its obvious importance, we have examined both morphological and genetic data for indications of hybridization in El Tobar.

In summer a very small proportion, 0.3%, of the individuals in our samples had one or two neck teeth located in a somewhat higher position than those described for *D. rosea* (in Margaritora, 1985 or Negrea, 1983). A very similar toothed form has been described for *D. longispina* by Johnson (1952) and Christie (1983). These neck teeth were found in some summer primiparae and males but were absent in other seasons and individuals. This rare toothed form is the only notable morphological variability seen in the course of our study. It shows no correspondence with the allozyme variation we studied. Our samples showed no indication of the morphological variability and intergradations that usually accompany interspecific hybridization and it is not possible to attempt to relate morphological variation with the observed genetic differences.

Our genetic data also exclude hybridization and other forms of population subdivision as an explanation for the observed departures from Hardy-Weinberg equilibrium. Wahlund's principle (see Crow & Kimura, 1970 for a discussion) is a basic tenet of population genetics and describes allele frequencies when the population is subdivided into isolated or partially isolated groups. Even when the separate units are in equilibrium, whenever a mixed assemblage is combined into a single sample because of unrecognized population subdivision, the Wahlund effect leads to a departure from Hardy-Weinberg frequencies. However, these deviations are always in the direction of a net deficiency of heterozygotes. While our samples did depart from Hardy-Weinberg equilibrium, the departure was in the direction of an excess of heterozygotes.

We must therefore turn to other mechanisms that might explain the observed deviations from Hardy-Weinberg equilibrium. The most likely explanation, particularly in view of the dramatic changes in clone frequencies as well as the described environmental changes over the period of our study, is that the *D. longispina* population in lake El Tobar is subject to intense seasonal selection of the type discussed by King (1972, 1977) for rotifers. When clones vary in their fitness responses to physical and biotic conditions, environmental changes may lead to a succession of different clonal types.

Paralleling the dramatic temporal changes in the physical, chemical and biological characteristics of lake El Tobar is a change in the genetic structure of the *Daphnia* population. In April, near the start of an annual cycle, both the number and diversity of clones in the lake is very high. Six clones were found only in April, and of these three had frequencies over 1% (up to 3.6%). Among all other collections, only the September collection had exclusive clones.

September marks the end of a period of relative stability in which the lake has stratified and attained a high degree of vertical heterogeneity. The diversity of the whole planktonic community is also higher (Miracle *et al.*, 1993). By November, the deep maximum of the *Daphnia* population has been lost, and both the demographic and clonal structure of the population have been dramatically altered. These changes demonstrate that the ecological as well as the genetic structure of the El Tobar *Daphnia* population undergoes substantial temporal modifications.

As reviewed by Hebert (1987), Wolf (1988), and Lynch (1987), Hardy-Weinberg equilibria occur in some pond and lake populations and are absent in others. Do these results demonstrate a contrast between the adaptive structures of the *Daphnia* populations, or are they due to environmental differences between the habitats studied?

The Hardy-Weinberg principle stands as a null hypothesis linking allele frequencies in the successive generations of a population of sexually reproducing organisms. During the diploid, ameiotic parthenogenetic reproduction of *Daphnia*, the genome of each clone is effectively congealed into a single unit. Allele frequencies, under these circumstances, are determined by a process of hitch-hiking in which the frequencies of alleles at non-selected loci will reflect the operation of selection on loci that are directly related to the clone's fitness characteristics. Allele frequencies, therefore, will be simple functions of clone frequencies and their relative rates of reproduction.

These factors complicate the genetic analysis of evolutionary processes in cyclical parthenogens. A more powerful approach is needed if we are to develop an understanding of the genetic structure of *Daphnia* and other cyclical parthenogens. The solution, we suggest, is to focus not on allele frequencies but on individual clones and the environmental factors that influence their frequencies.

Acknowledgments

We are most grateful to M. Dolores Boronat for her important assistance with the electrophoresis and the experimental part of this work. We thank also Ros Jagoe for her help when she was an Erasmus student in Spain and Manuel Serra for providing helpful comments on the manuscript. This research was supported by DGICYT grant NT 89-1124 to MRM and DGICYT grant for CEK on his sabbatical leave from Oregon State University. We also acknowledge the cooperation of University of Valencia and 'Consejeria de Agricultura de Castilla-La Mancha'.

References

Alonso, M., 1985. Las lagunas de la España peninsular: taxonomía, ecología y distribución de los cladóceros. Ph. D. thesis. University of Barcelona.

Armengol, J., 1978. Los crustáceos del plancton de los embalses españoles. Oecol. aquat. 3: 3–96.

Christie, P., 1983. A taxonomic reappraisal of the *Daphnia hyalina* complex (Crustacea, Cladocera): an experimental and ecological approach. J. Zool. Lond. 199: 75–100.

Crease, T. J., D. J. Stanton & P. D. N. Hebert, 1989. Polyphyletic origins of asexuality in *Daphnia pulex*. II Mitochondrial-DNA variation. Evolution 43: 1016–1026.

Crow, J. F. & M. Kimura, 1970. An introduction to population genetic theory. Harper and Row, New York.

Green, J., 1974. Parasites and epibionts of Cladocera. Trans. zool. Soc. Lond. 32: 417–515.

Hebert, P. D. N., 1987. Genotypic characteristics of the Cladocera. Hydrobiologia 145 (Dev. Hydrobiol. 35): 183–193.

Hebert, P. D. N. & W. J. Payne, 1985. Genetic variation in populations of the hermaphroditic flatworm *Mesostoma lingua* (Turbellaria, Rhabdocoela). Biol. Bull. 169: 143–151.

Hebert, P. D. N., S. S. Shwartz & J. Hrbacek, 1989. Patterns of genotypic diversity in Czechoslovakian *Daphnia*. Heredity 62: 207–216.

Hebert, P. D. N., R. D. Ward & L. J. Weider, 1988. Clonal-diversity patterns and breeding system variation in *Daphnia pulex*, an asexual-sexual complex. Evolution 42: 147–159.

Jacobs, J., 1990. Microevolution in predominantly clonal populations of pelagic *Daphnia* (Crustacea: Phyllopoda): Selection, exchange, and sex. J. Evol. Biol. 3: 257–282.

Johnson, D. S. 1952. The British species of the genus *Daphnia* (Crustacea, Cladocera) Proc. zool. Soc. Lond. 122: 435–462.

King, C. E., 1972. Adaptation of rotifers to seasonal variation. Ecology 53: 408–418.

King, C. E., 1977. Genetics of reproduction, variation, and adaptation in rotifers. Arch. Hydrobiol. Ergebn. Limnology 8: 187–201.

King, C. E., 1993. Random genetic drift during cyclical ameiotic parthenogenesis. Hydrobiologia 255/256 (Dev. Hydrobiol. 83): 205–212.

King, C.E. & M.R. Miracle, 1995. Diel vertical migration by *Daphnia longispina* in a Spanish lake: Genetic sources of distributional variation. Limnol. Oceanogr. 40(2): *in press*.

Lynch, M., 1987. The consequences of fluctuating selection on isozyme polymorphisms in *Daphnia*. Genetics 115: 657–669.

Margalef, R., 1953. Los crustáceos de las aguas continentales ibéricas. Inst. Forestal de Investigaciones y Experiencias. Madrid, 243 pp.

Margaritora, F., 1985. Cladocera. In Fauna de Italia. Edizioni Calderini. Bologna, 399 pp.

Miracle, M. R., 1978 Composición específica de las comunidades zooplanctónicas de 153 lagos de los Pirineos y su interés biogeográfico. Oecol. Aquat. 3: 167–191.

Miracle, M. R., J. Armengol-Díaz & M. J. Dasí, 1993. Extreme meromixis determines strong differential planktonic vertical distributions. Verh. int. Ver. Limnol. 25: 705–710.

Mort, M. A. & H. G. Wolf, 1986. The genetic structure of large–lake *Daphnia* populations. Evolution 40: 756–766.

Negrea, S., 1983. Crustacea Cladocera. Fauna republici socialiste Romania. Vol. 4, Fasc. 12 Acad. Rep. Soc. Romania, 399 pp.

Nei, M., 1977. F-statistics and analysis of gene diversity in subdivided populations. Ann. Hum. Genet. 41: 225–233.

Spaak, P., 1993. Genetical ecology of a coexisting *Daphnia* hybrid species complex. Ph.D. Thesis, University of Utrecht, 125 pp.

Taylor, D. J. & P. D. N. Hebert, 1992. *Daphnia galeata mendotae* as a cryptic species complex with interspecific hybrids. Limnol. Oceanogr. 37: 658–665.

Taylor, D. J. & P. D. N. Hebert, 1993. Habitat-dependent hybrid parentage and differential introgression between neighboringly sympatric *Daphnia* species. Proc. natn. Acad. Sci. USA. 90: 7079–7083.

Vicente, E., A. Camacho & M. A. Rodrigo, 1993. Morphometry and physicochemistry of the crenogenic meromictic lake El Tobar (Spain). Verh. Int. Ver. Limnol. 25 (in press).

Weider, L. J., 1985. Spatial and temporal genetic heterogeneity in a natural *Daphnia* population. J. Plankton Res. 7: 101–123.

Weider, L. J., 1993. Niche breadth and life history variation in a hybrid species complex of *Daphnia*. Ecology 74: 935–943.

Wolf, H. G., 1988. Differences in the genetic structure of pond-dwelling and lake-dwelling *Daphnia*. Verh. int. Ver. Limnol. 23: 2056–2059.

Wolf, H. G. & M. A. Mort, 1986. Inter-specific hybridization underlies phenotypic variability in *Daphnia* populations. Oecologia 68: 507–511.

Wright, S., 1965. The interpretation of population structure by F-statistics with special regard to systems of mating. Evolution 19: 393–420.

Hydrobiologia **307**: 25–32, 1995.
P. Larsson & L.J. Weider (eds), Cladocera as Model Organisms in Biology.

Differences in genetic structure and ecological diversity between parental forms and hybrids in a *Daphnia* species complex

Jakob Müller & Alfred Seitz
Department of Population Biology, Zoological Institute of the University, Saarstr. 21, 55099 Mainz, Germany

Key words: clonal diversity, *Daphnia*, hybrids, mode of reproduction, parentals, spatial heterogeneity, temporal heterogeneity

Abstract

Hybridization is a common phenomenon in *Daphnia* species complexes. Hybrids often dominate in *Daphnia* populations; therefore it is worthwhile to look for principal differences between parental and hybrid populations with respect to their genetic structure and clonal differentiation. We studied natural populations of members of the *Daphnia galeata/hyalina/cucullata* complex in three lakes. In one of these lakes, one parental species (*D. galeata*) and one hybrid (*D. galeata* × *cucullata*) were investigated more intensively. The frequency of sexual reproduction was higher in parental populations, whereas clonal diversity was higher in hybrid populations. Ecological differentiation among clonal groups was more pronounced in the *D. galeata* × *cucullata* hybrid compared to *D. galeata*, whereas selection intensity was weaker. These results are discussed with respect to stability of clonal groups, multiple hybridizations and selective constraints.

Introduction

From morphological studies, the occurrence of hybrids in some *Daphnia* groups has been long suspected (Brooks, 1957; Lieder, 1983). Genetic evidence has come from allozyme electrophoresis. Now, interspecific hybridization has been documented for many *Daphnia* species complexes such as *D. carinata/cephalata* (Hebert, 1985), *D. galeata/hyalina/cucullata* (Wolf & Mort, 1986), *D. pulex/pulicaria* (Hebert *et al.*, 1989) and *D. galeata mendotae/rosea* (Taylor & Hebert, 1992). In general, hybrid clones show intermediate life-history traits when compared with parentals (Weider & Wolf, 1991). However it appears that certain characters expressed by the hybrids resemble one of the two parentals more than the other. For example, the hybrid *D. galeata* × *cucullata* seems to combine the small size of *D. cucullata* with the high reproductive rate of *D. galeata* (Spaak & Hoekstra, 1995). In the field, hybrid populations dominate in intermediate habitats when compared with parentals (Taylor & Hebert, 1992) or they occur in high abundances together with only one of their parentals (Hebert, 1985; Wolf, 1987; Bast, 1993). For example, *D. galeata* × *cucullata* hybrids can be abundant together with *D. galeata* in the same lake at the same time, but often they do not coexist in great numbers with *D. cucullata* (Wolf, 1987; Bast, 1993).

The intermediacy of hybrid characters alone does not explain their successful distribution in most natural lakes. Therefore, some general differences are suspected in the genetic structure and ecological diversification among clones between hybrid and parental populations. Here we present comparisons of genetic structure among the members of the *D. galeata/hyalina/cucullata* complex. Concerning the spatial and temporal dynamics of clones, results from one parental (*D. galeata*) and one hybrid population (*D. galeata* × *cucullata*) are shown in more detail.

Methods

We studied *D. galeata*, *D. hyalina*, *D. cucullata*, and their three interspecific hybrids in three permanent lakes that are located in southwestern Germany:

Meerfelder Maar (eutrophic), Neuhofener Altrhein (eutrophic) and Hegbachsee (hypertrophic). *D. galeata* and *D. galeata* × *cucullata* (*D. gal* × *cuc*) in Neuhofener Altrhein were studied more intensively. We took samples at the deepest site by vertical hauls with a plankton net for the temporal survey (May 1989 until June 1991). During each sampling bout, secchi disk depths, vertical temperature profiles, population densities of each *Daphnia* form (Bast, 1993, counting method) and fish densities (Wagner, 1992, echo-sounding method) were measured. For the spatial survey, we took samples in vertical series at 1 m intervals in May, July, October 1990, and May 1991. This was done by pumping water from each meter depth interval through a plankton net. Separate samples were taken at noon and the following midnight. Sample sizes per *Daphnia* type per date (temporal survey) or per depth (spatial survey), respectively, were about 36 individuals.

We used cellulose acetate electrophoresis (Hebert & Beaton, 1989) to assay individual adults at six polymorphic allozyme loci: aldehyde oxidase (AO, EC 1.2.3.1), glutamate oxaloacetate transaminase (GOT, EC 2.6.1.2), malate dehydrogenase (MDH, EC 1.1.1.37), mannosephosphate isomerase (MPI, EC 5.3.1.8), phosphoglucomutase (PGM, EC 2.7.5.1) and phosphoglucose isomerase (PGI, EC 5.3.1.9). It has been shown previously that morphological discrimination (Flößner & Kraus, 1986; Flößner, 1993) among all *Daphnia* forms was in good agreement with the genetic data (Müller, 1993). Therefore, we used the diagnostic locus GOT, to discriminate between all *Daphnia* forms (Wolf & Mort, 1986).

Our *Daphnia* populations reproduce predominantly by parthenogenesis and only occasionally by sexual reproduction. Therefore, it is convenient to consider them as assemblages of clones during most of the year. Within each *Daphnia* taxon, individual daphnids were classified into clonal groups according to their three-locus genotypes (MPI, PGM and PGI). We used the combination of only three loci as markers of clonal groups to get sufficiently large sample sizes per clonal group for statistical tests. In the following, we use clonal groups and genotypes as synonyms.

G-tests (likelihood ratio tests) for agreements with Hardy-Weinberg expectations (rare alleles pooled) were performed for all loci (except the diagnostic locus GOT). The negative common logarithms of the probabilities of these G-tests ($-\log P(G)$) were taken as measures for the frequency of sexual reproduction within a population (Hebert *et al.*, 1988). Clonal diversity was calculated as the effective number of genotypes (D_{eff}):

$$D_{eff} = 1/(\sum p_i^2), \quad (1)$$

with p_i being the frequency of the ith genotype (Simpson, 1949). This measure gives more weight to frequent genotypes and is therefore less influenced by sample size (Parker, 1979). For the temporal survey, genotype-specific selection coefficients (s) were calculated:

$$s = \{\ln((1-p_0)/p_0) - \ln((1-p_1)/p_1)\}/T, \quad (2)$$

with p_0 and p_1 being the genotype frequencies at the beginning and at the end of a given time interval and T being the length of the time interval (Lynch, 1987). The mean absolute values of all selection coefficients were used as measures for selection intensities and their temporal fluctuations.

Results

In order to assess the mode of origin and reproduction in these *Daphnia* populations, we examined deviations from Hardy-Weinberg equilibria ($-\log P(G)$) and levels of heterozygosity (Fig. 1). Almost all of the *Daphnia* populations displayed significant Hardy-Weinberg deviations ($-\log P(G) > 1.3$). In all lakes, the hybrid populations displayed stronger deviations from Hardy-Weinberg equilibria than the parental populations. On average, hybrids deviated more strongly (mean ± standard error: 3.6 ± 1.5) than parental forms (2.0 ± 0.9) (t-test: $t = 4.3$, $df = 42$, $p < 0.01$). Similar patterns were observed for the heterozygosities with higher values for the hybrids (0.66 ± 0.14) and lower values for the parentals (0.40 ± 0.15) (t-test: $t = 6.4$, $df = 42$, $p < 0.01$; values prior to the test were arcsin square-root transformed). These results indicate that in populations of parental forms, sexual reproduction is rare but more frequent than in hybrid populations. This suggests that hybrid populations may consist mainly of F_1-hybrids, and is supported by their high heterozygosities.

There was also a difference in clonal diversity between hybrids and parentals. In the total spatial and temporal survey in Neuhofener Altrhein with comparable sample sizes (*D. galeata*, N = 1658; *D. galeata* × *cucullata*, N = 1793), we found more three-locus genotypes (50) in the *D. galeata* × *cucullata* hybrid population than in the *D. galeata* parental population (24). Clonal diversities and evenness for all *Daphnia*

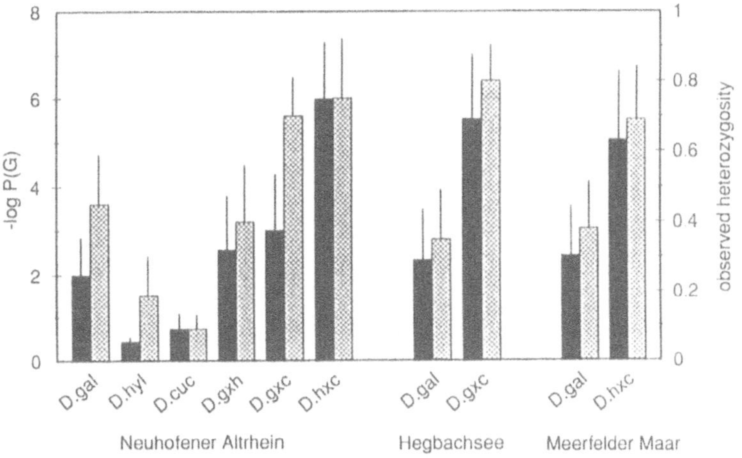

Fig. 1. Deviations from Hardy-Weinberg equilibria (-logP(G); black columns) and heterozygosities (hatched columns) of all *Daphnia* populations; means±standard errors; average values over the five loci AO, PGI, MPI, MDH, PGM and all dates of the temporal survey; D.gal = *D. galeata*, D.hyl = *D. hyalina*, D.cuc = *D. cucullata*, D.gxh = *D. galeata x hyalina*, D.gxc = *D. galeata x cucullata*, D.hxc = *D. hyalina x cucullata*.

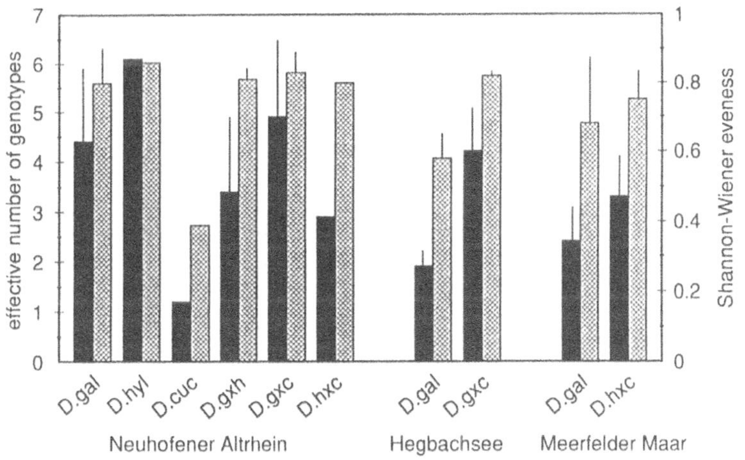

Fig. 2. Clonal diversity (effective number of genotypes; black columns) and clonal evenness (Shannon-Wiener evenness; hatched columns) of all *Daphnia* populations, based on three-locus genotypes; means±standard deviations are given, averaged over all dates of the temporal survey; for abbreviations see Fig. 1.

populations are shown (Fig. 2). On average, hybrids showed higher diversities (4.3 ± 1.4) than parentals (3.7 ± 1.7). This difference was only significant (t-test, 5% level) in the Hegbachsee. The same trend held true for the evenness with higher values for the hybrids (0.81 ± 0.05) and lower values for the parentals (0.74 ± 0.16).

In a previous study, we found that the vertical distributions of the most abundant genotypes of *D. galeata* and *D. galeata* × *cucullata* differed significantly during both day and night (Müller & Seitz, 1993). The most abundant genotypes were defined as those genotypes, that comprised 10% or more in at least one of the water depth intervals. Further, it appeared that diurnal vertical migration behaviour was more heterogeneous among hybrid genotypes than among *D. galeata* genotypes. For example, more genotypes with reverse migration behaviour were found for the hybrid. In Müller & Seitz (1993) we analysed five-locus genotypes. Here we present the results of G-tests of independence for vertical depth distributions of the most abundant three-locus genotypes (Table 1). One addi-

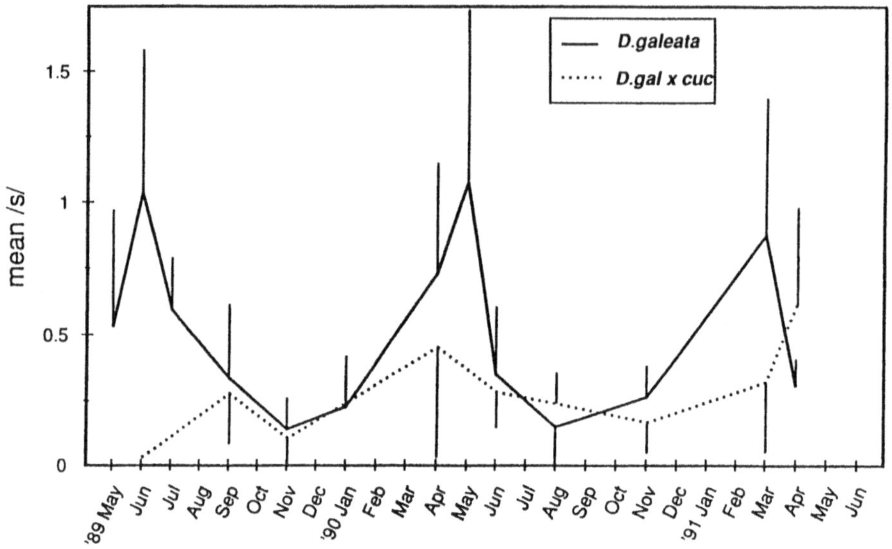

Fig. 3. Temporal dynamics of mean selection coefficients (absolute values) for the most abundant genotypes of *D. galeata* and *D. galeata x cucullata*; mean |s|±standard deviations.

Table 1. Probability values of RxC-tests (G-tests) for independence of vertical depth distributions of the most abundant genotypes; there were three to seven abundant three-locus genotypes per test; values in bold represent lower probability values.

Time	Date	D. galeata	D. gal x cuc
Noon	May 90	0.0475	**0.0143**
	July 90	0.1150	**0.0398**
	May 91	0.0029	**<0.0001**
Midnight	May 90	<0.0001	<0.0001
	July 90	**0.0331**	0.0413
	May 91	0.0805	**<0.0001**

tional date is included (July 1990). The probability values for the G-tests are given, which are in most cases (in particular during daytime) smaller for *D. galeata* × *cucullata* than for *D. galeata*. This corroborates the previous results that genotypic variation in vertical distribution in the hybrid population is more pronounced than in the parental population.

On a temporal scale, the situation seems to be just the opposite, with *D. galeata* genotypes being more variable. Figure 3 shows the temporal dynamics of the means of the absolute selection coefficients. For both *Daphnia* forms, a seasonal trend can be seen with maxima in spring and early summer. However, selection coefficients and therefore shifts in clonal groups, were more pronounced for the *D. galeata* population. The grand means over time and all genotypes were 0.51 ± 0.24 (S.E.) for *D. galeata* and 0.28 ± 0.13 (S.E.) for *D. galeata* × *cucullata*. This difference was significant ($t = 2.6$, $df = 20$, $p < 0.05$). However, selection coefficients can vary over short-time intervals and they can change their sign within a given sampling interval. Therefore the calculated coefficients depend on the lengths of the sampling intervals, which were not always the same for both *Daphnia* types. For a better comparison, we calculated the coefficients for corresponding time intervals for both *Daphnia* types. Even after this correction, *D. galeata* showed higher values, although not significantly higher than for the hybrid.

To find out which environmental parameters were relevant for the observed temporal fluctuations in selection intensities, we performed a correlation analysis with some simultaneously measured environmental factors (Table 2). Significant correlations were found only for *D. galeata* for fish densities and temperature range (i.e. difference between maximum and minimum values) for the inhabitable water strata (i.e. water strata > 1 mg/l oxygen content). When the temperature range is broad, then differential vertical distributions of genotypes can result in a divergence in life history parameters via temperature effects; for example growth, natality and temperature-dependent predation pressure.

Table 2. Correlation analysis (Pearson's correlation coefficients) between mean selection coefficients per month (mean |s|) and environmental factors; for *D. galeata*, 13 cases are included, except for fish density (11 cases); for *D. galeata x cucullata*, 9 cases are included, except for fish density (7 cases); * $p<0.05$.

Daphnia form	Transparency depth	Temperature Mean	Temperature Range	Population density	Fish density
D. galeata	0.046	0.406	0.549*	0.380	0.548*
D. gal x cuc	0.417	−0.087	0.391	−0.087	−0.020

Table 3 a. ANOVA (nested design) of lengths of adults of all abundant genotypes in July 90 (*D. galeata*) and July and October 90 (*D. gal x cuc*); **spatial survey**; noon- and midnight- samples; *$p<0.05$; **$p<0.01$; ***$p<0.001$.

Source	df	MS	F
D. galeata			
time (of day)	1	0.001	0.12
genotypes within time	8	0.009	1.50
water depths within time	6	0.029	**4.77****
genotypes × depths within time	23	0.010	1.76
within cells (error)	41	0.006	
D. gal x cuc			
date	1	0.003	0.02
time (of day) within date	2	0.014	2.92
genotypes within time × date	8	0.042	**8.38*****
water depths within time × date	20	0.009	**1.87***
genotypes × depths within time × date	40	0.010	**2.06*****
within cells (error)	345	0.005	

Table 3 b. ANOVA (two-way) of lengths of adults of all abundant genotypes from June 89 to June 91; **temporal survey**; *$p<0.05$; ***$p<0.001$.

Source	df	MS	F
D. galeata			
genotypes	3	0.034	2.39
date	10	0.165	**11.54*****
genotypes × date	23	0.017	1.20
error	156	0.014	
D. gal x cuc			
genotypes	4	0.061	**7.00*****
date	10	0.396	**45.12*****
genotypes × date	27	0.015	**1.74***
error	207	0.009	

Finally, we looked for genotypic differentiation in body length. Body length is a crucial parameter (e.g. for mortality via size-selective predation or for natality via clutch size, Seitz, 1980). Table 3 shows the results of an analysis of variance with length of adults as the dependent variable. Tests were done separately for the spatial (Table 3a) and temporal surveys (Table 3b). The body lengths of *D. galeata* varied significantly only with regard to water depth and between dates. In general, body length increased with water depth during daytime (Bast, 1993). For the *D. galeata* × *cucullata* hybrid, significant genotype and interaction effects also were found. In other words, the body lengths of the hybrid genotypes varied significantly and fluctuated in different ways with respect to water depth and time.

Discussion

In general, all the *Daphnia* populations examined in this study displayed significant deviations from Hardy-Weinberg expectations (Fig. 1). The more intensively studied populations of *D. galeata* and *D. galeata* × *cucullata* deviated significantly from Hardy-Weinberg expectations on all dates, and thus indicates that sexual reproduction within these pelagic populations is quite infrequent. Shifts to weaker deviations from Hardy-Weinberg expectations (interpreted as recruitment of sexually produced ephippia) were found during periods of maximal genotypic fitness differences (spring and early summer, see Müller & Seitz, 1994).

In comparison with hybrid populations, parental populations showed weaker deviations, i.e. interpreted as more frequent sexual reproduction. In terms of coexistence of males and ephippia-bearing females, the potential for sexual reproduction was also higher for *D. galeata* compared to *D. galeata* × *cucullata* (Müller & Seitz, 1994). There is some evidence that sexual reproduction (i.e. formation of F_2-generation or backcrossing) plays a minor part in hybrid populations and that these populations consist essentially of F_1-hybrids: (i) Hybrids form quite distinct morphological groups (Flößner & Kraus, 1986), except *D. galeata* × *hyalina*, which exhibits gradual morphological transitions that resemble *D. galeata*; (ii) Heterozygosity is significantly higher in hybrid populations. A test for independence of observed and expected heterozygosities in hybrid populations was not performed, because we sampled two parental populations (*D. hyalina* and *D. cucullata*) only once; (iii) Recombinant multilocus genotypes, which combine a typical hybrid genotype at one locus with a parental genotype at a second locus, or combine a genotype of one parent at the first locus with a genotype of the other parent at the second locus are rare (personal observations, Wolf & Mort, 1986). Despite the fact that sexual reproduction and introgression can be underestimated when only a few diagnostic loci are analysed, the three points mentioned above suggest that backcrosses and sexual reproduction within hybrid populations are rare events in the short-term. However, the long-term significance of introgression and selection for new advantageous recombinant genotypes cannot be ignored (Taylor & Hebert, 1993a, b).

In permanent pelagic *Daphnia* populations, clonal diversity is relatively high (Mort & Wolf, 1986). In most permanent hybrid populations, clonal diversity and richness is even higher. These results along with the greater number of rare genotypes found for *D. galeata* × *cucullata* provide indirect evidence that hybrid populations arose and still arise from multiple hybridization events (Vrijenhoek, 1990; Taylor & Hebert, 1992, 1993b). Direct evidence comes from *in situ* observation of hatching F_1-hybrids out of ephippia (Wolf & Carvalho, 1989). In the *D. galeata/hyalina/cucullata* species complex, often both parentals coexist with their hybrids in the same lake, although one parental is usually at a low density. In addition, it is known that hybridization between *D. galeata* and *D. cucullata* is likely to be unidirectional with *D. cucullata* being the mother (Schwenk, 1993). *D. cucullata* females and *D. galeata* males coexist in the Neuhofener Altrhein (Bast, 1993): ephippia-bearing *D. cucullata* females and *D. galeata* males coexisted during four out of the 26 months studied. During that time, they comprised on average 3.3% of the total population of both species with a mean sex ratio of two females to one male. On the other hand, no *D. cucullata* males were observed in the Neuhofener Altrhein, thus there is no evidence for potential hybridization in the other direction. During most of the year, *D. galeata* is more abundant than *D. cucullata* (Bast, 1993). This does not support the suggestion that hybridization most often involves females of the common species and males of the rare species in a given lake (Taylor & Hebert, 1993a). There was no significant correlation between the population density of *D. cucullata* and the production of *D. galeata* males or *D. cucullata* ephippia (Bast, pers. comm.). However, the analysis of interspecific sexual induction should consider the population dynamics of all coexisting *Daphnia* taxa, including the hybrids. Further studies will consider this point, as well as consider other potential environmental cues for sexual induction.

Additional evidence for multiple hybridizations comes from the finding that hybrid genotypes were quite differentiated with respect to vertical migration behaviour, body length and temporal dynamics. As was shown for vertical migration, these are polygenic traits (De Meester, 1991). It implies that clonal groups differ in whole genotype combinations, which cannot be easily ascribed to a single hybridization event in the past, followed by diversification via mutation (Parker, 1979).

In comparison with the *D. galeata* × *cucullata* hybrid, the parental *D. galeata* displayed less genotypic variation in ecologically relevant traits, like body length and vertical distribution. But on a temporal scale, parental genotypes showed more fluctuations. We attribute these genotypic shifts to selection because population sizes were large (estimated minimum size: 10^7 individuals) and therefore the influence of genetic drift should be negligible.

In general, there should be a positive correlation between the degree of variation and selection intensity. We will try to explain the contradictory results of genotypic differentiation and selection intensity for both *Daphnia* types in two ways. Firstly, one can never exclude the possibility that there is additional clonal variation in other selectively relevant parameters, which were not measured here (e.g. resistance to hunger or parasites). Further, the correlation between selection and variability should not be seen as too rigid. For example, in the case of *D. galeata*, size- and depth-selective fish predation could have a greater impact on this larger species than on the smaller *D. galeata* × *cucullata* (Jacobs, 1978). *D. galeata* × *cucullata* seems to be less selectively constrained and therefore pronounced genotypic diversification could be observed. Support for this hypothesis comes from the result that a positive correlation between the intensity of selection and fish density was found only for *D. galeata*. Secondly, clonal groups (marked by three-locus genotypes) seem to be less stable in parental populations than in hybrid populations. Sexual recombination, which is apparently more frequent in parental populations, can change the clonal composition within a clonal group and therefore can change the fitness of that clonal group as well. Whereas, the clonal and ecological variation in hybrid populations seems to be more-or-less frozen. Several clonal groups appear to be adapted to specific habitats and are relatively stable with regard to temporal fluctuations. The hybrid clones are presumably created by multiple hybridizations of two genetically variable parental gene pools, and after hybridization some well-adapted hybrid clones with non-overlapping niches are selected. This hypothesis is consistent with the frozen-niche variation model of Vrijenhoek (1984, 1990). In order to determine which of these two scenarios is more realistic, future investigations must be able to quantify the frequency of sexual reproduction within *Daphnia* populations.

Acknowledgements

We are grateful to a number of people for field assistance, to Susanne Rieder for technical assistance, to Siegmar Bast and Andreas Wagner for supplying the population dynamic data for daphnids and fishes, respectively, and to Hans Redlef Siegismund for the use of his population genetic program G-Stat. We thank Sabine Gießler, Piet Spaak and Larry Weider for helpful comments and criticisms on earlier versions of this manuscript. Financial support was received from the German Federal Ministry for Science and Technology (Project number: 0339200 D and 07OTX018).

References

Bast, S., 1993. Populationdynamik ausgewählter Cladocerenarten in Seen unterschiedlichen Trophiegrades – Ph.D. thesis, Faculty of Biology, J-Gutenberg University, Mainz.

Brooks, J. L., 1957. The species problem in freshwater animals. In Mayr, E. (ed.), The species problem. Am. Ass. Advmt. Sci., Washington: 81–123.

De Meester, L., 1991. An analysis of the phototactic behaviour of *Daphnia magna* clones and their sexual descendants. Hydrobiologia 225 (Dev. Hydrobiol. 71): 217–227.

Flößner, D., 1993. On the knowledge of some hybrids of *Daphnia*. Limnologica 23: 71–79.

Flößner, D. & K. Kraus, 1986. On the taxonomy of the *Daphnia hyalina – galeata* complex (Crustacea: Cladocera). Hydrobiologia 137: 97–115.

Hebert, P. D. N., 1985. Interspecific hybridization between cyclic parthenogens. Evolution 39: 216–220.

Hebert, P. D. N. & M. J. Beaton, 1989. Methodologies for allozyme analysis using Cellulose Acetate electrophoresis: a practical handbook. Helena Laboratories, Beaumont, Texas.

Hebert, P. D. N., R. D. Ward & L. J. Weider, 1988. Clonal-diversity patterns and breeding-system variation in *Daphnia pulex*, an asexual-sexual complex. Evolution 42: 147–159.

Hebert, P. D. N., M. J. Beaton, S. S. Schwartz & D. J. Stanton, 1989. Polyphyletic origins of asexuality in *Daphnia pulex*. I. Breeding-system variation and levels of clonal diversity. Evolution 43: 1004–1015.

Jacobs, J., 1978. Influence of prey size, light intensity, and alternative prey on the selectivity of plankton feeding fish. Verh. Int. Ver. Limnol. 20: 2461–2466.

Lieder, U., 1983. Introgression as a factor in the evolution of polytypical plankton Cladocera. Int. Revue ges. Hydrobiol. 68: 269–284.

Lynch, M., 1987. The consequences of fluctuating selection for isozyme polymorphisms in *Daphnia*. Genetics 115: 657–669.

Mort, M. A. & H. G. Wolf, 1986. The genetic structure of large-lake *Daphnia* populations. Evolution 40: 756–766.

Müller, J., 1993. Räumliche und zeitliche Variabilität der genetischen Struktur natürlicher Cladocerenpopulationen (Crustacea: Cladocera). Ph.D. thesis, Faculty of Biology, J-Gutenberg University, Mainz.

Müller, J. & A. Seitz, 1993. Habitat partitioning and differential vertical migration of some *Daphnia* genotypes in a lake. Arch. Hydrobiol. Beih. 39: 167–174.

Müller, J. & A. Seitz, 1994. Influence of differential natality and mortality on temporal fluctuations of *Daphnia* genotypes in natural populations. In Beaumont, A. R. (ed.), Genetics and evolution of aquatic organisms. Chapman & Hall, London: 342–350.

Parker, E. D., 1979. Ecological Implications of clonal diversity in parthenogenetic morphospecies. Am. Zool. 19: 753–762.

Schwenk, K., 1993. Interspecific hybridization in *Daphnia*: distinction and origin of hybrid matrilines. Mol. Biol. Evol. 10: 1289–1302.

Seitz, A., 1980. The coexistence of three species of *Daphnia* in the Klostersee. II. The stabilizing effect of selective mortality and conclusions for the stability of the system. Oecologia 47: 333–339.

Simpson, E. H., 1949. Measurement of Diversity. Nature 163: 688.

Spaak, P. & R. J. Hoekstra, 1995. Life history variation and the coexistence of a *Daphnia* hybrid with its parental species. Hydrobiologia (in press).

Taylor, D. J. & P. D. N. Hebert, 1992. *Daphnia galeata mendotae* as a cryptic species complex with interspecific hybrids. Limnol. Oceanogr. 37: 658–665.

Taylor, D. J. & P. D. N. Hebert, 1993a. Habitat-dependent hybrid parentage and differential introgression between neighboringly sympatric *Daphnia* species. Proc. Natl. Acad. Sci. USA 90: 7079–7083.

Taylor, D. J. & P. D. N. Hebert, 1993b. Cryptic intercontinental hybridization in *Daphnia* (Crustacea): the ghost of introductions past. Proc. R. Soc. Lond. B 254: 163–168.

Vrijenhoek, R. C., 1984. Ecological differentiation among clones: the frozen niche variation model. In Wöhrmann, K. & V. Loeschcke (eds), Population biology and evolution. Springer, Berlin: 217–231.

Vrijenhoek, R. C., 1990. Genetic diversity and the ecology of asexual populations. In Wöhrmann, K. & Jain (eds), Population Biology; ecological and evolutionary viewpoints. Springer, Berlin: 175–199.

Wagner, A. R., 1992. Untersuchungen zur Verteilungsdynamik pelagischer Fische in fünf Seen unterschiedlicher trophischer Struktur, unter besonderer Berücksichtigung der Populationsbiologie von Flußbarsch (Perca fluviatilis L., Pisces, Percidae) und Rotauge (Rutilis rutilis L., Pisces, Cyprinidae). Ph.D. thesis, Faculty of Biology, J-Gutenberg University, Mainz.

Weider, L. J. & H. G. Wolf, 1991. Life-history variation in a hybrid species complex of *Daphnia*. Oecologia 87: 506–513.

Wolf, H. G., 1987. Interspecific hybridization between *Daphnia hyalina*, *D. galeata*, and *D. cucullata* and seasonal abundances of these species and their hybrids. Hydrobiologia 145 (Dev. Hydrobiol. 35): 213–217.

Wolf, H.G. & M. A. Mort, 1986. Interspecific hybridization underlies phenotypic variability in *Daphnia* populations. Oecologia 68: 507–511.

Wolf, H. G. & G. R. Carvalho, 1989. Resting eggs of lake-*Daphnia*. II. In situ observations on the hatching of eggs and their contribution to population and community structure. Freshwat. Biol. 22: 471–478.

Hydrobiologia **307**: 33–41, 1995.
P. Larsson & L.J. Weider (eds), Cladocera as Model Organisms in Biology.

Effects of salinity on the life history and fitness of *Daphnia magna*: variability within and between populations

Martina Teschner
Max Planck Institute for Limnology, Department of Physiological Ecology, Postfach 165, D-24302 Plön, Germany

Key words: reaction norms, life history traits, *Daphnia magna*, clonal differences, environmental tolerance, local adaptation

Abstract

The life history traits of *Daphnia magna* were studied in laboratory experiments under freshwater and brackish (5‰ salinity) conditions. The variability of responses within and between populations was examined by comparing 11 clones from a brackish lake and 10 clones from a freshwater pond. Experimental clones were hatched from ephippia collected from the sediment and thus represent random samples of the clone banks of each population.

Most clones with a high salinity tolerance were from the population of the brackish habitat, but some were also found in the freshwater population. Thus, freshwater populations appear to have the potential to invade brackish habitats. A proportion of clones from the brackish population had very low fitness (measured as e^r) under freshwater conditions. This unexpected result means that freshwater adaptation can be lost by the freshwater cladoceran *Daphnia magna*. The effects of unfavourable conditions on growth and reproduction varied among clones and were not correlated. This clonal variation in growth and reproduction indicates that the environmental sensitivities of these traits are independent. The pattern of fitness reaction norms showed no trade-off between fitness under brackish and under freshwater conditions for either population. Thus, euryhaline generalists should be favoured in habitats with salinity fluctuations between freshwater and brackish conditions.

Introduction

The freshwater cladoceran *Daphnia magna* is known to have a relatively high salinity tolerance (Lagerspetz, 1955). Ranta (1979) stated that its 'niche' is defined by a broad salinity range compared to other *Daphnia* species. *Daphnia magna* occurs both in freshwater and brackish habitats and has been found in rockpools with salinities up to 12.5‰ (Lagerspetz, 1955). In a recent study Arnér & Koivisto (1993) measured higher fitness in *Daphnia magna* from a brackish rockpool under brackish than under freshwater conditions. Since it is known, from other *Daphnia* species, that clones and populations can differ in salinity tolerance (e.g. *D. pulex*: Weider & Hebert, 1987), the following questions arise: Is high salinity tolerance a general property of the species, or is it restricted to populations from brackish habitats? How adaptable are freshwater populations to brackish conditions? To answer these questions, I compared 11 clones from a brackish lake with 10 clones from a freshwater pond.

'A reaction norm is the mapping of the genotype onto the phenotype as a function of the environment' (Stearns, 1992). The concept of reaction norms allows a population to be represented by a bundle of reaction norms over a range of environments and therefore is a useful tool in combining ecophysiology and quantitative genetics (van Noordwijk, 1989). The results of this study are presented as reaction norms, which are plots of a clone's performance under varying environmental conditions. In this case, performance, including fitness, clutch size and body size and age at first reproduction, was measured under two experimental salinity levels (freshwater and brackish). In the figures of my study the reaction norms are shown as lines, connecting mean clonal performances obtained in each salinity treatment.

The examination of populations under extreme environmental conditions provides information not only about adaptability, but can also measure correlations between the responses of different life history traits to environmental change. This throws light on the possibilities and constraints of life history evolution, since linkages between traits (and between the expression of traits in different environments) can constrain the response to simultaneous selection (Via & Lande, 1985; Stearns, 1992). If a population is adapted to an extreme environment, the response of its life history traits should be optimal within the constraints that can be reflected by correlations between traits. These constraints might include trade-offs between the allocation of energy between growth and reproduction or between reproduction and survival. If the population is not adapted to the extreme environment, comparison of the vulnerabilities of different life history traits to stress can give information about genetical and developmental constraints. In general, life history traits related to fitness, such as survival, growth, and reproduction, are reduced by stress (Sibly & Calow, 1986; Hoffmann & Parsons, 1991). If the pattern of reduction in these characters in an adverse environment is similar for all clones, one can assume that these traits are linked and do not have easy options for independent evolution. If no common pattern exists, then the population possesses variability in the environmental sensitivities of different life history traits. This means open evolutionary options for the evolution of reaction norms.

Materials and methods

Life history experiments were performed with several clones from one brackish and one freshwater habitat, both located near Plön in northern Germany. The experimental clones were obtained from ephippia from sediment samples and are therefore representatives of the clone banks of each population. Since ephippia were produced sexually (unpubl. enzyme electrophoresis data), all clones differ in genotype.

The brackish lake Großer Binnensee was characterized by a fluctuating level of salinity due to a connection with the Baltic Sea, before 1991. Lampert (1991) studied the population dynamics of *Daphnia magna* in this lake and reported occurrence of *D. magna* at a salinity of 2.5‰. The connection with the Baltic Sea was closed in 1990 and the lake then became a freshwater lake in 1991. *D. magna* was still found after this change (Stolpe, 1993). The ephippia for the 11 experimental clones (clone 11–21) were collected in 1992. The freshwater population was isolated from Rixdorfer Pferdetränke, a small eutrophic pond. A total of 10 clones (clone 1–10) were hatched from ephippia that were collected in 1993.

Life history experiments

The experiments were run a few weeks after establishing the clones from hatched ephippia. Hatching occurred under constant light in 0.45 μm-filtered freshwater from Lake Schöhsee without additives. The experimental design included two salinity treatments, freshwater and brackish (5‰). The freshwater treatment used filtered (0.45 μm) water from Lake Schöhsee. Brackish conditions were created by adding sea-salt to filtered Schöhsee water at a salinity of 5‰. Single animals were reared in jars with 200 ml of water. They were fed daily with 1.5 mg C l^{-1} of the green algae *Scenedesmus acutus* from a chemostat culture. This food concentration is high, i.e. above the incipient limiting level (Lampert, 1987), for *D. magna*. The medium was renewed every other day. The temperature was 20 °C and the light was constant and dim.

The grandmothers and mothers of the experimental animals were acclimated to identical freshwater conditions to minimize maternal and grandmaternal effects (Falconer, 1981; Lynch & Ennis, 1983). Neonate size in *D. magna* is smaller in the first and second clutches (Ebert, 1991). To reduce the within-clone variance for body length at birth, the experimental animals and their mothers originated from the third clutch of the respective preceeding generation.

The experimental animals were transferred to experimental conditions within the first 12 h after birth. A total of 8–20 replicate animals from 2–5 mothers were tested for each clone and treatment. After the animals deposited their first clutch in the brood chamber, body length from head to base of tailspine was measured (size at first reproduction). Age at release of first clutch offspring was determined within 12 h (age at first reproduction). For each animal the growth rate 'gr' (mm d^{-1}) was calculated by the growth increment from birth until the first reproduction devided by the age at first reproduction:

$$gr = (s_{fr} - 0.97)a_{fr},$$

where s_{fr} is the size (mm) and a_{fr} the age (days) at first reproduction; 0.97 mm is the mean neonate length of a number of clones from both populations under the acclimation conditions (Teschner, unpubl. data). The number of eggs in the first clutch was counted (clutch size). In *Daphnia*, clutch size is dependent on body size. For comparisons of clutch sizes among animals with different body sizes, standardized clutch sizes were calculated from clutch size data with size at first reproduction as a covariate. Measurements were performed with a dissecting microscope (25 × or 40 × magnification) and a computer based image analysis system that allows measurements on digitized video images (Soft-Imaging Software, Münster, Germany).

From the age at first reproduction and the size of the first clutch, the intrinsic rate of population increase 'r' was calculated for each individual by the Euler equation

$$\sum e^{-rx} l_x m_x = 1,$$

where x is the age (days), l_x is the age specific probability of survival, and m_x the age specific fertility. The parameter r serves as a measure for fitness. Generally, r increases if one considers further clutches. However, I found a good linear relation between r calculated from the first clutch only and r calculated from the first three clutches in the Lake Großer Binnensee population data ($r_3 = 0.192 + 0.553\ r_1$, $r = 0.82$, $n = 182$). As the relative differences in r values remain, it was possible to take only the first clutch into account. Some animals did not reproduce, because they died as juveniles. The lack of reproduction results in an r-value of $-\infty$. In order to include all data I present the fitness values as e^r instead of r. e^r is the finite rate of population increase. For animals without offspring e^r equals zero.

Experiments with the two populations were run under the same conditions but were not performed simultaneously. To test for comparability, 38 animals, representing two clones from the Großer Binnensee, were retested during the experiments with the freshwater population. ANOVA's were performed on these two brackish clones for size and age at first reproduction, clutch size, and e^r; for each of the four possible clone-salinity combinations with time of experiment as a factor.

Results

Comparability of data sets

Experiments on two clones (19 and 20) from the Großer Binnensee population were repeated in 1993 parallel with experiments with the freshwater clones. In general the results were comparable. Significant differences between experiments in fitness were found for only one clone (20). This clone had under both treatments higher er values in 1993 than in 1992. Its fitness difference between experiments under freshwater conditions was relatively high (0.932 *vs* 1.313, $p = 0.060$) and mainly due to higher mortality in 1992 (18.8% *vs* 0%), whereas the fitness difference under brackish conditions was very low (1.242 *vs* 1.291) and only marginally significant ($p = 0.084$). In the brackish treatments, there were no significant differences between experiments in clutch size and size and age at first reproduction in either clone. In the freshwater treatments clone 20 reproduced one day earlier ($p = 0.005$) with a larger clutch size (12.9 *vs* 15.1), but no significant difference in body size, in 1993 compared to 1992. The only significant year effects detected for the other clone in the freshwater treatments, were an one day earlier reproduction ($p = 0.001$) at a 7.7% larger body size ($p = 0.005$) in 1993 than in 1992. But this did not result in significantly different fitness values.

Salinity tolerance

Figure 1 shows reaction norms for fitness, measured as e^r; under freshwater and brackish conditions for all clones from both populations. The brackish population showed both rising and falling fitness reaction norms between environments, whereas all clones of the freshwater population exhibited a tendency for reduced fitness under brackish conditions. Relatively high fitness in one environment was not generally accompanied by low fitness in the other. Thus, neither population showed a trade-off between fitness under brackish and under freshwater conditions. Under brackish conditions, the brackish population was on average fitter than the freshwater population (t-test: $p < 0.001$). This was due to fewer unfit clones in the brackish population. The single highest clonal value under brackish conditions was not significantly higher in the brackish population (clone 11: $e^r = 1.330$) than in the freshwater population (clone 9: $e^r = 1.318$; t-test: $p = 0.281$). Under freshwater conditions, the freshwater population was on average fitter than the brackish population (t-test: $p = 0.001$). This was due to fewer unfit clones and a higher maximum fitness in the freshwater population.

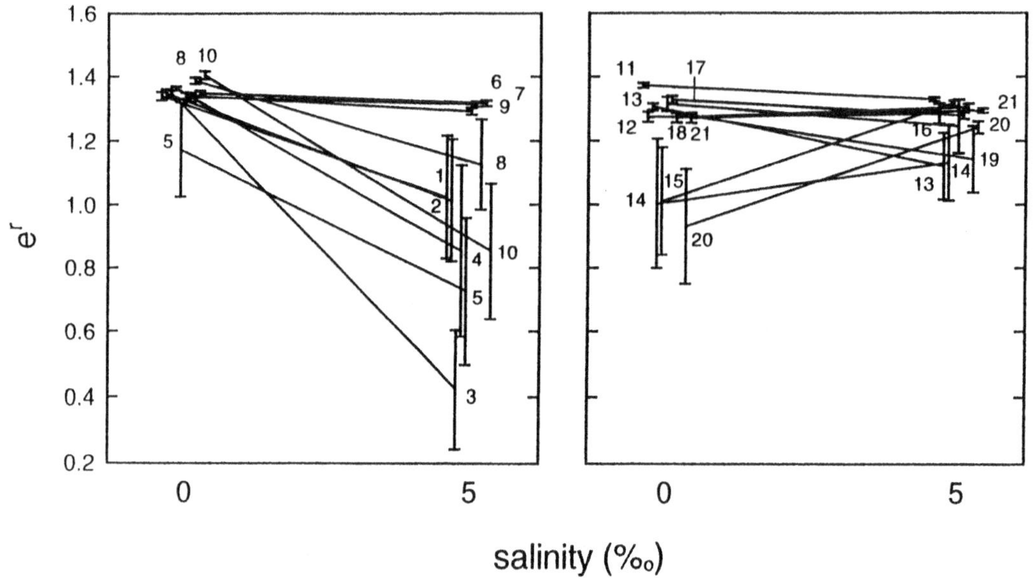

Fig. 1. Reaction norms for fitness under freshwater and brackish conditions in the freshwater population (left panel) and the brackish population (right panel). The reaction norms are shifted horizontally, in order to make the standard error bars more evident. Numbers indicate clones. Each line represents data for 8–20 replicate individuals from one clone.

Effects of salinity on life history traits

Age and size at maturity

Figure 2 shows reaction norms for size against age at first reproduction under freshwater and brackish conditions for clones of both populations. All clones had a larger size at first reproduction under freshwater conditions. The freshwater clones grew more rapidly under freshwater conditions, than under brackish treatments. Their reaction norms have negative slopes, because the animals matured on average later and at a smaller size under brackish conditions. There were deviations from this typical growth pattern in the brackish population: 5 of 11 clones (15, 18, 19, 20 and 21) showed reaction norms with positive slopes. They reproduced later and were larger in the freshwater treatments. However, the increase in size was relatively small, compared to the age difference. This indicates a relative reduction of growth under freshwater conditions in these clones compared to the general growth pattern demonstrated by all other clones. To quantify such deviations in growth pattern, growth rates were calculated (Fig. 3). If the difference between a clone's growth rate in freshwater minus its growth rate under brackish conditions lies above the average difference from all measurements (0.0397 mm d^{-1}), the clone has enhanced growth in freshwater and consequently reduced growth under the brackish treatment, when compared to the mean behaviour of all clones. All freshwater clones and two of 11 brackish clones showed a reduction in growth under the brackish treatment; nine of 11 brackish clones showed a reduction in growth under freshwater conditions (Fig. 7).

Clutch size

For a comparison of reproduction, it is necessary to take into account that clutch size is dependent on body size. Figure 4 shows reaction norms for clutch size against body size under freshwater and brackish conditions. Analogous to growth (see above) the reaction norms of clutch size against body size reaction norms followed a general pattern that can be used as a reference for comparisons. Most clones had larger clutches under freshwater conditions, which corresponded with their larger body size. The regression of clutch size (*cs*) against size at first reproduction (*sfr*) from all measurements is given in Fig. 3 ($cs = -7.98 + 7.70\,sfr$, $r = 0.38$, $n = 434$). To detect deviations from the general reproductive pattern, standardized clutch sizes were calculated from the clutch size data with size at first reproduction as a covariate (ANOVA: $p < 0.0001$). Figure 5 shows average standardized clutch sizes for all clone-treatment combinations. If the difference between a clone's standardized clutch size under brackish minus standardized clutch size under freshwater conditions lies above the average difference from all measure-

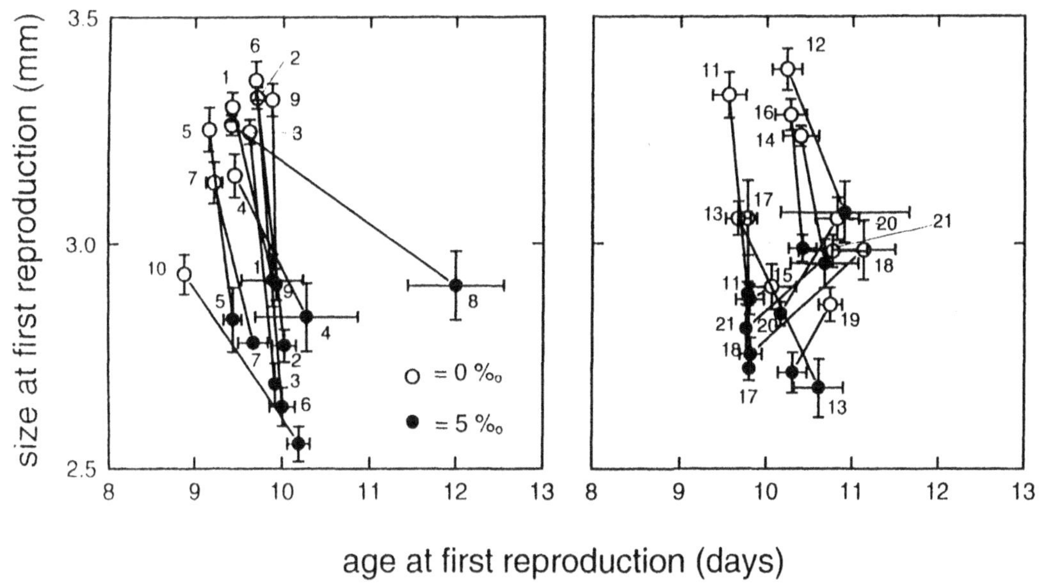

Fig. 2. Bivariate reaction norms for size *vs* age at first reproduction for the freshwater population (left panel) and the brackish population (right panel). The lines connect mean clonal performances (±S.E.) for each salinity treatment. Numbers indicate clones.

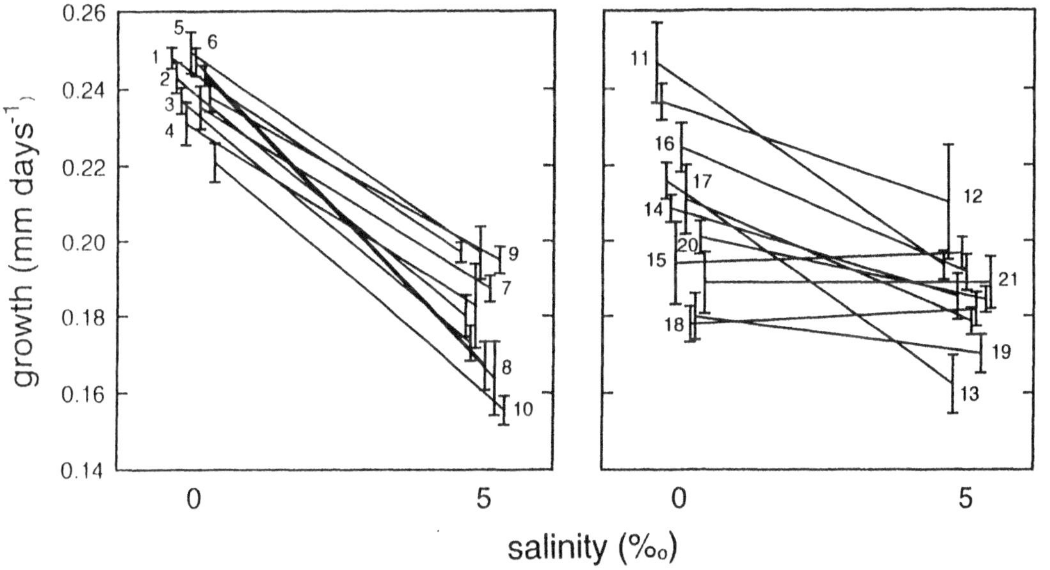

Fig. 3. Reaction norms for growth rate under freshwater and brackish conditions in the freshwater population (left panel) and the brackish population (right panel). The reaction norms are shifted horizontally in order to make the standard error bars more evident. Numbers indicate clones.

ments (2.504), the clone reduces reproduction in freshwater, and consequently enhances reproduction under brackish conditions, when compared to the average behaviour of all clones. The largest reduction in reproduction occurred in the brackish clone 14 under the freshwater treatment. Five of 11 brackish clones and four of 10 freshwater clones reduced standardized clutch size under freshwater conditions (i.e. relative enlargement of clutch sizes under brackish conditions) (Fig. 7).

Pre-reproductive mortality

Figure 6 shows fitness versus mortality. Low salinity tolerance of freshwater clones was expressed by high mortality; the highest mortalities occurred among the freshwater clones under the brackish treatment. In contrast, only three brackish clones with reduced freshwater adaptation showed increased mortality. The freshwater clones followed a clear pattern: either no mortality or a higher mortality under brackish compared

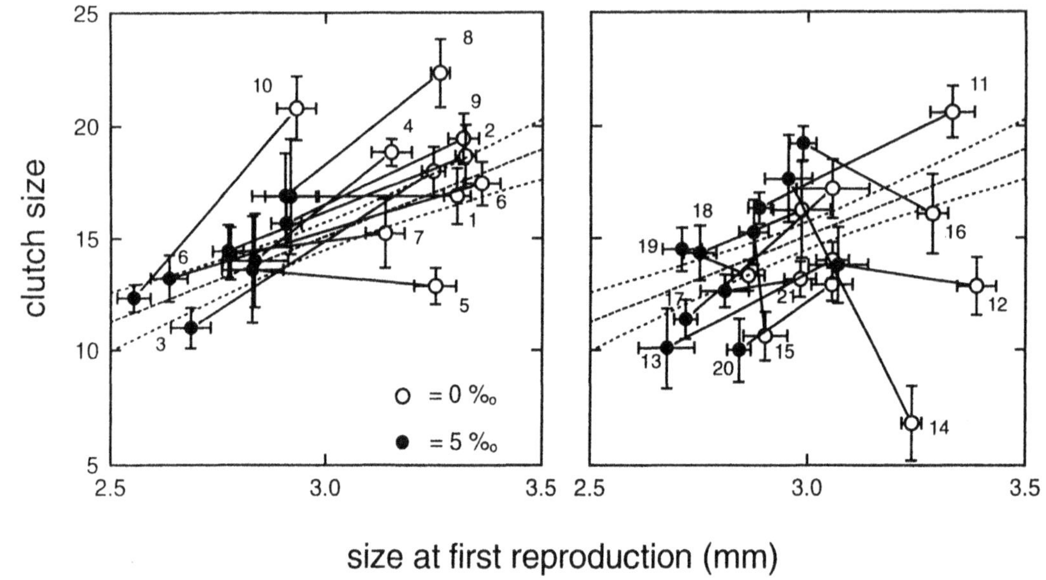

Fig. 4. Bivariate reaction norms for clutch size *vs* body size at first reproduction in the freshwater population (left panel) and the brackish population (right panel). The lines connect mean clonal performances (±S.E.) for each salinity treatment. Numbers indicate clones.

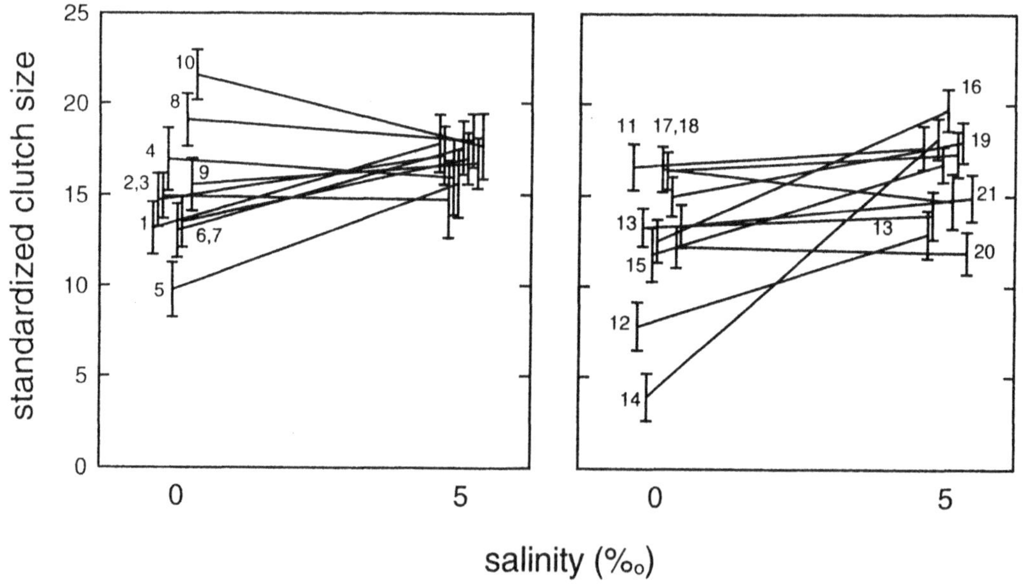

Fig. 5. Reaction norms for clutch size, standardized for body size, under freshwater and brackish conditions in the freshwater population (left panel) and the brackish population (right panel). The reaction norms are shifted horizontally in order to make the standard error bars more evident. Numbers indicate clones.

to the freshwater treatment. The reactions of brackish clones were more diverse; some showed higher mortality under brackish conditions, some under the freshwater treatment, and some showed no mortality in both treatments.

Clonal differences in relative sensitivities of growth and reproduction to adverse environmental conditions

The effects of unfavourable conditions on growth and reproduction varied among clones (Fig. 7). Among the clones of the freshwater population reduction in growth

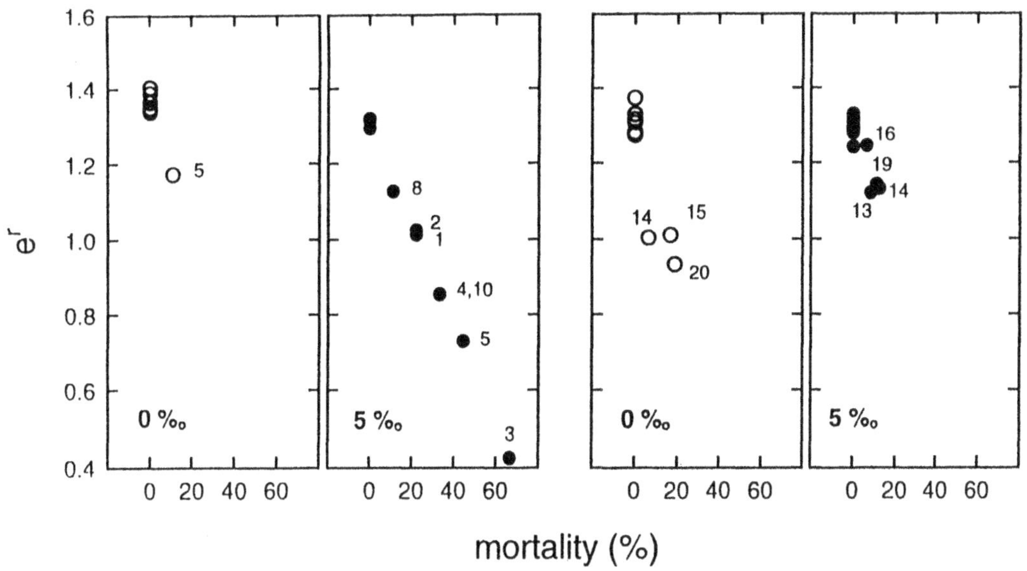

Fig. 6. Fitness *vs* mortality under freshwater (hollow circles) and brackish conditions (filled circles) in the freshwater clones (left panels) and the brackish clones (right panels). Numbers indicate clones.

and reproduction under brackish conditions were not correlated ($r = 0.203$, $p = 0.57$) and there was also no correlation between the sensitivities of these traits to freshwater stress among the brackish clones ($r = 0.017$, $p = 0.96$).

Discussion

Comparability of data sets

The results of repeated experiments with two clones suggest that data from the brackish treatments are comparable for both populations over time. Since the e^r value of one clone and the somatic growth rate of the other clone varied significantly in the freshwater treatments, data from the freshwater treatments may be less comparable. If the higher growth rate in 1993 was due to a change in clonal properties during one year of culturing in the lab, then the data are comparable because both experiments were performed shortly after hatching of ephippia. If the observed changes were due to alterations in the quality of the culture medium, the data are not strictly comparable. In the worst case, the measured values for the brackish clones under freshwater conditions would have to be elevated about 10% for growth and 20% for reproduction for a direct comparison with the freshwater population. However, this deviation is relatively low compared to the magnitude of measured differences between populations.

Salinity tolerance in populations from freshwater and from brackish habitats

In this study I have shown that salinity tolerance in *D. magna* is not restricted to populations from brackish habitats. Based on their performance under brackish conditions, it appears that some clones from freshwater populations have the potential to invade brackish habitats. However, this result is relative, because the experimental salinity (5‰) was somewhat higher than the salinity in the brackish lake (around 2.5‰). It is possible therefore, that populations from more saline waters than the brackish lake in this study would show higher fitness under the experimental brackish conditions. As expected, the proportion of clones with a low salinity tolerance was higher in the freshwater pond than in the brackish lake, where selection should act against low salinity tolerance (Fig. 1).

In my study, the comparison of reaction norms between populations suggests, that *Daphnia magna* can loose freshwater adaptation. Lower fitness under freshwater compared to a brackish (4‰) treatment was also found by Arnér & Koivisto (1993). They studied one *D. magna* clone, which came from a brackish rockpool. As the genus *Daphnia* is freshwater in origin and distribution, I did not expect to find reduction in fresh-

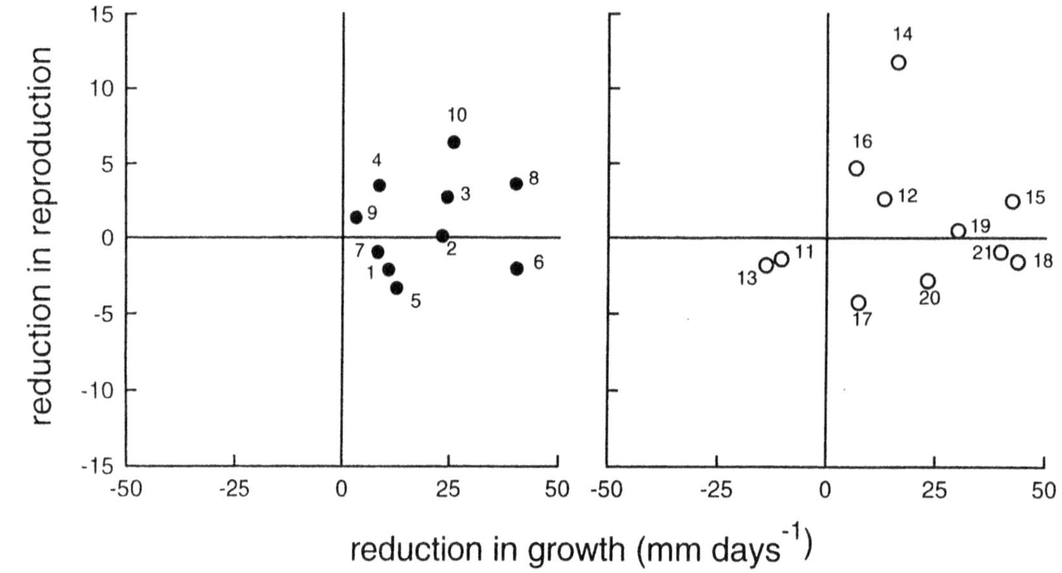

Fig. 7. Reduction in reproduction *vs* reduction in growth in the freshwater clones under brackish treatment (left panel) and in the brackish clones under freshwater treatment (right panel). Numbers indicate clones.

water adaptation. Since *D. magna* osmoregulates up to a salinity of about 5‰ in the external medium and above this salinity it is an osmoconformer (Fritsche, 1916; Aladin, 1991), it seems likely that a reduction in the ability to osmoregulate is involved in the loss of freshwater adaptation.

It is sometimes suggested that a trade-off might exist between fitness under favourable conditions and stress resistance (review in Hoffmann & Parsons, 1991). In my study, the pattern of fitness reaction norms showed no trade-off between fitness under brackish and under freshwater conditions for either population (Fig. 1). Thus, euryhaline generalists should be favoured in both populations, if salinity conditions change from being constant to short-term fluctuations between freshwater and brackish water. Rockpools are a common brackish habitat for *Daphnia magna*. Fluctuating salinity is probably the norm in most small brackish pools due to storms, rainfall and evaporation. Here one should expect euryhaline clones.

Indications for constraints on life history evolution

In the brackish treatment, larger clutch sizes and higher growth rates were expected for the brackish population than for the freshwater population. Since all clones from both populations showed the same pattern of somatic growth reduction (Fig. 2) and relatively high clutch sizes (Figs 4 and 5) in brackish water, growth under brackish conditions appears to be physiologically constrained. Reduced growth in brackish water was also observed in other studies of *Daphnia magna* (Fritsche, 1916; Arnér & Koivisto, 1993). Perhaps the experimental salinity of 5‰ in my study was so much beyond the salinity optima for both populations, that all clones were stressed and therefore showed reduced growth. But it is also possible that a physical mechanism, and not salinity tolerance per se constrains growth. Fritsche (1916) found that daphnids with a higher osmogradient from internal to external medium grow better. This might be explained in the physiology of the crustacean molt cycle: At ecdysis, expansion of the body volume occurs by rapid water uptake. During the following intermolt period, tissues grow to the volume established by tissue hydration (Drach, 1939, ref in Passano, 1960). The growth determining water uptake may be impeded for *Daphnia* in brackish water compared to freshwater.

While it is easy to imagine that the upper limit for growth under brackish conditions could be similarly physiologically constrained in both populations, it is less clear why the freshwater population showed relatively high clutch sizes in brackish water (Figs 4 and 5). Again, the possibility has to be considered, that a salinity of 5‰ is beyond the optimum and therefore similarly constrains reproduction in both populations. On the other hand, clutch sizes under brackish conditions were relatively large. In both populations, standardized clutch sizes were on average larger in the brackish treatment. Clones with low fitness under the brackish treatment showed high mortality, while reduced reproduction played a minor role. It is possible that the sustained high reproduction despite lethal salinity stress is adaptive, since life history theory predicts that high adult mortality rates favour high reproductive effort. Such an adaptive stress response should not be specific for salinity stress, as it would not evolve in a freshwater population that never experiences salinity pressure. It would be rather an unspecific response caused by severe (i.e. at the edge of lethality) stresses and should be found also in the presence of some other stressfactors that act on the population under natural conditions.

A comparison of the sensitivities of growth and reproduction to osmotic stress can provide additional information about constraints in *Daphnia*'s life history evolution. My data show that growth and relative clutch size are not necessarily both affected by stress. There is clonal diversity in the responses of these two life history traits to adverse environmental conditions. Thus, in *Daphnia*'s ontogeny, growth and reproductive effort appear to be largely independent of each other. These traits show different sensitivities to, respectively buffering against, environmental influences. This means, that plasticities of both traits have evolved and could continue to evolve independently.

It would be interesting to study the effects of other environmental factors in addition to salinity, because linkages between responses to different factors can also constrain evolution. Other ecologically and therefore also evolutionary important factors for cladocerans include, for example, food conditions and chemical cues from predators. Interactions between both factors and cladocerans are the topic of many interesting papers in this volume. The approach of comparing the reaction norms of different populations under a range of environmental conditions can contribute considerably to finding answers to the question: How do populations become adapted to their environment?

Acknowledgments

I thank Carola Peters, Hinnerk Borіß, and Uta Bohlen for their assistance during experimental work. Ralph Tollrian and Bill DeMott provided useful comments on earlier versions of this manuscript. Tad Kawecki came to my rescue with a critical comment on presentation of this material.

References

Aladin, N. V., 1991. Salinity tolerance and morphology of the osmoregulation organs in Cladocera with special reference to Cladocera from the Aral sea. Hydrobiologia 225 (Dev. Hydrobiol. 71): 291–299.

Arnér, M. & S. Koivisto, 1993. Effects of salinity on metabolism and life history characteristics of *Daphnia magna*. Hydrobiologia 259: 69–77.

Drach, P., 1939. Mue et cycle d'intermue chez les Crustacés Décapodes. Ann. inst. océanog. (Paris) [N.S.] 19: 103–391.

Ebert, D., 1991. The effect of size at birth, maturation threshold and genetic differences on the life-history of *Daphnia magna*. Oecologia 86: 243–250.

Falconer, D. S., 1981. Introduction to quantitative genetics. 2nd edition, Longman, New York.

Fritsche, H., 1916. Studien über die Schwankungen des osmotischen Druckes der Körperflüssigkeiten bei *Daphnia magna*. Int. Revue Hydrobiol. 8: 22–80, 125–203.

Hoffmann, A. A. & P. A. Parsons, 1991. Evolutionary genetics and environmental stress. Oxford University Press, Oxford.

Lagerspetz, K., 1955. Physiological studies on the brackish water tolerance of some species of *Daphnia*. Arch. Soc. Vanamo Suppl. 9: 138–143.

Lampert, W., 1987. Feeding and nutrition in *Daphnia*. Mem. Ist. ital. Idrobiol. 45: 143–192.

Lampert, W., 1991. The dynamics of *Daphnia magna* in a shallow lake. Verh. Int. Ver. Limnol. 24: 795–798.

Lynch, M. & R. Ennis, 1983. Resource availability, maternal effects, and longevity. Exp. Gerontol. 18: 147–165.

Passano, L. M., 1960. Molting and its control. In T. H. Watermann (ed.), The physiology of crustacea. Vol. 1. Academic Press, New York and London: 473–536.

Ranta, E., 1979. Niche of *Daphnia* species in rockpools. Arch. Hydrobiol. 87: 205–223.

Sibly, R. M. & P. Calow, 1986. Physiological ecology of animals: an evolutionary approach. Blackwell Scientific Publications, Oxford.

Stearns, S. C., 1992. The evolution of life histories. Oxford University Press, Oxford.

Stolpe, G., 1993. Populationsdynamik und biotische Interaktionen von *Daphnia magna* im Großen Binnensee. Diplom-thesis, Universität Kiel, Germany.

van Noordwijk, A. J., 1989. Reaction norms in genetical ecology. BioScience 39: 453–458.

Via, S. & R. Lande, 1985. Genotype-environment interaction and the evolution of phenotypic plasticity. Evolution 39: 505–522.

Weider, L. J. & P. D. N. Hebert, 1987. Ecological and physiological differentiation among low-arctic clones of *Daphnia pulex*. Ecology 68: 188–198.

Hydrobiologia **307**: 43–55, 1995.
P. Larsson & L.J. Weider (eds), Cladocera as Model Organisms in Biology.

Changing attitudes toward chydorid anomopods since 1769

David G. Frey[†]
Department of Biology, Jordan Hall, Indiana University, Bloomington, IN 47405, USA

Key words: Anomopod, biogeography, chydorids, cosmopolitanism, historical, speciation

Abstract

The first mention of the chydorids in the literature was by O.F. Müller in 1769, although he did not use any names for the taxa that he described. The first century after O. F. Müller's first paper was concerned almost exclusively with the cladocerans of Europe. The second century saw European scientists looking at cladocerans from other continents and frequently placing the European names on them. Species had not been described adequately, and the persons looking at the non-European taxa seldom made any detailed comparisons to document their claim of conspecificity. Consequently, the names became distributed everywhere for look-alike taxa, and scientists routinely regarded the chydorid anomopods as cosmopolitan. During the past two decades, scientists are finding through detailed comparisons, that cognate taxa of chydorids on different continents, or even on the same continent, usually are not the same. Attitudes are shifting away from cosmopolitanism (presumably brought about by the passive dispersal of resting eggs), to vicariant events as one of the major mechanisms affecting the evolution of the chydorids and involved in controlling their distributions.

Introduction

The Chydoridae constitute one of five families of the Anomopoda, which is one of four orders comprising the former order Cladocera (Fryer, 1987). It is by far the largest family, and contains more than half the species of all cladocerans, and about half the total number of genera. Chydorids are small, with a modal adult maximum size of 0.5 mm. They are members of the meiobenthos rather than the plankton, and hence, researchers who concentrate on true plankton in the pelagia, have little appreciation for their numbers or biodiversity. Yet the numbers of individuals commonly exceed more than 1×10^{-6} m^{-2} of littoral bottom, and species diversity often ranges between 10–25 taxa in a given small portion of a waterbody at any given time. Being small, inconspicuous, and not occurring in the pelagia, the chydorids have been largely neglected or ignored in many aquatic studies. This is despite the fact that they are the principal converters of littoral algae, detritus, and bacteria into secondary production, which can be utilized by higher trophic levels (e.g. fish).

Most cladocerans reproduce by cyclical parthenogenesis. During most of the year, a population consists strictly of parthenogenetic females, but at certain times of the year, males and ephippial females are produced, sexual reproduction occurs, with the subsequent production of resting eggs. Because these resting eggs in their enclosed ephippia have been believed to be distributed everywhere by wind, birds, insects, mammals, or by other means, scientists considered these animals to be capable of widespread dispersal. This lead to the belief that the same species occurred on different continents, and the idea of cosmopolitanism took hold. Except for some endemic species and genera confined to particular continents, the fauna of the World seemed to be much the same everywhere. During the past two decades, however, in which scientists have begun to examine the chydorids more critically by comparing cognate populations from different continents with type populations, significant differences have emerged. These differences indicate that the cognate taxa under closer examination are not conspecific.

[†] Contributed posthumously.

These observations suggest that perhaps most chydorid taxa with claimed intercontinental distributions, are really members of species groups, which have one or more different species represented on each continent (Frey 1982b, 1986).

The goal of this paper is to present an historical review of how we gravitated into the unwitting acceptance of the concept of cosmopolitanism, and what evidence is forcing us to change this attitude.

O. F. Müller: 1776 and before

The first mention of chydorids in the literature are the identical 1769 and 1770 notes of O. F. Müller, in which he wrote in Danish the same descriptions of the taxa that he later gave in Latin, but he did not use any names for the taxa. In 1776, however, Müller listed all the animals then known from Denmark and gave each a binomial scientific name if they didn't already have one. Members of the genus *Lynceus* constitute numbers 2392–2398 in this list, members of *Daphne* (changed to *Daphnia* in 1785) are numbers 2399–2406, and one species of *Polyphemus* is listed (number 2417). The 1776 paper really marks the beginning of work on the chydorids. Only about 20 papers published prior to 1776, are concerned with Cladocera (Baird, 1843, 1850), but only with daphniids. The papers by Swammerdam (1669, 1685) are the first to mention cladocerans. These papers and other earlier papers refer to the daphniids as waterfleas from the vernacular expressions, and it seems certain that *Daphnia* was actually regarded as a parasite, like a flea sucking out its victim's juices with its pointed beak (Frey, 1982b).

Müller's 1785 paper is more elaborate, and contains the same brief diagnostic descriptions as in his three earlier papers. However, Müller (1785) presents expanded descriptions of the various taxa and provides engravings that, although small, are mostly recognizable. Except for two species of *Daphnia* in this paper, all other taxa were shown later to be from different genera, and they now represent the type species of the following genera: *Sida, Latona, Daphnia, Bosmina, Eurycercus, Camptocercus, Alona, Pleuroxus, Peracantha,* and *Chydorus*. *Lynceus brachyurus*, the first species in this genus in the 1785 paper, but not in the 1776 paper, is a conchostracan, and is now the type species of the Lynceidae. Thus, it is obvious that Müller was seeing genera rather than species, even though he had grouped all these taxa into just three genera.

First century post Müller (1776–1878): Europe

From 1785 until 1878, scientists were concerned mostly with the Cladocera of Europe. By 1880 most of the species now known from Europe had been described and most of the genera had been established. Up until 1878, only about 20 papers had been published on non-European cladocerans (Richard, 1895, 1896). These were by European scientists, who made collections from India, extreme northern and eastern Siberia, Jerusulem, Egypt, Algeria, Turkestan, Kerguelensland, and South America. This list also included four significant papers by King (1853a, b, 1855, 1866) on the Cladocera of Australia, which were written while he resided there. This non-European effort, however, was so small that one can conveniently refer to the first century post-Müller as European. Workers were examining collections from various European countries. Often at first, they considered the taxa to be different, but then with further study, these supposed differences were rationalized into single species, sometimes with considerable intraspecific variation. P. E. Müller (1867) was uncertain of the taxonomic status of a number of taxa, and this uncertainty remains even today for most of these taxa (Frey, 1989). This indicates that little progress has been made during the past century in improving our understanding of the species in Europe. One of P. E. Müller's difficulties was that he considered all morphology as normal and unvarying. Consequently, he described four new species from single aberrant specimens.

During this period, the work of G. O. Sars is of particular significance. In six papers (Sars, 1862a, b, 1863, 1864, 1865, 1890), Sars described a total of 52 new species of cladocerans from Norway, of which 16 were chydorids. The published descriptions are in Latin, they were very short, and without any illustrations except for one species published in 1865. Yet, in his presentation of these papers, particularly the first two, Sars used a considerable number of illustrations, as evidenced by the title of the first paper (Sars, 1862a). These papers were found in the University of Oslo library, along with a large unpublished manuscript dating from 1861, which considers all these taxa in much greater detail and keys their morphology to the figures.

Subsequently, Sars' handwritten manuscript entitled, 'On the Freshwater Crustaceans occurring in the Vicinity of Christiana', has been translated into English, and has been published by the University of Bergen (Sars, 1993); it is now available for scholarly taxonomic studies.

Sars had intended to publish a monograph on the Cladocera of Norway, for which this 1865 paper was the first part. But he never progressed with the project any further, because he was unable to handle *Daphnia* to his satisfaction (Frey, 1982a). Few specimens that he actually used in drafting the descriptions are available in the museum collections in Oslo, although there are other specimens from Norway. He preferred to make his drawings from live specimens and then paint them to approximate their fresh appearance. The specimens so used were not saved, which means that his unpublished drawings and paintings are the substitute types. In spite of this poor documentation of Sars' species, most of them are claimed to occur on other continents as well. Yet, there is no way of comparing the cognate populations with specimens actually used by Sars except through his illustrations and colored portraits.

Second century post-Müller (1878–1973): Cosmopolitanism

In 1878, E. A. Birge received his Ph.D. from Harvard with a thesis entitled, 'On Crustacea Cladocera Collected in Fish Pond and Glacialis, Cambridge, Mass. 1876 and at Madison, Wis. 1877'. Because this was the first major study of cladocerans done in residence outside of Europe, Birge had little to guide him. He described a total of 16 new species and a new genus in his thesis, to which four more species were added when the paper was published a year later (Birge, 1879). In his next three papers concerned with Cladocera (Birge, 1892, 1893, 1910), Birge described 13 additional species and two new genera, all from North America. Thus, 33 new species and three genera are contained in his thesis and in his first four papers on Cladocera, but of these, only 14 species and one genus have survived inclusion in his monograph on the Cladocera of North America (Birge, 1918). The rest have vanished without any indication from Birge as to their supposed equivalence.

In Birge's monograph, there are 57 species of chydorids, 35 of which were originally described from Europe, nine from North America, six from South America, three from Sri Lanka and Australia, and one from Asia Minor. Thus, according to Birge, only 16% of the chydorids in North America had been described from that continent, and 61% had European names. All the species were well-illustrated, and unlike the various national treatments of Europe, Birge's monograph included almost a whole continent with a considerable number of tropical and subtropical taxa. It was used and still is being used (see the revision by Brooks, 1959) by scientists everywhere to name the taxa of their continents. Thus, the European names have been transferred readily and without critical appraisal from one continent to another, thereby helping to establish the occurrence of cognates and the derivative concept of cosmopolitanism.

Another important work is the monograph by Lilljeborg (1901) on the Cladocera of Sweden. This contains many illustrations of such high quality that investigators elsewhere, obviously convinced of the validity of cosmopolitanism, have borrowed them to illustrate the taxa on their continents, rather than attempt lesser quality illustrations of the local populations. Scientists had become careless. If they did not look meaningfully at their specimens, then one illustration was as good as another for representing them.

So, for 100 years after Birge's thesis, many investigators both in Europe and elsewhere, were considering the same cladoceran species as present on these continents. Many taxa were new and restricted in their distributions, but many others were given European names or names from elsewhere because of a general similarity. Detailed comparisons between type populations and presumed conspecific populations were not made, and hence one had to rely on the scientific reputation of the person making the decisions. Cosmopolitanism had become the 'Leitmotiv' of cladoceran distributions without anyone taking the time and energy necessary to document it or refute it.

Third century post-Müller (1973–): Non-cosmopolitanism?

In order to compare two cognate taxa adequately, large populations of each taxon, which contain all gamogenetic stages and instars as well as parthenogenetic females, are required. Type specimens by themselves are not adequate. Experience has shown that parthenogenetic females are less well-differentiated than males or ephippial females, and that the ontogeny of the three instars of males is also important. Hence, no taxon can be considered adequately described if males, ephippial females, and cast ephippia are not included. Yet, many chydorid taxa have been described from only one or just a few parthenogenetic females, or even from abnormal specimens. Even experts on the chydorids have paid little or no attention to the presence of males and ephippial females in their samples (e.g. Sars, 1901 in Brazil; Dady, 1905 in Paraguay; Frey, 1982b). Consequently, the chydorid taxa were often poorly or inadequately described, which made it difficult to determine the taxonomic equivalence of populations and their evolutionary relationships. Because of taxonomic carelessness and deceptively poor decisions, names in the literature, whether or not accompanied by descriptions and illustrations, are mostly unreliable for indicating the presence of particular species.

Since we are interested in phylogenetic relationships, related species should be included in the same genus, although many of the chydorid genera are in 'name only', without any formal descriptions. Table 1 shows the time sequence of establishing the genera that are in current use. Twenty of the total of 39 genera were described from Europe. Except for King's *Dunhevedia* from Australia, all the genera proposed through 1895 are European in origin. In the five years beginning in 1900, one other European genus was described, and Sars described three genera from South America and one from Australia. Then there comes a gap of 60 years with no new genera. Beginning in 1965 and continuing through 1991, seven genera were split off from other genera, a new genus was described from Lake Baikal, and nine presumably endemic genera were described from Australia. *Propleuroxus*, a genus of fossil chydorids from the Permian of Siberia, was also described.

But we are most interested in species, and whether or not some species occur on different continents as is implied by the widespread distribution of their names. The first study that seriously questioned cosmopolitanism was the description of a new species of *Eurycercus* (Frey, 1973); see Table 2 (#1) and the establishment of three subgenera (Frey, 1975). Previously, there were only two species of *Eurycercus* in the world – *E. glacialis* which was confined to the northern fringe of the northern continents, and *E. lamellatus*, which was widely distributed in the Holarctic region, with isolated populations in northern Argentina and southern Brazil, and in South Africa. *E. lamellatus* is now split into two subgenera, one presently with two species, and the other with four (Hann, 1990). The two widespread taxa in North America are in a different subgenus from the dominant *E. lamellatus* of Europe, and in fact, the latter taxon does not occur in North America at all. Thus, the formerly cosmopolitan taxon *E. lamellatus* now has been divided into six different species, and undoubtedly with more to come. Frey's 1973 paper represents, in my opinion, the beginning of the third century post-Müller in which detailed comparisons of populations containing all stages and instars are used to describe species and to determine if cognates are conspecific. One population must be of the named taxon from the locality of origin, and the others must represent the populations to be compared.

A couple of papers published prior to Frey (1973), which were done with less consideration to morphological details and with less critical evaluation of mis-identified taxa, seem to counter the idea of non-cosmopolitanism. Megard (1967; see #10 in Table 2) described two species of *Alona* from North America without giving any indication that these are what were probably previously called *Alona rectangula*. The latter taxon, with a purported widespread distribution on many continents, does not seem to occur either in North America, or on the sub-Antarctic islands as claimed by Heywood (1967). Flößner & Frey (1970) attempted to compare European and North American populations of *Alona costata* (Frey, 1965b) and *Alona rustica*. They concluded that these two taxa are the same species, with *A. rustica* differing only at the subspecific level on the two continents. On the basis of later work with additional species, I am now not satisfied with the conclusions of Flößner & Frey (1970). I believe that the study should be redone, with particular attention given to details that have turned out to be important in other species pairs.

The study by Flößner & Kraus (1977; #16 in Table 2), which compared *Pleuroxus denticulatus* in Europe and North America and which concluded that the taxa are the same, although highly variable, leaves some doubts. This is based on the fact that at least two (as yet undescribed) species of *P. denticulatus* exist in North America (personal observations) with considerable differences between them. It is not entirely clear what was being compared in this earlier study (Flößner & Kraus, 1977).

After Frey (1973), there is a whole series of papers that show that cognate populations on different continents can be different. The level of taxonomic resolution used in the past was too coarse to reveal these differences, and yet, once these differences were discovered and documented, there is no question about considering the taxa as being distinct. *Chydorus sphaeri-*

Table 1. Time sequence of description of currently recognized genera of chydorid anomopods and continent from which they were described.

Europe	Australia	South America	Asia	North America
1816 *Chydorus* Leach				
1843 *Pleuroxus* Baird				
Peracantha Baird				
Eurycercus Baird				
Camptocercus Baird				
Alona Baird				
Acroperus Baird				
	1853 *Dunhevedia* King			
1862 *Monospilus* Sars				
Graptoleberis Sars				
Anchistropus Sars				
Alonopsis Sars				
Alonella Sars				
1867 *Phrixura* P.E. Müller				
1874 *Leydigia* Kurz				
1895 *Kursia* Dybowski & Grochowski				
Oxyurella Dybowski & Grochowski				
		1901 *Dadaya* Sars		
		Euryalona Sars		
		Leydigiopsis Sars		
1903 *Rhynchotalona* Norman				
	1904 *Saycia* Sars			
1965 *Tretocephala* Frey			1966 *Indialona* Petkovski	
1968 *Pseudochydorus* Fryer			1969 *Kozhowia* Vasil'eva & Smirnov 1970	*Propleuroxus* Smirnov 1982 *Ephemeroporus* Frey
1971 *Biapertura* Smirnov				
	1983 *Archepleuroxus* Smirnov & Timms			
	Australochydorus Smirnov & Timms			
	Monope Smirnov & Timms			
	Rak Smirnov & Timms			
	Rhynchochydorus Smirnov & Timms		1987 *Notoalona* Rapjapaksa & Fernando	
	1989 *Leberis* Smirnov			
	1991 *Celsinotum* Frey			
	Planicirclus Frey			
	Plurispina Frey			

cus (#3 in Table 2) was regarded as the most common and widespread of all the cladocerans, and occurred on all continents, at nearly all latitudes, and at elevations ranging from sea-level into the highest mountains. But *C. sphaericus* is now known to be a complex of at least two species groups, each with an unknown number of species. *C. sphaericus* sensu stricto possibly does not occur in North America at all.

Ephemeroperus barroisi (Table 2, #5) is a species described from Syria, which has four teeth or notches on the labral keel and one tooth on the shell. Around the world, there are many taxa that resemble *E. barroisi* in these characters, including two species from Spain, and three undescribed species from Florida. Other continents still have further variations on this central theme.

Chydorus faviformis (Table 2, #8) was described from North America by Birge (1893). Birge was so impressed by the honeycombed shell and head, that he could say relatively little else about its morphology. Honeycombed taxa have since turned up on other continents, and have been called *C. faviformis*. Yet, when

Table 2. Results of detailed morphological comparisons of cognate taxa of chydorid anomopods on different continents. Each taxon is in italics for the continent from which it was described, and the resolution of the originally cognate taxa are not in italics. Many, but not all, of these originally cognate taxa are now considered members of species groups.

	Europe	North America	Asia	Australia	South America	Africa	References
1)	*Eurycercus lamellatus*		E. macracanthus				Frey 1973
	E. pompholygodes						Frey 1975
		E. microdontus					Frey 1978
		E. vernalis					Hann 1982
2)	*Alonopsis elongata*	A. americana					Kubersky 1977
3)	*Chydorus sphaericus*	C. brevilabris					Frey 1980
		C. biovatus					Frey 1985
4)		C. linguilabris	*Chydorus reticulatus*				Frey 1982d
			C. invaginatus				
5)		3 species in Florida	*Ephemeroporus barroisi* other species	other species	other species	other species	Frey 1982c
	E. margalefi						Alonso 1987a
	E. episphantioi						
6)	*Oxyurella tenuicaudis*	O. brevicaudis					Michael & Frey 1983
7)	*Phrixura rostrata*	P. leei					Michael & Frey 1984
8)		*Chydorus faviformis*	C. sinensis	C. obscurirostris	C. parvireticulatus		Frey 1987
			C. angustirostris	C. opacus			
			C. obscurirostris tasekberae				
9)	different species				*Alona verrucosa*		Alonso 1987b
10)	*Alona rectangula*	A. setulosa					Megard 1967
		A. circumfimbriata					
				A. weinecki			Frey 1988b
11)	*Pleuroxus laevis*	P. straminius					Frey 1988a
		P. chiangi					Frey 1988a
12)	A. iberica			*Alona karua*			Alonso & Pretus 1989
13)	*Pleuroxus aduncus*			various endemic species			Frey 1991a,b
14)	*Alona costata*	A. costata					Flössner & Frey 1970
15)	*Alona rustica*	A. rustica americana					Flössner & Frey 1970
16)	*P. denticulatus*	*Pleuroxus denticulatus*					Flössner & Kraus 1977
17)		D. macrops	*Daday macrops*	D. macrops	D. macrops		Rajapasa & Fernando 1982
18)		K. longirostris	*Kursia longirostris* K. brevilabris	K. longirostris	K. longirostris		Rajapasa & Fernando 1986
19)		E. orientalis	*Euryalona orientalis*	E. orientalis	E. orientalis		Rajapaksa & Fernando 1987b
20)		N. freyi	*Notoalona globulosa*	N. globulosa australiensis	N. globulosa		Rajapaksa & Fernando 1987a
21)		D. americana	*Dunhevedia serrata*				Rajapaksa & Fernando 1987c

populations of these taxa were closely compared, they were all found to be distinctly different at the species level. The taxon from Malaysia is possibly a subspecies of one of the two taxa from Australia, but no definite conclusion can be reached until males are described from this taxon.

Pleuroxus aduncus (Table 2, #13) from Europe has been claimed to occur on all continents. A study of the *Pleuroxus* species in Australia, first had to present a detailed description of *P. aduncus* from Scandinavia, and then compare this with the Australian taxa. *P. aduncus*, sensu stricto, was not found anywhere in Australia. All the Australian taxa are endemic. Moreover, Harding (1955) grouped quite a number of distinct species from four continents under the name *P. aduncus*. This

Table 3. Tabulation of chydorid anomopods by genus, number of species in each genus, continent of description, and endemicity for four continents. The data for Europe are based on the summary by Hrbáček et al. (1978), plus the subsequent papers by Alonso (1987a, 1988), Alonso & Pretus (1988), Brancelj (1990, 1991), Frenzel & Alonso (1988), and Frey (1975, 1989). The data for North America are based on Brooks' (1959) version of Birge's monograph (1918), plus the subsequent papers of Chengalath & Hann (1981a, 1981b), Flössner & Frey (1970), Flössner & Kraus (1977), Frey (1961, 1978, 1982c, 1982d, 1982e, 1985, 1988a), Hann (1982), Hann & Chengalath (1981), Kubersky (1977), Megard (1967), Michael & Frey (1983, 1984), Rajapaksa & Fernando (1987a, 1987c), Roen (1973, 1987), and Smirnov (1987). The data for Australia are based on the monograph by Smirnov & Timms (1983), plus the subsequent papers by Frey (1991a, 1991b), Rajapaksa & Fernando (1987a), and Smirnov (1989a, 1989b). The data for New Zealand are based on the rough summary in Chapman & Lewis (1976), plus the subsequent papers of Frey (1980, 1991a), and the extensive fieldwork by Frey in 1987.

Subfamily	Genus	Europe			North America			Australia			New Zealand		
		Total species	From Europe	Endemic	Total species	From North America	Endemic	Total species	From Australia	Endemic	Total species	From New Zealand	Endemic
Eurycercinae	Eurycercus	3	2	1	4	3	3	-	-	-	-	-	-
Sayciinae	Saycia	-	-	-	-	-	-	1	1	1	1	1	1
Chydorinae	Alonella	4	3	0	5	2	1	4	2	1	3	1	1
	Phrixura	1	1	0	3	3	1	-	-	-	-	-	-
	Ephemeroporus	4	3	2	6	5	5	2	0	0	1	1	1
	Rak	-	-	-	-	-	-	2	2	2	-	-	-
	Dunhevedia	1	0	0	2	1	1	1	1	0	1	0	0
	Daday	-	-	-	1	0	0	1	0	0	-	-	-
	Monope	-	-	-	-	-	-	1	1	1	-	-	-
	Peracantha	1	1	0	-	-	-	-	-	-	-	-	-
	Fleuroxus	9	6	1	7	4	3	7	4	3	2	2	2
	Archepleuroxus	-	-	-	-	-	-	1	1	1	-	-	-
	Flanicirclus	-	-	-	-	-	-	1	1	1	-	-	-
	Flurispina	-	-	-	-	-	-	2	2	2	-	-	-
	Pseudochydorus	1	1	0	1	0	0	1	0	0	1	0	0
	Chydorus	6	6	1	14	8	7	5	1	1	3	2	2
	Australochydorus	-	-	-	-	-	-	1	1	1	-	-	-
	Rhynchochydorus	-	-	-	-	-	-	1	1	1	-	-	-
	Anchistropus	1	1	0	1	1	1	-	-	-	-	-	-
Aloninae	Camptocercus	5	4	0	1	1	1	1	1	0	1	0	0
	Acroperus	4	4	0	1	0	0	4	1	1	-	-	-
	Alonopsis	1	1	0	1	1	1	-	-	-	-	-	-
	Kursia	1	1	0	2	0	0	2	0	0	-	-	-
	Euryalona	-	-	-	1	0	0	1	0	0	-	-	-
	Oxyurella	1	1	0	2	2	1	2	0	0	1	0	0
	Leydigia	2	2	0	3	0	0	5	2	1	1	0	0
	Tretocephala	1	1	0	-	-	-	-	-	-	-	-	-
	Alona	19	17	10	16	9	7	21	11	8	8	3	0
	Biapertura	4	2	0	4	0	0	13	9	8	7	3	0
	Celsinotum	-	-	-	-	-	-	3	3	3	-	-	-
	Notoalona	-	-	-	1	1	1	1	1	1	-	-	-
	Bryospilus	-	-	-	1	1	0	-	-	-	2	1	1
	Leberis	-	-	-	-	-	-	1	1	1	-	-	-
	Graptoleberis	1	1	0	1	0	0	1	0	0	1	0	0
	Rhynchotalona	1	1	0	2	1	1	-	-	-	-	-	-
	Monospilus	1	1	0	1	0	0	2[a]	2	2	1	0	0
Totals		72	60	15	81	43	34	88	49	41	34	14	8
% of totals			83	21	.	53	42		56	47		41	24
Number of genera		22			24			28			15		
Continent of description %													
Europe		60	83.3		22	27.2		21	23.9		10	29.4	
North America		2	2.8		43	53.1		1	1.1		2	5.9	
Australia		3	4.2		3	3.7		49	55.7		7	20.6	
New Zealand		1	4.4		0	—		4	4.5		14	41.2	
Asia		2	2.8		6	6.2		7	8.0		0	—	
Africa		2	2.8		1	1.2		2	2.3		1	2.9	
South America		2	2.8		7	8.6		4	4.5		0	—	

[a] These two species probably do not belong in *Monospilus*

has lead to further claims of cosmopolitanism, and this matter will eventually have to be clarified as well.

Other comparisons between taxa named from Europe and their cognates in North America have shown the cognate taxa to be differents species (Table 2, #2, 6, 7, and 11), and taxa named from Asia, Africa, and South America (Table 2, #4, 9, and 12) to be different from their cognates in North America and Spain. In one instance (Table 2, #2), the morphological differences at the species level have been substantiated by the inability to cross European and North American populations of the cognates.

Thus, *Eurycercus lamellatus*, *Alonopsis elongata*, *Oxyurella tenuicaudis*, *Phrixura rostrata*, *Alona rectangula*, *Pleuroxus laevis*, and possibly *Chydorus sphaericus* all described from Europe and claimed to occur in North America, definitely do not occur in North America. Instead, they are represented by other members of these species groups. *Dunhevedia serrata* and *Alona globulosa*, described originally from Sri Lanka, have been replaced by other species names in North America, and for *A. globulosa*, also by a new genus name for its world-wide distribution. *Ephemeropterus barroisi* from Syria does not occur in North America, although several related species do. These are all documented examples of non-cosmopolitanism, which have involved the detailed comparison of type populations with the cognate populations.

In addition, there are many examples that remain to be documented. Examination of hundreds of collections from different regions of North America has revealed no equivalents of the european *Camptocercus rectirostris*, *C. macrurus*, *Acroperus harpae*, *Pleuroxus uncinatus*, and *Peracantha truncata*. Equivalent populations (i.e. cognates) in North America of other taxa from Europe and from other continents differ from the named populations more subtly, but probably no less realistically. Differences are so consistently present that is is impossible to say in many instances what species really occur in North America, or on any other continent except Europe, from which most of them were described.

While this evidence of non-cosmopolitanism in the chydorids seems almost overwhelming, some other studies suggest that at least certain tropical-subtropical species can be intercontinental and cosmopolitan. Rajapaksa (1986) in her Ph.D. dissertation at the University of Waterloo, Canada, sought to determine if cognate populations on other continents, of species described from Sri Lanka by Daday (1898) are the same or different. Of the five species considered, (Table 2, #17–21) two are undifferentiated on any of the five continents, one is uniformly distributed but has an additional species in Asia, while the fourth is differentiated as a new species in North America and a new subspecies in Australia. The fifth species, which involves a comparison of material from Asia, Africa, and North America, has a new differentiated species in North America. Rajapaksa and Fernando (1982, 1986, 1987a, b, c) discuss the presence of pantropical species, which some of these seem to be and which implies cosmopolitanism. Such differences in distribution of tropical as compared with temperate species were suggested earlier by Frey (1971). On the other hand, Alonso & Pretus (1988), who were also working with tropical and subtropical taxa from other continents (Table 2, #5, 9 and 12), found distinctly different, although related taxa in Spain. Thus, temperate taxa seem to be quite consistently differentiated as separate species; some tropical and subtropical taxa are also differentiated, while others are not.

These 21 detailed comparisons between named taxa on one continent and their cognates on other continents are the kind of comparisons that must be made to determine the likelihood of conspecificity. But there is another less satisfying way of defining the magnitude of the problem. Three continents – Europe, North America, and Australia – have reasonable lists of their chydorid taxa, which are based on attempts in the past to monograph the species present on each continent. These lists are supplemented with information from individual research papers that have been published subsequently. In addition, New Zealand has a 'rough' unpublished list, which has been supplemented by the extensive research that I did there in 1987. For each of these continents (Table 3), the number of taxa in each genus is given, plus the number that are endemic to each continent. Obviously, this is a constantly changing array of data as new taxa are found, and as detailed comparisons are made to establish if conspecificity (which is implied by cognates) is indeed a fact. At the bottom of the table, all taxa that are presently recognized from each continent are tabulated according to the continent from which they were described. Summaries of the chydorids present on the other three continents have not yet been published (e.g. the work of J. C. Paggi in South America), or are not yet organized, even though attempts have been made to summarize the faunas of individual countries (Africa: South Africa – Sars, 1916, the Sahelo-Soudan Zone – Rey and St. Jeans, 1980; Asia: Japan – Ueno, 1937, China – Chiang and Du, 1979, Sri Lanka – Daday, 1898, India – Michael and Sharma, in press, Malaysia – Idris, 1983).

Considering genera first, presently there are 39 genera of chydorids recognized world-wide. The four not included in Table 3 are the Permian *Propleuroxus* from Siberia, *Kozhowia* and *Indialona* from Asia, and *Leydigiopsis* from South America. A few of the genera are mostly Holarctic in distribution – *Eurycercus*, *Phrixura*, *Anchistropus* (although there is now a species recorded from South America), *Alonopsis*, and *Rhynchotalona*. On the other hand, *Saycia* is confined to Australia and New Zealand. Australia is unique among all the continents in that it has nine highly distinctive genera (10 if one includes *Saycia*) that are presently confined to it. This reflects most likely the long isolation of Australia from any other continental land mass. Thus, the distribution of genera, even though many of them are poorly defined or even not defined at all, indicates that chydorids as a whole, are not distributed intercontinentally, as required by the concept of cosmopolitanism.

On all other continents, as tabulated here, 53% of the genera presently are represented by just single taxa, and more than three-quarters have no more than three taxa each. Only the genera *Pleuroxus*, *Chydorus*, *Alona*, and *Biapertura* have seven or more taxa on individual continents. These are the species rich genera. North America seems to have a relatively greater number of *Chydorus* species, while Australia and New Zealand have more species of *Biaperatura*. In Europe, 80% of all chydorid taxa there were described from Europe. On the other three continents, about 50% of the species were described from that continent, and about 25% were described from Europe. This demonstrates how easy it has been to use European names for look-alike taxa on other continents. Many of these, as well as many of the few taxa named from other countries, are members of species groups that still need to be studied in detail, and differentiated into their component species. When this is achieved, the number of species of chydorids will have increased considerably.

Distributions of species on single continents likewise, are often more restricted than previously assumed. North America is the best example thus far (Frey, 1986). Here, cognate populations in the northern States and Canada, and in the south can be differentiated at the level of full species (e.g. *Eurycercus longirostris* and *E. vernalis*, Hann, 1982; *Pleuroxus denticulatus*, Shan & Frey, 1968), probably as a result of their southward displacement during the last glaciation. Southern species have moved northward and southward along the Atlantic Coast in response to major changes in climate, such as in the mid-Holocene warm period. Moreover, in the southern States, many related taxa, such as the various species of *Ephemeroperus*, have very restricted distributions. The species that can become established in any waterbody are probably controlled mostly by the fine structure of the hydrological, climatic, chemical, and biological environments, and not by the geographic location of the waterbody.

Speculation concerning the origin of the species groups

Demonstrating that cosmopolitanism, at least among temperate-zone taxa, is much less common than formerly believed and that many species considered intercontinental, even cosmopolitan, are really members of species groups is only the first (and possibly the easiest) part of the problem to tackle.

The major question now is: How did these species groups come into being, over what time periods, and why do we have so many of them?

The cladocerans are an ancient group, with scattered fossil records dating from the Permian (Smirnov, 1970), Jurassic (Smirnov, 1971), and Cretaceous (G. Fryer, personal communication), and with more abundant records, especially of daphniid ephippia, from the Tertiary (Frey, 1964). At least one of the latter, cannot be distinguished from ephippia produced by a present-day species. Moreover, 25 species of chydorids from a single lake from the Eemian interglacial of Denmark, a mere 100,000 years ago, are indistinguishable morphologically from the same species occurring in Denmark today. Their relative abundance (with respect to each other) does not appear to have changed much from the Eemian (Frey, 1962). This sparse, but consistent evidence indicates that the cladocerans are very old and it suggests that they have been stable evolutionarily for tens of millions of years. They are old enough and diversified enough to have been present on Gondwanaland and Pangaea before breakup of these primordial continents into daughter continents, and before the evolution of birds as possible dispersal vectors.

In general, new species arise either by dispersal to some previously unoccupied region, or by the breaking up of a formerly continuous range by the creation of some impassable barrier, such as an ocean, mountain range, or desert. In either case, the mother and daughter populations, or the two sister populations must be kept separated from one another long enough for genetic differentiation to lead to genetic incompatibility. For new species to evolve by dispersalism, the passive distribution of resting eggs is the key process. For new species to evolve by range fragmentation (i.e. vicariance), the creation of impassable barriers is the key. Here the separation and movement of daughter continents and continental fragments from both Pangaea and Gondwanaland are critical, with the oceans separating them, gradually becoming complete barriers to passive dispersal.

It should be obvious from the discussion to this point, that we really know too little about the chydorid cladocerans to engage in any meaningful speculation about their evolutionary history. Much work still needs to be done in the analysis of genera and species. Nevertheless, there are many anecdotal examples that implicate vicariance as the dominant process. Dispersalism may well be the initiator of many of the troubling species groups, but the dispersal appears to have occurred many millions of years ago (based on the apparently very slow rate of evolution among these crustaceans).

Some examples include: northern Asia seems to have most of the species (or species groups) that occur in Europe, although some taxa, such as *Alonopsis elongata*, *Tretocephala ambigua*, *Peracantha truncata*, *Alona weinecki*, and others seem confined to Europe or else, they extend into Asia for only a short distance beyond the Ural Mountains. Europe and North America share many cognates. Where these have been studied by detailed analysis of populations (Table 2), there seems to be some effect of the vicariant separation of Europe and North America. Evidence suggests that the two cognate populations were likely to have been the same before continental separation. In fact, it appears that North America was quite isolated from the rest of the world, as suggested by the three contrasting species of honeycombed *Chydorus* there, and the presence even of unique tropical species of *Dunhevedia* and *Notoalona*. *Eurycercus* is an interesting example. Europe contains a dominant species of the subgenus *Eurycercus*, and a minor species of the subgenus *Bullatifrons* is found only in northern Sweden and Finland. On the other hand, North America has two dominant species of the subgenus *Bullatifrons*, and only one very minor species of *Eurycercus*, which is restricted to the southern Atlantic Coast and into Florida. The genus *Eurycercus* is largely holarctic, but with disjunct populations in northern Argentina and southern Brazil, and in South Africa. Each disjunct population belongs to the dominant subgenus in North America and Europe, respectively. The genus *Tretocephala*, with one species in Europe and the other in South Africa, is another example of north–south discontinuity.

Australia is a unique continent for chydorids as well as for the more conspicuous marsupials. The polyembryonic genus *Saycia* occurs only in Australia and New Zealand. Furthermore, Australia has nine recently described genera that are considered endemic. The species in these genera are unique morphologically, with no related species known from any of the other continents. Moreover, the proportion of two-pored *Alona*-like taxa (e.g. *Biapertura*) found in Australia is much greater than on any other continent.

The unresolved genus *Pleuroxus* is composed of four subgenera, plus three more from Australia. *Pleuroxus trigonellus* is the type species. *P. uncinatus* is closely related. *Peracantha truncata* is a good species, and is easily separated from other congeneric species; a strict dichotomy in morphology exists – one species group is related to *P. aduncus* of Europe, while the other group is related to *P. laevis*, also from Europe. The three new Australian genera (*Archepleuroxus*, *Planicirclus*, and *Plurispina*) are in the *aduncus* group. Species in this group are mainly from the Southern Hemisphere, while species in the *laevis* group are found mainly in the Northern Hemisphere. This is the kind of information that eventually will help us understand the evolution and distribution of the chydorids. Most available information to date, is too coarse-grained to be very useful.

Direct evidence for the influence of the breakup of Gondwanaland on the chydorids is rather meager at present. South Africa contains only one species of the *Ephemeroperus hybridus*-species group, which hitherto has been restricted to South America and subtropical North America. A new genus for South Africa will be created for the species, *Chydorus tilhoi* (Rey & St. Jean, 1969). I hope other examples of Gondwanian affinity will emerge as the faunas of the southern continents are analyzed in more detail.

This study has attempted to document the changing attitudes toward the chydorid anomopods over the past 220 years (i.e. since the first mention of these organisms in the literature). From an early uncritical acceptance that the chydorids could be cosmopolitan, the attitude over the past 20 years has now changed, with many researchers believing that most previous claims of cosmopolitanism are incorrect. Short-term dispersalism does not seem to be a significant factor affecting intercontinental species distributions. Vicariance events, which resulted from plate tectonics, seem to have been a more important factor.

References

Alonso, M., 1987a. Morphological differentiation of two new *Ephemeroporus* species (Cladocera, Chydoridae) belonging to the *barroisi* complex: *E. margalefi* and *E. epiaphantoii*, in Spain. Hydrobiologia 145 (Dev. Hydrobiol. 35): 141–146.

Alonso, M., 1987b. Aportaciones a la idea del localismo de los Cladoceros: algunas especies espanoles como paradigma. Actas IV Congr. Espanol Limnol., Sevilla: 369–378.

Alonso, M., 1988. *Chydorus pizarri* sp. nov. A new chydorid (Cladocera) from western Spain. Limnetica 3: 27–39.

Alonso, M. V. & J. L. Pretus, 1989. *Alona iberica*, new species: first evidence of non-cosmopolitanism with the *A. karua* complex (Cladocera: Chydoridae). J. Crust. Biol. 9: 459–476.

Baird, W., 1843. The Natural History of British Entomostraca, no. VI. Ann. Mag. nat. Hist. 11: 81–95.

Baird, W., 1850. The Natural History of British Entomostraca. The Ray Society, London.

Birge, E. A., 1878. Crustacea Cladocera collected in Fish Pond and Glacialis. Cambridge, Mass. 1876, and at Madison, Wis. 1877. Ph.D. thesis, Harvard University.

Birge, E. A., 1879. Notes on Cladocera. Trans. Wis. Acad. Sci. Arts Lett. 4: 77–109.

Birge, E. A., 1892. List of Crustacea Cladocera from Madison, Wisconsin. Trans. Wis. Acad. Sci. Arts Lett. 8: 379–398.

Birge, E. A., 1893. Notes on Cladocera, III. Trans. Wis. Acad. Sci. Arts Lett. 9: 275–317.

Birge, E. A., 1910. Notes on Cladocera, IV. Trans. Wis. Acad. Sci. Arts Lett. 16: 1017–1066.

Birge, E. A., 1918. The water fleas (Cladocera). In Ward, H. B. & G. C. Whipple (eds), Freshwater Biology, Wiley, New York: 676–740.

Brancelj, A., 1990. *Alona hercegovinae* n. sp. (Cladocera: Chydoridae), a blind cave inhabiting cladoceran from Hercegovinia (Yugoslavia). Hydrobiologia 199: 7–16.

Brancelj, A., 1992. *Alona sketi* sp. n. (Cladocera: Chydoridae), the second cave-inhabiting Cladocera from former Yugoslavia. Hydrobiologia 248: 105–114.

Brooks, J. L., 1959. *Cladocera*. In Edmondson, W. T. (ed.), Freshwater Biology, Wiley, New York: 587–656.

Chapman, M. A. & M. W. Lewis, 1976. An Introduction to the Freshwater Crustacea of New Zealand, with a special chapter on the Arachnida by V. M. Stout. Collins, Aukland: 261 pp.

Chengalath, R. & B. J. Hann, 1981a. A new species of *Chydorus* (Cladocera: Chydoridae) from Ontario, Canada. Trans. am. Micros. Soc. 100: 229–238.

Chengalath, R. & B. J. Hann, 1981b. Two new species of *Alona* (Chydoridae, Cladocera) from western Canada. Can. J. Zool. 59: 377–389.

Chiang, S.-C. & N.-S. Du, 1979. *Freshwater Cladocera*. In Fauna Sinica, Crustacea. Academia Sinica, Peking, 297 pp.

Daday, E., 1898. Mikroskopische Süßwassertiere aus Ceylon. Termes. Füzetek. Anhangsheft. 21: 1–123.

Daday, E., 1905. Untersuchungen über die Sußwasser-Mikrofauna Paraguays. Zoologica (Stuttgart) 18: 1–374.

Dybowski, B. & M. Grochowski, 1895. Spis systematyczny Wioslarek (Cladocera) krajowych. Kosmos polski (Lemberg) 320: 139–165.

Flößner, D. & D. G. Frey, 1970. A morphological comparison of European and North American populations of *Alona rustica* Scott and *Alona costata* Sars (Cladocera, Chydoridae). Limnologica (Berlin) 7: 325–337.

Flößner, D. & K. Kraus, 1977. On the variability and taxonomy of *Pleuroxus denticulatus* Birge (Cladocera: Chydoridae). J. Fish. Res. Bd Can. 34: 463–476.

Frenzel, P. & M. Alonso, 1988. *Alona azorica* sp. nov., a new chydorid (Cladocera) from the western Palearctic. Arch. Hydrobiol. 112: 449–465.

Frey, D. G., 1961. Differentiation of *Alonella acutirostris* (Birge, 1897) and *Alonella rostrata* (Koch, 1841) (Cladocera,Chydoridae). Trans. am. Micros. Soc. 79: 129–140.

Frey, D. G., 1962. Cladocera from the Eemian Interglacial of Denmark. J. Paleont. 36: 1133–1154.

Frey, D. G., 1964. Remains of animals in Quaternary lake and bog sediments and their interpretation. Arch. Hydrobiol., suppl. Erg. Limnol. 2: 1–116.

Frey, D. G., 1965a. A new genus Chydoridae. Int. Rev. ges. Hydrobiol. 50: 153–168.

Frey, D. G., 1965b. Differentiation of *Alona costata* Sars from two related species (Cladocera, Chydoridae). Crustaceana 8: 169–173.

Frey, D. G., 1971. Zoogeography and Evolution (Abstract). In Frey, D. G. (ed.), Symposium on Ecology of Cladocera. Trans. am. Micros. Soc. 90: 119–121.

Frey, D. G., 1973. Comparative morphology and biology of three species of *Eurycercus* (Cladocera,Chydoridae) with a description of *Eurycercus macracanthus* sp. nov. Int. Rev. ges. Hydrobiol. 58: 221–267.

Frey, D. G., 1975. Subgeneric differentiation within *Eurycercus* (Cladocera, Chydoridae) and a new species from northern Sweden. Hydrobiologia 46: 263–300.

Frey, D. G., 1978. A new species of *Eurycercus* (Cladocera, Chydoridae) from the southern United States. Tulane Stud. Zool. Bot. 20: 1–26.

Frey, D. G., 1980a. On the plurality of *Chydorus sphaericus* (O.F. Müller) (Cladocera, Chydoridae), and designation of a neotype from Sjaelso, Denmark. Hydrobiologia 69: 83–123.

Frey, D. G., 1980b. The non-swimming chydorid Cladocera of wet forests, with descriptions of a new genus and two new species. Int. Rev. ges. Hydrobiol. 65: 613–641.

Frey, D. G., 1982a. G. O. Sars and the Norwegian Cladocera: a continuing frustration. Hydrobiologia 96: 267–293.

Frey, D. G., 1982b. Questions concerning cosmopolitanism in Cladocera. Arch. Hydrobiol. 93: 484–502.

Frey, D. G., 1982c. Relocation of *Chydorus barroisi* and related species (Cladocera,Chydoridae) to a new genus and description of two new species. Hydrobiologia 86: 231–269.

Frey, D. G., 1982d. The reticulated species of *Chydorus* (Cladocera, Chydoridae): two new species with suggestions of convergence. Hydrobiologia 93: 255–279.

Frey, D. G., 1982e. The honeycombed species of *Chydorus* (Cladocera,Chydoridae): comparison of *bicornutus* and *bicollaris* n. sp. with some preliminary comments on *faviformis*. Can. J. Zool. 60: 1892–1916.

Frey, D. G., 1985. A new species of the *Chydorus sphaericus* group (Cladocera:Chydoridae) from western Montana. Int. Rev. ges. Hydrobiol. 69: 3–20.

Frey, D. G., 1986. The non-cosmopolitanism of chydorid Cladocera: Implications for biogeography and evolution. In Gore, R. H. & K. L. Heck (eds), Crustacean Biogeography, Balkema, Rotterdam: 237–256.

Frey, D. G., 1987. The North American *Chydorus faviformis* (Cladocera,Chydoridae) and the honeycombed taxa of other continents. Phil. Trans. r. Soc., Lond., Ser. B. 315: 353–404.

Frey, D. G., 1988a. Separation of *Pleuroxus laevis* Sars, 1861 from two resembling species in North America: *Pleuroxus straminius* Birge, 1897 and *Pleuroxus chiangi* n. sp. (Cladocera,Chydoridae). Can. J. Zool. 88: 2534–2563.

Frey, D. G., 1988b. *Alona weinecki* Studer on the subantarctic islands, not *Alona rectangula* Sars (Chydoridae,Cladocera). Limnol. Oceanogr. 33: 1386–1411.

Frey, D. G., 1989. P.E. Müller's chydorid Cladocera from Denmark. Steenstrupia 15: 105–134.

Frey, D. G., 1991a. The species of *Pleuroxus* and of three related genera (Anomopoda,Chydoridae) in southern Australia and New Zealand. Rec. Austral. Mus. (in press).

Frey, D. G., 1991b. A new genus of alonine chydorid cladocerans from athalassic saline waters of new South Wales, Australia. Hydrobiologia 224: 11–48.

Fryer, G., 1968. Evolution and adaptive radiation in the Chydoridae (Crustacea: Cladocera): a study in comparative functional morphology and ecology. Phil. Trans. r. Soc., Lond., Ser. B., 254: 221–385.

Fryer, G., 1987. A new classification of the branchiopod Crustacea. Zool. J. Linn. Soc. 91: 357–383.

Hann, B. J., 1982. Two new species of *Eurycercus* (*Bullatifrons*) from eastern North America (Chydoridae, Cladocera). Taxonomy, ontogeny, and biology. Int. Rev. ges. Hydrobiol. 67: 585–610.

Hann, B. J., 1990. Redescription of *Eurycercus* (*Teretifrons*) *glacialis* (Cladocera,Chydoridae), and description of a new species, *E.* (*T.*) *nigracanthus*, from Newfoundland, Canada. Can. J. Zool. 68: 2146–2157.

Hann, B. J. & R. Chengalath, 1981. Redescription of *Alonella pulchella* Herrick, 1884 (Cladocera,Chydoridae), and a description of the male. Crustaceana 41: 249–262.

Harding, J. P., 1955. The Percy Sladen Trust Expedition to Lake Titicaca in 1937: Crustacean: Cladocera. Trans. Linn. Soc., Lond. 1: 329–354.

Heywood, R. B., 1967. Antarctic ecosystems. The freshwater lakes of Signy Island and their fauna. Phil. Trans. r. Soc. Lond., Ser. B., 261: 347–362.

Hrbacek, J., V. Korinek & D. G. Frey, 1978. *Cladocera*. In Illies, J. (ed), *Limnofauna Europaea*, Fischer, Stuttgart: 189–195.

Idris, B. A. G., 1983. Freshwater Zooplankton of Malaysia (Crustacea: Cladocera). Penerbit Universiti Pertanian Malaysia, Serdang, Selangor. Syarikat Percetakan Selangor: Kuala Lumpur. vii, 153 pp.

King, R. L., 1853a. On some species of Daphniadae found in New South Wales. Pap. Proc. R. Soc. Van Diemen's Land. 2:243–253.

King, R. L., 1853b. On some Australian Entomostraca – in continuation. Pap. Proc. R. Soc. Van Diemen's Land. 2:253–263.

King, R. L., 1855. On Australian Entomostracans. Pap. Proc. R. Soc. Van Diemen's Land. 3: 56–75.

King, R. L., 1866. On the anatomy of certain forms of Australian Entomostraca. Trans. Ent. Soc. N. S. Wales (Sydney). 1: 162–166.

Kubersky, E. S., 1977. World-wide distribution and ecology of *Alonopsis* (Cladocera:Chydoridae) with a description of *Alonopsis americana* sp. nov. Int. Rev. ges. Hydrobiol. 62: 649–685.

Kurz, W., 1874. Dodekas neuer Cladoceren nebst einer kurzen Übersicht der Cladocerenfauna Böhmens. Sitzb. K. Akad. Wissensch. 70: 7–88.

Leach, W. E., 1816. Annulosa. Encycloped. Brit., 4th edn., Suppl., pp. 406.

Lilljeborg, W., 1901. Cladocera Sueciae, oder Beiträge zur Kenntniss der in Schweden lebenden Krebstiere von der Ordnung der Branchiopoden und der Unterordnung der Cladoceren. Nova Acta reg. Soc. Sci. Upps. Ser. 3: i–vi, 1–701.

Megard, R. O., 1967. Three new species of *Alona* (Cladocera, Chydoridae) from the United States. Int. Rev. ges. Hydrobiol. 53: 37–50.

Michael, R. G. & D. G. Frey, 1983. Assumed amphi-atlantic distribution of *Oxyurella tenuicaudis* (Cladocera, Chydoridae) denied by a new species from North America. Hydrobiologia 106: 3–35.

Michael, R. G. & D. G. Frey, 1984. Separation of *Disparalona leei* (Chien, 1970) in North America from *D. rostrata* (Koch, 1841) in Europe (Cladocera,Chydoridae). Hydrobiologia 114: 81–108.

Michael, R. G. & B. K. Sharma, 1988. Indian Cladocera (Crustacea: Branchiopoda: Cladocera). Zool. Surv. India, Calcutta. xvii, 262 pp.

Müller, O. F., 1769. Efterretning om ubekiendte Een-Oyer i vore faerske Vande. Kiobenhavnske Efterretninger om laerde Sager for Aar 1769, Andet Haefte: 848–851.

Müller, O. F., 1770. Folgende Efterretning om ubekiednte Een-Oyer i vore Vande. Kiobenhavnske Kongl. privil. Addressecontoirs Kritiske Journal, 1770, No. 3: 17–20.

Müller, O. F., 1776. *Zoologiae Danicae Prodromus seu Animalium Daniae et Norwegiae indigenarum characteres, nomina, et synonyma imprimis popularium.* Havniae: 1–282.

Müller, O. F., 1785. *Entomostraca seu Insecta Testacea, quae in aquis Daniae et Norvegiae reperit, descripsit et iconibus illustravit.* Lipsiae et Havniae: 1–135.

Müller, P. E., 1867. Danmarks Cladocera. Naturhistisk Tidsskrift, Kjöbenhavn. 3: 53–240.

Norman, A. M., 1903. New generic names for some Entomostraca and Cirripedia. Ann. Mag. nat. Hist. Ser. 7, 11: 367–369.

Petkovski, T. K., 1966. Eine neue Cladoceren-Gattung aus dem Westen Indiens *Indialona ganapati* n. gen. et n. sp. Fragmenta Balcanica, Skopje 5(22): 157–165.

Rajapaksa, R., 1986. A contribution to the taxonomy, biogeography and gamogenesis of freshwater Cladocera, with special reference to tropical region. Ph.D. thesis, Univ. Waterloo. xiv, 221, 21 appendix.

Rajapaksa, R. & C. H. Fernando, 1982. The first description of the male and ephippial female of *Dadaya macrops* (Daday, 1898) (Cladocera, Chydoridae), with additional notes on this common tropical species. Can. J. Zool. 60: 1841–1850.

Rajapaksa, R. & C. H. Fernando, 1986. Tropical species of *Kurzia* (Crustacea, Cladocera) with a description of *Kurzia brevilabris* sp. nov. Can. J. Zool. 64: 2590–2602.

Rajapaksa, R. & C. H. Fernando, 1987a. Redescription and assignment of *Alona globulosa* Daday, 1898 to a new genus *Notoalona* and a description of *Notoalona freyi* sp. nov. Hydrobiologia 144: 131–153.

Rajapaksa, R. & C. H. Fernando, 1987b. A redescription of *Euryalona orientalis* (Daday, 1898), with a consideration of other species in the genus *Euryalona* (Cladocera:Chydoridae). Hydrobiologia 150: 75–90.

Rajapaksa, R. & C. H. Fernando, 1987c. Redescription of *Dunhevedia serrata* Daday, 1898 (Cladocera, Chydoridae) and a description of *Dunhevedia americana* sp. nov. from America. Can. J. Zool. 65: 432–440.

Rey, J. & L. Saint-Jean, 1969. Les Cladoceres (Crustaces Branchiopodes) du Tchad (Deuxieme note). Cah. O.R.S.T.O.M., ser. Hydrobiol. 3: 21–42.

Rey, J. & L. Saint-Jean, 1980. Branchiopodes (Cladoceres). In Durand, J.-R. & C. Lévêque (eds), Flore et Fauna Aquatiques de l'Afrique Sahelo-Soudanienna, Tome I: 307–332.

Richard, J., 1895. Revision des Cladocere. Ann. Sci. nat. Zool., Paris (7e ser.) 18: 279–389.

Richard, J., 1896. Revision des Cladoceres. Deuxieme Partie. Ann. Sci. nat. Zool., Paris (8e ser.) 2: 187–363.

Roen, U., 1973. *Rhynchotalona kistarae* sp. n. from south Greenland (Crustacea, Cladocera, Chydoridae, Aloninae). Steenstrupia 3: 89–92.

Roen, U., 1987. *Chydorus arcticus* n. sp., a new cladoceran crustacean (Chydoridae: Chydorinae) from the North Atlantic arctic and subarctic areas. Hydrobiologia 145 (Dev. Hydrobiol. 35): 125–130.

Sars, G. O., 1862a. (1861 in separate). Hr. Studios med. G.O. Sars medeelte en af talrige Afbildninger ledsaget Oversigt af de af ham i Omegnen af Christiania iagttagne Crustacea cladocera. Forh. Vidensk.-Selskab. Christiania, Aar 1861: 144–167.

Sars, G. O., 1862b. Hr. Studios med. G.O. Sars fortsatte sit Foredrag over de af ham i Omegnen af Christiania iagttagne Crustacea cladocera. Forh. Vidensk.-Selskab. Christiania, Aar 1861: 250–302.

Sars, G. O., 1863. Beretning om en i Sommeren 1862 foretagen zoologisk Reise i Christianias og Trondhjems Stifter. Nyt Magazin for Naturvideskaberne. 12: 193–252.

Sars, G. O., 1864. Beretning om en i Sommeren 1863 foretagen zoologisk Reise i Christianias Stift. Nyt Magazin for Naturvidenskaberne. 13: 225–260.

Sars, G. O., 1865. Norges Ferskvandskrebsdyr. Forste Afsnit Branchiopoda I. Cladodera Ctenopoda (Fam. Sididae & Holopedidae). Brogger & Christie's Bogtrykkeri, Christiania, viii, 71 pp.

Sars, G. O., 1890. Oversigt af Norges Crustaceer med forelobige Bemaerkninger over de nye eller mindre bekjendte Arter. II. (Branchiopoda – Ostracoda – Cirripedia). Forh. Vidensk.-Selskab. Christiania, Aar 1890, 1: 1–80.

Sars, G. O., 1901. Contributions to the knowledge of the fresh-water Entomostraca of South America, as shown by artificial hatching from dried material. Part I. Cladocera. Arch. Math. Naturvidensk. 23: 1–101.

Sars, G. O., 1904. On a remarkable new chydorid, *Saycia orbicularis* G.O. Sars, from Victoria, Australia. Arch. Math. Naturvidensk. 26: 3–15.

Sars, G. O., 1916. The freshwater Entomostraca of Cape Province (Union of South Africa). Part I: Cladocera. Ann. S. Afr. Mus. 15: 303–351.

Sars, G. O., 1993. *On the Freshwater Crustaceans Occurring in the Vicinity of Christiania*, M.E. Christiansen, J.-A. Eie, G. Halvorsen, A. Hobaek & P. Larsson (eds), University of Bergen Press, Bergen, 197 pp.

Shan, R. K. & D. G. Frey, 1968. Induced interbreeding between two stocks of a chydorid cladoceran. BioScience 18: 203–205.

Smirnov, N. N., 1970. Cladocera (Crustacea) iz permskikh otlozheniy Vostochnogo Kazakhstana. Paleontol. Zhurnal 3: 95–100 (in Russian).

Smirnov, N. N., 1971. A new species of *Archedaphnia* (Cladocera, Crustacea) from the Jurassic deposits of Tranbaikalia. Paleontol. Zhurnal. 1971: 119–121 (in Russian).

Smirnov, N. N., 1987. Cladocera (Crustacea) from Nicaragua. Hydrobiologia 160: 63–77.

Smirnov, N. N., 1989a. Tropicheckiye Cladocera. 1. Novyye vidy rodov *Alona* i *Biapertura* (Aloninae,Chydoridae) tropicheskoy Avstralii. Zool. Zhurnal. 68: 135–140 (in Russian).

Smirnov, N. N., 1989b. Tropicheskiye Cladocera. 2. Novyye vidy semeystv Chydoridae, Macrothricidae i Moinidae tropicheskoy Avstralii. Zool. Zhurnal. 68: 51–59 (in Russian).

Smirnov, N. N. & B. V. Timms, 1983. A revision of the Australian Cladocera (Crustacea). Rec. Austral. Mus., Suppl. 1: 1–32.

Swammerdam, J., 1669 (1685). Historie generale des Insectes. Ou l'on expose clairment la maniere lente & presq'insensible de l'acrroissement de leurs membres, & ou l'on decouvre evidemment l'Erreur ou l'on tombe d'ordinaire au sujet de leur pretendue transformation. Autrecht, Chez Jean Ribbius. (The original version was published in Latin in 1669.)

Ueno, M., 1937. Order Branchiopoda (Class Crustacea). Fauna Nipponica 9: 1–135.

Vasil'yeva, G. L. & N. N. Smirnov, 1969. Chydoridae (Cladocera) Baykala. Zool. Zhurnal. 48: 184–196 (in Russian).

Hydrobiologia **307**: 57–68, 1995.
P. Larsson & L.J. Weider (eds), Cladocera as Model Organisms in Biology.

Phylogeny and adaptive radiation within the Anomopoda: a preliminary exploration

Geoffrey Fryer
Institute of Environmental and Biological Sciences, University of Lancaster

Key words: Anomopoda, evolution, phylogeny, adaptive radiation, morphology, ecology

Abstract

The distinctness of the Anomopoda and the polyphyletic nature of the so-called Cladocera are emphasized.

An attempt is made to reconstruct the ancestral anomopod, which probably lived in Palaeozoic times. This task is facilitated by the availability of detailed information on extant forms, which includes functional as well as purely morphological considerations and enables us to understand the means whereby complex mechanisms were transformed during evolution. Comparative studies on the ecology and habits of extant forms also throw light on the probable way of life of the ancestral anomopod.

Adaptive radiation within the Anomopoda is briefly surveyed and an outline of the suggested phylogeny of the order is indicated.

Introduction

The Anomopoda is a well-defined order of the Branchiopoda, separated by many trenchant characters from the other three orders with which it was for long united in a polyphyletic assemblage under the name Cladocera (Fryer, 1987a, 1987c). Diagnostic features of the order are listed elsewhere (Fryer, 1987c). Differences between the Anomopoda and the superficially similar Ctenopoda, with which it shares certain primitive branchiopod features, but from which it is distinguished in many ways, are discussed in Fryer (1987a). Here it is sufficient to note that a fundamental distinction lies in the nature of the trunk limbs. In the Anomopoda exploitation of diversity within the series of limbs for many different means of collecting and manipulating the food has been an outstanding feature of adaptive radiation. This has been accompanied by related modifications of the carapace, antennae and post-abdomen, in ways that have no parallel in the Ctenopoda, whose trunk limbs are characterised by serial uniformity and display much more restrained modification. These two orders also differ fundamentally in other features, including the nature of the maxillules, the arrangement and details of the ovaries and the means of protecting the resting eggs.

Apart from its implications for Branchiopod evolution as a whole (to be considered elsewhere) recognition of Anomopod distinctness invites examination of affinities within the order, certain aspects of which are explored here. Opinions differ as to the number of families recognised. The Macrothricidae, Chydoridae, Bosminidae and Daphniidae are universally accepted. From the Daphniidae Goulden (1968) split off the Moinidae and this taxon has been widely recognised though it seems not to merit more than sub-familial rank (Fryer, 1991b), and the same applies to the proposed Ilyocryptidae. Likewise Smirnov (1992) allocated what are much the most informative anomopod fossils yet discovered to a new family, the Prochydoridae. These Early Cretaceous animals certainly differ from all modern chydorids in well-defined ways but whether, especially in the absence of information on the nature of the trunk limbs, this justifies familial separation is arguable. Neither of these matters, however, presents difficulties when affinities within the order are considered.

The search for the ancestral anomopod

The Anomopoda is an ancient group. Just how ancient is unknown. As long ago as early Cretaceous times (120–130 m y ago) ephippia were produced that are not only similar to those of modern daphniids but can be assigned to the extant genera *Daphnia*, *Simocephalus* and probably *Moina* (Fryer, 1991; Smirnov, 1992). Others are known of Eocene (Lai & Li, 1987), Oligocene (Heydon, 1862) and (*Moina*) Miocene (Goulden, 1968) age. Daphniid ephippia are the most specialised of all such structures within the Anomopoda and owe their preservation to their robust protective envelopes. The more delicate, more primitive, and often very small ephippia of other families are less likely to be preserved but chydorid ephippia may yet be found in a fossil state. The inference from fossil daphniid ephippia is that they were produced by animals that had other attributes of these genera, but this is unproven. Geographical evidence also indicates that *Daphnia* is an ancient genus (Benzie, 1987). If even extant genera existed so long ago, the order must be much older. If phylogenetic reasoning to be put forward elsewhere is correct, it must have originated earlier than the Devonian.

The claim by Tasch (1963), based on curious logic, that the origin of 'cladocerans' can be restricted to post-Cretaceous, pre-Oligocene times was never convincing and can now be forgotten, as too, incidentally, can his completely untenable, and contradictory, claim that the Haplopoda probably arose in 'the old Pleistocene'. This too is clearly an ancient group.

We recently acquired for the first time information on some features of the morphology of anomopods that flourished in early Cretaceous times (Smirnov, 1992). Whether these are regarded as primitive members of the Chydoridae or as representatives of a distinct family, the Prochydoridae, is unimportant: their significance is that they show that animals with many features of chydorid organisation were abundant some 130 m y ago. These extinct forms also reveal interesting differences from modern chydorids – e.g. they had more antennal setae and some of them apparently had an unconvoluted alimentary canal – and provide valuable information on certain evolutionary trends within the chydorid line and on the time span over which changes of antennal armature may have occurred. In various respects, however, they were already too specialised to throw much light on the nature of the ancestral anomopod.

Reconstruction of the anatomy and way of life of this organism must therefore be based largely on comparative studies of extant forms, particularly within, but also outside, the order. This is a perfectly legitimate approach which, as a result of studies made during recent years, can include functional as well as purely morphological considerations. Indeed without functional knowledge of the mechanisms involved (and there are many), not only is it impossible to understand how one mechanism might have been derived from another, but meaningful comparisons between families are impossible, and it is very difficult to appreciate what the primitive condition may have been. Information on the ecology and habits of extant forms is also extremely helpful. As Patterson (1981) has said, knowledge of the recent biota is the starting-point for all phylogenetic work.

Some attributes can be ascribed to the ancestral anomopod with certainty. For example, when a feature characteristic of, and unique to, the order, such as the nature of the maxillule, or the formation of an ephippium, is shared by all species irrespective of their diverse specialisations, this feature was clearly present in the common ancestor. The only, and less likely, alternative is that it evolved independently several times. This does not mean that certain features have not evolved more than once during the adaptive radiation of the order. Convergence has certainly occurred at times.

The ancestral anomopod: its structure and way of life

The suggested gross morphology of a female of the ancestral anomopod is shown in Figs 1 and 2. Save for a pronounced ventral gape, the trunk was completely enveloped by the carapace. This was derived from a dorsal shield of still earlier ancestors that later folded and extended ventrally on each side to give a functionally bivalved structure while remaining as a single hingeless entity. Its margins were more or less straight and unelaborated but probably bore a few short setae. The carapace was double-walled, the inner wall being very thin, the outer more robust but by no means thick, and scarcely sclerotized. Between the walls was a space, spanned by fibrils, within which blood circulated, and which, anteriorly, housed the tubules of the maxillary glands. The trunk was composed of only six limb-bearing segments and a much abbreviated region posterior to them where it was flexed ventrally and perhaps somewhat anteriorly to form a simple structure, later to become elaborated into the post-abdomen, that terminated in a pair of spines. A pair of sensory setae arose in the region of flexure. The posterior region had some mobility but was not articulated with more anterior parts, nor was it provided with specialised marginal armature though a few spinules were present on the morphologically dorsal surface.

The short head lacked a headshield. The eyes were 'internalised', having sunk beneath the cuticle and were already fused to give a single median structure or were completing the process. An ocellus was present.

The movable antennules were long, slender and of two segments, terminated in a tuft of sensillae, and had a more proximal sensory seta.

The antennae (Fig. 3) were large with a straight protopodite, and bore a four-segmented endopodite and a three segmented exopodite. Fossil evidence produced by Smirnov (1992) shows that in the chydorid lineage there were more setae (up to five) on the basal segment of the exopodite of certain early Cretaceous forms than in modern species that have one or none. As even primitive macrothricids never have more than one long seta or spine on the segment in question, this may have been an experiment in an offshoot of the chydorid lineage, but it remains an open question whether the ancestral anomopod had more exopodite setae than any extant species. Other branchiopod orders that retain locomotory antennae in the adult have more such setae than anomopods, and this is also the case in anostracan nauplii, so the precursor of the Anomopoda may also have had more setae than modern anomopods, but the basic condition suggests that reduction occurred early in their evolution. Only fossils can resolve this question.

The mandibles were of the basic rolling, crushing, branchiopod type and lacked a palp. Already they displayed slight asymmetry, both skeletal-and muscular.

The maxillules (Figs 5 and 6) were similar to those of modern forms, being small and with armature reduced to three spines, which, save for occasional reductions, is a conservative feature in the order and very different from, for example, the arrangement in the Ctenopoda.

It is probable that the maxilla was a simple papilla on which the duct of the maxillary gland terminated and opened. As an alternative hypothesis, the similarity of the inner lobe of trunk limb 1 of certain macrothricids and chydorids to the maxilla, or to the similar lobe borne by this appendage, in the Anostraca and Lipostraca, raises the possibility that it is in fact the maxilla, of which there is otherwise no trace in, for example, the chydorid *Eurycercus* (Fryer, 1968). Even the possibility that what has long been called the first trunk limb is in fact the maxilla needs consideration. There are, for example, beguiling similarities between the first trunk limb of macrothricids and chydorids and the maxilla of such an anostracan as *Branchinecta paludosa* (Fig. in Cannon 1933). If this were the case, trunk limb 2 of anomopods would be the homologue of trunk limb 1 of the lipostracan *Lepidocaris*, to which it bears striking structural, and obviously functional, similarities. However, as Grobben (1879) showed long ago for *Moina* (in which there is no maxilla: only a setose mound), both maxillules and maxillae are evident in the embryo, and the same is true of the macrothricid *Acantholeberis* in which the maxilla is distinct in the embryo (Figs 7and 8) and persists in the adult as a small papilla in addition to the conspicuous inner lobe of trunk limb 1 (Fryer, 1974). Furthermore, Kotov (1995) has shown that the maxilla makes a transitory appearance in the embryo of chydorids, bosminids and daphniids in which there is no trace of this appendage in the adult.

A large median fleshy labrum, with no ventral elaboration, extended posteriorly from the head, covering the mandibles and maxillules ventrally and forming, with the adjacent body wall, a tube bridging the gap between the anterior end of the food groove (see below) and the atrium oris (Figs 5 and 6). Within the labrum were labral glands that discharged entangling secretions.

The mid ventral region of the essentially cylindrical trunk was excavated into a well-marked food groove that extended posteriorly to the level of the fifth pair of trunk limbs.

The trunk limbs lacked joints, nor was there a joint between limb and trunk. Although form was defined by a cuticle, rigidity of the limb was in large measure provided by turgor pressure of the haemocoelic blood system. The trunk limbs already differed among themselves, the last pair being little more than simple flaps. Comparisons of extant species enable a generalised pattern to be determined for each limb and such are shown in Fig. 5. Limbs 3, 4 and 5 bore flap-like exopodites laterally and a swollen epipodite also arose laterally from each limb. A horizontal slice through the mouthparts and trunk limbs is shown in Fig. 6. The proximal portions of the second to fifth pairs of limbs extended as small gnathobases into the median food groove (Fig. 4). This was of great functional significance: remotion (posterior swing) of the limbs result-

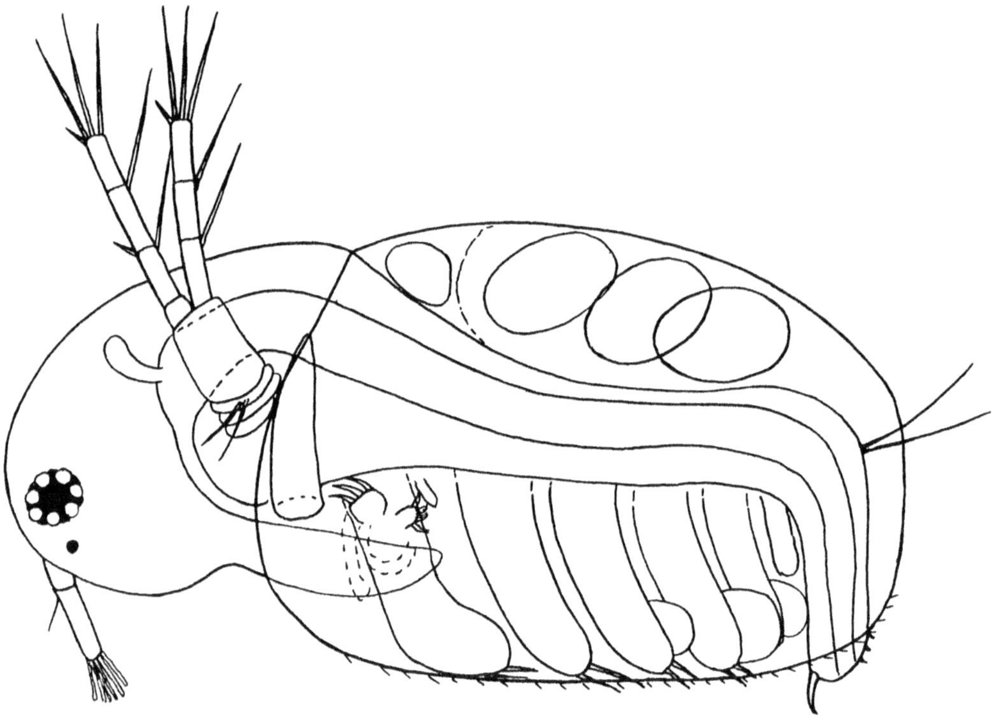

Fig. 1. The ancestral anomopod, lateral. The carapace is shown as if transparent. The trunk limbs are shown as simple outlines only: epipodites are omitted. For details, see Fig. 3.

ed in promotion (anterior swing) of the gnathobases which were concerned with the forward movement of food material.

The posteriorly-facing mouth opened into a short, muscular oesophagus that curved dorsally to open into the wider mid-gut which extended into the head. Apart from the necessary sinuosity to curve towards the trunk and, posteriorly, to accommodate the flexure of the proto-postabdomen, the mid gut passed directly from the head along the length of the trunk. There were no convolutions. Anteriorly a pair of simple caeca arose from the mid-gut and extended anteriorly.

From close to its anterior origin the carapace was free from the trunk, leaving a dorsal space which was already used as a receptacle for eggs, both parthenogenetic and sexual. Parthenogenetic eggs were already produced, though precursors of the ancestral anomopod produced only eggs that required fertilization, which probably underwent diapause. This is typical of many primitive extant branchiopods though there have been a few experimental departures, e.g. in species of *Artemia* (Anostraca). All extant Spinicaudata and Laevicaudata except *Cyclestheria* produce only fertilized resting eggs, which they protect beneath the carapace, though not (except in *Cyclestheria*) in a dorsal brood pouch as is the case in the Anomopoda. *Cyclestheria* exemplifies the independent acquisition of the ability to reproduce by parthenogenesis, an ability that was also acquired independently by the Anomopoda and other 'cladoceran' orders. Development of both parthenogenetic and sexual eggs was direct, the nauplius larva having been eliminated from the life cycle.

As in modern anomopods, whether parthenogenetic or sexual eggs were produced was dependent on ecological conditions. Sexual (resting) eggs were produced in small clutches whose number was not fixed, as is still the case in such primitive macrothricids as *Acantholeberis* and *Ophryoxus* and primitive chydorids such as *Eurycercus* and *Saycia*. Little or no modification of the carapace took place for the reception of resting eggs save perhaps for the laying down of a thickened dorsal ridge, but the moulted carapace was used as a container for such eggs and formed a primitive ephippium (or proto-ephippium), a structure diagnostic of the Anomopoda. The thickened dorsal ridge acted as a spring which pulled the valves together when the carapace was moulted and secured the eggs within, offering a measure of protection.

Small secondary sexual differences were apparent. The first trunk limbs played only a minor part

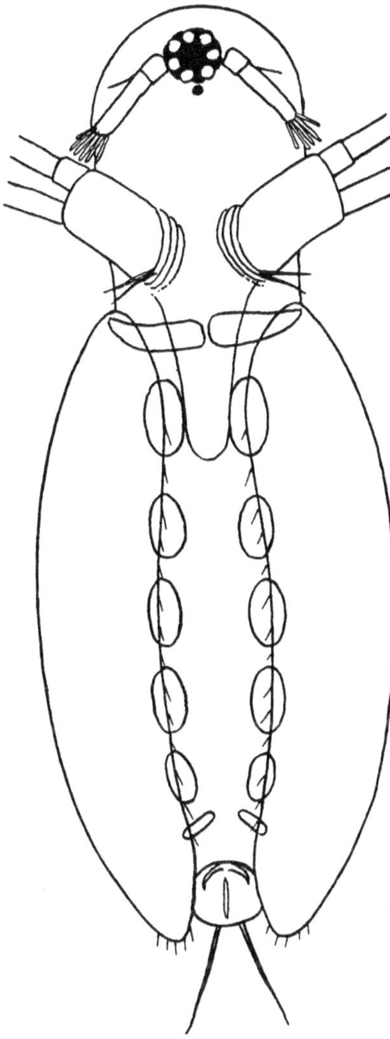

2

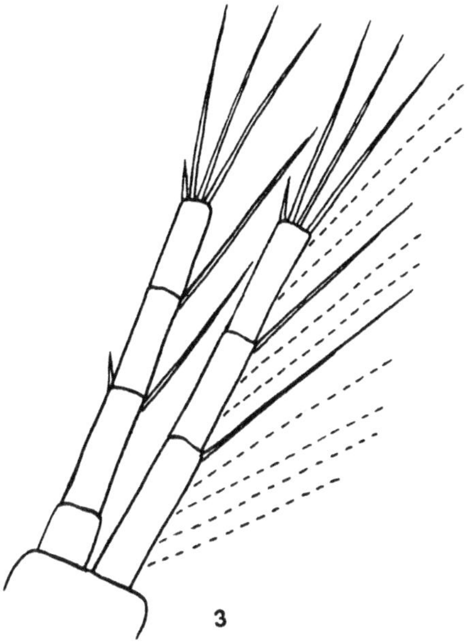

3

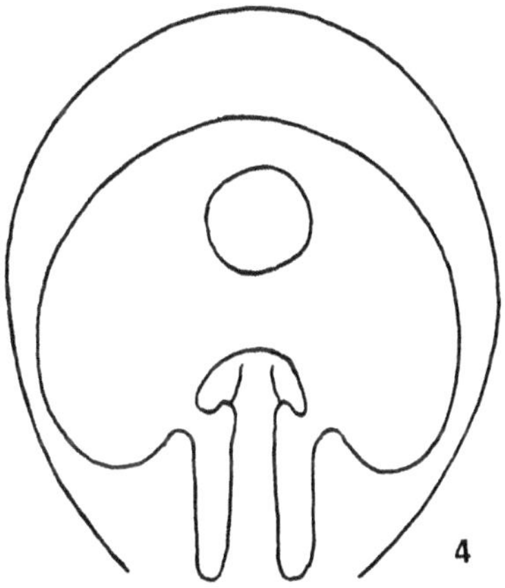

4

Fig. 2. The ancestral anomopod, ventral. Only the positions of the trunk limbs are indicated. *Fig. 3.* Antenna. Dashed lines on the exopodite indicate the possible presence of additional setae, such as were present in certain Early Cretaceous members of the chydorid line. Whether these are indicative of the ancestral state or of an experiment in that lineage remains unproven. *Fig. 4.* Diagrammatic transverse section through the trunk and a pair of limbs, showing the trunk/limb union and how the gnathobasic spines extend into the food groove and how, when the limbs remote, the gnathobase promotes.

in food collection, were not particularly specialised for other functions, were conveniently located, and acquired simple hooks in the male that helped it to grasp the female. The vasa deferentia opened on the post-abdomen, that resembled that of the female. From such a generalised anomopod it is possible to derive all extant forms including such different animals as *Daphnia*, *Bosmina*, *Eurycercus*, *Pseudochydorus*, *Graptoleberis*, *Ilyocryptus* and *Lathonura*, to mention only a few of the many widely divergent genera of all families. That these derivations are based on perfectly feasible morphological transformations, not only of individual appendages but of entire, and complex, mechanical arrangements, gives one confidence that the reconstruction is realistic and that it serves as at least a first approximation of what the ancestral anomopod was like. Furthermore the trends that can be recognised are not based on structure alone: recently acquired knowledge of the functional morphology and ecology of many very different anomopods makes it possible to see how one mechanism could be transformed into or derived from another and to correlate the transforma-

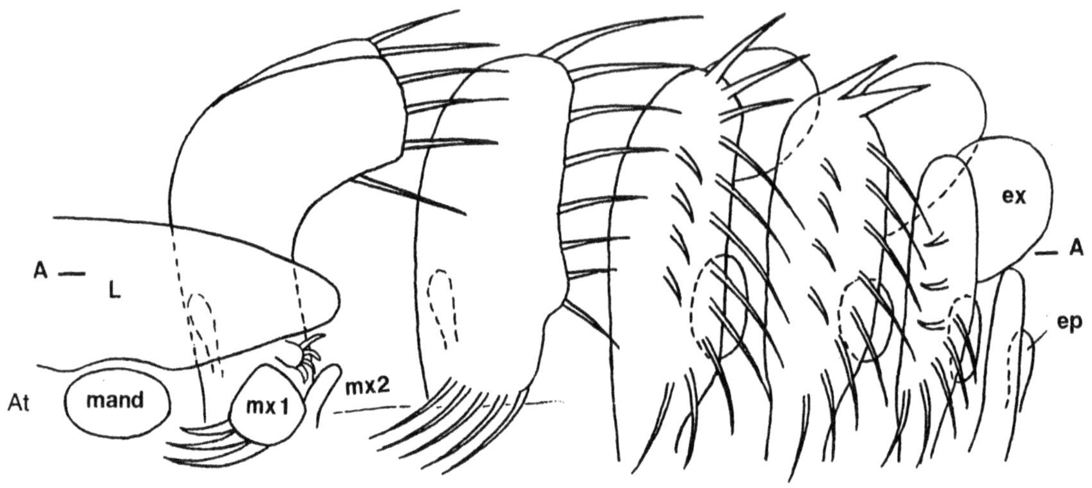

5

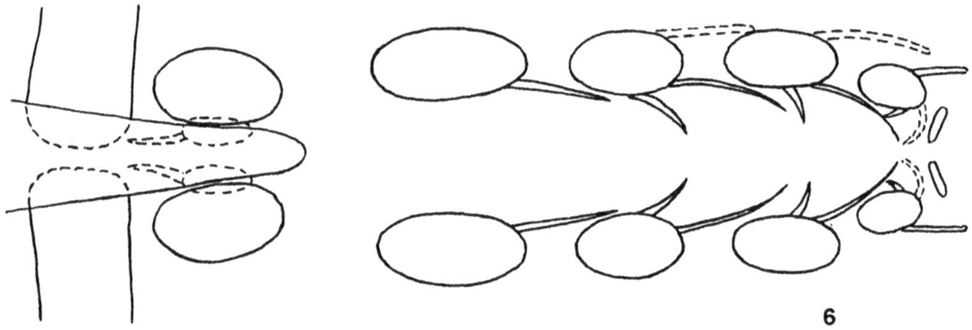

6

Fig. 5. The ancestral anomopod. Arrangement of the trunk limbs, medial aspect, with the labrum bisected longitudinally. At = atrium oris; mand = Mandible; m×1 = Maxillule, m×2 = Maxilla; ex = exopodite; ep = epipodite; L = labrum. *Fig. 6.* The same. Horizontal slice through the trunk limbs at about the level indicated by A–A in Fig. 5, ventral. The exopodites of trunk limbs 3 and 4, as they appear more ventrally, and a gnathobasic spine of trunk limb 5 as it appears more dorsally, are indicated by dashed lines. Although the sixth trunk limbs are dorsal to the level of the section, their position is indicated, as are the more dorsally lying maxillules (shown by dashed lines) and parts of the mandibles.

tions with shifts in ecology and behaviour. They also make it possible to suggest not only how the ancestral anomopod was constructed but how it worked at the mechanical level – which in these animals is a supremely important level. Only fossil finds can verify or modify these deductions. The sort of transformations that have taken place during the evolution of trunk limb 5, which assumes diverse forms, are shown in Fig. 9 as an example of how continuity was maintained from the ancestral condition. Transitions in limb 4, whether to the basic macrothicid/chydorid type or the daphnid type with its enormous filter – are even easier to envisage.

Habits were essentially benthic. Orientation was horizontal and locomotion was by swimming, usually in short bursts, using the antennae for propulsion. Whenever the antennae ceased to beat the animal sank to the adjacent bottom and there rested on the ventral margins of the carapace. Balance was not very efficient but was facilitated by the gape between the carapace valves and the width of the carapace and was perhaps assisted at times by the outstretched antennae, though these were generally held clear of the bottom and directed somewhat anteriorly and dorsally as well as being extended laterally.

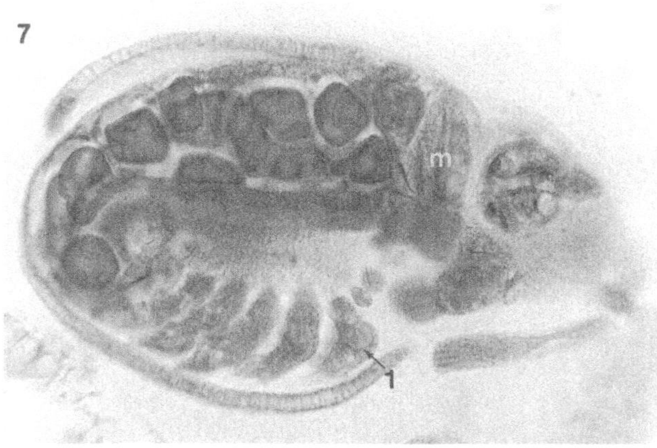

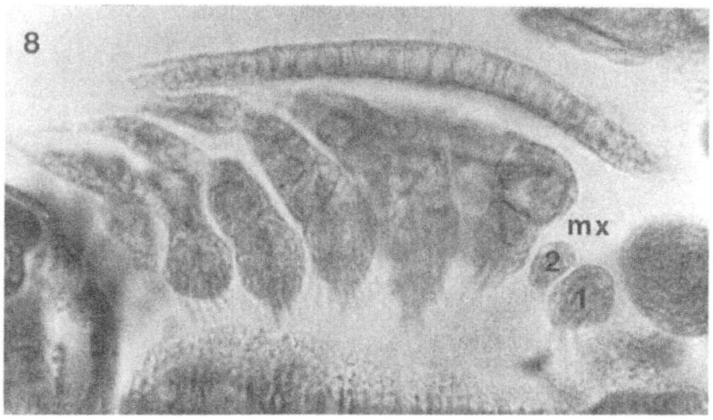

Fig. 7. Median longitudinal section of an embryo of *Acantholeberis curvirostris* showing the presence of a well-defined maxilla between the maxillule and first trunk limb. m = mandible; 1 = trunk limb one. *Fig. 8.* The same, other side, more highly magnified. For ease of comparison Fig. 7 is shown dorsal side uppermost; Fig. 8 ventral side uppermost. mx 1 & 2 = maxillule and maxilla respectively.

In a settled individual the tips of the first three pairs of trunk limbs lay in close proximity to the substratum. Those of the first pair may have been used occasionally for grasping objects, which would assist balance or anchorage, but were not specialised for this. By simple promotion and remotion, assisted by some flexure and extension, the distal spines of each would reach and scratch the substratum and, on remotion, sweep coarse particles posteriorly and dorsally. Further transport of these particles was assisted by suction set up as trunk limb 4 drew away from limb 5, thereby expanding the interlimb space 4/5 towards the end of promotion, to be followed by limb 3 that enlarged interlimb space 3/4, a process that continued as limb 5 began remotion before limb 4. Suction was greatly enhanced by the flap-like exopodites of limbs 3, 4 and 5 which formed a simple pump, later to be modified in different ways in different lineages.

Trunk limb 2 had already acquired a uniseriate row of spines that extended from its tip to more proximal regions. These spines helped to sweep particles dorsally. Particles collected and drawn in by these various means were then trapped in a cage made up of rows of spines, borne on the inner posterior faces of limbs 3, 4 and 5. The rudimentary sixth limb helped to block the posterior end of the cage. Although coarse-meshed, the cage contained the precursors of a filter-feeding device, but true filter-feeding did not occur, and very fine particles could not be retained. The orientation of the cage-forming spines ensured that collected material passed dorsally as well as posteriorly, and, assisted by more anteriorly located spines on the inner faces of limbs 3 and 4, ultimately arrived within the food groove. Passage forward was ensured by the spines of the simple gnathobases of limbs 3, 4 and 5 which swung anteriorly as the limb remoted, as with greater effect did the more elaborately armed gnathobasic region of limb 2 that showed specialisation in this respect from an early stage in evolution, and swept material towards the maxillules which assisted its passage to the mandibles that operated in the basic branchiopod manner.

Adaptive radiation and anomopod phylogeny

Macrothricidae

Of extant anomopods the Macrothricidae retains the most primitive features. I originally thought that the primitive macrothricid was probably a filter feeder, and we now know that true filters, remarkably similar to those of modern branchiopods, had evolved in small arthropods even by early Cambrian times (Butterfield, 1994). The most likely owners of these filters were branchiopod crustaceans. Nevertheless, subsequent studies and comparisons lead me to conclude that several species of the Macrothricidae probably never acquired the ability to filter. In this they differ from non-filter feeding chydorids in which this condition represents a secondary loss. It is much easier to envisage the evolution of the trunk limbs of macrothricids such as *Streblocerus* and *Drepanothrix* from the postulated ancestral condition by reduction and specialisation for grasping and pushing, in minute animals that retained the ancestral benthic habits, and continued to move by means of large antennae, than via the development of filters and their subsequent loss. Likewise it is difficult to see how the substratum-hugging *Lathonura* could have been derived via a filter feeding stage, but its morphological specialisations can be

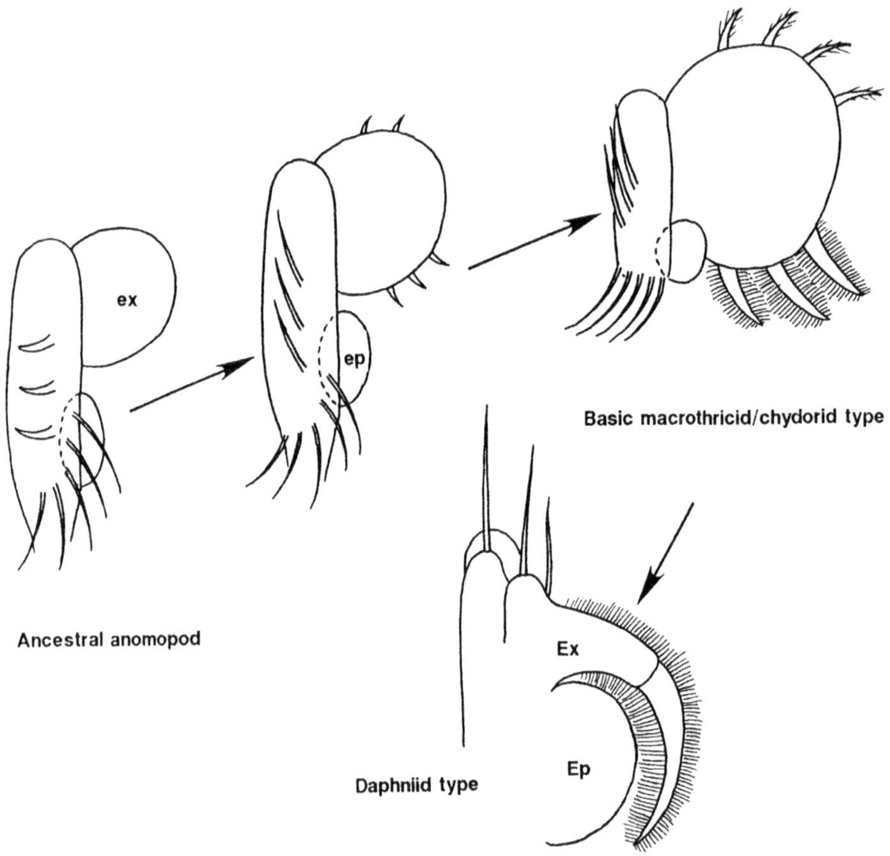

Fig. 9. Transformations of trunk limb 5 from the condition in the ancestral anomopod to that of modern families.

understood if derivation was direct from the condition postulated for the ancestral anomopod.

All macrothricids retain the ancestral association with the bottom or, in a few cases (*Iheringula*, *Macrothrix triserialis*, *Bunops*, *Onchobunops*) with vegetation. Some (*Ilyocryptus*, *Neothrix*) have become specialised burrowers, and others (*Streblocerus*, *Drepanothrix*) readily penetrate flocculent deposits. Most species swim for short distances only but *Ophryoxus* has become a more persistent swimmer, though still over the bottom.

Modifications of the ancestral arrangement of limbs are related to changes in habits, which are reflected in many other morphological features. Primitively the tips of trunk limbs 1 to 3 (and possibly 4) were involved in scratching and sweeping food from the substratum, but limbs 3 and 4 gradually became specialised for dealing with the material so collected and ceased to be involved in food collection. While trunk limb 1 continues to contribute to this process in a few cases (*Acantholeberis*, *Streblocerus*) this role was generally taken over largely or entirely by limb 2, whose armature of spines became more specialised and in many cases developed into a row of scrapers whose form is intimately related to the circumstances in which food is collected. This left the first trunk limbs free to take on other functions, e.g. grasping in *Onchobunops*, or organs of true locomotion, primitive in *Ophryoxus*, more specialised in *Macrothrix triserialis* and *Lathonura*. Limbs 3, 4 and 5 became specialised for handling the material collected by the anterior limbs and were modified in many different ways. Trunk limb 6, originally serving to close the posterior exit from the chamber formed by the trunk limbs, was in most cases eliminated and this role was assumed by limb 5 that, primitively, served a basically similar function to limb 4.

The primitive arrangement of the armature of limbs 3 to 5 was such that it formed a cage that retained collected material. A major, and probably very early, trend was the conversion of the posterior row of spines of these limbs into filter setae, which also called for improved efficiency in the exopodite pump that drew water through the sieve so formed. In several, but not all, cases where filter feeding was adopted, trunk limb 5 ceased to contribute and filtration was concentrated on limbs 3 and 4. In the super-specialised *Ilyocryptus* filtration became so effective that the animal can subsist purely by this process, no scraping by trunk limb 2 being necessary, and that limb has lost its primitive function of food collection (and its armature of scrapers). Trunk limb 1 has also become specialised to provide a screen that excludes the ingress of coarse particles to the filter chamber. These developments went hand in hand with improvements in the gnathobases of the limbs that pass food forward, this being especially the case in trunk limb 2 that came to play a major role in this process.

At the same time, many other changes occurred in different macrothricids. These included diversification of the shape of the carapace, sometimes the development of a headshield, or coiling of the gut, or flexing of the antennae, elaborations of the postabdomen, specialisations of the still primitive ephippium, and other morphological changes, as well as physiological specialisations that led eventually to such abilities as the toleration of strongly acidic conditions (*Acantholeberis*) or of low oxygen levels via the development of haemoglobin (*Ilyocryptus*).

Daphniidae

The daphniid feeding mechanism (Fryer, 1991b) can easily be derived from that of a generalised macrothricid by the expansion of the filtering area of trunk limbs 3 and 4, and improvements in the pumping mechanism, which involves exopodites 3 and 4 and modifications of trunk limb 5 to form a seal. These developments obviate the need to collect particles by scraping and enable them to be obtained entirely from suspension. Furthermore, an extant macrothricid, *Ophryoxus gracilis*, provides a clear indication of the route whereby the Daphniidae arose from the macrothricid stem. *Ophryoxus* is the most persistent swimmer of all the studied macrothricids and its antennae are very similar to those of *Daphnia* (which have an extra exopodite seta). What is more, *Ophryoxus* sometimes swims along vertical objects, its orientation then being the same as that typical of *Daphnia*. It is also a filter feeder (though its feeding mechanism involves scraping) and is one of the macrothricids that has acquired a headshield. It is also, as it happens, the only macrothricid whose alimentary canal has daphniid-like anterior caeca. While not the direct ancestor of modern daphniids, *Ophryoxus* nevertheless throws much light on that ancestry. In essence, if one turns it through 90 ° and changes its feeding mechanism to filtering with trunk limbs 3 and 4, with no preliminary scraping one has a basic daphniid. Such an origin of daphniids explains the large postabdomen, whose derivation is from a structure that served for pushing in their benthic ancestors, but not from one so specialised as to have acquired an articulation.

Adaptive radiation within the Daphniidae has been considerable, but into fewer niches than have been exploited by either the Macrothricidae or Chydoridae. Specialisations include association with the surface film (*Scapholeberis* and *Megafenestra*), filtering while attached to objects (*Simocephalus*), and especially emancipation from the bottom, (*Moina*, *Ceriodaphnia*, *Daphnia*, *Daphniopsis*). Open-water living has led to a truly planktonic lifestyle in at least some members of all the genera that practise it. The exploitation of different size ranges by *Daphnia* and *Ceriodaphnia* has helped to partition resources in what is superficially a uniform habitat. There have also been physiological specialisations, including tolerance of waters of high ionic concentration (especially *Daphniopsis* and several species of *Moina*), and modifications of the reproductive process have been exploited, including obligate parthenogenesis by certain species of *Daphnia*.

Chydoridae

The Chydoridae, with which for the time being I include Smirnov's Prochydoridae, has retained the ancestral association with substrata. Such habits offer the greatest opportunities to the anomopod Bauplan and the Chydoridae has acquired an enormous range of specialisations. Association with bottom deposits or vegetation and the exploitation of scrambling, and especially crawling, of which the family has many specialist exponents, has enabled chydorids to develop thicker carapaces than those of macrothricids – or of daphniids where this trait is incompatible with efficient persistent swimming and has been exploited only to a minor extent in the semi-sedentary *Simocephalus*. Thicker carapaces have in turn facilitated elaboration of their ventral margins which display a wide range of modifications related to association with substrata, some of them exceedingly complex. In the most specialised case, that of *Graptoleberis*, which glides over surfaces like a minute snail, mechanical integration between the structures concerned and the surfaces with which they make contact achieves a level of intricacy that can have few rivals in the animal kingdom.

A thickened cuticle, besides being well-suited to a benthic existence, offers protection against predators, which has been exploited to a striking extent in *Pseudochydorus*, and in a remarkable manner as protection against the nematocysts of *Hydra* by *Anchistropus*. A novel way of obtaining protection, achieved by *Alonopsis elongata* and *Monospilus dispar*, is retention of one or more carapaces after the moult. In the case of the former, an individual seized by a small predator, such as a cyclopoid copepod, can slip out of the old carapace and escape. Remarkable cuticular elaboration, unique to chydorids, has arisen in the so-called honeycombed species of *Chydorus*, of which the most detailed account is given by Frey (1987).

Chydorids have reduced the size of their antennae, which beat rapidly when swimming is necessary, and both rami are reduced to three segments. They have also exploited the basic anomopod trunk limb arrangement in a variety of ways and their feeding mechanisms are diverse. Most are particle feeders, many of which employ filtration, and as many as four pairs of limbs may bear filters. In several genera of different subfamilies, not only are filters borne on limbs 3, 4 and 5, but the gnathobase of limb 2 is filtratory, examples being found in *Eurycercus*, *Alona* and *Chydorus* to mention only three genera. Filtering may, however, be restricted to only two pairs of limbs. For example, in *Alonopsis* filtering is carried out largely by trunk limb 3, with a minor contribution from limb 4. Particles to be filtered are first collected by scrapers borne on trunk limb 2, assisted in some cases (e.g. *Alonopsis elongata*, *Alona affinis*) by distal spines of trunk limb 3.

Some chydorids have abandoned filtration. In the case of *Leydigia*, this is clearly a secondary loss, elements of a filter feeding mechanism being employed. The functional series that leads to the super-specialised *Graptoleberis* also suggests derivation from filter-feeding ancestors, as do certain morphological features, though the mechanism has been drastically modified during the course of its evolutionary history. Morphological considerations also suggest that *Pseudochydorus* (a scavenger) and *Anchistropus* (a parasite of *Hydra*) are derived from filter feeding ancestors. It therefore seems likely that the chydorid line was derived from an ancestral filter feeder.

Of all anomopods the Chydoridae has most fully exploited the potential of the postabdomen whose efficiency in levering, and sometimes its use in ways unknown in other families, has been enhanced by the acquisition of a true articulation between it and the trunk.

Chydorid organisation has permitted the colonisation of habitats inaccessible to other anomopods. Most remarkable has been the exploitation of moisture-holding clumps of bryophytes in wet forests. For life in the thin film of water present in such sites *Bryospilus* has short antennae, no longer used for swimming, of which it is incapable, and is a slow-moving crawler that has lost the compound eye but retains the ocellus (Frey, 1980). Chydorids have also colonised groundwaters, a logical extension of benthic habits. Of the few known species of such habitats *Alona phreatica* has lost the ability to swim and locomotion is restricted to crawling or scrambling (Dumont, 1983). *A. smirnovi* from L. Ohrid also appears to be of interstitial habits, having been found among sand on a beach, and perhaps frequents subterranean springs that debouch there. Its compound eye has lost its pigment (Petkovski & Flössner, 1972). *A. hercegovina* is a true cave-frequenter and has lost the compound eye and the pigment from the ocellus (Brancelj, 1990).

A specialisation of many chydorids (but not of the most primitive) is that they have reduced the clutch and brood size to two, implying a heavier investment per offspring than is usually the case in other anomopods – perhaps exploiting the protection conferred by a more robust carapace? The evolutionary trend in the chydorids has not been towards increased fecundity – and runs counter to theories which claim that to maximise fecundity always has selective advantages – but the implications of the situation remain to be explored (see discussion in Fryer, 1988).

Smirnov's (1992) recent discoveries show that animals similar in many respects to modern chydorids were abundant in Early Cretaceous times. The general arrangement of the headshield, carapace and antennae, both of whose rami had three segments, was remarkably similar to that in modern forms. As to be expected, the mandibles were, like those of modern chydorids, of the typical branchiopod type, and the post-abdomen, of which details are vague, certainly terminated in two spines like those of modern forms. The extra antennal setae, one of the distinguishing features of these animals, may indicate that they tended to swim over bottom deposits rather than scramble or crawl. Conditions in some of these fossils are interpreted by Smirnov as indicating that the gut was not convoluted. If this was so, it represents the retention of an ancestral feature not seen in any extant chydorid. It is, however, difficult to recognise the gut in most of the fossils and it is possible that both the convoluted and non-convoluted condition prevailed in different species. Additional evidence is needed. A coiled gut is a specialisation of extant chydorids. Bottom- or vegetation-frequenting detritus eaters often collect large quantities of low quality food and often much indigestible material, including even inorganic particles. A coiled, and therefore extended, gut is advantageous here, and any increase in density (probably slight) so acquired is no burden and may even be advantageous. Open-water forms, such as daphniids, usually feed on more nutritious material, including living algae, and have less need of a long gut, which would be an inconvenience to animals that have to combat a continuous tendency to sink. They have also developed certain specialisations that improve the efficient utilisation of what is collected (Fryer, 1970). (Macrothricids display both trends. Some retain a primitive condition: others, such as *Drepanothrix dentata*, have a specialised, coiled gut.)

Bosminidae

The Bosminidae arose from a chydorid-like ancestor that, like the Daphniidae, adopted an open-water way of life based on persistent swimming and the filtration of suspended particles. The dichotomy evidently arose before the Chydoridae had lost one antennal segment from the exopodite as the importance of swimming declined. Ancestry is clearly revealed by the trunk limbs. While a statement that bosminids can be regarded as swimming chydorids that have adopted true filtration carried out predominantly by trunk limb 3 is much too crude, it has useful descriptive value and contains a considerable element of truth.

Apart from a pioneer effort by Graf (1930), which established certain points but left some problems unanswered, no real attempt has been made to describe the feeding mechanism of bosminids, of which certain specialisations remain unrecorded. Early statements by Naumann (1921) are incorrect, and while DeMott (1982, 1985) and DeMott and Kerfoot (1982) have produced some useful information, certain functional details have yet to be described. It is, however, clear that an enormous expansion of the filtering surface of trunk limb 3 has been the key to their success as filter feeders. Morphological relics of former scraping by trunk limb 2 in the chydorid manner remain. DeMott and Kerfoot (1982) and DeMott (1985) have emphasised the use of these spines in, as they put it, seizing and manipulating individual algal cells, and regard *Bosmina* as having a 'dual option' feeding mechanism. However, these spines can only manipulate material that has been brought into the filter chamber by currents. They do not collect any material themselves and therefore do not add to the diversity of food collected. *Bosmina* is purely a suspension feeder. That it preferentially retains algal cells rather than bacteria is interesting, but this does not mean that it can collect the former independently of the filtering process. DeMott's demonstration (1982) that it took *Chlamydomonas* and *Aerobacter* from a mixed suspension but hardly removed the latter from a pure offering is explained by the fact that it *had* to take some *Aerobacter* in order to obtain *Chlamydomonas* (which it readily removed from a pure culture), but when offered *Aerobacter* alone, it either let it pass through the filter chamber or, more likely, simply stopped pumping. Respiration-wise, *Bosmina* can probably afford not to pump for long periods. Several benthic macrothricids and chydorids meet their respiratory demands in situations that are often less well oxygenated than those frequented by *Bosmina* without pumping (some have no pump), or by very occasional pumping. Even the actively pumping globular *Chydorus sphaericus* can close up its carapace completely as a defence mechanism and so remain for at least 2 hours while a predatory cyclopoid copepod attempts to gain access.

Possession of delicate cuticle and the lack of association with substrata have granted both daphniids and bosminids freedom to modify external shape to a degree denied to the benthic macrothricids and chydorids. As well as much experimental divergence in form in both families, the adaptive significance of which in many cases remains to be explained, both have been able to exploit cyclomorphosis, which has various ecological advantages.

Phylogeny

Figure 10 shows in a simple manner the suggested phyletic relationships within the Anomopoda. To refine this by including subfamilies (as which the Moininae is here regarded) is best delayed until the taxonomic arrangement has been more precisely analysed. It is, however, already evident that, for example, the Eurycercinae belongs nearer to the base of the chydorid stem than other subfamilies and that what may eventually be defined as the Prochydorinae also belongs near to the base of this lineage so far as it is yet known. As for the early origins of the Anomopoda,

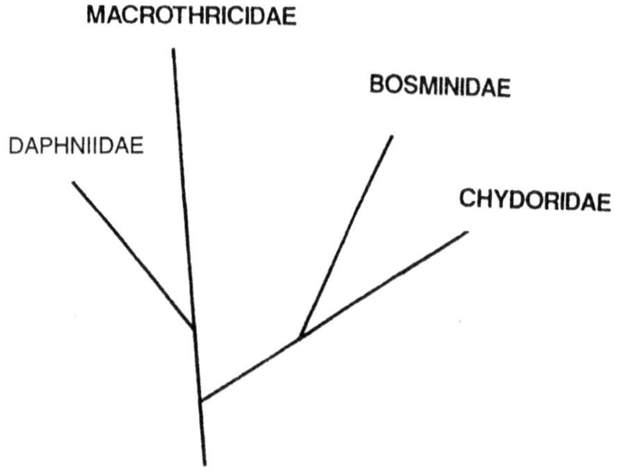

Fig. 10. A suggested phylogeny of the Anomopoda. For comments see text.

these clearly lie far back in time. Discoveries such as the chert-entombed *Lepidocaris* of the Devonian, the phosphatised microcrustaceans of the Cambrian Orsten deposits, and minute fragments, including filters, even in the early Cambrian, give grounds for hope that actual remains of early anomopods or their relatives may yet be found.

Factual information on which much of this analysis is based is presented in Fryer (1963, 1968, 1970, 1972, 1974, 1987b, 1991b).

References

Benzie, J. A. H., 1987. The biogeography of Australian *Daphnia*: clues of an ancient (>70 m y) origin for the genus. Hydrobiologia 145 (Dev. Hydrobiol. 35): 51–65.

Brancelj, A., 1990. *Alona hercegovinae* n.sp. (Cladocera: Chydoridae), a blind cave-inhabiting Cladoceran from Herzegovina (Yugoslavia). Hydrobiologia 192: 7–16.

Butterfield, N. J., 1994. Burgess Shale-type fossils from a Lower Cambrian shallow-shelf sequence in northwestern Canada. Nature Lond. 369: 477–479.

Cannon, H. G., 1933. On the feeding mechanisms of the Branchiopoda. Phil. Trans. r. Soc. B. 222: 267–352.

DeMott, W. R., 1982. Feeding selectivities and relative ingestion rates of *Daphnia* and *Bosmina*. Limnol. Oceanogr. 27: 518–527.

DeMott, W. R., 1985. Relations between filter mesh-size, feeding mode and capture efficiency for cladocerans feeding on ultrafine particles. Arch. Hydrobiol. Beih. Ergebn. Limnol. 21: 125–134.

DeMott, W. R. & W. C. Kerfoot, 1982. Competition among cladocerans: nature of the interaction between *Bosmina* and *Daphnia*. Ecology 63: 1949–1963.

Dumont, H. J., 1983. Discovery of groundwater-inhabiting Chydoridae (Crustacea: Cladocera), with a description of two new species. Hydrobiologia 106: 97–106.

Frey, D. G., 1980. The non-swimming chydorid Cladocera of wet forests, with descriptions of a new genus and two new species. Int. Rev. ges. Hydrobiol 65: 613–641.

Frey, D. G., 1987. The North American *Chydorus faviformis* (Cladocera, Chydoridae) and the honeycombed taxa of other continents. Phil. Trans. r. Soc. B 315: 353–402.

Fryer, G., 1963. The functional morphology and feeding mechanism of the chydorid cladoceran *Eurycercus lamellatus* (O. F. Müller). Trans. r. Soc. Edinb. 65: 335–381.

Fryer, G., 1968. Evolution and adaptive radiation in the Chydoridae (Crustacea: Cladocera): a study in comparative functional morphology and ecology. Phil. Trans. r. Soc. B 254: 221–385.

Fryer, G., 1970. Defaecation in some macrothricid and chydorid cladocerans, and some problems of water intake and digestion in the Anomopoda. Zool. J. Linn. Soc. 49: 255–270.

Fryer, G., 1972. Observations on the ephippia of certain macrothricid cladocerans. Zool. J. Linn. Soc. 51: 79–96.

Fryer, G., 1974. Evolution and adaptive radiation in the Macrothricidae (Crustacea: Cladocera): a study in comparative functional morphology and ecology. Phil. Trans. r. Soc. B. 269: 137–274.

Fryer, G., 1987a. Morphology and the classification of the so-called Cladocera. Hydrobiologia 145 (Dev. Hydrobiol. 35): 19–28.

Fryer, G., 1987b. The feeding mechanisms of the Daphniidae (Crustacea: Cladocera): recent suggestions and neglected considerations. J. Plankton Res. 9: 419–432.

Fryer, G., 1987c. A new classification of the branchiopod Crustacea. Zool. J. Linn. Soc. 91: 357–383.

Fryer, G., 1988. Functional morphology and functional ecology. Funct. Ecol. 2: 270–275.

Fryer, G., 1991a. A daphniid ephippium (Branchiopoda: Anomopoda) of Cretaceous age. Zool. J. Linn. Soc. 102: 163–167.

Fryer, G., 1991b. Functional morphology and the adaptive radiation of the Daphniidae (Branchiopoda: Anomopoda). Phil. Trans. r. Soc. B 331: 1–99.

Goulden, C. E., 1968. The systematics and evolution of the Moinidae. Trans. am. phil. Soc. 58: 1–101.

Graf, H., 1930. Der Fangapparat von *Bosmina*. Z. Morph. Okol. Tiere 19: 381–396.

Grobben, C., 1879. Die Entwicklungsgeschichte der *Moina rectirostris*. Zugleich ein Beitrag zur Kenntniss der Anatomie der Phyllopoden. Arb. Zool. Inst. Univ. Wien 2: 203–268.

Heydon, C. von, 1862. Gliederthiere aus der Braunkohle des Niederhein's, der Wetterau und der Rohn. Palaeontographica 10: 62–82.

Kotov, A.A. 1995. The fate of the second maxilla in the later embryogenesis of some Anomopoda (Branchiopoda: Crustacea). Zool. J. Linn. Soc. (in press).

Lai, Xing-rong & Ying-pei Li, 1987. Ephippia of Cladocera from the Tertiary of China. Acta Palaeontol. Sin. 26: 171–180. [In Chinese and English]

Naumann, E., 1921. Spezielle Untersuchungen über die Ernahrungsbiologie des tierischen Limnoplanktons. 1 Über die Technik des Nahrungserwerbes bei den Cladoceren und ihre Bedeutung für die Biologie der Gewassertypen. Lunds Univ. Arsskr. N.F. Ayd. 17: 1–22.

Patterson, C., 1981. Significance of fossils in determining evolutionary relationships. Ann. Rev. Ecol. Syst. 12: 195–223.

Petkovski, T & D. Flössner, 1972. Eine Neue *Alona* – Art (Crustacea: Cladocera) aus dem Ohridsee. Fragmenta Balcanica 9: 97–106.

Smirnov, N. N., 1992. Mesozoic Anomopoda (Crustacea) from Mongolia. Zool. J. Linn. Soc. 104: 97–116.

Tasch, P., 1963. Evolution of the Branchiopoda. In Phylogeny and Evolution of Crustacea. Mus. Compar. Zool. Special Pub: pp 145–157.

Hydrobiologia **307**: 69–74, 1995.
P. Larsson & L.J. Weider (eds), Cladocera as Model Organisms in Biology.

The evolution of groundwater Cladocera

Henri J. Dumont
Institute of Ecology, State University of Ghent, K.L. Ledeganckstraat 35, B-9000 Gent, Belgium

Key words: groundwater, evolution, Cladocera, Alona, conserved species

Abstract

Cladocera occur in various types of groundwater, but are most common in the underflow of rivers. Numerous surface water species occasionally venture into groundwater; some chydorids are more common in groundwaters than in surface waters; at least three groups within *Alona*, finally, have evolved exclusive groundwater species. The latter show few obvious adaptations to the subterranean habitat, except loss of an eye or total blindness. Some, however, have conserved an array of primitive characters (e.g. on the end-claw of the postabdomen, and the setation of the valve rims) which suggest that the physical protection and relative constancy of the hyporheic has permitted the survival of some ancient taxa.

Introduction

Since 1983, (Dumont, 1983) attention has refocused on the existence of groundwater-dwelling Cladocera, and the field has expanded considerably. Of the numerous older records, which had not received the attention they deserve, quite a few have been reinterpreted as genuine (see Dumont, 1983, 1987; Brancelj & Sket, 1990; Schmidt, 1994, and many older references therein). What has advanced the field most is the finding that the groundwater fauna is composed, besides many surface-water (stygoxenic) species (the list now stands at *c.* 35 spp. – Table 1), of an array of *c.* 10 species entirely limited to a groundwater (stygobitic) mode of life. In addition, *c.* 10 spp. with transitory (stygophilic) ways of life exist (Table 1). Also, the population dynamics and phenology of one river-gravel dweller, *Alona phreatica* Dumont, has now been studied (Sabater, 1987). There is thus little doubt that systematic attention to Cladocera in all sources of groundwater will reveal plenty of undescribed taxa in the future, with possibly quite a few surprises. Available evidence indicates that Cladocera are, on the whole, a much less common structural element of hypogean faunas than, say, the Copepods and the Ostracods. Yet, they may at times be quite abundant, even if often of strictly local occurrence. Thus, of two samples taken a few kilometres apart in the Wadi Surdud, a river in central Yemen, one contained a rich population of *Alona alsafadii*, an undescribed primitive *Alona* species (Dumont & Brancelj, 1994), while the second contained no Cladocera at all. Quite a bit of research will be needed to show whether such localized populations are the rule rather than the exception, and why they are distributed in such a patchy way. Extended surveys using quantitative or, at least, semi-quantitative methods such as that of Beladjal *et al.* (1992), may hopefully be widely applied in the near future.

The aim of the present paper is to explore possible pathways behind the evolution of a groundwater mode of life in the Cladocera.

Taxonomic incidence

Table 1 illustrates convincingly the fact that, while the majority of families of so-called Cladocera are represented in groundwater, it is the chydorids that virtually hold a monopoly on a strict stygobitic mode of life. Within the Chydoridae, there are quite a few genera (*Chydorus*, *Disparalona*, *Rhychotalona*) with species that might be stygophilic, and are therefore only rarely recorded in lakes and rivers (where they are often – and not by chance – found over sandy bottoms, or in sphagnum). There is also a suggestion that some predominantly or exclusively stygobitic species, traditionally placed in *Chydorus* or in *Pleuroxus* (e.g. '*Chydorus*' *piger* and *pigroides*) might in fact better be transferred

Table 1. List of Cladocera (Anomopoda and Ctenopoda) recorded from various types of groundwater (up to October 1993) Categories : X - stygo<u>X</u>enic; ? - stygo<u>P</u>hilic; B - stygo<u>B</u>itic.

	Category	Occurrence	Key reference
Daphniidae			
Daphnia longispina O.F.M.	x	Yugoslavia, France	Brancelj & Sket (1990); Schmidt (1994)
D. hyalina Leydig	x	France (Rhone Valley)	Schmidt (1994)
D. cucullata Sars	x	Sweden, France (Rhone Valley)	Schmidt (1994)
D. galeata Sars	x	France (Rhone Valley)	Schmidt (1994)
Ceriodaphnia megops Sars	x	France (Rhone Valley)	Schmidt (1994)
C. reticulata (Jurine)	x	France (Rhone Valley)	Schmidt (1994)
C. dubia Rich.	x	Yugoslavia	Brancelj & Sket (1990)
Scapholeberis mucronata (O.F.M.)	x	France (Rhone Valley)	Schmidt (1994)
S.rammneri Dumont & Pensaert	x	South China cave	personal observation
Simocephalus vetulus (O.F.M.)	x	Yugoslavia; France	Brancelj & Sket (1990); Schmidt (1994)
S. expinosus (Koch)	x	France (Rhone Valley)	Schmidt (1994)
Moinidae			
Moina macrocopa Straus	x	France	Pacaud (1952)
Macrothricidae			
Macrothrix laticornis (Jurine)	P	Yugoslavia	Brancelj & Sket (1990)
M.bialatus Motas & Orgh.	B	Rumania	Motas & Orghidan (1948)
Neothrix sp.	P(?)	Australia	Dumont (1987)
Bosminidae			
Bosmina longirostris (O.F.M)	x	Yugoslavia; France	Brancelj & Sket (1990); Schmidt (1994)
Sididae			
Sida crystallina (O.F.M.)	x	France (Rhone valley)	Schmidt (1994)
Ilyocryptidae			
Ilyocryptus sordidus (Liévin)	x	Rumania, Yugoslavia, France	Dumont (1987); Schmidt (1994)
I. agilis (Kurz)	x	Rumania, Yugoslavia, France	Dumont (1987); Schmidt (1984)
Chydoridae			
Eurycercus lamellatus O.F.M.	x	Yugoslavia; Wales	Brancelj & Sket (1990); personal observation
Acroperus harpae Baird	x	Yugoslavia; France	Brancelj & Sket (1990); Schmidt (1994)
Alona affinis (Leydig)	P	Rumania, Ireland elsewhere in Europe	Motas & Orghidan (1948); Schmidt (1994); Hazelton (1974)

to a separate (sub)genus (Frey, 1993). Yet, at this point in time, the preponderance of *Alona* among the strict stygobitic species is absolutely overwhelming. While at first sight more Macrothricids, and especially Ilyocryptids might have been expected, it should be borne in mind that most – if not all – of these bottom-living species are really mud dwellers, which makes them quite distinct from true groundwater species. The latter are rather to be looked for among species adjusted to scraping the surface of solid substrates, an adaptation quite common among the chydorids (Fryer, 1968).

Morphological distinctiveness

Within *Alona*, we presently know of three species groups that contain an 'evolutionary line' connecting strictly groundwater to surface-water dwelling species: the *Alona protzi*-group (three species), the *Alona*

Table 1. (continued).

	Category	Occurrence	Key reference
A.quadrangularis (O.F.M.)	P	Widespread in Europe	Beladjal *et al.* (1992)
A.macrops Motas & Orghidan	B	Rumania	Motas & Orgidan (1948)
A.rectangula Sars	x	Sweden & elsewhere in Europe	Husmann & Teschner (1970)
A.guttata group			
A. cf guttata	P	Sweden, Belgium, Australia	Dumont (1992); Beladjal *et al.* (1992)
A.bessei Dumont	B	Central Massif, France	Dumont (1983)
A.intermedia Sars	x	Sweden	Husmann & Teschner (1970)
Alona protzi-group			
(A. protzi Hartwig	P	Rhone gravels, France	Dumont (1987)
(A. smirnovi Petk. & Flössner	B	Springs at Lake Ohrid	Petkovski & Flössner (1972)
(A. phreatica Dumont	B	France, Catalonia	Dumont (1983); Sabater (1987)
Alona diaphana-group			
(A. diaphana Sars	x	Slovenia	Brancelj (1990)
(A. sketi Brancelj	B	Slovenia	Brancelj (1990)
(A. hercegovinae Brancelj	B	Bosnia-Hercegovina	Brancelj (1992)
Alona karua-group			
A. alsafadii Dum. & Branc.	B	Yemen, Wadi Surdud	Dumont & Brancelj (1994)
A. rustica Scott	?	Europe; Brazil	Dumont (1987); Beladjal *et al.* (1992)
Disparalona rostrata (Koch)	x	France, Rhone valley; Slovenia	Brancelj & Sket (1990); Schmidt (1994)
Rhynchotalona falcata Sars	P	Lake Llyndanas (Wales)	leg J. Mertens (1991), unpublished
Chydorus sphaericus (O.F.M.) group	x	Widespread in Europe; Australia	Dumont (1987); Schmidt (1994)
C. ovalis Kurz	x	Slovenia; Croatia	Brancelj & Sket (1990)
C. parvus Daday	P	Sandy shores of Lake Malawi at Senga Bay	personal observation
"Chydorus" piger[1] Sars	P	Slovenia; probably widespread in Europe	Brancelj & Sket (1990)
"Chydorus" pigroides Lillj[1]	B(?)	Uppsala, Sweden	Frey (1976)
Pseudochydorus globus (Baird)	x	France, Rhone valley	Schmidt (1994)
Pleuroxus uncinatus Baird	x	France, Rhone valley	Schmidt (1994)
Pleuroxus aduncus (Jurine)	x	Slovenia	Brancelj & Sket (1990)
P. laevis Sars	x	Slovenia	Brancelj & Sket (1990)
P. denticulatus Birge	x	France, Rhone valley	Schmidt (1994)
Pleuroxus piger[2] (sensu Dumont, 1983)	B	Auvergne, France	Dumont (1983)
Alonopsis elongata Sars	x	Sweden	Husmann & Teschner (1970)
Tretocephala ambigua (Lillj)	x	Croatia	Brancelj & Sket (1990)
Monospilus dispar Sars	x	Rhone valley, France	Schmidt (1994)
Leydigia leydigi (Schoedler)	x	Rumania, Yugoslavia, France	Dumont (1987); Schmidt (1994)
L. acanthocercoides (Fischer)	x	France, Rhone valley	Schmidt (1994)
Camptocercus rectirostris (Schoedler)	x	France, Rhone valley	Schmidt (1994)

[1] Generic allocation likely incorrect.
[2] Not the same species as "Chydorus' piger.

arua-group (two species), and the *Alona diaphana*-roup (three species).

From a combination of the character states encoun-red in these eight taxa, we can establish that:

1. *eye and ocellus pigmentation is lost* (Figs 1–3): in *all* species that, at least, at times occur in surface waters, both an eye and an ocellus occur. Two groundwater forms (*A. smirnovi*, *A. alsafadii*)

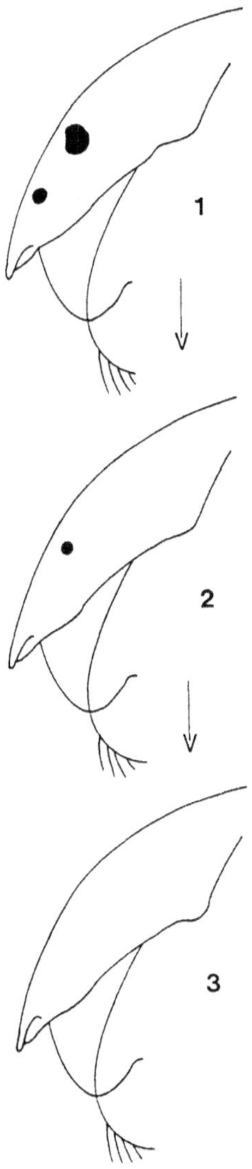

Figs 1–3. Progressive eye-pigment reduction in *Alona* as an adaptation to subterranean life. 1. surface-living species (e.g. *A. karua*); 2. groundwater species (e.g. *A. smirnovi, A. alsafadii*); 3. cave-dwelling species (e.g. *A. sketi, A. hercegovinae*).

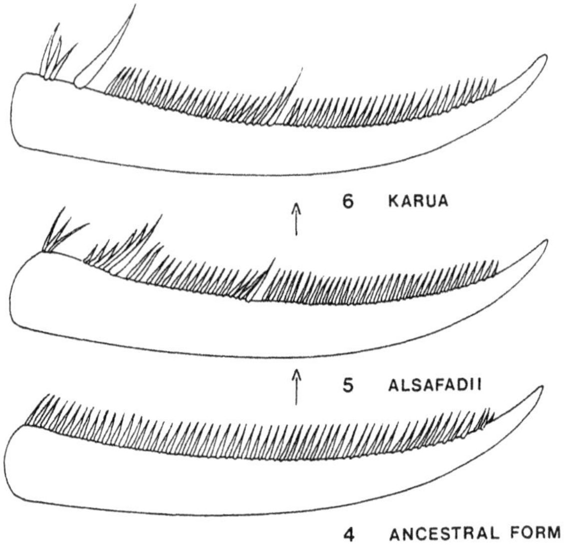

Figs 4–6. The evolution of the external pecten of the end-claw in the *A. karua*-group. The ancestral form is hypothetical.

only show a pigmented ocellus. Two blind cave-dwellers (*A. sketi, A. hercegovina*) have neither.

2. *feeding appendages*: in the two hypogean species in which the fine structure of the trunk limbs has been described in sufficient detail (*A. sketi* and *A. alsafadii*) only minute differences in structure with their surface-dwelling relatives were observed. These could not be related to differences in habitat, but rather represented species-specific characters (Brancelj, 1990, 1992; Dumont & Brancelj, 1994).

3. *Postabdomen*: all species-groups with hypogean representatives have, in addition to the usual dorsal spines, fans of lateral fascicles of stiff setae on the postabdomen. The significance of these to a groundwater mode of life is, at present, unclear. The endclaw of the postabdomen is adorned with a number of pectens, and a basal spine in all cases. Most of these are trendless, but an interesting exception is offered by *A. karua* (and *A. hispanica*) and *A. alsafadii*, where the basal spine is seen (Figs 4–6) to be composed of a number of fused (fusion incomplete in *alsafadii*) denticles derived from the basal outer pecten.

4. *valve rim adornments*: like in the preceding case, the *A. karua* group is the only one to contain a trend: the ventro-posterior rim in *A. karua* is differentiated into a zone with setae, a series of strong, sclerotized teeth, and a row of setules of haphazard length moving up the distal valve margin. In *A. alsafadii*, it is clear (Figs 7–9) that these structures are all modified setae.

Discussion

Adaptation versus conservation

Loss of optical perception (eyes, ocelli) is widespread in a lightless environment. However, it remains unclear

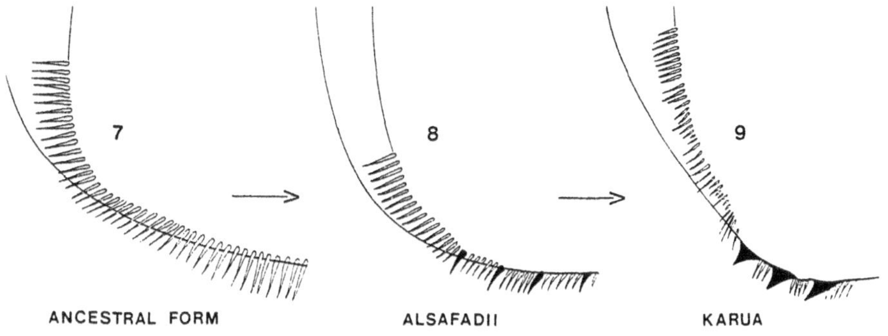

Figs 7–9. The evolution of the setation along the posterior section of the carapace valves. The ancestral form is hypothetical.

whether this is irreversible (genetic) or reversible (phenetic). Cases of – possibly reversible – loss of eyes and of body pigmentation have indeed been documented from other groups, e.g. fishes. (Thinès & Proudlove, 1986).

At present, we have no means of determining whether in Cladocera, the loss of eye and/or ocellus pigmentation is a fixed trait. The fact that eye loss does occur, however, confirms that Cladocera in a cave environment respond to environmental selective pressures exactly as any other animal would. Other structural adaptations, such as in the trunk limbs, are doubtful, and of the same order as 'any' interspecific differences, as stated earlier.

In addition, the adornment of the postabdominal claw in *A. alsafadii*, as well as the setation along its valve rim, cannot be interpreted as an adaptation to subterranean life. Yet, both sets of characters clearly suggest a conserved character state, which throws a new light on the origin of the basal spine of the postabdomen, as well as the valve setation and denticulation. The most primitive character state appears to have been a postabdominal claw with a uniform external longitudinal pecten, which, by a process of spinule fusion, gradually transformed into a basal spine, some isolated basal spinules, and two separate external pectens (Figs 4–6). Likewise, the valve setation may originally have consisted of a row of uniform setae, which later specialized into ventral setae of different length, fused among each other and with the carapace to become carapace teeth, or spinules and setules of variable length running up along the inner distal margin of the valves (Figs 7–9).

Surprisingly, therefore, some of the truly groundwater *Alona* appear to be conserved primitive species, rather than highly specialized ones. Because many more such species, as yet undiscovered, may exist, it is anticipated that continued surveys of different types of groundwater may significantly advance our insight into the origin and evolution of these Anomopoda.

The survival of such ancient taxa in the Cladocera is probably for the same reason as in other crustacean groups (e.g. the Syncarida) (Schminke, 1986): the physical protection and relative constancy of the hyporheic environment acted as a refuge in which otherwise vulnerable ancient species found enough shelter to survive.

References

Beladjal, L., J. Mertens & H. J. Dumont, 1992. A simple basket trap for estimating relative abundances of some components of hyporheic faunas: application to the Cladocera. Stygologia 7: 193–195.

Brancelj, A., 1990. *Alona hercegovinae* n. sp. (Cladocera: Chydoridae), a blind cave-inhabiting cladoceran from Hercegovina (Yugoslavia). Hydrobiologia 199: 7–16.

Brancelj, A., 1992. *Alona sketi* sp. n. (Cladocera: Chydoridae), the second cave-inhabiting cladoceran from former Yugoslavia. Hydrobiologia 248: 105–114.

Brancelj, A. & B. Sket, 1990. Occurrence of Cladocera (Crustacea) in subterranean waters in Yugoslavia. Hydrobiologia 199: 17–20.

Dumont, H. J., 1983. Discovery of groundwater-inhabiting Chydoridae (Crustacea: Cladocera) with the description of two new species. Hydrobiologia 106: 97–106.

Dumont, H. J., 1987. Groundwater Cladocera. A synopsis. Hydrobiologia 145 (Dev. Hydrobiol. 35): 169–173.

Dumont, H. J. & A. Brancelj, 1994. *Alona alsafadii* n. sp. from Yemen, a primitive groundwater-dwelling member of the *A. karua*-group. Hydrobiologia 281: 57–64.

Frey, D. G., 1976. Redescription of *Chydorus pigroides* Lilljeborg, 1900 and allocation to Pleuroxus (Cladocera, Chydoridae). Crustaceana 30: 89–97.

Frey, G. D., 1993. Subdivision of the genus Pleuroxus (Anomopoda, Chydoridae) into subgenera worldwide. Hydrobiologia 262: 133–144.

Fryer, G., 1968. Evolution and adaptive radiation in the Chydoridae (Crustacea: Cladocera). A study in comparative functional morphology and ecology. Phil. Trans. r. Soc. B 254: 221–335.

Hazelton, M., 1974. Hypogean fauna recorded from Ireland 1952–1971. Trans. Cave Res. Group Great Britain 15: 225–252.

Husmann, S. & D. Teschner, 1970. Subterrane Wassermilben aus Schweden. Arch. Hydrobiol. 67: 242–267.

Motas, C. & T. Orghidan, 1948. Quelques Cladocères recueillis dans l'eau phreatique. Notat. biol. Bucarest 6: 123–129.

Pacaud, A., 1952. Remarque sur la systématique du genre Moina Baird (Cladocères) et sur sa distribution autour du bassin occidental de la Méditerrannée. Vie Milieu 3: 68–76.

Petkovski, T. K. & D. Flössner, 1972. Eine neue Alona-Art (Crustacea: Cladocera) aus dem Ohridsee. Fragm. Balc. Mus. maced. Sci. Nat. 9: 97–106.

Sabater, F., 1987. On the interstitial Cladocera of the River Ter (Catalonia, NE Spain), with a description of the male of *Alona phreatica*. Hydrobiologia 144: 51–62.

Schmidt, C. M., 1994. Distribution in space and time of Cladocera (Crustacea) in the alluvial sediments of the river Rhône (France). Hydrobiologia 291: 131–140.

Schminke, H. K., 1986. Syncarida. In L. Botosaneanu (ed.), Stygofauna Mundi. Brill., Leiden: 389–404.

Thinès, G. & G. Proudlove, 1986. Pisces. In L. Botosaneanu (ed.), Stygofauna Mundi. Brill., Leiden: 709–733.

Hydrobiologia **307**: 75–84, 1995.
P. Larsson & L.J. Weider (eds), Cladocera as Model Organisms in Biology.

Altitudinal distribution of tropical planktonic Cladocera

J. Green
17 King Edwards Grove, Teddington, Middx. TW11 9LY, UK

Key words: Cladocera, Africa, South America, tropics, altitude

Abstract

The altitudinal distributions of planktonic Cladocera in Africa in three latitudinal bands: 0–10 °, 10–20 ° and 28–31 ° are described. Some species have altitudinal ranges of 2000 m within a latitudinal band of 10 °. It is possible to distinguish tropical species which do not extend up to high altitudes, particularly at higher latitudes. At the lowest latitudes, some species are found only at high altitudes, but occur at progressively lower altitudes with increasing latitude: *Daphnia pulex* and *D. obtusa* are good examples. A study in South America, with a smaller database, gives similar results.

Introduction

In a recent paper, Green & Kling (1988) reported the presence of *Daphnia obtusa* in a small crater lake in Cameroon at an altitude of 1900 m. Its occurrence at high altitudes in East Africa has been recorded by Löffler (1968), who also found other species of *Daphnia* at similar altitudes. These occurrences led to a consideration of the altitudinal distribution of planktonic Cladocera in tropical Africa. Several questions were posed:

To what altitudes do tropical species ascend, and are these altitudes reduced at higher latitudes ?

How low can essentially temperate species, such as *D. obtusa*, descend in the tropics, and how does this vary with latitude?

The data are derived mainly from my own collections, with additions from the literature where possible (see references). The papers of Löffler (1964, 1968, 1978) are particularly relevant and useful. There are many other records of tropical Cladocera in the literature that I have not been able to use because they were not accompanied by altitudinal data.

Materials and methods

Altitudes of localities were taken from the most recent available maps, and in some cases were approximately confirmed using a Lufft altitude aneroid barometer, calibrated at a known altitude on the day of collection. African localities are shown in Fig. 1, and Neotropical localities in Fig. 5.

Plankton samples were collected with 55 μm and 250 μm meshed nets and were preserved in 5% formaldehyde. Only planktonic Cladocera were considered (Tables 1 & 3), and only those species that occurred in a reasonable number of localities; for instance the few records of *Diaphanosoma mongolianum* and *D. sarsi* in tropical Africa are not included.

Results

Altitudinal distributions between latitudes 0–10 ° in Africa

Typical tropical species, such as *Ceriodaphnia cornuta* and *Diaphanosoma excisum* range from sea level up to 2000 m or slightly more (Fig. 2). *Daphnia barbata*, often quoted as one of the few tropical lowland species, ranges up to 1680 m. Of the species one might regard as temperate, *Daphnia rosea* extends down from 2220 m to 680 m, and *D. pulex* occupies a middle range, from 2920 m to 1890 m. Species occurring at higher altitudes are *Daphnia obtusa* (1900 to 4100 m) and *Daphnia dolichocephala* (3480 to 4320 m). The only other

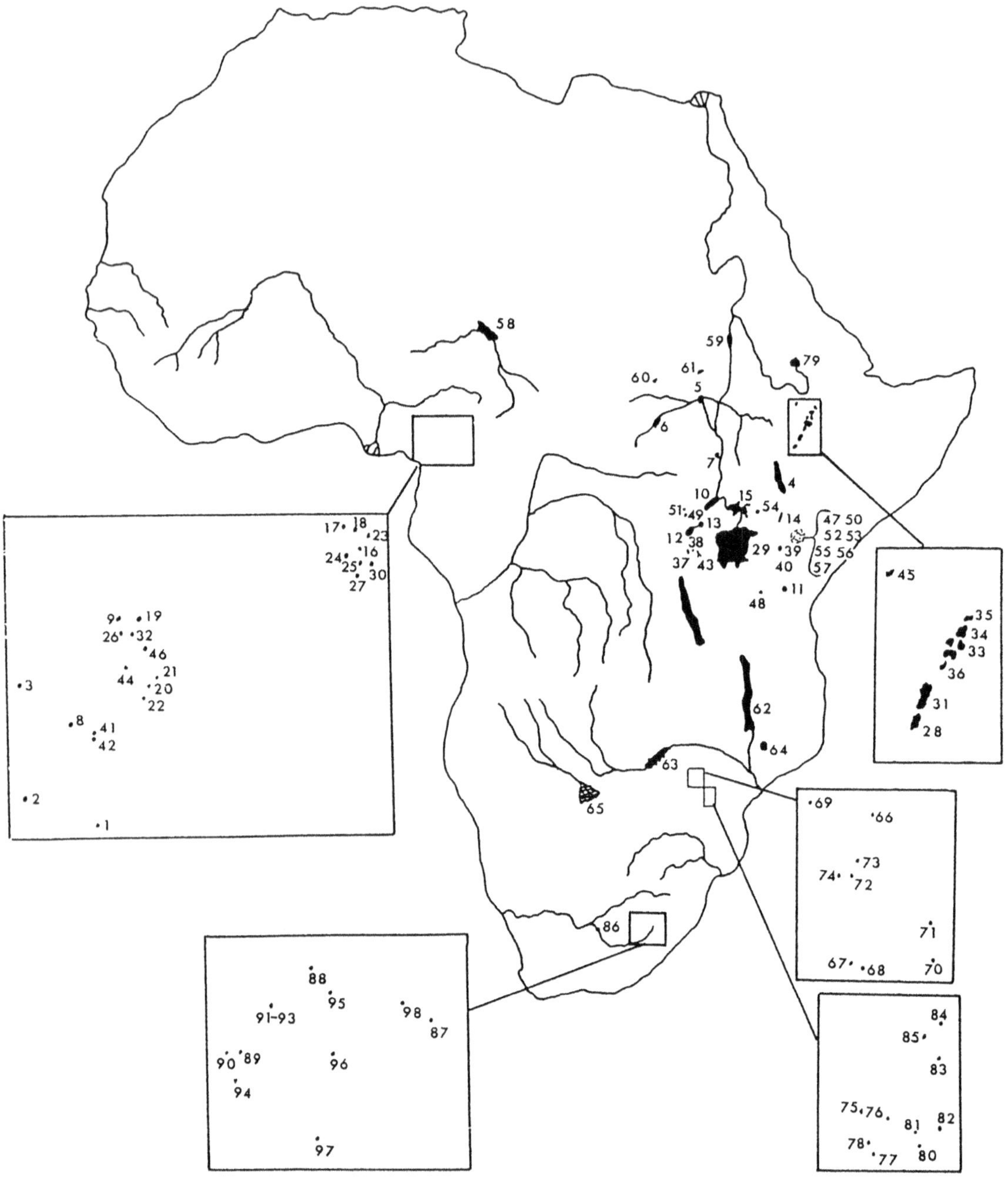

Fig. 1. African water bodies mentioned in the text; altitudes are given in Table 2.

species occurring over 3000 m is *Daphnia curvirostris*, which also extends downwards to 1780 m.

Figure 2 makes it clear that while one can recognise lowland and upland species, there is a continuous spectrum of overlapping species, and it is notable that more species (15) were found between 1000 and 2000 m than between 0 and 1000 m (12). Above 2000 m the number of species (8) drops rapidly.

Altitudinal distributions between latitudes 10–20 ° in Africa

The expectation was that some of the tropical species would not extend to the same altitude at these latitudes as at 0–10 °. The highest records for *Moina micrura*, *Ceriodaphnia cornuta* and *Diaphanosoma excisum* are all at least 150 m below their highest at 0–10 °

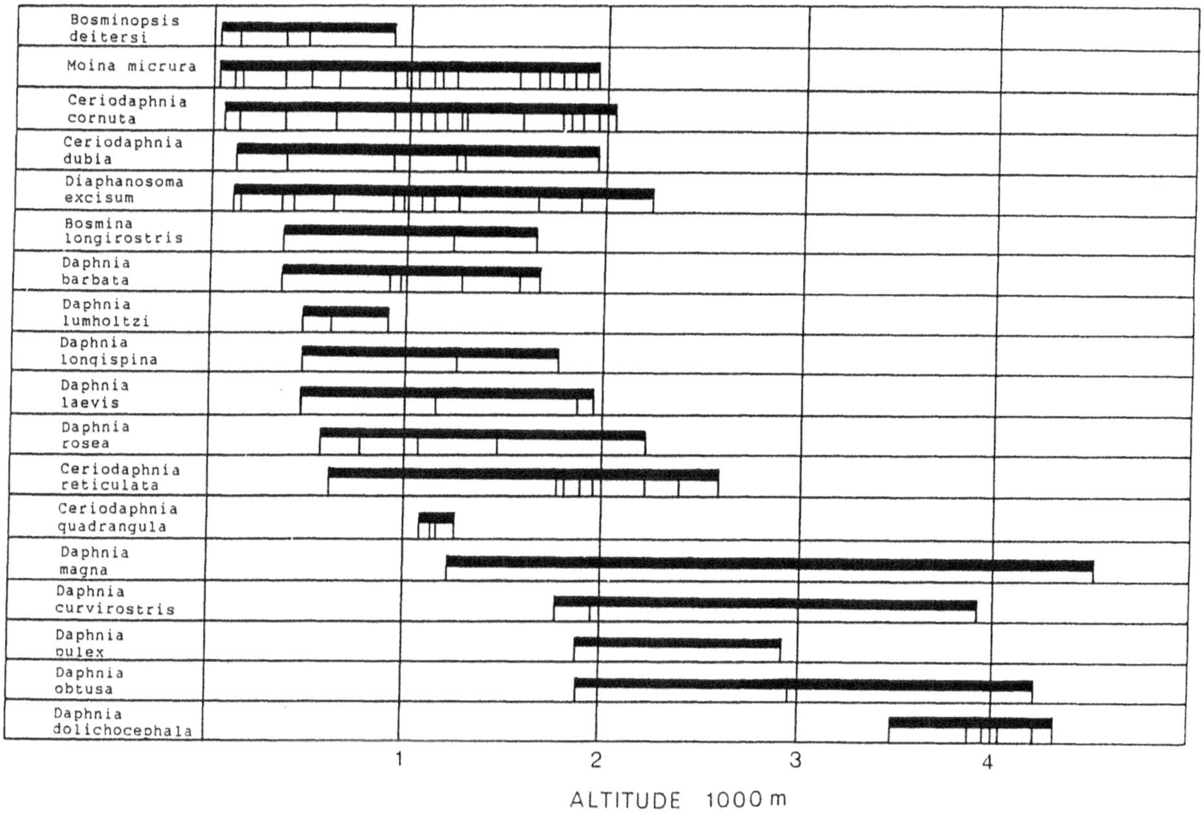

Fig. 2. Altitudinal distribution of planktonic Cladocera in Africa between latitudes 0 and 10 °.

Table 1. List of species considered in Africa.

Diaphanosoma excisum Sars
Daphnia barbata Weltner
Daphnia curvirostris Eylmann
Daphnia dolichocephala Sars
Daphnia laevis Birge
Daphnia lonqispina O.F. Muller
Daphnia lumholtzi Sars
Daphnia magna Straus
Daphnia obtusa (Kurz) em. Scourfield
Daphnia pulex Leydig em. Scourfield
Daphnia rosea Sars
Ceriodaphnia cornuta Sars
Ceriodaphnia dubia Richard
Ceriodaphnia quadrangula (O.F. Muller)
Ceriodaphnia reticulata (Jurine)
Moina micrura Kurz
Bosmina longirostris (O.F. Muller)
Bosminopsis deitersi Richard

(Fig. 3). But this figure also illustrates the vagaries of field data: *Daphnia lumholtzi* is recorded at higher altitude at these latitudes than at 0–10 °. This is an indication that various other factors are influencing the distribution of planktonic Cladocera, and we must not be too rigid in our interpretations of altitudinal distributions. Among the factors causing variation in the data are unequal seasonal sampling coverage, and sometimes simply the presence or absence of suitable localities.

Unfortunately the data did not contain any localities above 2500 m in latitudes 10–20 °, and had only half the localities of the 0–10 ° range. But the same general distribution is found, with a good variety of species up to 2000 m and a rapid fall in number above that level. In both latitudinal ranges *Bosminopsis deitersi* is the species with the lowest maximum altitude.

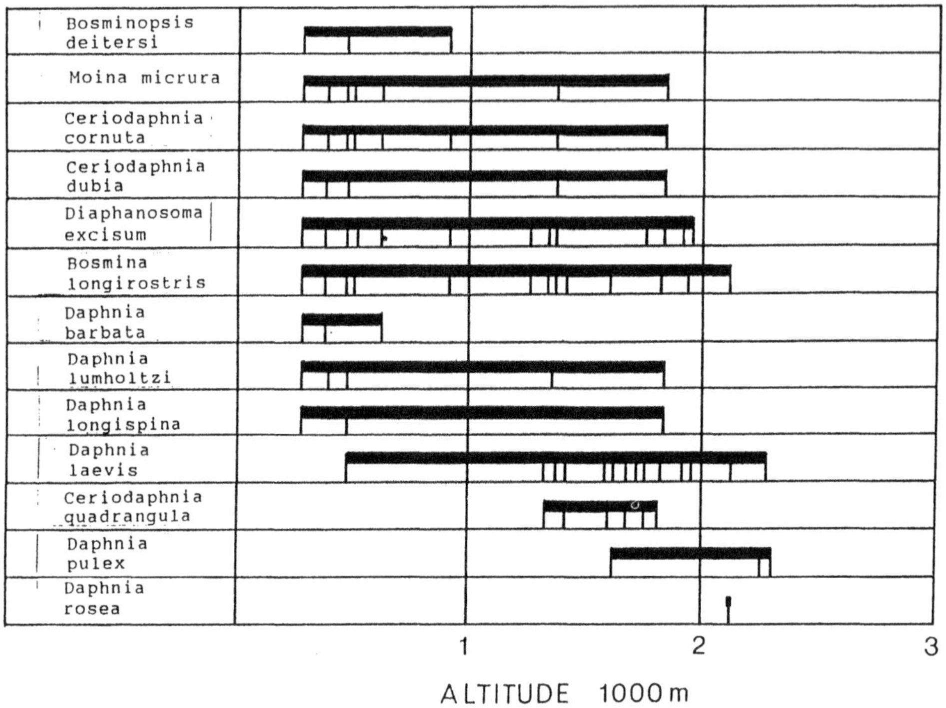

Fig. 3. Altitudinal distribution of planktonic Cladocera in Africa between latitudes 10 and 20 °.

Altitudinal distributions between latitudes 28–31 ° in Africa

The data in this section are derived mostly from my collections in Lesotho, with the addition of Lake Le Roux (Hart, 1987). They are given for comparative purposes, to indicate how the latitudinal range may change as one moves out of the tropics.

Figure 4 shows that *Daphnia barbata* extends up to 1661 m, which is close to its upper limit at latitudes 0–10 °.

Both *Ceriodaphnia cornuta* and *Moina micrura* do not extend above 1700 m at these latitudes, thus showing a further small decrease in their upper limits compared with latitudes 10–20 °. Of the temperate species, *Daphnia pulex* ranged from 2947 m down to 1300 m, which is about 300 m lower than its lower limit at latitudes 10–20 °.

Daphnia dolichocephala, not found below 3400 m at latitudes 0–10 °, is now found down to 1585 m. In the region of Tunis (37 °N) it occurs near sea level in temporary brackish water (Gauthier, 1928). The original description of this species (Sars, 1895) was from dried mud collected at Knysna, South Africa (34 °S) close to sea level.

Altitudinal distributions in the Neotropics

The localities included in this survey are shown in Fig. 5; references in Table 4. The data are less satisfactory than those for Africa, as there are no records for altitudes between 1000 and 2000 m. Nevertheless the data show some parallels to those from Africa. In Fig. 6 *Bosminopsis deitersi, Bosmina hagmanni, Ceriodaphnia cornuta, Moina micrura* and *Diaphanosoma spinulosum* form a clear cut tropical lowland group. *Bosmina huaronensis, Daphnia peruviana* and *Daphnia longispina* were found only above 3000 m; while *Daphnia pulex, Ceriodaphnia dubia* and *Daphnia obtusa* extend progressively down below 3000 towards 2000 m. The extraordinary range of *Ceriodaphnia quadrangula* raises the question of whether more than one species is involved.

Discussion

Table 5 summarises the distribution of planktonic species richness in relation to altitude in Africa. Both the number of species found and the number of species per locality decrease above 2000 m. Interpretation of such a table is not simple because numerous factors

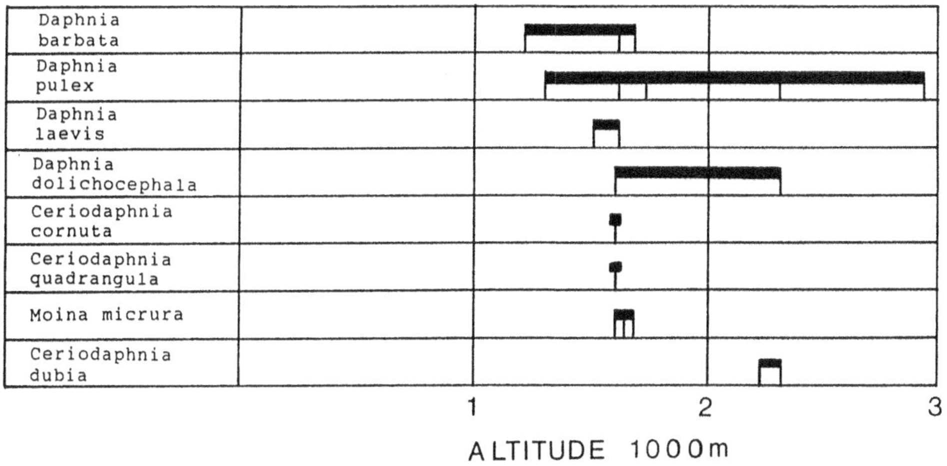

Fig. 4. Altitudinal distribution of planktonic Cladocera in Africa between latitudes 28 and 31 °S. There are no localities in Lesotho under 1000 m.

influence the number of species in a locality. The size of a water body is important. The high number of species per locality for latitude 10–20 ° and altitude 0–1000 m is a result of several large lakes falling into this latitudinal and altitudinal range.

The highest recorded planktonic species in tropical Africa is *Daphnia magna* from Mawenzi Tarn (4400 m). This record is shown in Fig. 2, but the locality is not included in the list because the record is not based on a plankton sample, but on cultures from dried mud (Löffler, 1968). *Daphnia dolichocephala* is the most consistently recorded species at high altitude in tropical Africa, and extends up to 4330 m in Emerald Tarn (Löffler, 1968).

Three Neotropical species: *Ceriodaphnia dubia*, *Daphnia longispina* and *Bosmina huaronensis* occur at altitudes over 5000 m. The highest recorded altitude for a Neotropical species is for *Daphniopsis chilensis* in Licancabur Crater Lake (5883 m) (Hann, 1986), its only known locality. A *Daphniopsis* sp. occurs in fresh pools around the Laguna Colorada (4278 m) in nearby Bolivia (pers. obs.).

The altitudinal distributions of 18 species in Africa are shown in Figs 2–4, and of 12 Neotropical species in Fig. 6. Eight species are common to both regions. Comparisons are hampered by the lack of data between 1000 and 2000 m in the Neotropics, but in both regions *Bosminopsis* has not been recorded above 1000 m. In both regions *Daphnia pulex* does not descend below 1600 m. *Daphnia obtusa* is also a high altitude species in both regions.

There are some discrepancies in the distributions of some species in the two regions. *Daphnia longispina* in Africa descends down below 500 m, but the Neotropical data show none below 3500 m. *Ceriodaphnia dubia* occurs at low altitudes in tropical Africa, but in my data not below 2000 m in the Neotropics. These distributions raise the question as to whether more than one species are covered by each of these names.

The data for Africa show that a typical specific altitudinal range, within the tropics, is about 2000 m. Only *Daphnia magna* exceeds this by more than 500 m in the present data. The Neotropical data are scanty, but several species have altitudinal ranges of about 2000 m, and the most notable exception is *Ceriodaphnia quadrangula*, extending from sea level to 4700 m, again raising the question of possible sibling species.

Based on the generally accepted lapse rate of 6 °C per 1000 m for moist air, an altitudinal range of 2000 m is approximately equivalent to a mean temperature difference of 12 °C. In Fig. 7 the mean minimum and mean maximum air temperatures for twelve stations between latitudes 0 and 10 ° in Africa are plotted (data from Griffiths, 1972), together with the data of Talling (1993) for near bottom temperatures of tropical African water bodies. Talling's line gives a lapse rate of 6.4 °C per 1000 m, and is somewhat nearer to the mean maximum air temperature than the mean minimum.

The altitudinal ranges of many tropical Cladocera are roughly equivalent to a temperature range of about 13 °C. This is much smaller than the ranges of some temperate species. In the London area for instance *Simocephalus vetulus* can be found reproducing at 2 °C

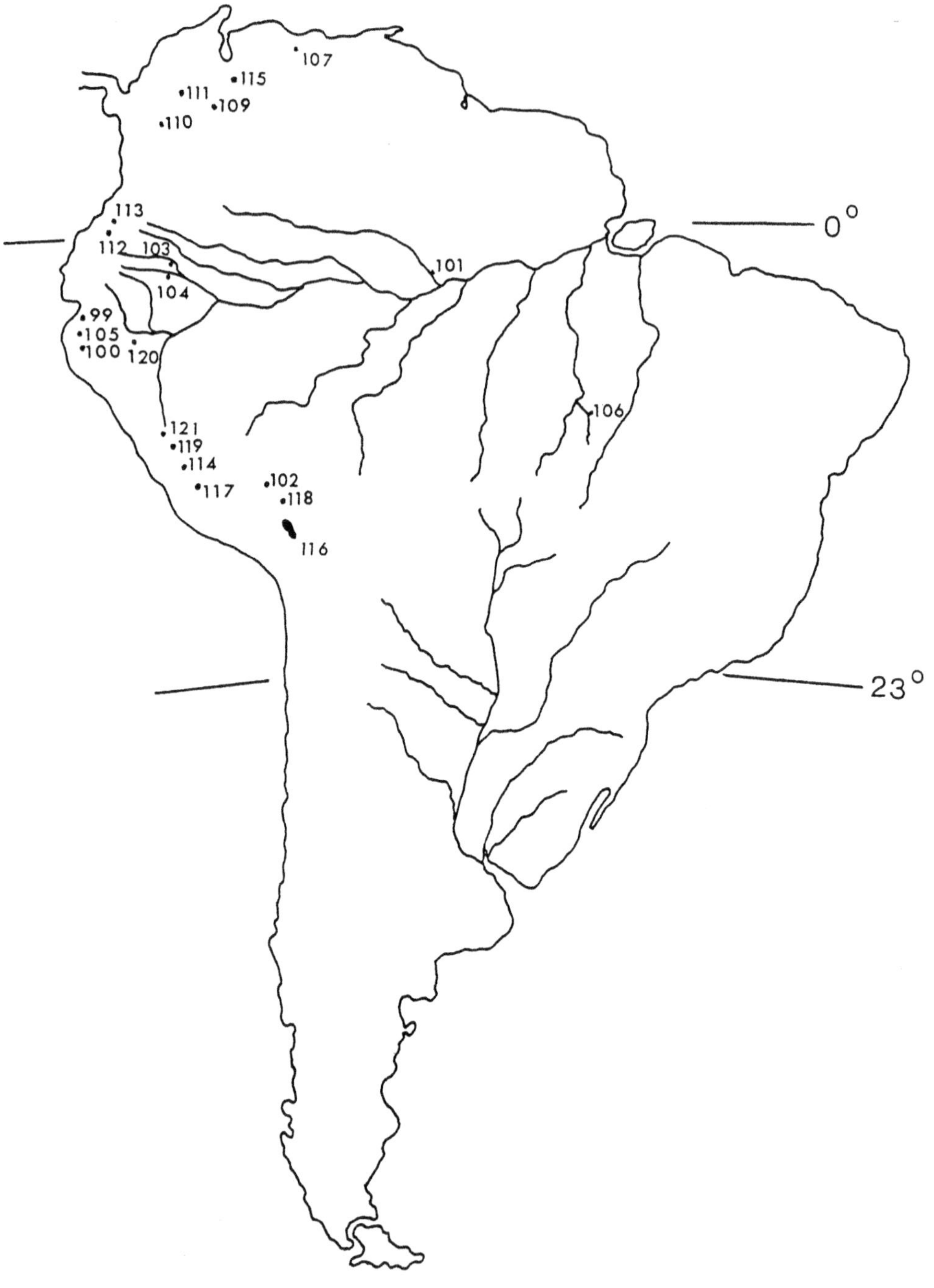

Fig. 5. South American water bodies mentioned in the text; altitudes are given in Table 4.

and at 22 °C (Green, 1966). In the same area a number of warm water species, such as *Ceriodaphnia reticulata* and *Scapholeberis mucronata*, do not become active until the water temperature reaches about 10 °C in the springtime. The normal temperatures experienced by these species will be from 10 to 23 °C, giving a range of 13 °C. They are sometimes found at temperatures below 10 °C in the autumn, but they do not survive for any significant length of time. There is thus a reasonable agreement between the temperature range attributable to altitude experienced by tropical lowland Cladocera, and the range of temperature normally experienced by warm water species in temperate regions.

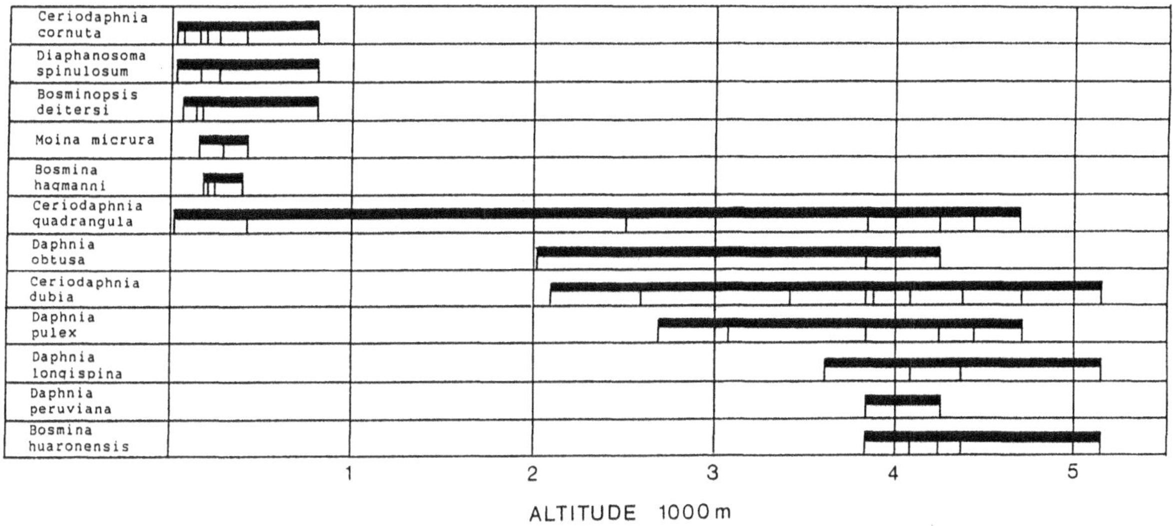

Fig. 6. Altitudinal distribution of planktonic Cladocera in the Neotropics.

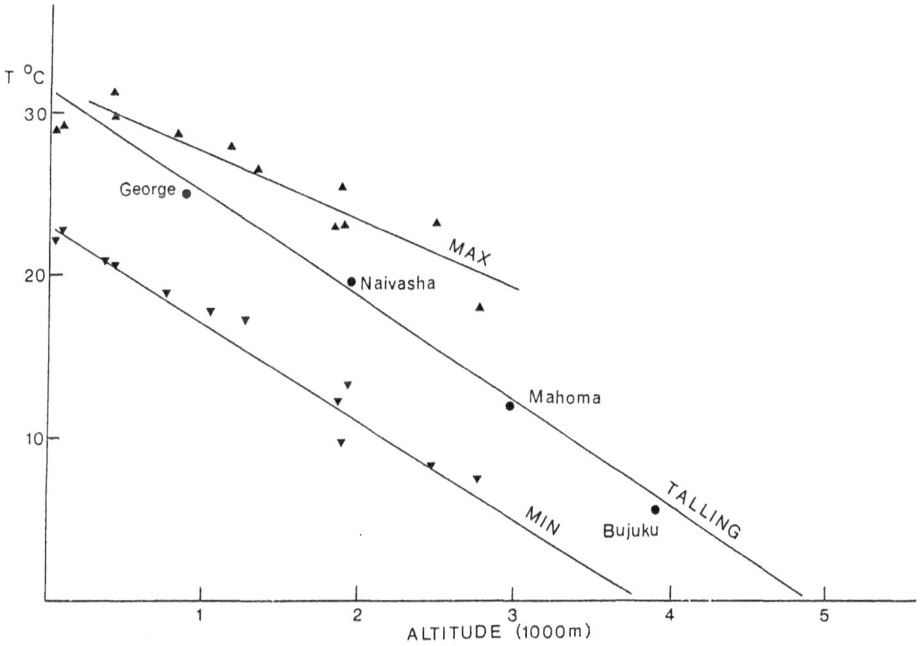

Fig. 7. Mean maximum and mean minimum air temperatures in relation to altitude at twelve climatic stations in Africa between latitudes 0 and 10 °, together with a line fitted to Talling's (1993) data for near bottom temperatures of water bodies in tropical Africa. Only four of Talling's 14 points are shown. The line fitted to the mean minimum temperatures is the theoretical lapse rate of 6 °C per 1000 m. The line through the mean maximum temperatures has been fitted by eye, and gives a lapse rate of 4.1 °C per 1000 m.

Within one species the relationship to temperature is not simple. Thermal clones were studied by Wood & Banta (1929), Banta (1939) and Johnson (1952) described a thermal race of *Daphnia atkinsoni* which had a lethal temperature significantly lower than other races of the same species. Loaring & Hebert (1981) showed that different clones of *Daphnia pulex* varied in their abilities to coexist at different temperatures. The temperature range of a species may thus be a combination of the temperature tolerances of a number of clones, none of which has the range of the whole species (cf. Carvalho, 1987). If this situation was extensively developed one could envisage that all species could occur at all temperatures, and there

Table 2. Localities in Africa (cf Fig. 1)

A	Latitudes 0–10°	Altitude (m)	Reference
1.	Ossa, Cameroon	1	D
2.	Debundsha, Cameroon	54	D
3.	Ejagham, Cameroon	100	D
4.	Turkana, Kenya	365	A
5.	No, Sudan	385	A
6.	Ambadi, Sudan	390	A
7.	Shambe, Sudan	406	A
8.	Beme, Cameroon	500	D
9.	Benakuma, Cameroon	576	D
10.	Albert, Uganda/Zaire	616	B
11.	Chala, Kenya/Tanzania	842	A
12.	Edward, Uganda/Zaire	913	B
13.	George, Uganda	913	B
14.	Baringo, Kenya	985	A
15.	Kyoga, Uganda	1034	B
16.	Massot, Cameroon	1054	D
17.	Bini, Cameroon	1079	D
18.	Dang, Cameroon	1079	D
19.	Nyos, Cameroon	1094	D
20.	Monoun, Cameroon	1100	D
21.	Petponoun, Cameroon	1120	D
22.	Negop Ghang, Cameroon	1120	D
23.	Mbalang, Cameroon	1130	D
24.	Tizong, Cameroon	1160	D
25.	Ngaoundaba, Cameroon	1160	D
26.	Wum, Cameroon	1177	D
27.	Gegouba, Cameroon	1180	D
28.	Chamo, Ethiopia	1233	A
29.	Victoria, Uganda	1240	B
30.	Baledjem, Cameroon	1249	D
31.	Abaya, Ethiopia	1285	A
32.	Nyi, Cameroon	1495	D
33.	Langano, Ethiopia	1582	A
34.	Ziway, Ethiopia	1636	A
35.	Koka, Ethiopia	1660	A
36.	Awasa, Ethiopia	1680	A
37.	Mutanda, Uganda	1791	C
38.	Mulehe, Uganda	1803	C
39.	Naivasha, Kenya	1889	A
40.	Oloidian, Kenya	1889	A

Table 2. (continued).

A	Latitudes 0–10°	Altitude (m)	Reference
41.	Manengouba, man, Camer.	1900	D
42.	Manengouba, woman,Camer.	1920	D
43.	Bunyonyi, Uganda	1973	C
44.	Bambuluwe, Cameroon	2053	D
45.	Fincha, Ethiopia	2225	A
46.	Oku, Cameroon	2227	D
47.	Sacred, Kenya	2400	G
48.	Mt Meru, Tanzania	2900	F
49.	Mahoma, Ruwenzori Mts.	2960	G
50.	Ellis, Mt Kenya	3475	G
51.	Bujuku, Ruwenzori Mts.	3920	G
52.	Michaelson, Mt Kenya	3970	G
53.	Hook, Mt Kenya	3970	G
54.	Mt Elgon, Uganda	4100	E
55.	Hohnel, Mt Kenya	4210	G
56.	Enchanted, Mt Kenya	4220	G
57.	Emerald, Mt Kenya	4330	G
B	**Latitudes 10–20°**	**Altitude (m)**	**Reference**
58.	Chad	282	H
59.	Gebel Aulia, Sudan	395	N
60.	Kundi, Sudan	450	J
61.	Keilak, Sudan	450	J
62.	Malawi	475	M
63.	Kariba	485	K
64.	Chilwa, Malawi	622	0
65.	Okavango, Botswana	930	A
66.	Sweet Valley, Zimbabwe	1260	L
67.	Mtemwa, Zimbabwe	1340	L
68.	Edinburgh, Zimbabwe	1350	L
69.	McIlwaine, Zimbabwe	1368	P
70.	Joberini, Zimbabwe	1410	L
71.	Gwaai Farm, Zimbabwe	1420	L
72.	Jo Denton, Zimbabwe	1580	L
73.	Nyambuya, Zimbabwe	1600	L
74.	Mariangwe, Zimbabwe	1620	L
75.	Udu, Zimbabwe	1676	L
76.	Rhodes, Zimbabwe	1707	L
77.	Moodie, Zimbabwe	1768	L
78.	Claremont, Zimbabwe	1813	L

would be no latitudinal or altitudinal differentiation. In nature this does not happen, and the present data indicate a relatively consistent range for many species, at least within the tropics. Outside the tropics altitudinal ranges are likely to be more variable. Extreme continental climates can lead to enormous seasonal variation in temperature, and even at one latitude and altitude the mean annual temperature can show very big

Table 2. (continued).

B	Latitudes 10–20°	Altitude (m)	Reference
79.	Tana, Ethiopia	1829	A
80.	Purdon, Zimbabwe	1920	L
81.	Mare, Zimbabwe	1951	L
82.	Gulliver, Zimbabwe	1951	L
83.	Troutbeck, Zimbabwe	2134	L
84.	Large Connemara, Zimbabwe	2255	L
85.	Small Connemara, Zimbabwe	2270	L
C.	Latitudes 28–31° S.	Altitude (m)	Reference
86.	Le Roux , S. Africa	1200	Q
87.	Drackensberg Sun, S. Africa	1300	A
88.	Leribe, Lesotho	1501	A
89.	Bolaea Tau, Lesotho	1585	A
90.	Linyane, Lesotho	1585	A
91.	University Pond 1, Lesotho	1605	A
92.	University pond 2, Lesotho	1610	A
93.	University pond 3, Lesotho	1615	A
94.	Tso Litiama, Lesotho	1661	A
95.	Collett Dam, Lesotho	1722	A
96.	Semonkong Dam, Lesotho	2210	A
97.	Letseng la Letsie, Lesotho	2317	A
98.	Letseng la Terae, Lesotho	2947	A

References: A-own observations. B-Green (1971). C-Green (1976). D-Kling & Green (in prep.). E-Brehm (1935). F-Brehm (1960). G-Loffler (1968). H-Rey & Saint-Jean (1968,1969). J-Green, Moghraby & Ali (1984). K-Green (1985). L-Green (1990). M-Fryer (1957). N-Brook & Rzoska (1954). O-Kalk *et al.* (1979). P-Thornton (1982). Q-Hart (1987).

Table 3. List of species considered in the Neotropics.

Diaphanosoma spinulosum Herbst
Daphnia longispina O.F. Muller
Daphnia obtusa (Kurz) em Scourfield
Daphnia peruviana Harding
Daphnia pulex Leydig em Scourfield
Ceriodaphnia cornuta Sars
Ceriodaphnia dubia Richard
Ceriodaphnia quadrangula (O.F. Muller)
Moina micrura Kurz
Bosmina hagmanni Stingelin
Bosmina huaronensis (Delachaux)
Bosminopsis deitersi Richard

Table 4. Localities in the Neotropics (cf Fig. 5)

		Altitude (m)	Reference
99	Rio Tumbes, Peru	0	R
100	Laguna Ramon Grande, Peru	20	R
101	Laguna Camalaeo, Brazil	44	S
102	Laguna Yarinacochas, Peru	150	R
103	Imuya Lake, Ecuador	190	A
104	Lake Zancudo, Ecuador	200	A
105	San Lorenzo, Peru	260	R
106	Suia Missu Lakes, Brazil	400	T
107	Lake Valencia, Venezuela	415	W
108	Pacaya Samiria, Peru	800	R
109	Laguna Pedropalo, Colombia	2000	X
110	Laguna nr Medellin, Colombia	2504	X
111	Laguna I, Colombia	2585	X
112	Laguna San Pablo, Ecuador	2661	A
113	Cuicocha, Ecuador	3068	A
114	Laguna de Paca, Peru	3400	R
115	Lago Macubayi, Venezuela	3600	U
116	Lake Titicaca, Bolivia/Peru	3809	V
117	Lago Junin, Peru	4080	R
118	Laguna Arapa & others, Peru	3810–4230	R
119	Laguna Antocotto & others, Peru	4425–4700	R
120	Laguna Suches, Peru	4452	R
121	Laguna Huaron & others, Peru	4360–5140	R
122	Licancabur Crater Lake, Chile	5833	Y

References: A - own observations. R - Valdivia-Villar (1988). S - Hardy, Robertson & Koste (1984). T - Green (1972). U - Brehm (1956). V - Rey (1992). W - Infante & Riehl (1984) X - Stingelin (1913). Y - Hann (1986).

Table 5. Numbers of species of planktonic Cladocera in relation to altitude at three latitudinal ranges in Africa.

Latitude	Altitude (1000 m)	No. localities	Total no. spp.	Mean no. spp/locality
0–10	0–1	14	12	2.86
	1–2	29	15	2.83
	2–3	6	6	1.17
	3–4	4	2	1.00
	4–5	4	3	1.00
10–20	0–1	8	10	6.0
	1–2	17	10	3.12
	2–3	3	4	2.0
28–31	0–1	0	–	–
	1–2	10	9	2.0
	2–3	3	3	1.67

differences in relation to geographical position within a large land mass.

Acknowledgments

My work in Africa has been financed at various times by the Royal Society, the Leverhume Trust, the Central Research Fund of London University, the Inter-University Council, the British Council, and the Canadian International Development Agency. I have received great hospitality from the universities of Khartoum, Makerere (Uganda), Ife (Nigeria), Addis Ababa and Lesotho, and the Helminthiasis Research Unit at Kumba, Cameroon. In Brazil I was a member of the Xavantina-Cachimbo Expedition, which enabled me to study the Suia Missu lakes; I am grateful to the Royal Society for this privilege.

References

Banta, A. M., 1939. Studies on the physiology, genetics and evolution of some Cladocera. Pap. Dept. Genet. Carneg. Instn, No. 39.

Brehm, V., 1935. Crustacea I. Cladocera und Euphyllopoda. Mission Scientifique de l'Omo 2: 141–166.

Brehm, V., 1956. Cladocera aus Venezuela. Ergeb. Deutsch. Limnol. Venezuela-Exped. 1: 217–232.

Brehm, V., 1960. Crustacea Cladocera. Mission zoologique de l'I.R.S.A.C. en Afrique orientale (P. Basilewsky & N. Leleup, 1957). Ann. Mus. Congo Tervuren, in 8 °, Zool. 81: 27.

Brook, A. J. & J. Rzoska, 1954. The influence of the Gebel Aulyia Dam on the development of Nile plankton. J. anim. Ecol. 23: 101–114.

Carvalho, G. R., 1987. The clonal ecology of *Daphnia magna* (Crustacea: Cladocera) II. Thermal differentiation among seasonal clones. J. anim. Ecol. 56: 469–478.

Fryer, G., 1957. Freeliving freshwater Crustacea from Lake Nyasa and adjoining waters. Part II. Cladocera and Conchostraca. Arch. Hydrobiol. 53: 223–239.

Gauthier, H., 1928. Recherches sur la faune des eaux continentales de l'Algerie et de la Tunisie. Minerva; Alger, 419 pp.

Green, J., 1966. Seasonal variation in egg production by Cladocera. J. anim. Ecol. 35: 77–104.

Green, J., 1971. Associations of Cladocera in the zooplankton of the lake sources of the White Nile. J. Zool., Lond. 165: 373–414.

Green, J., 1972. Freshwater ecology in the Mato Grosso, Central Brazil II. Associations of Cladocera in meander lakes of the Rio Suia Missu. J. nat. Hist. 6: 215–227.

Green, J., 1976. Changes in the zooplankton of Lakes Mutanda, Bunyonyi and Mulehe (Uganda). Freshwat. Biol. 6: 433–436.

Green, J., 1985. Horizontal variations in associations of zooplankton in Lake Kariba. J. Zool., Lond. 206: 225–239.

Green, J., 1990. Zooplankton associations in Zimbabwe. J. Zool., Lond. 222: 259–283.

Green, J. & G. Kling, 1988. The genus *Daphnia* in Cameroon, West Africa. Hydrobiologia 160: 257–261.

Green, J., A. I. Moghraby & O. M. M. Ali, 1984. A faunistic reconnaisance of Lakes Kundi and Keilak, Western Sudan. Hydrobiologia 110: 33–44.

Griffiths, J., 1972. World survey of Climatology Volume 10. Climates of Africa. Amsterdam: Elsevier, 604 pp.

Hann, B. J., 1986. Revision of the genus *Daphniopsis* Sars,1903 (Cladocera: Daphniidae) and a description of *Daphniopsis chilensis*, new species from South America. J. Crust. Biol. 6: 246–263.

Hardy, E. R., B. Robertson & W. Koste, 1984. About the relationship between the zooplankton and fluctuating water levels of Lago Camaleao, a central Amazonian varzea lake. Amazoniana 9: 43–52.

Hart, R. C., 1987. Population dynamics and production of five crustacean zooplankters in a subtropical reservoir during years of contrasting turbidity. Freshwat. Biol. 18: 287–318.

Infante, A. & W. Riehl, 1984. The effect of Cyanophyta upon zooplankton in a eutrophic tropical lake (Lake Valencia, Venezuela). Hydrobiologia 113 (Dev. Hydrobiol. 23): 293–298.

Johnson, D. S., 1952. A thermal race of *Daphnia atkinsoni* Baird and its distributional significance. J. anim. Ecol. 21: 118–119.

Kalk, M., A. J. McLachlan & C. Howard-Williams, 1979. Lake Chilwa. Monographiae Biologicae 35: 1–462.

Loaring, J. M. & P. D. N. Hebert, 1981. Ecological differences among clones of *Daphnia pulex* Leydig. Oecologia 51: 161–168.

Löffler, H., 1964. The limnology of tropical high-mountain lakes. Verh. int. Ver. Limnol. 15: 176–193.

Löffler, H., 1968. Die Crustaceenfauna der Binnengewässer Ostafrikanischer Hochberge. Hochgebirgsforschung 1: 107–170.

Löffler, H., 1978. Limnological and paleolimnological data on the Bale Mountain Lakes (Ethiopia). Verh. int. Ver. Limnol. 20: 1131–1138.

Rey, J., 1992. The Cladocera; pp 254–267 in Dejoux, C. & A. Iltis (eds) Lake Titicaca. Monographiae Biologicae 68: 1–573.

Rey, J. & L. Saint-Jean, 1968. Les Cladocères (Crustacés, Branchiopodes) du Tchad (Première note). Cah. O.R.S.T.O.M. ser Hydrobiol. 2: 79–118.

Rey, J. & L. Saint-Jean, 1969. Les Cladocères (Crustacés, Branchiopodes) du Tchad (Deuxième note). Cah. O.R.S.T.O.M. ser Hydrobiol. 3: 21–42.

Sars, G. O., 1895. On some South-African Entomostraca raised from dried mud. Vidensk. Selsk. Skr. 1895: 1–56.

Stingelin, T., 1913. Cladocera. O. Fuhrmann & E. Mayor Voyage d'exploration scientifique en Colombie. Mem. Soc. neuchatel. Sci. nat. 5: 600–638.

Talling, J. F., 1993. Environmental regulation in African shallow lakes and wetlands. Rev. Hydrobiol. trop. 25: 87–144.

Thornton, J. A., 1982. Lake McIlwaine. The eutrophication and recovery of a tropical African man-made lake. Monographiae Biologicae 49: 1–251.

Valdivia Villar, R. S., 1988. Lista de Cladoceros dulceacuicolas del Peru. Amazoniana 10: 283–297.

Wood, T. R. & A. M. Banta, 1929. Physiological studies on thermal clones. Carnegie Instn Year Book 28: 43–44.

Hydrobiologia **307**: 85–97, 1995.
P. Larsson & L.J. Weider (eds), Cladocera as Model Organisms in Biology.

The conservation ecology of the Podonidae from the Caspian and Aral seas *

N. V. Aladin
Zoological Institute, Russian Academy of Sciences, St. Petersburg, 199034, Russia

Key words: Caspian and Aral seas, Podonidae, salinity and temperature tolerance

Abstract

Results of five expeditions to the Caspian Sea and 17 expeditions to the Aral Sea from 1976–1993 are presented. In these two large salt lakes, nine species of Podonidae were found, all of which were fairly euryhaline, except *Caspievadne maximowitschi*. Further, all of them were thermophilic, except *Cornigerius bicornis*, *Caspievadne maximowitschi*, and *Pleopis polyphemoides*. Sexual reproduction of the Podonidae in these two salt lakes appears to be insignificant. The possibility exists that parthenogenetic females of some species can live and reproduce even during the winter in the southern parts of the lakes. The complete disappearance of the Podonidae from the Aral Sea in 1989–1990 and the later reappearance of *Podonevadne camptonyx* in the north Aral in 1991, demonstrates that today it is possible to rehabilitate and conserve a small part of this lake.

Introduction

The Family Podonidae includes 15 species from seven genera: *Pleopis*, *Pseudoevadne*, *Evadne*, *Podon*, *Podonevadne*, *Cornigerius*, and *Caspievadne*. At present, nine species are known from the Caspian and Aral Seas, and only seven species from marine habitats. The systematics of the Podonidae was well-studied by Sars (1897, 1902, 1927), and Mordukhai-Boltovskoi (1968a, b, 1974). Unfortunately, very little is known about the ecology of the Podonidae from these two large lakes. The present study summarizes results from numerous expeditions to the Caspian and Aral Seas in the period from 1976 to 1993. It forms part of a more extensive study that covers various aspects of the ecology of the Podonidae from the Caspian and Aral Seas, and summarizes some earlier published papers (Aladin, 1982a, b, 1983, 1989a, b, 1991a, b; Aladin & Andreev, 1984; Aladin & Valdivia Villar, 1987; Andreev *et al.*, 1992).

Fig. 1. Map showing the locations of the Caspian and Aral Seas. The black dots in each basin represent the main sampling stations from which the Podonidae were collected.

* This paper is dedicated to the memory of David Frey, who gave the author the idea for this study, and discussed its main conclusions.

Materials and methods

Animals were collected by the author mainly in the summer from various sites in these large salt lakes during five expeditions to the Caspian Sea and 17 expeditions to the Aral Sea (Fig. 1). Some samples were collected in the absence of the author (mainly in spring and autumn) by crews of the research vessels and staff of the Russian Space Center in Kazakhstan. Only the main sampling stations are shown in Fig. 1. During this study, collections of plankton and Cladocera from the Zoological Institute, Russian Academy of Sciences, including the G. O. Sars collection (St. Petersburg) were used to aid in species identifications. Plankton collections of CaspNIRKH (Caspian Fisheries Research Institute in Astrkhan), and one from KazNIRKH (Kazakhstan Fisheries Research Institute in Aralsk) also were examined. Experimental studies of salinity and thermal tolerances of the Podonidae were conducted under laboratory conditions at the Zoological Institute (St. Petersburg), CaspNIRKH, KazNIRKH and also on board the Research vessels 'Bakuvi', 'Biolog', and 'Alma-Ata'. Some experiments were conducted at the Shikhovo Biological Station (Caspian Sea, near Baku), and on the Isle of Barsakelmes Wildlife Preservation Station (Central Aral Sea). Methods of sampling and experimental design are given in earlier papers (Aladin, 1982a, b; Aladin & Andreev, 1984).

Results

Nine species of Podonidae were found in the Caspian Sea and four species were detected in the Aral Sea. Originally, eight species of Podonidae were described from the Caspian Sea: *Evadne anonyx* (G. O. Sars, 1897); *Podonevadne trigona* (G. O. Sars, 1897); *Podonevadne camptonyx* (G. O. Sars, 1987); *Podonevadne angusta* (G. O. Sars, 1902); *Cornigerius maeoticus* (Pengo, 1879); *Cornigerius arvidi* (Mordukhai-Boltovskoi, 1967); *Cornigerius bicornis* (Zernov, 1901); and *Caspievadne maximowitschi* (G. O. Sars, 1902). Rivier (1966b) considered that in the Caspian Sea, one additional species of Podonidae occurs – *Evadne prolongata* (Behning, 1938). Behning (1938) described this taxon as a subspecies of *E. anonyx*. Because the limbs of *E. prolongata* and *E. anonyx* are identical, and the only difference is the length of the carapace covering the brood chamber, here I consider this organism as a subspecies of *E. anonyx*.

In the Aral Sea, there were originally only two species of Podonidae: *P. camptonyx* and *E. anonyx*. Later, two additional species, *P. angusta* and *P. trigona*, were found. *P. angusta* was found for the first time in 1968 (Mordukhai-Boltovskoi, 1974), while *P. trigona* was first detected in 1980 (Aladin & Andreev, 1984). It is unclear how these two species entered the lake. They could have been introduced from the Caspian Sea either naturally by migrating birds, or artificially by man. It is also possible that *P. angusta* and *P. trigona* were always in the Aral Sea, but for various reasons (e.g. low population densities), they remained undetected for many years. By 1989–1990, all Podonidae had disappeared from the Aral Sea. This happened when the salinity reached a concentration of 30–32 g l^{-1} (Aladin, 1991b). This disappearance was predicted by and agrees with earlier laboratory experiments with Aral Sea Podonidae (Aladin, 1982b). Later, in the autumn of 1991, *P. camptonyx* was collected again in the northern part of the Aral Sea (Plotnikov, 1993). Since that time, this single species of Podonidae has persisted in the north Aral; however, no Podonidae have been detected in the southern part of the Aral. Only two specimens of *E. camptonyx* were detected in the south Aral in autumn 1989 (Andreev *et al.*, 1992).

Species distributions

Evadne anonyx. This species now lives only in the Caspian Sea. Originally, it was plentiful in the Aral Sea (Kortunova, 1958, 1975; Lukonina, 1960; Kortunova & Lukonina, 1970; Aladin & Andreev, 1984; Andreev, 1989). Some specimens have also been recorded from river estuaries in the Black and Azov Seas (Mordukhai-Boltovskoi & Negrea, 1965). The reason for its disappearance from the Aral Sea was because of an increase in salinity; its disappearance from the Black and Azov Seas may have been related to increased pollution. *E. anonyx* was last found in the Aral Sea in summer 1987 in the coastal waters near the Isle of Barsakelmes (Aladin, 1989b) Salinity at that time was 24.7 g l^{-1}.

Podonevadne trigona. This species now lives in the Caspian and Azov Seas, lagoons of the Black Sea (Dneprovskyi, Dneprobugskyi), freshwater reservoirs of the River Dnepr (Kakhovskoe and others), and River Don (Tsimlyanskoe) (Mordukhai-Boltovskoi, 1965a, b, 1979; Aladin, 1982b). Earlier, *P. trigona* was found in the Aral Sea and in the salt lake Chelkar, which is located about 250 km north-west of the Aral (Behning, 1938, 1940; Aladin & Andreev, 1984). Its disappearance from the Aral can be attributed to increased salinization, while its disappearance from the Chelkar may be due to increased pollution. *P. trigona* was last found in the Aral Sea in summer 1984 in the coastal waters near the Isle of Barsakelmes, when the salinity was 21.5 g l^{-1} (Aladin, 1989b).

Podonevadne camptonyx. This species is endemic to the Caspian and Aral Seas. It was completely absent in the Aral Sea by 1990, but it reappeared again (probably from diapausing eggs) in 1991 in the northern part of the lake (Plotnikov, 1993).

Podonevadne angusta. This species now lives only in the Caspian Sea. Originally, *P. angusta* was abundant in the Aral Sea (Lukonina, 1960; Kortunova & Lukonina, 1970; Kortunova, 1975; Aladin & Andreev, 1984). The reason for its disappearance from the Aral Sea can once again be attributed to salinity increase. This species was last found in the Aral Sea in summer 1987 in Bolshoy Sarychaganak Bay, when the salinity was 28.9 g l^{-1} (Aladin, 1991a).

Cornigerius maeoticus. *C. maeoticus* has a similar distribution to *P. trigona*. It now lives in the Caspian and Azov Seas, lagoons of the Black Sea (Dneprovskyi, Dneprobugskyi), the delta of the River Danube, and freshwater reservoirs of the Rivers Dnepr (Kakhovskoye) and Don (Tsimlyanskoe). *C. maeoticus* has never been detected in the Aral Sea, but under laboratory conditions, it can survive in Aral Sea water.

Cornigerius arvidi. This species is endemic to the Caspian Sea, and was collected from the central and southern parts of this lake. In the laboratory, it can survive in water from the Aral Sea.

Cornigerius bicornis. *C. bicornis* has a similar distribution to both *P. trigona* and *C. maeoticus*. It has been collected from the Caspian and Azov Seas, and the freshwater canals and reservoirs of the Rivers Dnepr (Kakhovskoe) and Don (Tsimlyanskoe). It has never been detected from the Aral Sea, but under laboratory conditions, it can survive in Aral Sea water.

Caspievadne maximowitschi. Like *C. arvidi*, this species is endemic to the Caspian Sea, and has been detected at extremely low densities (7 ind. m^{-3}, August 1978) in the central and southern parts of this lake, when the surface temperature was 22.4 °C and salinity was 12.1 g l^{-1}. Mean abundance was slightly greater than 1 ind. m^{-3}. Because of the low densities, it was not possible to conduct any laboratory experiments with this species.

Pleopis polyphemoides. This species is found in coastal marine waters, mainly in the Atlantic and Pacific Oceans between 65–70° N and 40° S (Mordukhai-Boltovskoi & Rivier, 1987). It is also numerous in the seas connecting these two ocean basins. It is a very common species in the Black and Azov Seas, and it first appeared in the Caspian Sea in 1957, after the opening of the Volga-Don canal (Mordukhai-Boltovskoi, 1962b). *P. polyphemoides* has never been detected in the Aral Sea, but under laboratory conditions, it can survive in Aral Sea water.

Salinity and temperature tolerance, and population fluctuations

Laboratory salinity and temperature tolerance experiments (Figs. 2 and 3) using water from the Caspian and Aral Seas showed that most species could tolerate fairly broad ranges of salinities and temperatures (As mentioned above, we were unable to conduct laboratory experiments on *C. maximowitschi* because of its rarity). I will summarize the data for all nine species as follows:

Evadne anonyx. Our experimental data on salinity tolerance of *E. anonyx* indicated that under laboratory conditions, it can tolerate salinities of 25 and 30 g l^{-1} in Caspian and Aral Sea water, respectively (Aladin, 1989a). Its lower limit of salinity tolerance is approximately 4 g l^{-1}, and this lower limit is similar for water with Caspian and Aral salt compositions.

In the Caspian Sea, this species is most numerous (i.e. mean abundance greater than 150 ind. m^{-3}) within the salinity range from 9.2–11.7 g l^{-1} (Fig. 2). The highest and lowest salinities where we found *E. anonyx* in this lake, were 12.5 and 8.0 g l^{-1}, respectively (Aladin, 1982b). In the Aral Sea, it was most numerous (mean abundance greater than 100 ind. m^{-3}) within a salinity range of 15.9–21.5 g l^{-1} (Fig. 2). The highest and lowest salinities where we found *E. anonyx* in this lake, were 24.7 and 15.4 g l^{-1}, respectively (Aladin & Andreev, 1984; Aladin, 1989b).

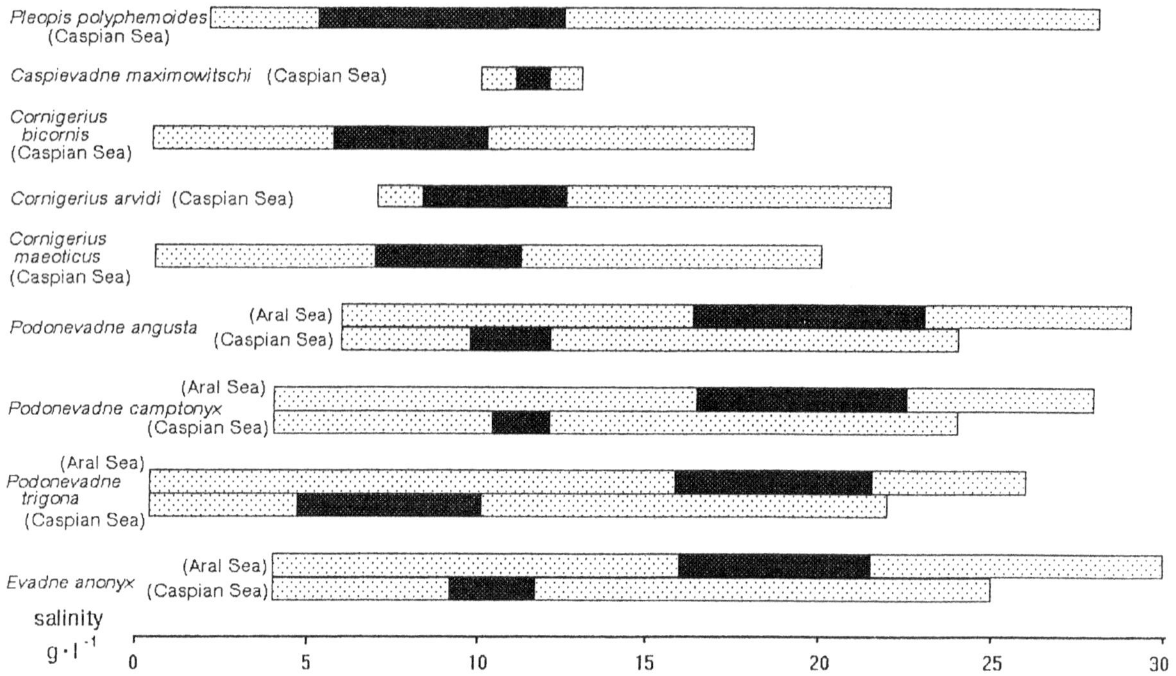

Fig. 2. Optimum (darkened) and tolerant (stippled) salinity ranges for parthenogenetic females of the nine species of Podonidae from the Caspian and Aral Seas.

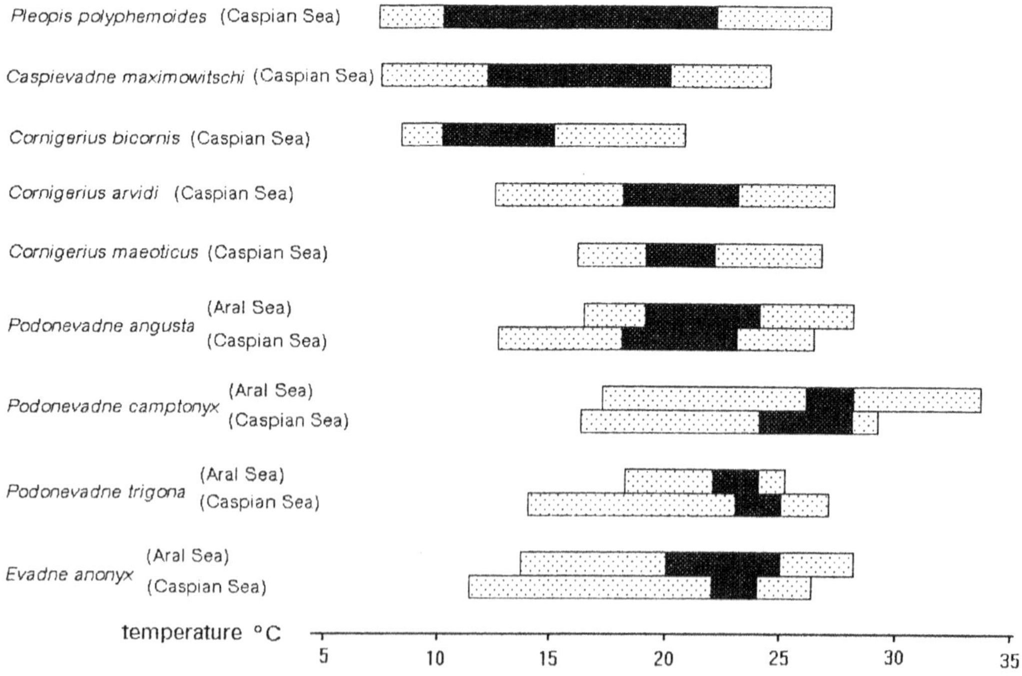

Fig. 3. Optimum (darkened) and tolerant (stippled) temperature ranges for parthenogenetic females of the nine species of Podonidae from the Caspian and Aral Seas.

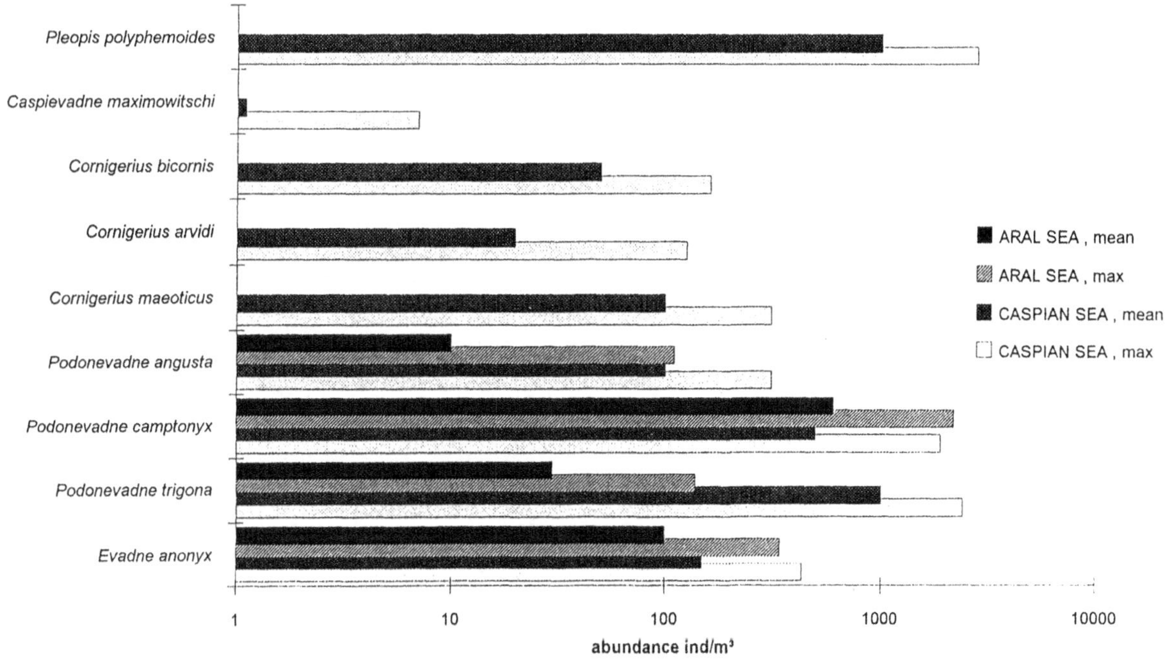

Fig. 4. Population density estimates (mean and maximum abundances) for parthenogenetic females of the nine species of Podonidae from the Caspian and Aral Seas.

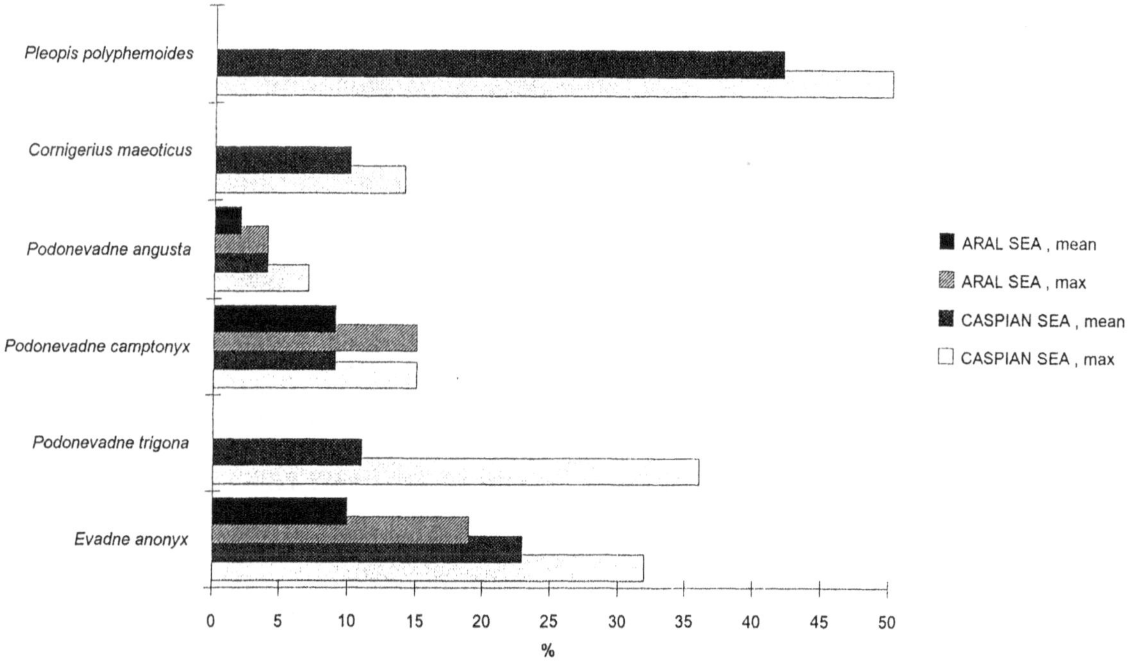

Fig. 5. Maximum and mean percentages of gametogenetic (sexual) forms of six species of Podonidae from the Caspian and Aral Seas.

This species is also thermophilic (Fig. 3), and reaches maximum abundance in July-August at water temperatures ranging from 20–25 °C. In samples from both the Caspian and Aral Seas, *E. anonyx*, when found within the optimal salinities mentioned above, usually had an abundance of 200–300 ind. m^{-3} when the surface temperature was about 22–24 °C. The maximum abundance of 473 ind. m^{-3} was recorded in the southern part of the Caspian Sea in July 1977 (Fig. 4), when the surface temperature was 23.7 °C and salinity was 11.6 g l^{-1}. The highest and lowest temperatures where we found parthenogenetic females of *E. anonyx* in the Caspian Sea, were 26.4° and 11.4 °C, respectively. The corresponding values for the Aral Sea were 28.2° and 13.7 °C, respectively (Fig. 3).

The first sexual forms of *E. anonyx* appear in September in the Caspian Sea, and August-September in the Aral Sea. They reach a maximum adundance of 32% (mean of 23%) of the total population in October in the Caspian Sea, and a maximum abundance of 19% (mean of 10%) in September-October in the Aral Sea (Fig. 5). As a rule, the number of males exceeds the number of sexual females. At the end of October 1977, some males were collected in the middle part of the Caspian Sea when the surface temperature was 8.2 °C, and salinity was 10.1 g l^{-1}. At the beginning of October 1978, some sexual females were also collected in the northern middle part of the Caspian Sea when the temperature was 6.8 °C and salinity was 8.7 g l^{-1}.

Podonevadne trigona. The upper limits of salinity tolerance of *P. trigona* under laboratory conditions are 22 and 26 g l^{-1} in Caspian and Aral water, respectively (Fig. 2). This species can survive in freshwater (0.4 g l^{-1}) in the laboratory, and thus must be considered euryhaline.

In samples from the Caspian Sea, this species is most numerous (mean abundance greater than 1000 ind. m^{-3}) within the salinity range of 4.7–10.1 g l^{-1}. The highest and lowest salinities where we found *P. trigona* in this lake were 12.5 and 2.8 g l^{-1}, respectively.

This species was not abundant in the Aral Sea (Fig. 4); it was first detected there in July 1980, but was found only at nine stations, with minimum and maximum densities of 16 and 139 ind. m^{-3}, respectively (Aladin & Andreev, 1984). *P. trigona* showed stable low summer population densities until 1983. It was last detected in the Aral Sea in May-August 1984, with a mean abundance of 2–6 ind. m^{-3}. The highest and lowest salinities where we found it in this lake were 21.5 and 15.8 g l^{-1}, respectively (Aladin & Andreev, 1984).

P. trigona is very common in the northern part of the Caspian Sea in July-August, when surface water temperatures are above 20 °C. The maximum population density recorded for this species was 2407 ind. m^{-3} in July 1978 (Fig. 4), when the surface water temperature was 25.5 °C and salinity was 5.6 g l^{-1}. The highest and lowest temperatures where we found parthenogenetic females in the Caspian Sea were 27.1° and 14.0 °C, respectively. The corresponding temperatures in the Aral Sea were 25.2° and 18.2 °C, respectively.

The first sexual forms of *P. trigona* appear at the end of August in the Caspian Sea. Sexual females and males have never been detected in the Aral Sea. The sexual forms reach a maximum adundance of 36% (mean of 11%) of the total population in September-October in the Caspian Sea (Fig. 5). Generally, sexual females are more abundant than males. At the beginning of October 1977, both sexual forms were collected in the northern part of the Caspian Sea when the temperature was 4.9 °C and salinity, 3.7 g l^{-1}.

Podonevadne camptonyx. The salinity tolerance of *P. camptonyx* to water from the Caspian and Aral Seas is different for high salinities, but similar for low salinities (Fig. 2). In the laboratory, this species is able to survive at 24 and 28 g l^{-1} in Caspian and Aral Sea water, respectively (Aladin, 1989a). It can tolerate salinities as low as 4 g l^{-1}.

In the Caspian Sea, this species is most numerous (mean abundance greater than 500 ind. m^{-3}) within the salinity range of 10.4–12.1 g l^{-1}. The highest and lowest salinities where we found *P. camptonyx* in the Caspian were 13.2 and 6.2 g l^{-1}, respectively. In the Aral Sea, it was most numerous (mean abundance greater than 600 ind. m^{-3}) at salinities ranging from 16.4–22.5 g l^{-1}. It has been detected in the Aral Sea at salinities as high as 28.0 g l^{-1} (Andreev *et al.*, 1992).

P. camptonyx is the most thermophilic of all the species studied (Fig. 3), and is able to reach maximum abundances in June-August at surface water temperatures ranging from 24–28 °C. At optimal salinities in the Caspian Sea, it has been detected at a density of 1500–1600 ind. m^{-3} (Fig. 4), when the temperature was about 25 °C. At optimal salinities in the Aral Sea, it was found at a density of 1200–1300 ind. m^{-3} (Fig. 4), when temperatures ranged from 26–28 °C. It was found at a maximum density of 2192 ind. m^{-3} in the south-western part of the Aral Sea in July 1980, when the temperature was 27.6 °C and salinity was 15.9 g l^{-1}. Its upper and lower temperature limits in the Caspian Sea are 29.2° and 16.3 °C, respectively. The corresponding values for the Aral Sea are 33.7° and 17.2 °C, respectively.

The first sexual forms of *P. camptonyx* appear in August in the Caspian Sea, and in September in the Aral Sea. The sexual forms reach a maximum adundance of 15% (mean of 9%) of the total population in each lake (Fig. 5). The number of sexual females in both lakes generally exceeds that of the males. In some autumn samples from both the Caspian and Aral Seas, males have been completely absent. At the end of September 1977, a few males and sexual females were collected in the northern part of the Caspian Sea when the temperature was only 12.0 °C and salinity, 8.2 g l^{-1}.

Podonevadne angusta. In the laboratory, *P. angusta* can withstand salinities as high as 24 and 29 g l^{-1} in Caspian and Aral Sea water, respectively (Fig. 2). Its lower limit of salinity tolerance is 6 g l^{-1} in water with both Caspian and Aral salt compositions.

In the Caspian Sea, it is most numerous (mean abundance greater than 100 ind. m^{-3}) within a salinity range of 9.7–12.1 g l^{-1}. The highest and lowest salinities where we found it in this lake were 12.8 and 7.0 g l^{-1}, respectively. In the Aral Sea, this species was not abundant. It was first detected there in August 1968, although only in the south-eastern part of the lake (Mordukhai-Boltovskoi, 1974). From 1980 to 1986, the mean summer abundance was stable at about 10 ind. m^{-3} (Fig. 4), also mainly in the south-eastern section of the lake, but some specimens were regularly found in the north-eastern part as well. *P. angusta* was last detected in the Aral Sea in July-August 1987, when its mean abundance was reduced to about 2 ind. m^{-3} (maximum abundance of 4 ind. m^{-3}). The highest and lowest salinities where we found it in this lake were 28.9 and 16.3 g l^{-1}.

P. angusta is abundant in the central and southern parts of the Caspian Sea during summer at water temperatures above 17 °C. A maximum abundance of 312 ind. m^{-3} was recorded here in July 1978 (Fig. 4), when the surface temperature was 26.6 °C and salinity was 11.3 g l^{-1}. The highest and lowest temperatures where we found parthenogenetic females in the Caspian Sea were 26.4° and 12.6 °C, respectively. The corresponding temperatures in the Aral Sea were 28.1° and 16.4 °C, respectively.

The first sexual forms of this species appear in the Caspian Sea at the end of September. In the Aral Sea, sexual forms were very rare, even in late October. The sexual forms reach a maximum adundance of 7% (mean of 4%) of the total population in the Caspian Sea in September-October (Fig. 5). The corresponding figures for the Aral Sea are 4% and 2%, respectively. Generally, the number of sexual females exceeded that of the males by twofold. A few sexual females were found in October 1980 in the central part of the Caspian Sea when the temperature was only 5.1 °C and salinity, 11.0 g l^{-1}. Two sexual females were found in October 1985 in the central part of the Aral Sea when the temperature was 4.7 °C and salinity was 22.0 g l^{-1}.

Cornigerius maeoticus. Under laboratory conditions, *C. maeoticus* shows upper limits of salinity tolerance of 20 and 24 g l^{-1} in Caspian and Aral Sea water, respectively. It is able to survive in freshwater (0.5 g l^{-1}) as well.

In the Caspian Sea, it is most abundant (mean adundance greater than 100 ind. m^{-3}) within the salinity range of 7.9–11.2 g l^{-1} (Fig. 2). The highest and lowest salinities where we found it in this lake were 12.5 and 6.0 g l^{-1}, respectively (Aladin, 1982b). *C. maeoticus* was detected at a maximum abundance of 311 ind. m^{-3} in the central part of the Caspian Sea in July 1977 (Fig. 4), when the surface temperature was 25.3 °C and salinity, 10.3 g l^{-1}.

The highest and lowest temperatures where we found parthenogenetic females in the Caspian Sea, were 26.7° and 16.1 °C, respectively (Fig. 3).

Sexual forms of *C. maeoticus* are very rare in the Caspian Sea. Some sexual forms have been found in several samples collected in September-October. The sexual forms reach a maximum adundance of 14% (mean of 10%) of the total population in the Caspian Sea at these times (Fig. 5). Generally, the number of sexual females exceeds that of the males by threefold. Five sexual females were collected in October 1977 in the northern part of the Caspian Sea when the temperature was 4.2 °C and salinity was 8.2 g l^{-1}.

Cornigerius arvidi. The upper limit of salinity tolerance in this species is unclear, because we did not have enough specimens for our laboratory experiments. In the Caspian Sea, one should consider a realistic upper limit of 22 g l^{-1} because four individuals survived at this concentration. The lower limit of salinity tolerance in the laboratory is 7 g l^{-1}.

C. arvidi has never been numerous in the Caspian Sea. Its maximum abundance was 126 ind. m^{-3} in the southern part of this lake in July 1979 (Fig. 4), when the surface temperature was 25.2 °C, and salinity was 11.9 g l^{-1}. Mean abundance in summer is about 20 ind. m^{-3} at a salinity of 8.3–12.5 g l^{-1}. The highest and lowest temperatures where we found parthenogenetic females were 27.2° and 12.4 °C, respectively. Sexual forms of *C. arvidi* are not known (Mordukhai-Boltovskoi & Rivier, 1971, 1987).

Cornigerius bicornis. In the laboratory, the upper limits of salinity tolerance for this species are 18 and 22 g l^{-1} in Caspian and Aral Sea water, respectively. It is also able to survive in freshwater (0.4 g l^{-1}).

In the Caspian Sea, it is most abundant (mean abundance greater than 50 ind. m^{-3}) within the salinity range of 5.7–10.2 g l^{-1}. The highest and lowest salinities where we found *C. bicornis* in this lake were 11.0 and 3.1 g l^{-1}. It was detected at a maximum density of 162 ind. m^{-3} in the central part of the Caspian Sea in June 1976 (Fig. 4), when the surface temperature was only 16.4 °C, and salinity was 9.8 g l^{-1}. The highest and lowest temperatures where we found parthenogenetic females were 20.7° and 8.2 °C, respectively. Based on laboratory experiments (Fig. 3), this organism prefers temperatures ranging from 10–15 °C, but not temperatures higher than 20–22 °C. Sexual forms of this species are not known.

Caspievadne maximowitschi. Due to its rarity, it was not possible to carry out any laboratory experiments. A maximum density of 7 ind. m^{-3} was recorded in the southern part of the Caspian Sea in August 1978, when the surface temperature was 22.4 °C, and salinity, 12.1 g l^{-1}. Mean abundance was slightly greater than 1 ind. m^{-3} (Fig. 4). The highest and lowest temperatures where we found parthenogenetic females were 22.4° and 7.3 °C, respectively. This species is probably fairly eurythermic. Sexual forms of this species are not known.

Pleopis polyphemoides. Experimental data on the salinity tolerance of *P. polyphemoides* to Caspian and Aral Sea water differed. Under laboratory conditions (Fig. 2), this species can survive up to 28 g l^{-1} in Caspian (Aladin, 1982b) and 32 g l^{-1} Aral Sea water. Its lower limit of salinity tolerance is 2 g l^{-1}, both in water with Caspian and Aral salt compositions. Based on these results, this species is the most euryhaline of the nine species studied.

In the Caspian Sea, it is most numerous (mean abundance greater than 1000 ind. m^{-3}) within the salinity range of 5.2–12.4 g l^{-1}. The highest and lowest salinities where we found *P. polyphemoides* in the lake, were 12.8 and 3.9 g l^{-1}, respectively.

This species is the most eurythermic of the nine species studied. It reaches a maximum abundance in May-September at water temperatures ranging from 10–22 °C. Within optimal salinities in the Caspian Sea, it had a mean abundance of 1600 ind. m^{-3} when temperatures ranged from 8–25 °C. Its maximum abundance of 2804 ind. m^{-3} was recorded in the southern part of the Caspian Sea in June 1978 (Fig. 4), when the surface temperature was 21.2 °C, and the salinity was 11.8 g l^{-1}. The highest and lowest temperatures where we found parthenogenetic females in the Caspian Sea were 27.0° and 7.2 °C, respectively.

The first sexual forms of *P. polyphemoides* appear in August-September in the Caspian Sea, and their appearance was not immediately connected with an increase in water temperature. The sexual forms reach a maximum adundance of 50% (mean of 42%) of the total population in August in the Caspian Sea (Fig. 5). The number of males slightly exceeds the number of sexual females. At the end of September 1978, some sexual females were collected in the northern part of the Caspian Sea when the temperature was only 6.3 °C and the salinity was as low as 3.7 g l^{-1}. At the same time, some males were found in the central part of the lake when the temperature was 10.4 °C, and the salinity was 9.2 g l^{-1}.

Discussion

Comparative analyses of the data and those available in the scientific literature (Mordukhai-Boltovskoi, 1962a, b, 1965a, b, 1966, 1967a, b, 1968a, b, 1974; Mordukhai-Boltovskoi & Rivier, 1970, 1971, 1987; Rivier, 1966a, b, 1968, 1969; Rivier & Mordukhai-Boltovskoi, 1966) make it possible to understand the salinity and temperature tolerances of the Podonidae from the Caspian and Aral Seas.

From Fig. 2, it is clear that all the Podonidae from the Caspian and Aral Seas are fairly euryhaline, except *Caspievadne maximowitschi.* At high salinities (about 30 g l^{-1}), four species (*E. anonyx, P. camptonyx, P. angusta, P. polyphemoides*) can survive. Under freshwater conditions (0.4–0.5 g l^{-1}), three species (*P. trigona, C. maeoticus, C. bicornis*) can survive quite well. In the Aral Sea, the upper limits of salinity tolerance of the Podonidae are always higher than in the Caspian Sea (Fig. 2), and this is probably because in the Aral Sea, the dominance of sodium and chloride ions was somewhat less than in the Caspian Sea (Aladin, 1983).

All species of Podonidae from the Caspian and Aral Seas are thermophilic except *C. bicornis, C.*

maximowitschi, and *P. polyphemoides* (Fig. 3). Only parthenogenetic females of *P. camptonyx* can survive high temperatures (around 35 °C). Sexual forms usually appeared at temperatures less than 20 °C. At low temperatures (5 °C), sexual forms of four species (*E. anonyx*, *P. angusta*, *C. maeoticus*, *P. polyphemoides*) can survive.

From Fig. 4, it is clear that all Podonidae studied except *P. trigona*, *P. camptonyx*, and *P. polyphemoides*, were never very abundant in the plankton of the Caspian and Aral Seas during the past 20 years. Only parthenogenetic females of the three aforesaid species reached maximum abundances of more than 1000 ind. m^{-3}. The rarest species is *C. maximowitschi* with a maximum abundance of less than 10 ind. m^{-3}. It is important to note that *P. polyphemoides* was a very successful colonist to the Caspian Sea. Its mean abundance increased steadily from its first appearance in 1957 until 1972; since then, its abundance has stabilized. All data included in Fig. 4 cover the period after 1975, thus the *P. polyphemoides* data were collected only for the stable period of population increase.

Some members of the Podonidae from the Caspian and Aral Seas, as with other members of the Cladocera, have two modes of reproduction: parthenogenetic and gametogenetic (sexual). Sexual forms have never been detected for *C. arvidi*, *C. bicornis*, and *C. maximowitschi*. Sexual reproduction (Fig. 5), appears to be very ineffective in the Caspian and Aral Seas because often times the sexual forms constitute less than 35% of the total population. Only *P. polyphemoides*, with maximum and mean percentages of sexual forms approaching 50%, has effective sexual reproduction. There is a high probability that in the southern parts of the Caspian and Aral Seas, parthenogenetic females could live and reproduce even in winter, and this may reduce the importance of sexual reproduction. Because of this potential for year round parthenogenesis, future systematic, taxonomic, and ecological studies should conduct population genetic analyses of the Podonidae from the Caspian and Aral Seas.

Earlier studies of the physiology of osmotic regulation of the Podonidae (Khlebovich & Aladin, 1976; Potts & Durning, 1980; Aladin, 1982a, b, 1991b) demonstrated that all members of this group are secondarily marine, with a freshwater origin. The low osmotic concentration of hemolymph in marine Podonidae reflects the presumed ancestral freshwater state. From the marine environment, only a maximum of seven species of Podonidae (Mordukhai-Boltovskoi, 1978) is known, and this number is lower than from the Caspian Sea. I hypothesize that the Podonidae first evolved from a freshwater cladoceran ancestor in a large and slightly brackish lake in the Caspian Sea depression and surrounding region. Later, during salinization of this lake, the Podonidae gradually evolved hypo-osmotic regulation of the hemolymph as an adaptation to a highly saline environment. This type of osmotic regulation permitted the Podonidae to spread from this ancient lake into the marine environment.

Conservation ecology of the Aral sea

The present study of the ecology of the Podonidae from the Caspian and Aral Seas has lead us to deal unexpectedly with both theoretical and applied issues related to the conservation ecology of these lakes. The complete disappearance of the Podonidae from the Aral Sea and the later reappearance of *P. camptonyx* in the northern Aral has demonstrated that there are still possibilities to rehabilitate and conserve at least a small portion of this massive inland sea. In addition, it serves as a warning that we must protect other large water bodies (e.g. Caspian and Balkhash Seas) from experiencing similar fates.

At this time, I believe it is necessary to present a brief historical look at developments around the Aral Sea for the past 30–40 years in order to put things into a conservation biology perspective. Historically, the Aral Sea has consisted of two main basins: the large southern basin (hereafter termed the Large Sea), and a smaller northern basin (hereafter termed the Small Sea; Fig. 6). From 1961 to 1990, the rates of decrease in water level in the two basins diverged. Over time, the Small Sea experienced a net increase in water level because of increased input via precipitation and groundwater flow. This caused water to discharge from the Small Sea to the Large Sea. The Small Sea was partially separated from the Large Sea by Kokaral island along an east-west transect (Fig. 6). On the western side between this island and the mainland, there was a shallow strait, Auzykokaral, with a maximum depth of about 2 m; this strait dried up in 1968. On the eastern side there was the relatively deep (13 m) Berg's strait. In 1989–1990, because of the decrease in water level of the Aral Sea as a result of water diversion for agricultural purposes, Berg's strait dried up, and the Small and Large Seas became completely separate basins.

Prior to this time, the average salinities in the Large and Small Seas differed to some extent. These differences were insignificant during the period from 1961 to

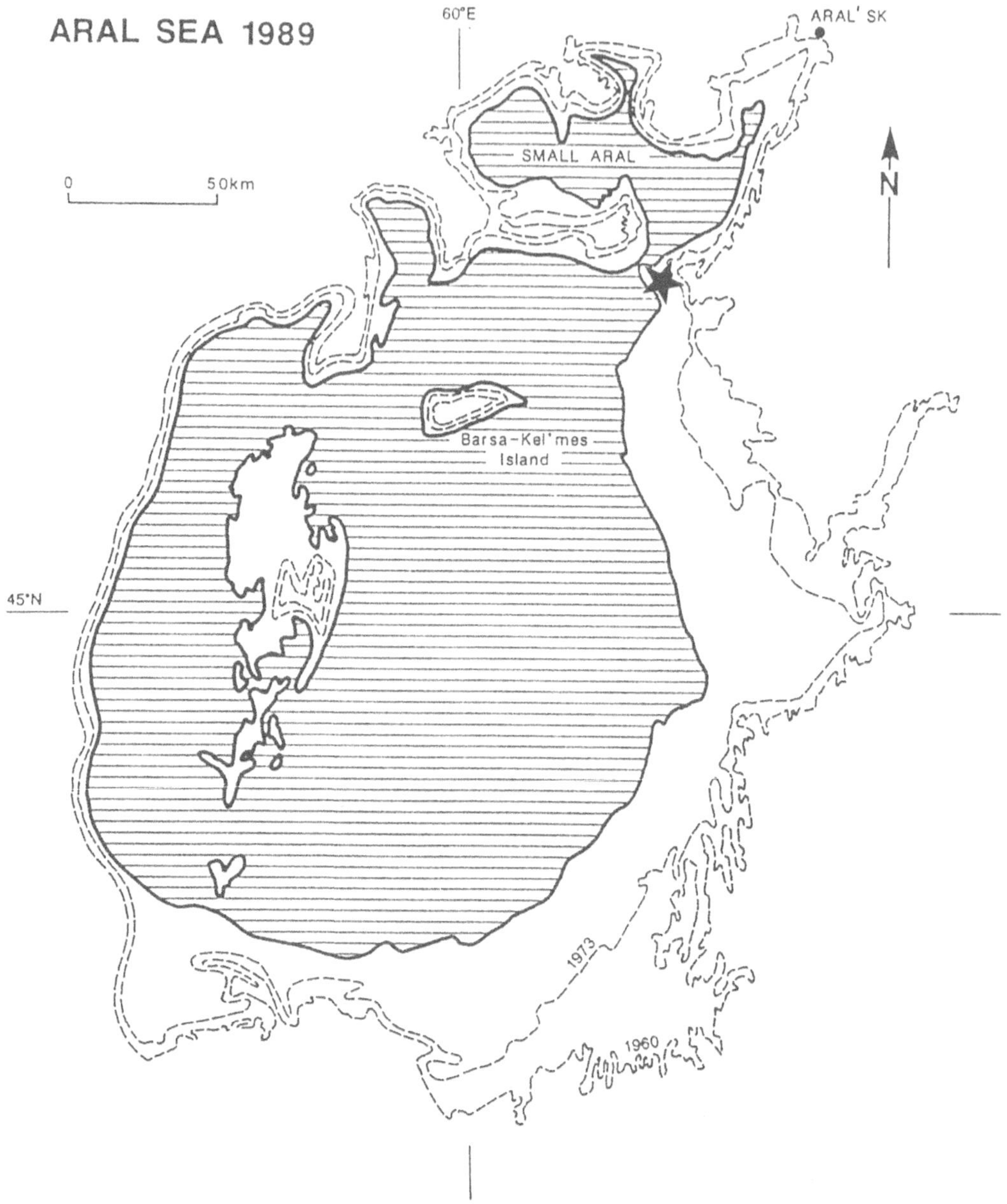

Fig. 6. Map showing the shoreline of the Small Aral and Large Aral Seas in 1989. The location of Berg's Strait and subsequent dam is depicted by a black star.

1970 because of sufficient water exchange between the two basins. The correspondence between water level and salinity in the Aral Sea before any separation of the two basins is given (Fig. 7). From 1971 to 1985 after the Auzykokaral strait dried up, and Berg's strait became shallower, the average salinity in the Small Sea rose 1.5–3.0 g l^{-1} higher than in the Large Sea. However, in 1986 the difference in average salinity between the Small and Large Seas decreased again, primarily because of a change in the direction of flow of the Syrdarya River, which resulted in the diversion of more freshwater into the Small Sea. Since 1988, the average salinity in the Small Sea has decreased by 1.5–2.0 g l^{-1} when compared with the Large Sea. Subsequently, the complete drying of Berg's strait in spring 1990, has resulted in still sharper differences in salinity concentrations between the two basins (Fig. 8).

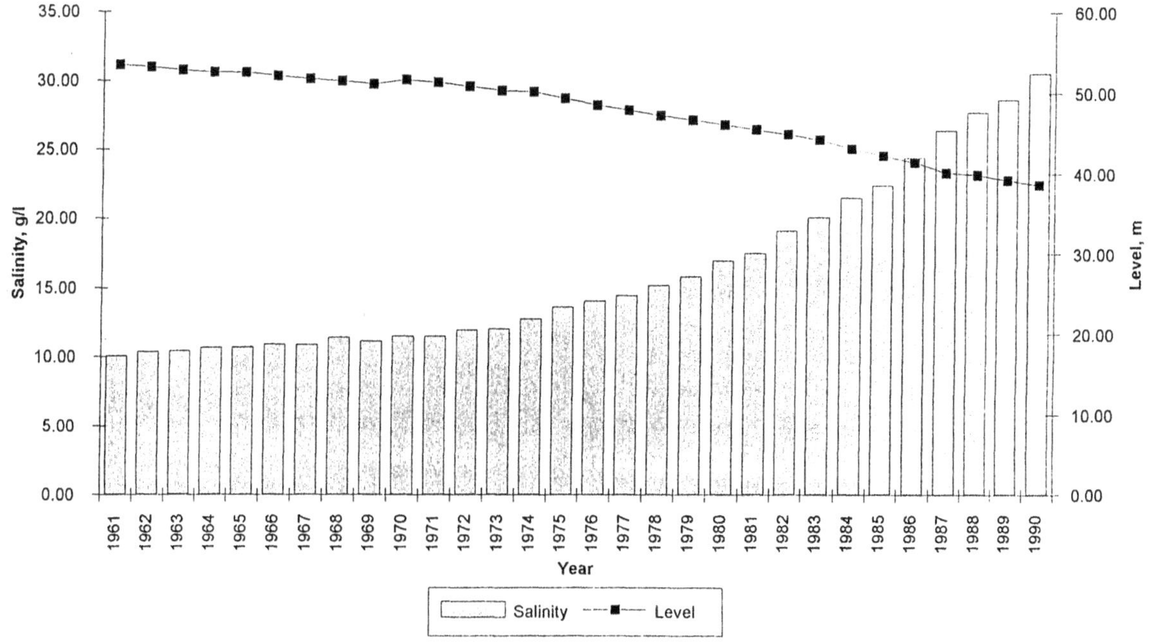

Fig. 7. Water level and salinity changes in the Aral Sea from 1961–1990.

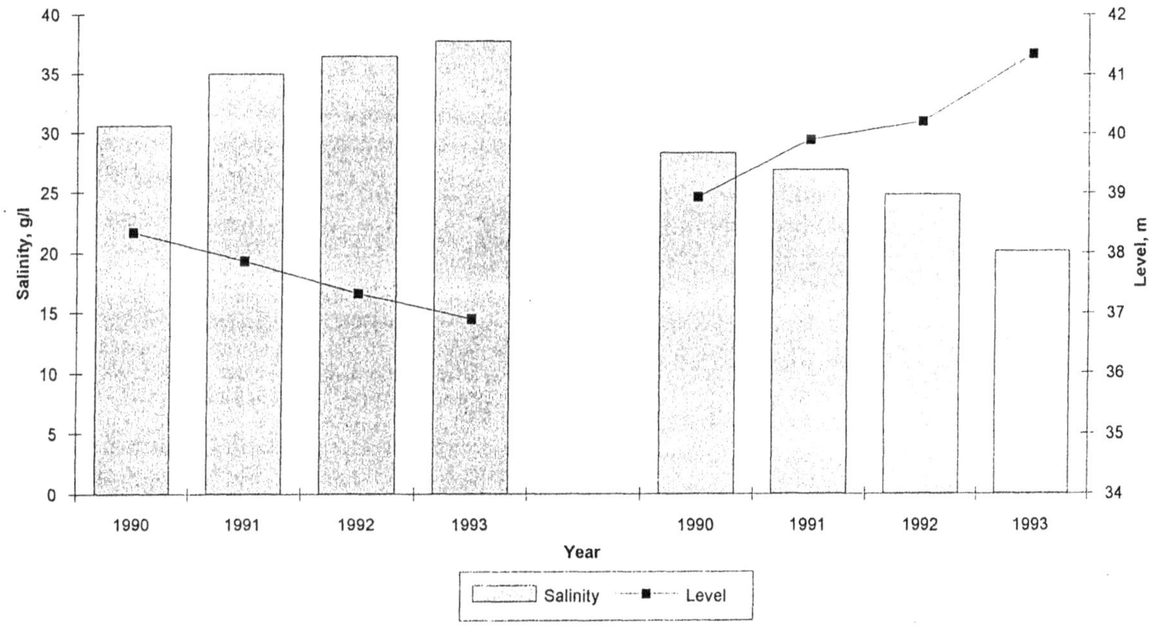

Fig. 8. Water level and salinity changes in the Large Aral Sea (left panel) and Small Aral Sea (right panel) from 1990–1993.

Sometime between spring 1990 and July 1992, a small but strong flowing channel/canal appeared in the area of Berg's strait, and it once again allowed water flow from the Small to the Large Sea. This channel was formed as a result of a rise in water level in the Small Sea during spring 1990. (In spring 1992, the flow rate from the Small to the Large Sea averaged 100 m^3 s^{-1}.)

In July 1992, this channel was filled up with sand to build a 1 m high dam/causeway, as part of an attempt to control the water level in the Small Sea, and as an attempt to rehabilitate this basin. The first attempt to construct a dam met with failure as the water pressure was too great, and the dam was breached. A second dam was constructed in late July/early August 1992, and it was in place for 9 months. However in March 1993, the water level in the Small Sea rose more than 1 m, and resulted in the destruction of the second dam.

Even though this second dam was only in place for a short (9 month) period, we are still able to draw some conclusions. The cessation of water flow from the Small to the Large Sea resulted in a significant enlargement of the estuarine zone of the Syrdarya River. Before damming, nearly all the estuarine water eventually discharged into the Large Sea. Now, the mixing zone of riverine and lake water is quite close to the Syrdarya River delta. This has resulted in a very smooth salinity gradient extending out from the delta into the lake. This gradient, with its very slow salinity increase, has made it possible for freshwater fishes to leave the delta and feed in the lake. Such an event has not been recorded by local people for more than 10–15 years.

While the first experimental dam, which was constructed in July 1992 on our recommendation, was partially destroyed, it confirmed that this location in Berg's strait was suitable, but obviously, a stronger dam is required for proper regulation of the water level in the Small Aral Sea. We anticipate that both local human populations and wildlife populations of the northern Aral region will benefit greatly from this project, while the damage to the southern Aral region will be minimal. We predict that not only *P. camptonyx*, but also other members of the Podonidae will eventually reappear in the Small Aral Sea.

Acknowledgements

I thank Professor P. Larsson and Dr L. Weider for editorial assistance, Drs J. Green and G. Fryer for important scientific comments, Dr I. Plotnikov, Zoological Institute, Russian Academy of Sciences for computer program assistance, and Miss O. Rusakova for typing the manuscript. I wish to thank Captain E. Pogrebniak of the research vessel 'Biolog', Dr N. Andreev, Kazakhstan Fisheries Research Institute, Mr. P. Popovich and Mr. G. Beregovoi, Russian Space Center in Kazakhstan for assistance in sampling during spring and autumn.

References

Aladin, N. V., 1982a. Salinity adaptations and osmoregulation abilities of the Cladocera. 1. Forms from the open seas and oceans. Zool. Zhurnal. 61: 341–351 (in Russian).

Aladin, N. V., 1982b. Salinity adaptations and osmoregulation abilities of the Cladocera. 2. Forms from Caspian and Aral Seas. Zool. Zhurnal. 61: 507–514 (in Russian).

Aladin, N. V., 1983. On displacement of the critical salinity barrier in the Caspian and Aral Seas, the Branchiopoda and Ostracoda taken as examples. Zool. Zhurnal. 62: 689–694 (in Russian).

Aladin, N. V., 1989a. Critical character of the biological influence of the Caspian water with salinity 7–11% and Aral water with salinity 8–13%. Trudy Zoologicheskogo Instituta Akademii nauk (Leningrad). 196: 12–21 (in Russian).

Aladin, N. V., 1989b. Zooplankton and zoobenthos in the coastal waters near the Isle of Barsakelmes, the Aral Sea. Trudy Zoologicheskogo Instituta Akademia nauk (Leningrad). 199: 110–114 (in Russian).

Aladin, N. V., 1991a. Studies of the influence of the water salinity in separating bays on the Aral Sea hydrobionts. Trudy Zoologicheskogo Instituta Akademia nauk (Leningrad). 237: 4–13 (in Russian).

Aladin, N. V., 1991b. Salinity tolerance and morphology of osmoregulation organs in Cladocera with special reference to Cladocera from the Aral Sea. Hydrobiologia 225 (Dev. Hydrobiol. 71): 191–199.

Aladin, N. V. & N. I. Andreev. 1984. Influence of the Aral Sea salinity on changes in the composition of the Cladocera fauna. Gidrobiol. Zhurnal. 20: 23–28 (in Russian).

Aladin, N. V. & R. S. Valdivia Villar. 1987. Microcryoscopic studies of liquid from eggs and embryos of Cladocera. Gidrobiol. Zhurnal. 23: 93–97 (in Russian).

Andreev, N. I., 1989. Zooplankton of the Aral Sea in the initial period of its salinization. Trudy Zoologicheskogo Instituta Akademii nauk (Leningrad). 199: 26–52 (in Russian).

Andreev, N. I., I. S. Plotnikov & N. V. Aladin. 1992. The fauna of the Aral Sea in 1989. 2. The zooplankton. Int. J. Salt Lake Res. 1: 111–116.

Behning, A. L., 1938. On the winter zooplankton of the Caspian Sea. Trudy Kompleks. izuch. Kasp. moria. 5: 7–97 (in Russian).

Behning, A. L., 1940. On the zooplankton of the gulfs of Komsomolets (Mertviy Kultuk) and Kaydak. Trudy komissii Kompleks. izuch. Kasp. moria. 3: 64–80 (in Russian).

Khlebovich, V. V. & N. V. Aladin. 1976. Hypotonic regulation in marine cladocerans *Evadne nordmanni* and *Podon leuckarti*. J. Comp. Biochem. Physiol. 12: 591–592 (in Russian).

Kortunova, T. A., 1958. On the influence of acclimatization (fish and invertebrates) and river inflow regulation on the zooplankton of the open part of the Aral Sea. In: Biological Fundamentals of Fisheries in Water-bodies of Central Asia and Kazakhstan. Fergana. Abstracts of Conference: 99–100 (in Russian).

Kortunova, T. A., 1975. On changes in the zooplankton of the Aral Sea in 1959–1968. Zool. Zhurnal. 54: 657–669 (in Russian).

Kortunova, T. A. & N. K. Lukonina, 1970. Quantitative characteristics of the Aral Sea zooplankton. Rybniye resursy vodoemov Kazakhstana i ikh ispolzovanie (Alma-Ata). 6: 52–60 (in Russian).

Lukonina, N. K., 1960. Zooplankton of the Aral Sea. Trudy Vsesoyuznogo Nauchno-Issledovatel'skogo Ribnogo. 43: 177–179 (in Russian).

Mordukhai-Boltovskoi, Ph. D., 1962a. Pontocaspian polyphemids (taxonomy, geographical distribution, biology). Bull. Mosk. Obshch. Ispyt. Prir. (biol.) 67: 131–132 (in Russian).

Mordukhai-Boltovskoi, Ph. D., 1962b. Appearance of a representative of Mediterranean polyphemids in the Caspian Sea. Zool. Zhurnal. 41: 289–290 (in Russian).

Mordukhai-Boltovskoi, Ph. D., 1965a. Polyphemidae of the Pontocaspian basin. Hydrobiologia 25: 212–219.

Mordukhai-Boltovskoi, Ph. D., 1965b. Caspian Polyphemidae in the Don and Dniepr reservoirs. Trudy Inst. Biol. vnutren. vod. 8: 37–43 (in Russian).

Mordukhai-Boltovskoi, Ph. D., 1966. Distribution, biology, and morphology of Polyphemidae in fresh and brackish waters of the Pontocaspian basin. Verh. int. Ver. Limnol. 16: 1677–1683.

Mordukhai-Boltovskoi, Ph. D., 1967a. On the taxonomy of the genus *Cornigerius* (Cladocera,Polyphemidae). Crustaceana 12: 74–86.

Mordukhai-Boltovskoi, Ph. D., 1967b. On the males and gamogenetic females of the Caspian Polyphemidae (Cladocera). Crustaceana 12: 113–123.

Mordukhai-Boltovskoi, Ph. D., 1968a. On the taxonomy of the Polyphemidae. Crustaceana 14: 197–209.

Mordukhai-Boltovskoi, Ph. D., 1968b. Order Cladocera. In Birstein, J. *et al.* (eds), *Atlas bespozbonochnych Kaspiiskogo moria*, Moscow: 120–160 (in Russian).

Mordukhai-Boltovskoi, Ph. D., 1974. Subclass Branchiopoda. In Mordukhai-Boltovskoi, Ph. D. (ed.), *Atlas of Aral Sea Invertebrates*, Moscow: 112–134 (in Russian).

Mordukhai-Boltovskoi, Ph. D., 1978. On the taxa of marine Podonidae. Zool. Zhurnal. 57: 523–529 (in Russian).

Mordukhai-Boltovskoi, Ph. D., 1979. Composition and distribution of Caspian fauna in the light of modern data. Int. Rev. ges. Hydrobiol. 64: 1–38.

Mordukhai-Boltovskoi, Ph. D. & St. Negrea, 1965. Date noi asurpa poliphemidelor casspice din bazinelo Dunarii si Niprului. St. si cerc. biol. Bucuresti (Ser. Zool.). 17: 29–38.

Mordukhai-Boltovskoi, Ph. D. & I. K. Rivier, 1971. A brief survey of the ecology and biology of the Caspian Polyphemoidea. Mar. Biol. 8: 160–169.

Mordukhai-Boltovskoi, Ph. D. & I. K. Rivier, 1987. Carnivorous Cladocera Podonidae, Polyphemidae, Cercopagidae of World Fauna. Opredeliteli po faune SSSR, vol. 148, Leningrad, 183 pp. (in Russian).

Plotnikov, I. S., 1993. The Aral Sea zooplankton (the Small Sea and Butakov Bay) in September 1991. Proc. Zool. Inst. 1993, vol. 250 (in Russian).

Potts, W. T. W. & C. T. Durning, 1980. Physiological evolution in the Branchiopoda. Comp. Biochem. Physiol. 67B: 475–484.

Rivier, I. K., 1966a. On the taxonomy of *Evadne anonyx* Sars. Trudy Inst. Biol. vnutren. vod. 12: 151–158 (in Russian).

Rivier, I. K., 1966b. On the biology of the Caspian Polyphemoidea. Vopr. morskoi Biol. Izd. Nauk. Dumka, Kiev: 112–114 (in Russian).

Rivier, I. K., 1968. On the feeding and diurnal vertical migration of the Caspian polyphemids (Podonidae). Trudy Inst. Biol. vnutren. vod. 17: 58–69 (in Russian).

Rivier, I. K., 1969. On the diurnal rhythmicity in the reproduction of Caspian polyphemids. Trudy Inst. Biol. vnutren vod. 19: 128–136 (in Russian).

Rivier, I. K. & Ph. D. Mordukhai-Boltovskoi, 1966. Contributions to the biology of Caspian Polyphemidae. Trudy Inst. Biol. vnutren. vod. 12: 159–169 (in Russian).

Sars, G. O., 1897. Pelagic Entomostraca of the Caspian Sea. Ezhegodn. Zool. muz. Akad. Nauk. 2: 1–73.

Sars, G. O., 1902. On the Polyphemidae of the Caspian Sea. Ezhegodn. Zool. muz. Akad. Nauk. 7: 31–54.

Sars, G. O., 1927. Notes on the crustacean fauna of the Caspian Sea. Sbornik v chest' prof. N. M. Knipovicha, Moscow.

Hydrobiologia **307**: 99–107, 1995.
P. Larsson & L.J. Weider (eds), Cladocera as Model Organisms in Biology.

Distribution and biology of marine cladocerans in the coastal waters of southern China

K. W. Tang[1], Q. C. Chen[2] & C. K. Wong[1]*
[1]*Department of Biology, The Chinese University of Hong Kong, Shatin, N. T., Hong Kong*
[2]*South China Sea Institute of Oceanology, Academia Sinica, Guangzhou, The People's Republic of China*
(Author for correspondence)*

Key words: marine cladocerans, distribution, South China Sea

Abstract

Geographical and seasonal distributions of marine cladocerans in the coastal waters of southern China were studied. *Penilia avirostris* was the most common species, followed by *Evadne tergestina* and *Podon schmackeri. P. avirostris* and *E. tergestina* were most common during summer. *P. schmackeri*, found only in a small bay northeast of Hong Kong, showed no clear seasonal pattern of occurrence. *P. avirostris* and *E. tergestina* were found at temperatures ranging from 16–32 °C and salinity ranging from 7.3–37.2 ‰. *P. schmackeri* was restricted to a temperature range of 17–29 °C and a salinity range of 31.0–37.2 ‰. No significant relationships between marine cladoceran abundance and chlorophyll *a* concentration were found in samples taken from Tolo Harbour. Parthenogenetic brood size of *P. avirostris* and *E. tergestina* ranged from 1 to 14, while *P. schmackeri* was found to carry up to 19 embryos per brood. No geographical trend in fecundity patterns was observed. No correlation was found between body length and brood size. The occurrence of females with resting eggs was rare.

Introduction

Cladocerans are small crustaceans which live almost exclusively in freshwater. Onbé (1977) noted that only eight cladoceran species from the families Sididae and Podonidae are truly marine. Because of their low diversity and small population size, the biology and ecology of marine cladocerans have received relatively little attention. However, recent studies have shown that marine cladocerans may sometimes form an important component of the zooplankton in productive coastal and estuarine waters (Bosch & Taylor, 1968; Gieskes, 1971; Della Croce & Venugopal, 1972; Onbé, 1985; Yoo & Kim, 1987). Cheng & Chao (1982) reported five species of marine cladocerans in the coastal waters of eastern China and suggested that marine cladocerans may be important food for carnivorous zooplankton and pelagic fish and fish larvae. In this paper, we first report on the geographical and seasonal distribution of marine cladocerans in the coastal waters of southern China; then we discuss geographical variations in body size and fecundity of marine cladocerans in our study region.

Materials and methods

Marine cladocerans were collected from the Zhujiang River estuary, Tolo Harbour, Mir Bay, Da Peng Ao and the coastal waters of Shan Wei (Fig. 1) between 1980 and 1991. Water depth in the study areas ranged from 5–30 m. Conical nets of 0.125 mm mesh size and 0.5–1.0 m mouth diameter were used. The nets were hauled vertically by hand from near the bottom to the surface. All samples were preserved in 4% formaldehyde. During the study in Tolo Harbour, water samples were collected from 1 m after each net haul for determination of chlorophyll *a* concentrations with a Turner-112 fluorometer.

In the laboratory, marine cladocerans were counted by species under a dissecting microscope and the number of animals in each net sample was converted into number per cubic meter of water. Immediately after

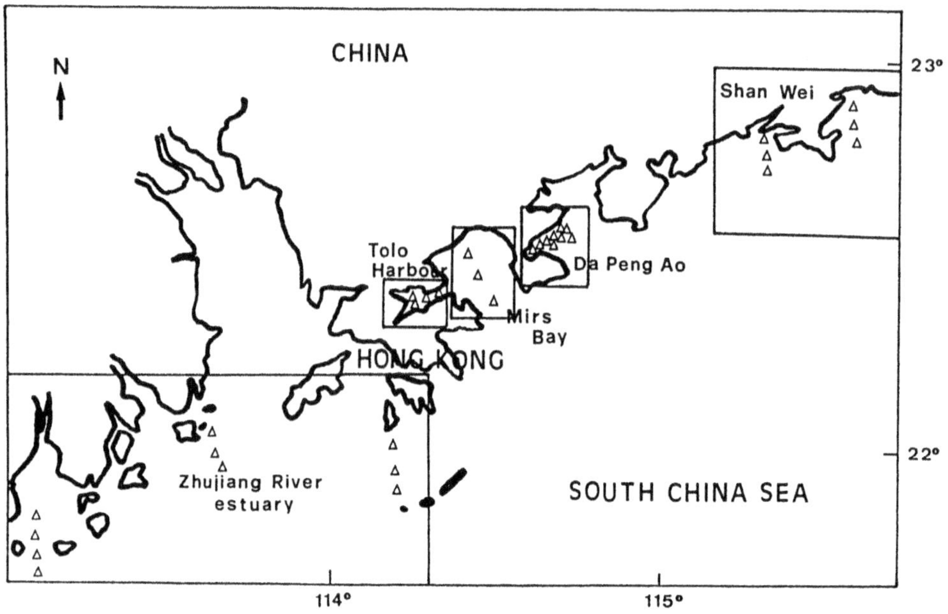

Fig. 1. Map of southern China showing study areas and sampling sites (triangles).

counting, animals were sorted randomly from each net sample for measurement of body length and brood size. Body length of parthenogenetic females was measured from the tip of the head to the dorso-posterior edge of the carapace before the shell spine. Each animal was carefully dissected to determine the number of eggs or embryos in the brood chamber.

Results

Climate in the study region is subtropical. Temperature is highest during the summer months from May to September (Fig. 2). Maximum surface temperature of 31.6 °C was recorded during our study in Da Peng Ao in July 1991. The rain bearing S.E. monsoon increases river flow during summer and exerts a strong influence on the salinity of coastal waters (Fig. 2). Discharge from the Zhujiang River creates an estuarine environment with an area of more than 10 000 km^2. The influence of the Zhujiang River diminishes towards the east of Hong Kong. Coastal waters in Mirs Bay, Da Peng Ao and Shan Wei are predominantly influenced by oceanic currents. Tolo Harbour opens into Mirs Bay, which is fully exposed to currents of the South China Sea from the south. Salinity in Tolo Harbour was close to natural seawater because rivers and streams in the catchment have been diverted to a reservoir and freshwater input was negligible. The N.E. monsoon blows during the winter months from October to March, resulting in a cold and dry climate. Surface water in coastal areas is characterised by low temperatures (16–20 °C) and high salinities (32-34 ‰).

Three species of marine cladocerans were found in the study area. *Penilia avirostris* was the most common species, followed by *Evadne tergestina* and *Podon schmackeri* (Fig. 3). The abundance of *P. avirostris* was highest in Tolo Harbour. Densities >2000 ind. m^{-3} were found in May 1990, November 1990 and March 1991. To the east of Hong Kong, from Mirs Bay and Da Peng Ao to Shan Wei, densities were lower and seldom exceeded 1000 ind. m^{-3}. *P. avirostris* was rare in the estuarine waters of the Zhujiang River estuary, attaining a maximum density of only about 35 ind. m^{-3} in the summer. *P. avirostris* did not show marked seasonality in occurrence. Abundance was lowest in winter and spring and highest in summer. Appearance in Tolo Harbour was sporadic. When data from all sampling stations were pooled (Fig. 4), it can be seen that *P. avirostris* occurred across a wide range of temperatures (16–32 °C) and salinities (7.3–37.2 ‰). Maximum densities were found at a temperature range of 24–28 °C and salinity range of 30–35 ‰. Distribution of *P. avirostris* extended to lower salinities at temperatures higher than 29 °C.

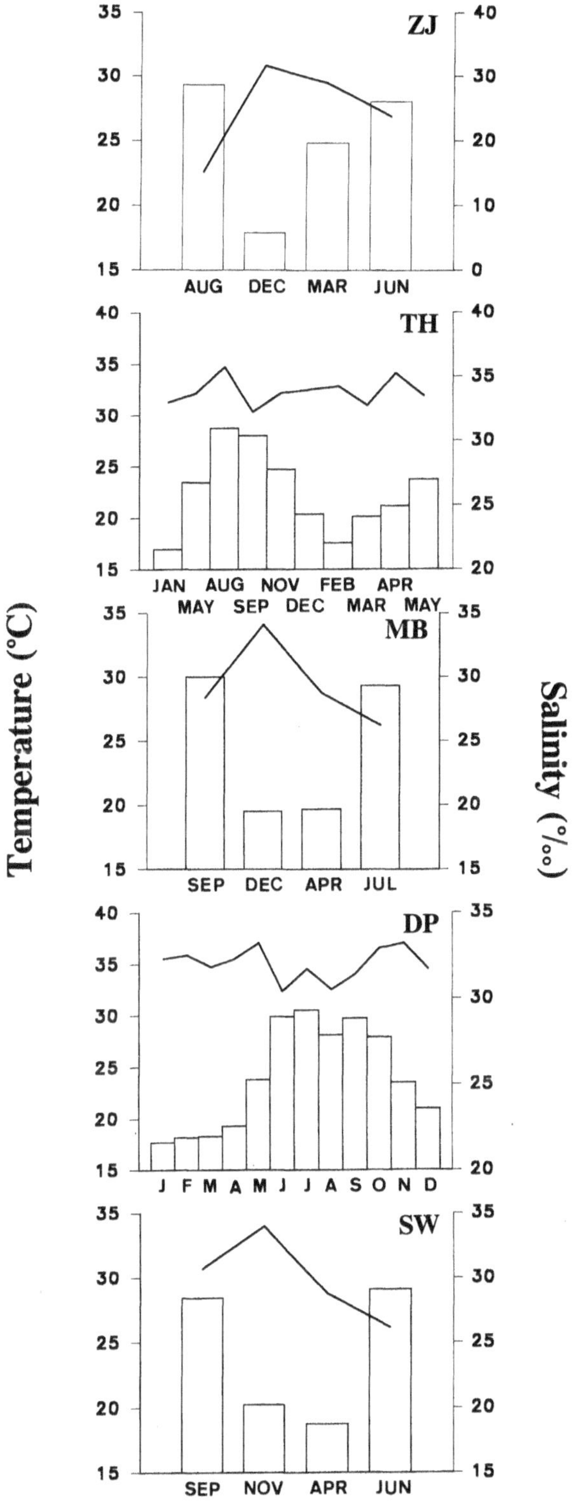

Fig. 2. Seasonal changes in surface temperature (bars) and salinity (lines) in the Zhujiang River estuary (ZJ), Tolo Harbour (TH), Mirs Bay (MB), Da Peng Ao (DP) and Shan Wei (SW).

Evadne tergestina occurred throughout the study area but dense populations were found only in Tolo Harbour (Fig. 3). Density in the Zhujiang River estuary, Da Peng Ao and Shan Wei was always <10 ind. m^{-3}. *E. tergestina* did not show a clear seasonal pattern of occurrence in Tolo Harbour. A maximum density of 1100 ind. m^{-3} was recorded in May 1990. Between November 1990 and April 1991, *E. tergestina* occurred in small numbers in the plankton. A second peak was observed in May 1991. In the Zhujiang River estuary, Mirs Bay, Da Peng Ao and Shan Wei, *E. tergestina* appeared during summer and tended to be rare or absent during winter and spring. Temperature and salinity ranges for *E. tergestina* were quite similar to those of *P. avirostris* (Fig. 4). The highest densities were recorded at around 24 °C and 30 ‰.

Podon schmackeri occurred in Tolo Harbour during the spring and early summer of 1991 (Fig. 3) at a temperature range of 17–29 °C and salinity range of 31.0–37.2 ‰. The species was first observed in March at a density of 1300 ind. m^{-3}. The population increased rapidly to >5000 ind. m^{-3} in April. By May, however, the population had declined to <300 ind. m^{-3}. *P. schmackeri* was not found in the other areas.

Since no clear seasonal patterns of marine cladoceran abundance were found, regression analysis was performed to determine if the observed abundance patterns in Tolo Harbour could be explained by changes in food levels. No significant relationships were found between cladoceran abundance and chlorophyll *a* concentration (Fig. 5). Cladoceran density was usually low at sampling sites where chlorophyll *a* concentrations were high. The low values of the coefficient of determination for the regressions indicated that variations in cladoceran abundance were not explained by changes in chlorophyll *a* concentration.

The body length of *P. avirostris* ranged from 0.36–1.04 mm (Fig. 6). Individuals smaller than 0.6 mm constituted <1% of the population in the Da Peng Ao. In contrast, >10% of the population were smaller than 0.6 mm in Tolo Harbour and the Zhujiang River Estuary. The brood size of parthenogenetic females ranged from 1–14 with a mean of 4.4 (Fig. 6). A modal brood size of four was observed at all five study areas. No individuals were found to carry resting eggs. No positive correlation was found between brood size and body length.

The body size and fecundity of *E. tergestina* were studied in Tolo Harbour, Mirs Bay and Shan Wei (Fig. 6). Number of animals from the Zhujiang River estuary and Da Peng Ao were too small for quantitative

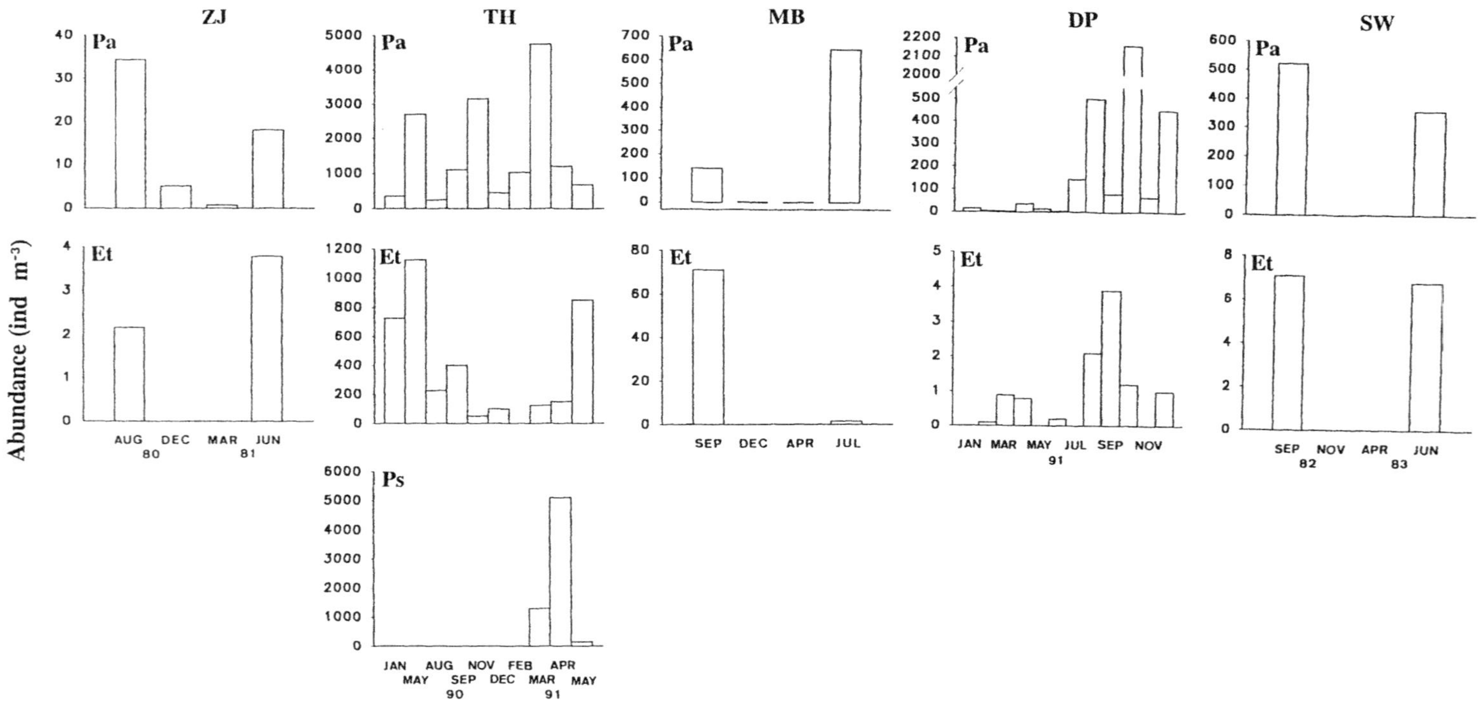

Fig. 3 Seasonal changes in abundance of *Penilia avirostris* (Pa), *Evadne tergestina* (Et) and *Podon schmackeri* (Ps) in the Zhujiang River estuary (ZJ), Tolo Harbour (TH), Mirs Bay (MB), Da Peng Ao (DP) and Shan Wei (SW).

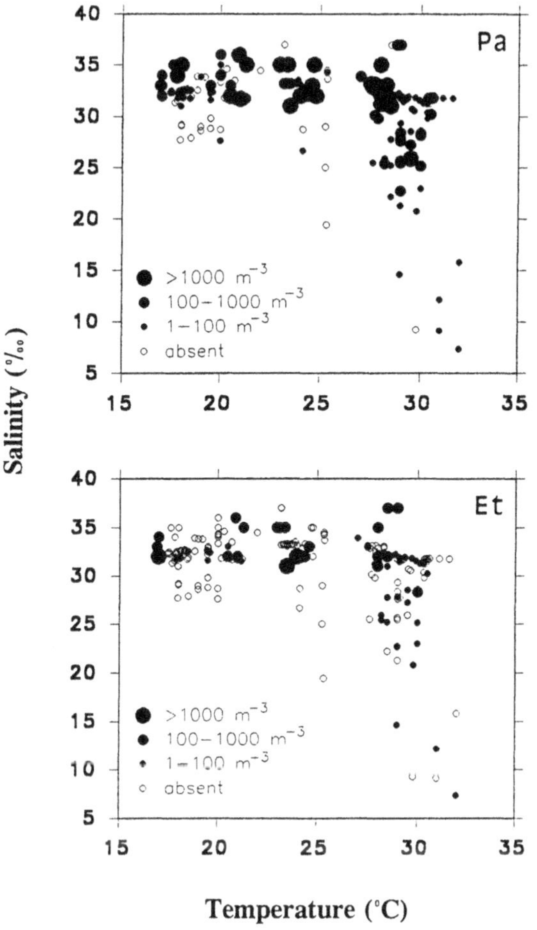

Fig. 4. Range of temperature and salinity for the occurrence of *Penilia avirostris* (Pa) and *Evadne tergestina* (Et) in the coastal waters of southern China.

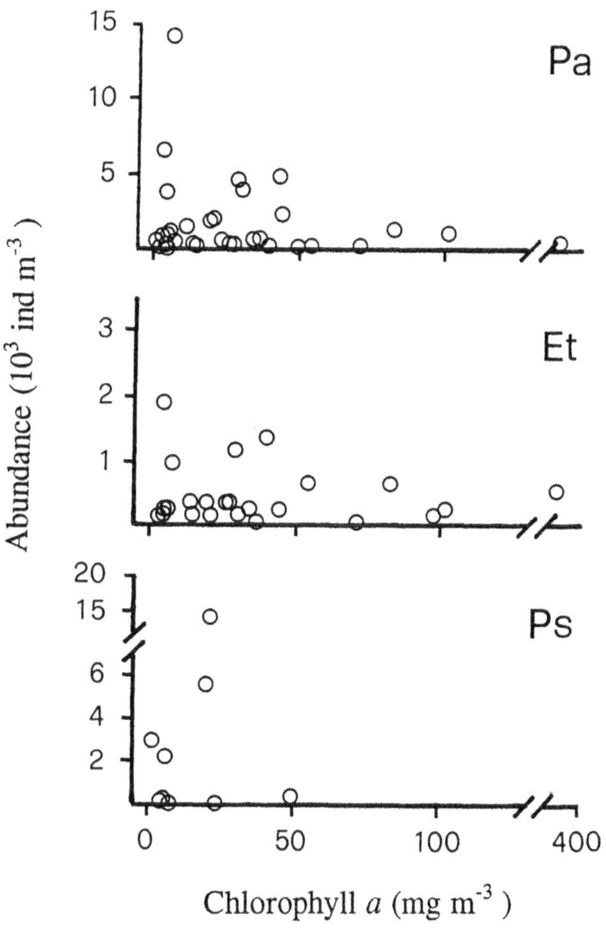

Fig. 5. Relationship between chlorophyll *a* concentration and abundance of *Penilia avirostris* (Pa), *Evadne tergestina* (Et) and *Podon schmackeri* (Ps) in Tolo Harbour.

analysis. The body length of *E. tergestina* ranged from 0.24–0.56 mm. Animals from Tolo Harbour appeared to be smaller than animals from Mirs Bay and Shan Wei. About 10% of the animals taken from Tolo Harbour were in the 0.2–0.3 mm size class. Only one individual was larger than 0.5 mm. In contrast, animals taken from Mirs Bay and Shan Wei were all larger than 0.3 mm. Thirty percent of the animals from Mirs Bay were larger than 0.5 mm. Parthenogenetic brood size of *E. tergestina* ranged from 1–14 with a mean of 4.6. Most females carried 4–6 embryos. Females with resting eggs were found in Shan Wei in August 1982 and in Tolo Harbour in September 1990. No positive correlation was found between brood size and body length.

The body length of *P. schmackeri* in Tolo Harbour ranged from 0.18–0.45 mm (Fig. 6). Ninety-seven percent of the animals were between 0.2 and 0.3 mm. The number of embryos per female ranged from 2–19 with a mean of 4.8 and a mode of 3. Most females carried 3–4 eggs or embryos in the brood chamber. Resting eggs were found in about 12% of the females. No positive correlation was found between brood size and body length.

Discussion

The distribution, size composition and fecundity of the marine cladocerans *Penilia avirostris, Evadne tergestina* and *Podon schmackeri* were studied in the subtropical coastal waters of southern China. *Penilia avirostris* and *E. tergestina* occurred throughout the study area in a temperature range of 16–32 °C. In waters of the northwestern Pacific, the occurrence of these species is mostly restricted to the summer (Onbé, 1977; Yoo & Kim, 1987). In tropical and subtropical waters, however, marine cladocerans may occur in the zooplankton throughout the year (Pillai & Pillai, 1975). Yoo & Kim (1987) reported that *P. avirostris* and *E. tergestina* remained in the plankton community in Cilinhae Bay, Korea, at temperatures as low as 10 °C. Results of this study confirm that the occurrence of *P. avirostris* and *E. tergestina* in subtropical waters is not restricted to the warmer seasons.

Penilia avirostris and *E. tergestina* occurred in a salinity range of 30–35 ‰, but their lower limit was extended to about 7 ‰ by the presence of small populations in the mouth of the Zhujiang River. Salinity does not appear to have a major effect on the fecundity of *P. avirostris* because individuals from the Zhujiang River estuary carried about the same number of embryos as individuals from the other study areas. *E. tergestina* occurred in the same range of temperature and salinity as *P. avirostris*, but dense populations of *E. tergestina* were encountered only in Tolo Harbour. Mullin & Onbé (1992) reported that the abundances of *P. avirostris* and *E. tergestina* in the Inland Sea of Japan were positively correlated over time. *P. avirostris* was more abundant than *E. tergestina* in most of our samples. The same was true in samples from the coastal waters of Korea (Yoo & Kim, 1987), Chesapeake Bay (Bosch & Taylor, 1967), the Inland Sea of Japan (Onbé, 1977, 1985) and the coastal waters of the Gulf of Mexico (Mullin & Onbé, 1992).

Podon schmackeri was found only in Tolo Harbour during the spring and early summer of 1991. *P. schmackeri* was not found in a survey of the marine zooplankton in the nearshore waters of Hong Kong by Chen (1980), but the occurrence of this species has previously been recorded in Da Peng Ao (Chen *et al.*, 1989) and in the surface waters off Hong Kong (Kim & Onbé, 1989a; Angelino & Della Croce, 1992). Distribution of *P. schmackeri* in the northwestern Pacific was mainly in waters off the Korean Peninsula and southern Japan (Kim & Onbé, 1989a). Its occurrence in the coastal waters of southern China and the South China Sea was sporadic. Kim & Onbé (1989a) believed that *P. schmackeri* was common in waters relatively distant from the coast and rare in nearshore waters. Its sporadic occurrence suggests that animals in our samples were transported into Tolo Harbour by ocean currents.

Maximum densities for all three species were found in Tolo Harbour, where marine cladocerans occasionally formed a significant part of the zooplankton community. The semi-enclosed topography of Tolo Harbour might limit water movements and help to maintain the population of marine cladocerans in the harbour. No clear relationship was found between marine cladocerans and chlorophyll *a* in samples taken from Tolo Harbour. Estimates of the grazing impact of *P. avirostris* on phytoplankton in Tolo Harbour averaged about 1% (Wong *et al.*, 1992), suggesting that the abundance of marine cladocerans in the highly eutrophic waters of our study area was not limited by food availability.

Penilia avirostris was the largest in body length, followed by *E. tergestina* and *P. schmackeri*. The body size of *P. avirostris* in our samples ranged from 0.36–1.04 mm and varied among study areas. Geographical variations were also observed in the size composition of *E. tergestina* which ranged from 0.24–0.56 mm. Geographical variation in body size has been reported for various species of marine cladocerans (Yoo & Kim, 1987; Kim & Onbé, 1989a). Yoo & Kim (1987) reported body size ranges of 0.32–1.12 mm for *P. avirostris* and 0.20–0.57 mm for *E. tergestina* in the northwestern Pacific and suggested that the body size of some marine cladocerans showed a tendency to increase at lower temperatures. McLaren (1963) proposed that crustacean zooplankton such as copepods in northern localities develop more slowly and reach a larger size than their counterparts in warmer waters. The size composition of *P. avirostris* and *E. tergestina* in our study region varied with locality, but animals in our samples were generally not smaller than their counterparts in the colder waters of the northwestern Pacific. Body size of *P. schmackeri* from waters of the northwestern Pacific ranged from 0.34–0.65 mm (Kim & Onbé, 1989a, 1989b). While animals from Tolo Harbour where size range was 0.18–0.45 mm appeared to be noticeably smaller, the largest individual from Tolo Harbour was as large as the largest individual collected by Onbé (1983) from the much colder waters off the south-eastern coast of Shikiku, Japan. These observations suggest that the body size of marine cladocerans may vary locally as well as on a broad latitudinal scale. Latitudinal differences in water temperature may exert an influence on adult size by affecting growth rates. Existence of substantial genetic variation for body size has been documented within small populations of *Daphnia* (Tessier *et al.*, 1992). There is a large literature on phenotypic plasticity among freshwater cladocerans (eg. Havel, 1987). While starvation has been known to cause a smaller size at maturity in some species (Lynch, 1989), the role of predators has received the most attention. Many species of freshwa-

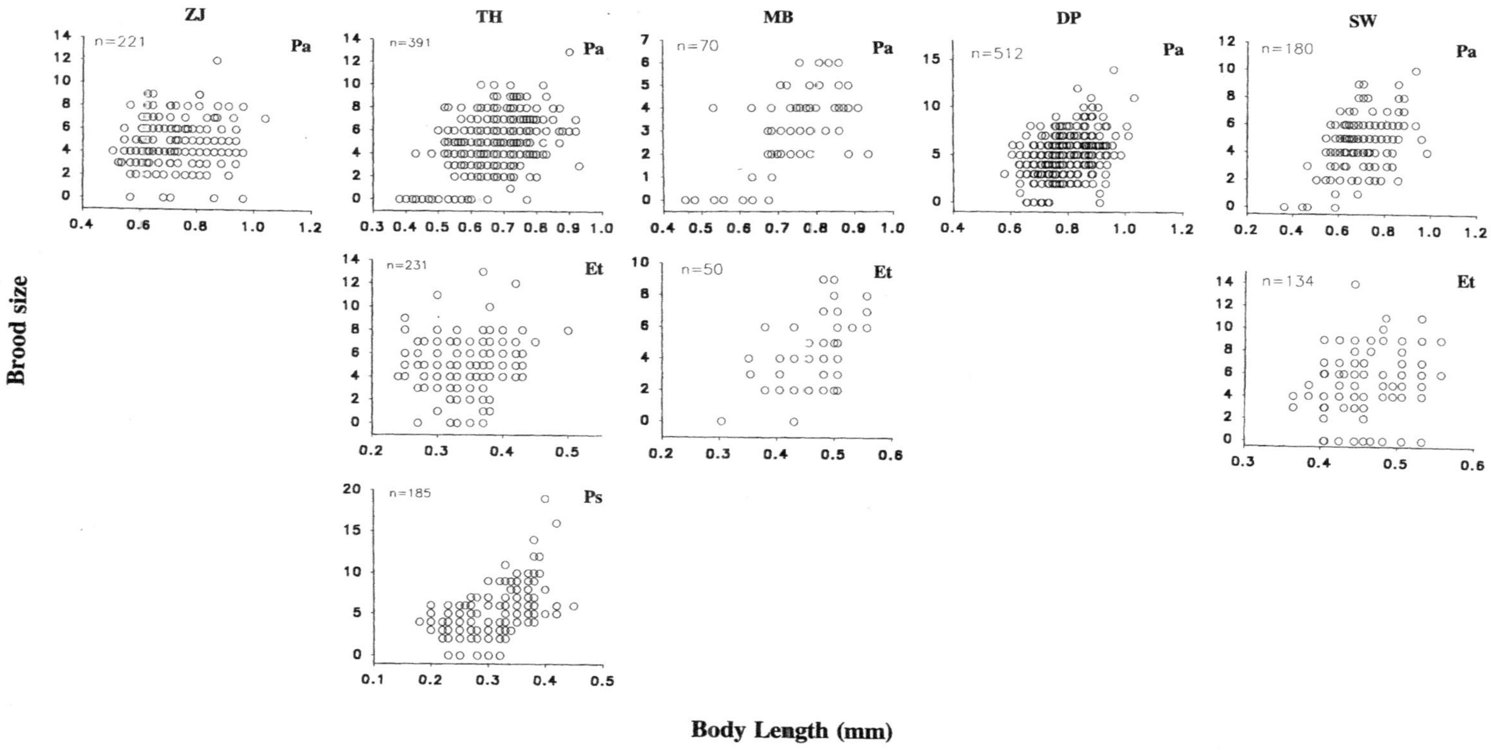

Fig. 6. Relationship between body length and brood size for *Penilia avirostris* (Pa), *Evadne tergestina* (Et) and *Podon schmackeri* (Ps) in the Zhujiang River estuary (ZJ), Tolo Harbour (TH), Mirs Bay (MB), Da Peng Ao (DP) and Shan Wei (Sw).

ter cladocerans change annually their size as defense against predators (Kerfoot, 1974; Dodson & Havel, 1988). The change is triggered by chemical signals produced by the predators and is influenced by food supply to the prey (Dodson & Havel, 1988).

Our data show no geographical variations in fecundity among cladocerans. This is in contrast to the result of Kim & Onbé (1987a), which showed marked geographical variations in fecundity among *P. schmackeri* populations in the northwestern Pacific. In general, comparison on a broad geographical scale reveals no significant trends in variations in fecundity. For *P. schmackeri*, the brood size in our samples was about equal to those found in the northern waters (Kim & Onbé, 1989a) where the animals were larger. For *P. avirostris*, the modal brood size in our samples, four, was equal to that found by Murakami & Onbé (1967) in the coastal waters of Japan. In the case of *E. tergestina*, the mean number of embryos per female in our samples was lower than those for animals in Chesapeake Bay (Bryan, 1979) but the range was about equal to those in Chesapeake Bay as well as those in Japanese waters (Bryan, 1979; Murakami & Onbé, 1967). There is general agreement among workers who study freshwater cladocerans that food level is an important determinant of fecundity (Lampert, 1978; Vanni, 1987; Lynch, 1989; Guisande & Gliwicz, 1992). The coastal waters of Hong Kong and southern China are characterized by high chlorophyll concentrations throughout the year (Chan & Hodgkiss, 1987; Huang *et al.*, 1989). Chlorophyll *a* levels at Tolo Harbour and Da Peng Ao were higher than those obtained from some bays in Japan (Huang *et al.*, 1989). In addition to resource levels, temperature has been discussed as a cause of variations in brood production in *Daphnia* (Hall, 1964). It is therefore reasonable to suggest that high temperature and favourable food conditions are the major causes of high fecundity in marine cladocerans in the coastal waters of southern China.

The population size of *P. avirostris*, the most dominant cladoceran in our samples, were substantially smaller than those reported by Yoo & Kim (1987) from the temperate waters of the northwestern Pacific and by Grahame (1976) from the tropical waters of Kingston Harbour, Jamaica. The scarcity of females with resting eggs suggests that environmental conditions in the warm subtropical waters of our study area are favourable to *P. avirostris* and the populations can propagate mainly by parthenogenesis. Throughout our study area, the density of small planktivorous fish is higher in inner bays and shallow near-shore waters than in waters relatively distant from the coast (C. K. Wong, per.obser.). Although we have no direct evidence, we consider the low numbers of marine cladocerans in our samples to be the result of intense predation by planktivorous fish.

Acknowledgments

We thank Mr K. C. Ng and Miss W. T. Foo for their assistance in the field and laboratory.

References

Angelino, M. I. & N. Della-Croce, 1992. Marine Cladocera in Hong Kong waters. Bol. Soc. Biol. Concepcion 63: 25–26.

Bosch, H. F. & W. R. Taylor, 1968. Marine cladocerans in the Chesapeake Bay estuary. Crustaceana 15: 161–164.

Bryan, B. B., 1979. The diurnal reproductive cycle of *Evadne tergestina* Claus (Cladocera, Podonidae) in Chesapeake Bay, USA. Crustaceana 36: 229–236.

Chan, B. S. S. & I. J. Hodgkiss, 1987. Phytoplankton productivity in Tolo Harbour. Asian mar. Biol. 4: 79–90.

Chen, Q. C., 1980. The marine zooplankton of Hong Kong. In: Morton, B. & C. K. Tseng, (ed.), The marine flora and fauna of Hong Kong and Southern China. Hong Kong University Press, Hong Kong: 789–799.

Chen, X. M., J. Q. Yin, G. X. Zhang, D. S. Cai, X. Y. Chen & Q. C. Chen, 1989. Zooplankton. In Xu, G. Z., (ed.), Environments and Resources of Daya Bay. An Huai Science Press, China: 175–192 (in Chinese).

Cheng, C. & W. C. Chao, 1982. Studies on the marine Cladocera of China II. Distribution. Acta Oceanol. Sinica 4: 731–742 (in Chinese).

Della Croce, N. & P. Venugopal, 1972. Distribution of marine cladocerans in the Indian Ocean. Mar. Biol. 15: 132–138.

Dodson, S. I. & J. E. Havel, 1988. Indirect prey effects: some morphological and life history responses of *Daphnia pulex* exposed to *Notonecta undulata*. Limnol. Oceanogr. 33: 1274–1285.

Gieskes, W. W. C., 1971. Ecology of the Cladocera of the north Atlantic and the North Sea. Neth. J. Sea Res. 5: 342–376.

Grahame, J., 1976. Zooplankton of a tropical harbour: the numbers, composition and response to physical factors of zooplankton in Kingston Harbour, Jamaica. J. exp. mar. Biol. Ecol. 25: 219–237.

Guisande, C. & Z. M. Gliwicz, 1992. Egg size and clutch size in two *Daphnia* species grown at different food levels. J. Plankton Res. 14: 997–1007.

Hall, D. J., 1964. An experimental approach to the dynamics of a natural population of *Daphnia galeata mendotae*. Ecology 45: 94–112.

Havel, J. E., 1987. Predator-induced defenses: a review. In Kerfoot, W. C. & A. Sih, (ed.), Predation: direct and indirect impacts on aquatic communities. University Press of New England, New Hampshire, USA: 263–278.

Huang, L. M., Q. C. Chen & W. B. Yuan, 1989. Characteristics of chlorophyll distribution and estimation of primary productivity in Daya Bay. Asian mar. Biol. 6: 115–128.

Kerfoot, W. C., 1974. Egg-size cycle of a cladoceran. Ecology 55: 1259–1270.

Kim, S. W. & T. Onbé, 1989a. Distribution and zoogeography of the marine cladoceran *Podon schmackeri* in the northwestern Pacific. Mar. Biol. 102: 203–210.

Kim, S. W. & T. Onbé, 1989b. Observations on the biology of the marine cladoceran *Podon schmackeri*. J. Crust. Biol. 9: 54–59.

Lampert, W., 1978. A field study of the dependence of the fecundity of *Daphnia* spec. on food concentration. Oecologia 36: 363–369.

Lynch, M., 1989. The life history consequences of resource depression in *Daphnia pulex*. Ecology 70: 246–256.

McLaren, I. A., 1963. Generation lengths of some temperate marine copepods: estimation, prediction, and implications. J. Fish. Res. Bd Can. 35: 1330–1342.

Mullin, M. M. & T. Onbé, 1992. Diel reproduction and vertical distributions of the marine cladocerans, *Evadne tergestina* and *Penilia avirostris*, in contrasting coastal environments. J. Plankton Res. 14: 41–59.

Murakami, Y. & T. Onbé, 1967. Notes on the marine cladocerans observed at a definite station near Tomo Marine Biological Station. Inform. Bull. Planktol. Japan, Comm. No. Dr. Matsue: 123–127.

Onbé, T., 1977. The biology of marine cladoceran in a warm temperate water. Symp. Warm Water Zoopl., Spec. Pub. UNESCO/NIO (Goa): 383–398.

Onbé, T., 1983. Preliminary observations on the biology of marine cladoceran *Pleopis ('Podon') schmackeri* (Poppe). J. Fac. Appl. Biol. Sci., Hiroshima Univ. 22: 55–64.

Onbé, T., 1985. Seasonal fluctuations in the abundance of populations of marine cladocerans and their resting eggs in the Inland Sea of Japan. Mar. Biol. 87: 83–88.

Pillai, P. P. & M. A. Pillai, 1975. Ecology of the cladocerans of the plankton community in the Cochin blackwater. Bull. Dept. Mar. Sci. Uni. Cochin 7: 127–136.

Tessier, A. J., A. Young & M. Leibold, 1992. Population dynamics and body-size selection in *Daphnia*. Limnol. Oceanogr. 37: 1–13.

Vanni, M. J., 1987. Effects of food availability and fish predation on a zooplankton community. Ecol. Monogr. 57: 61–88.

Wong, C. K., A. L. C. Chan & K. W. Tang, 1992. Natural ingestion rates and grazing impact of the marine cladoceran *Penilia avirostris* Dana in Tolo Harbour, Hong Kong. J. Plankton Res. 14: 1757–1765.

Yoo, K. I. & S. W. Kim, 1987. Seasonal distribution of marine cladocerans in Chinhae Bay, Korea. J. Oceanol. Soc. Korea 22: 80–86.

Hydrobiologia **307**: 109–115, 1995.
P. Larsson & L.J. Weider (eds), Cladocera as Model Organisms in Biology.

Synergistic effects of different food species on life-history traits of *Daphnia galeata*

Maarten Boersma[1] & Jacobus Vijverberg
Netherlands Institute of Ecology, Centre for Limnology, Rijksstraatweg 6, 3631 AC Nieuwersluis, The Netherlands
[1] *Present address: Max. Planck. Institut für Limnologie, Postfach 165, D-24302 Plön, Germany*
e-mail: boersma@mpil-ploen.mpg.d400.de

Key words: Daphnia, food quality, life history, standard carbon content

Abstract

In this paper we describe the life history consequences of feeding *Daphnia galeata* with different food types in different concentrations. We fed the animals with four concentrations of two green algae *Scenedesmus obliquus* and *Chlamydomonas globosa*, given separately as well as in a 1:1 mixture. Growth, reproduction and condition of the animals and consequently the intrinsic rate of population increase, *r*, increased with the concentration of the food for all three experimental series. The life history parameters measured were lowest in the animals raised on *C. globosa*. Animals fed on *S. obliquus* showed higher values for *r*. We observed a distinct synergistic effect of the two algal species: growth, reproduction, condition and *r* were significantly higher for the animals raised on the mixed media as compared to the animals raised on the mono-algal media. We conclude that the two algal species represented complementary food resources.

Introduction

In aquatic communities, filter feeders are among the most generalized herbivores, often ingesting all particles within a given size range. This implies that these animals are faced with a large spectrum of different quality food resources. Since the ability to select for given food types is usually limited, especially in species of the genus *Daphnia* (e.g. Kerfoot & Kirk, 1991), differences in food quality are likely to be of paramount importance for filter feeders.

In ecological studies of *Daphnia*, the effect of food quantity on life history characteristics has been addressed by many workers (e.g. Lampert, 1977; Duncan *et al.*, 1985). Most of these experiments have been carried out with high quality food items, such as Chlorophyceae and Cryptophyceae. Influences of food quality have drawn less attention, although several authors have emphasized the role of differences in quality between algae (e.g. Ahlgren *et al.*, 1990; Lundstedt & Brett, 1991). Most of the investigators compared the quality of Chlorophyceae and/or Cryptophyceae, groups which are usually considered as high quality food, with Cyanobacteria as a low quality resource. Possible toxicity of some Cyanobacteria has also been mentioned as a reason for the lower quality of Cyanobacteria (DeMott *et al.*, 1991). Recently, quality differences within food species as a result of the culturing conditions of the algae have been reported (Müller-Navarra, 1993; Sterner *et al.*, 1993; van Donk & Hessen, 1993).

The effect of mixtures of different food types on life histories of zooplankton has received considerably less attention, although approximately half of the studies reviewed by Vijverberg (1989) on growth and reproduction in zooplankton were carried out with a combination of food types. The differences in the nutritional value of two or more species of algae given separately, versus their value in combination, are not very well studied. Most authors, who have studied the nutritive value of different food species in mixed diets, considered mixtures of palatable and non-edible algae, either to show differences in uptake (Hartmann & Kunkel, 1991) or to investigate inhibitory effects of the 'inedible' species, usually blue-greens, on the feeding and growth of the zooplankton (Gliwicz & Lampert, 1990).

We are aware of only a few published works which deal with synergistic effects of different food types fed to herbivorous zooplankton in combination (Jamieson, 1980; Lewis & Maki, 1981; Hanazato & Yasuno, 1984; Roman, 1984; Cowgill *et al.*, 1985; Lundstedt & Brett, 1991; Vanni & Lampert, 1992).

In this paper we present the results of a study on the life history consequences for *Daphnia galeata*, fed with different concentrations of two green algae, *Scenedesmus obliquus*, and *Chlamydomonas globosa*. These were given separately as well as in a 1:1 mixture. Both algal species have been used extensively as food items for *Daphnia* species, and are considered to be high quality food (e.g. Vijverberg, 1989).

Material and methods

Algae

Chlamydomonas globosa and *Scenedesmus obliquus* were cultured axenically in a two-litre flow-through system on a medium as described elsewhere (Boersma & Vijverberg, 1994a). The algae were harvested daily from the overflow bottle of the continuous culture. The culture medium of the algae was removed by centrifuging twice for 20 min at 3000 rpm, and by subsequent washing of the algae with distilled water. The algae were resuspended in 0.45 μm filtered lake water from the Tjeukemeer. The algal density was measured using a haemacytometer, counting a minimum of 500 cells. The carbon content of the algae was measured using a UNICARB carbon analyser (Salonen, 1979). Algal dry weight was established after drying a known volume of a known concentration at 70 °C, and weighing the sample on a microbalance afterwards. Individual cells of *C. globosa* contained 2.52×10^{-2} ng C. The carbon content of *S. obliquus* cells was 2.30×10^{-2} ng C cell^{-1}, resulting in carbon to dry weight ratios of 0.53 and 0.50, respectively. *S. obliquus* was usually unicellular. Both algal species had a maximum length of around 15 μm.

Animals

D. galeata individuals were collected with a 120 μm tow net from the Tjeukemeer, a shallow eutrophic lake in the northern part of the Netherlands. These 'field' animals were placed individually in 100 ml test tubes and fed on a 1:1 mixture of *Chlamydomonas globosa* and *Scenedesmus obliquus*. The algal carbon content of this culture medium was 1 mg C l^{-1}. The incubation temperature used was 17.5 °C, and the light-dark period was 16:8 hours. As soon as the 'field' animals produced newborns (first generation), the 'field' animals were removed. The first generation animals were reared to maturity on the same medium as their mothers. The offspring produced by these animals (second generation) were used for the experiments described below.

Experimental design

Different amounts of *Chlamydomonas globosa* and *Scenedesmus obliquus* were added to 0.45 μm filtered Tjeukemeer water to obtain four food levels, with carbon contents of 0.13, 0.25, 0.50 and 2.5 mg C l^{-1}, respectively. Three types of media were prepared: 1) a series of pure *C. globosa* media (CHLAM), 2) a series of *S. obliquus* media (SCENE), and 3) a series of *C. globosa* and *S. obliquus* in a 1:1 mixture (CH_SC). All experiments were carried out at the same time.

The temperature and photoperiod used in the experiments were the same as during the acclimatization period. For each series, 40 second generation neonates were collected within 12 hours of birth, and placed individually in 100 ml test tubes. The animals were selected in such a way that the clonal composition of all series was the same. The animals were examined daily for moulting by inspection of the medium for exuviae. Then they were transferred to clean tubes with fresh medium. To avoid possible increased mortality due to handling we measured two second generation newborn from each brood, which were then discarded. The rest of the batch was used for the experiments. The time needed to reach maturity and the number of juvenile instars of the animals were recorded. The onset of adulthood is defined as the first appearance of eggs in the brood pouch, although development of the reproductive organs starts earlier. Once the second generation animals reached maturity, the number of eggs they produced was counted and the length of the individuals was recorded for each adult instar. The animals were measured from the upper edge of the eye to the base of the tail spine. The number of offspring produced by the second generation individuals was recorded, their lengths were measured, and subsequently they were removed from the tubes. Growth and reproduction were measured until the animals reached the fourth adult instar, that is approximately 20 days on average. The animals were then discarded, because under natural conditions, judging by the sizes found

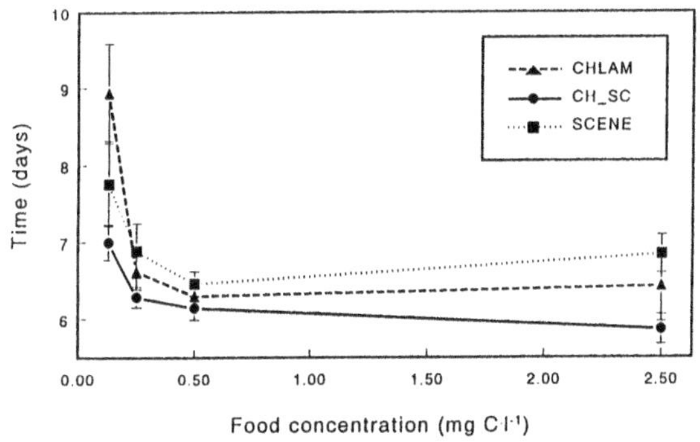

Fig. 1. Time needed for neonates to reach maturity at different food levels, and different food types. The food types used were *Chlamydomonas globosa* (CHLAM, dashed line, triangles), a 1:1 mixture of *C. globosa* and *Scenedesmus obliquus* (CH_SC, solid line, circles), and *S. obliquus* (SCENE, dotted line, squares). Error bars indicate 95% confidence limits of the means.

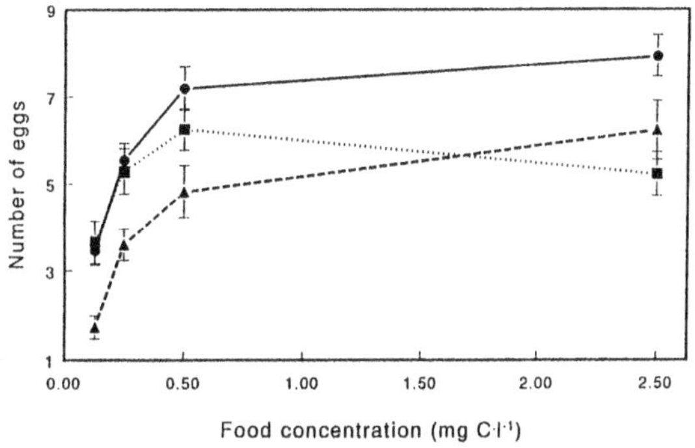

Fig. 2. Size at maturity of *D. galeata* raised at different food levels and different food types. The food types used were CHLAM (dashed line, triangles), CH_SC (solid line, circles), and SCENE (dotted line, squares). Error bars as in Fig. 1.

in the field, animals rarely develop further than the fourth adult instar. The volume of the test tubes was large enough in relation to the grazing pressure of the daphnids to guarantee only a small decrease in food concentration in 24 h. This decrease in algal concentration was never more than 20%. The intrinsic rate of increase (r) was estimated using the Euler equation:

$$1 = \sum_{x=0}^{N} e^{-rx}\, l_x\, m_x, \qquad (1)$$

where r = per capita rate of increase for the population (d^{-1}), x = age class (0,1....N) in days, l_x = probability of surviving to age x, and m_x = fecundity at age x. Since r is a population parameter it was not possible to directly compute standard errors. Hence standard errors for r were computed using a Jackknifing method (Meyer *et al.*, 1986). As only one value of r was available per treatment, it was impossible to compute the significance of the interaction between food type and food concentration. To overcome this problem, we split the series to obtain two independent values for r for each series. These values were analysed using an analysis of variance.

Carbon contents

From each of the adult instars, a number of individuals were taken at random and analysed for carbon content. Since changes in weight (carbon) can be substantial during an intermoult period (Tessier & Goulden, 1982), all animals that were analysed carried eggs which were so young that the first egg membrane

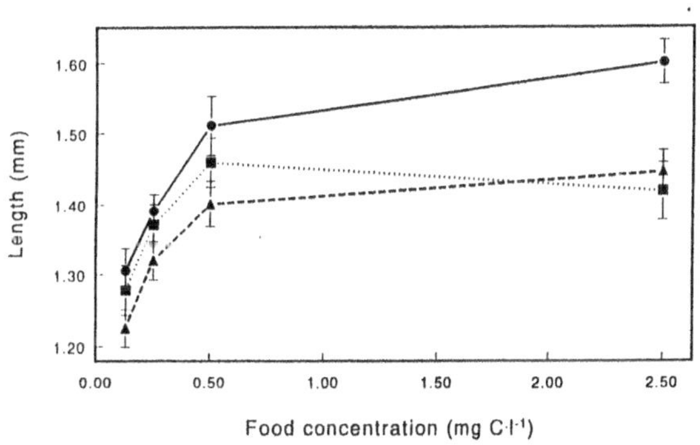

Fig. 3. Mean number of eggs per primiparous female of *D. galeata* raised at different food levels and different food types. The food types used were CHLAM (dashed line, triangles), CH_SC (solid line, circles), and SCENE (dotted line, squares). Error bars as in Fig. 1.

was not yet cast off. All animals were taken out of the cultures, washed in distilled water, and immediately analysed (alive) for carbon content (Boersma & Vijverberg, 1994b). Regression lines were computed with the log-transformed values of length (mm) as the independent variable and of carbon content (μg) of the animals as the dependent variable. Regression lines derived in laboratory cultures usually differ in intercept but not in slope (Duncan *et al.*, 1985). Hence, the exponent of the intercept of these regression lines, which represents the carbon content of a 1 mm individual, the Standard Carbon Content, SCC, (Boersma & Vijverberg, 1994c) will yield a direct measurement of the condition of the animals.

Table 1. F-values of two-way ANOVAs for selected life history characteristics of *Daphnia galeata.* The numbers of eggs per female were square-root transformed and carbon contents were log-transformed

Trait	Error	Food			Conc			F × C		
	df	df	F	*p*	df	F	*p*	df	F	*p*
Time to maturity	444	2	24.4	<0.001	3	60.8	<0.001	6	7.7	<0.001
Size at maturity	447	2	43.9	<0.001	3	112.2	<0.001	6	5.5	<0.001
Egg number	448	2	67.2	<0.001	3	127.9	<0.001	6	8.5	<0.001
r	6	2	11.1	0.009	3	17.2	–0.002	–	–	–
split-*r*	12	2	32.8	<0.001	3	50.2	0.001	6	3.1	0.045
Standard Carbon Content	206	2	32.2	<0.001	3	6.9	0.001	6	5.8	<0.001

Table 2. Probabilities of Tukey HSD tests between the different food sources for selected life history characteristics of *Daphnia galeata*

Trait	CH_SC-CHLAM	CH_SC-SCENE	SCENE-CHLAM
Time to maturity	<0.001	<0.001	0.770
Size at maturity	<0.001	<0.001	0.009
Egg number	<0.001	<0.001	<0.001
r	–	–	–
split-*r*	<0.001	0.002	<0.009
Standard Carbon Content	<0.001	<0.001	<0.001

Results

An effect of both food concentration and food type on the development, growth and reproduction of *D. galeata* was found. Higher food concentrations resulted in a shorter time to reach maturity (Fig. 1). The size at maturity was also influenced (Fig. 2). Animals fed with a mixture of *Chlamydomonas* and *Scenedesmus* had the largest size at maturity, whereas the *Chlamydomonas* fed animals were the smallest (Table 1).

The number of eggs produced by the first adult instar increased with increasing food levels, except for the highest *Scenedesmus* concentration (Fig. 3). This resulted in a significant food type, concentration and interaction effect (Table 1). The decrease in egg numbers from the 0.50 mg to the 2.50 mg *Scenedesmus* treatment was, however, not significant, as revealed by a Tukey Post-hoc comparison test. Animals fed on the mixture produced the highest number of eggs, although pair-wise comparisons showed that, with the exception of the highest food level, the differences between the mixture- and *Scenedesmus* fed animals were not significant. The differences between the animals reared on the mixture and the animals reared on

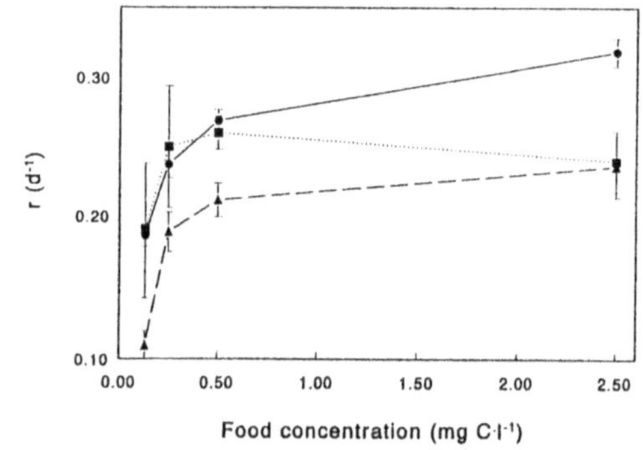

Fig. 4. Intrinsic rate of increase (*r*) of *D. galeata* raised at different food levels and different food types. The food types used were CHLAM (dashed line, triangles), CH_SC (solid line, circles), and SCENE (dotted line, squares). Error bars as in Fig. 1.

the CHLAM media were always significant. Post-hoc comparisons showed that overall differences between the three food types were significant for both the size and time at maturity and the number of eggs produced (Table 2).

Since both growth and reproductive rates were higher at higher food concentrations, the rate of population increase (*r*) increased with food concentration

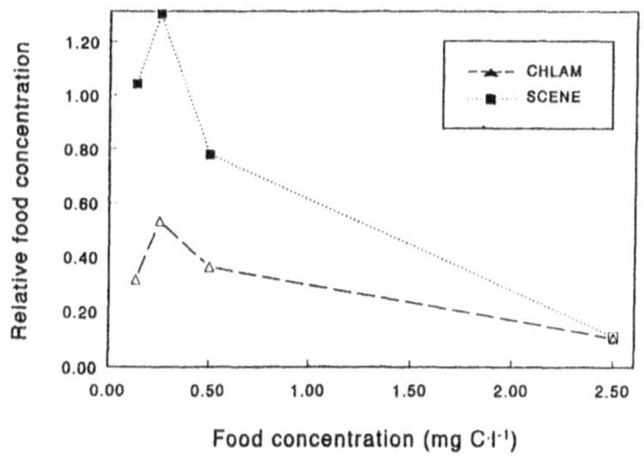

Fig. 5. Food concentration of the mono-algal cultures expressed as the relative concentration of the CH_SC series at which the animals have the same value for *r*. Open symbols indicate significant differences with the mixture, at the 5% level, closed symbols were not significantly different from the mixture.

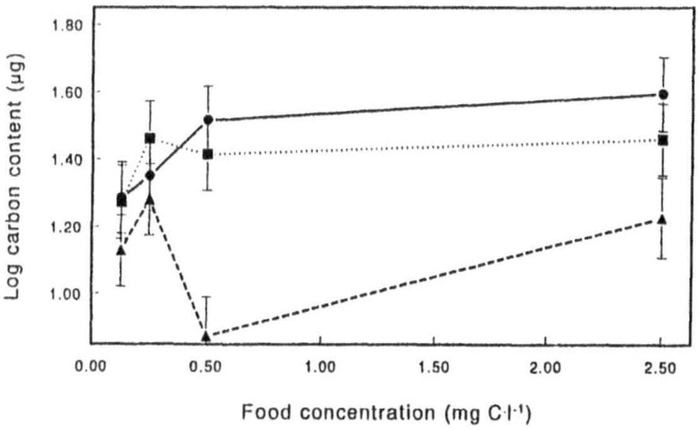

Fig. 6. Natural logarithm of the carbon content (μg) of a 1 mm animal ln(SCC) under the different feeding conditions. The food types used were CHLAM (dashed line, triangles), CH_SC (solid line, circles), and SCENE (dotted line, squares). Error bars indicate the 95% confidence intervals of the carbon contents of the grand mean.

for all three series (Fig. 4). Table 1 shows the results of two analyses of variance with *r*. The single measurements of *r* were analysed first, with the effects of the treatments tested using the interaction term as the error term. Both food type and concentration showed a significant effect on *r*. However, the analysis of split-*r* values showed the existence of a significant interaction, making the analysis of the single *r*-values less interpretable. As indicated by the data on growth and reproduction, the values for *r* were highest for the animals cultured on the mixture in most cases, although the difference with the *Scenedesmus* fed animals was only significant at the highest food concentration (Tukey post-hoc comparison; $p < 0.001$). Animals reared on the CHLAM series showed the lowest values for *r* in all cases. As was the case with the single life history traits, the overall differences between the three food types were highly significant (Table 2).

The calculation of *r* enabled us to compare the different series, and we estimated the 'nutritional value' of the mono-algal media relative to the mixture. This was done by computing the food concentration of the mixture at which the *r*-value would have been the same for each observation of *r* from the mono-algal cultures. This concentration was then expressed as a proportion of the mono-algal concentration with the same *r* value (Fig. 5). At the highest food concentration, which is well above the incipient limiting level, of the mono-algal media, we observed *r*-values comparable with *r*-values reached in mixed food at 0.25 mg Cl^{-1}, which is only 10% of the carbon content of the mono-algal media.

The differences in the slopes of the regression lines of the logarithm of the length versus the logarithm of

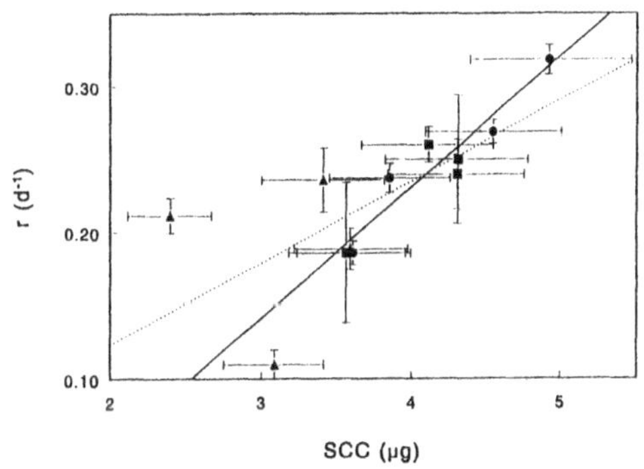

Fig. 7. Relation between the Standard Carbon Content (μg), with 95% confidence intervals, and *r* (d^{-1}). The dotted line indicates the regression line, with all points included, the solid line represents the regression line when the outlying CHLAM series is excluded. The different symbols indicate the different food sources (CHLAM: triangles; CH_SC: circles; SCENE: squares).

the carbon content were marginally significant ($F_{11,195} = 1.9$; $p = 0.04$), so we decided to ignore this non-parallelism, and analysed the data with an analysis of covariance, and a common slope of 2.78. Again, food type, food concentration and the interaction showed significant effects (Table 1). Figure 6 shows the intercepts of these regression lines, i.e. the natural logarithm of the carbon content of a 1 mm animal. The ln(SCC)-values roughly followed the same pattern as all the other measurements, the only exception being the ln(SCC)-value for the animals grown at *C. globosa* at 0.50 mg Cl^{-1}. This was also the series causing the non-parallelism in the regression lines. A combination of the two values measuring the quality of the food, i.e. *r*, and SCC (Fig. 7) showed a significant correlation ($r^2 = 0.53$; $p = 0.01$; N = 12). The correlation became even stronger when the outlying CHLAM point was omitted ($r^2 = 0.79$; $p < 0.001$; N = 11).

Discussion

Our results show that both growth and reproduction in *D. galeata* were influenced by the quantity and the quality of the available food sources, with higher values found on the pure *S. obliquus* media than on the pure *C. globosa* media (see also Ahlgren *et al.*, 1990; Lundstedt & Brett, 1991). Both *Chlamydomonas spp.* and *Scenedesmus spp.* are widely used as food for cladocerans and are usually regarded as high quality food items (e.g. Lampert, 1977; Vijverberg, 1989; Ahlgren *et al.*, 1990). Some reports, however, show the inadequacy of *Chlamydomonas reinhardtii* (Taub & Dollar, 1968) and *Scenedesmus obliquus* (Boersma & Vijverberg, 1994a) as food for *Daphnia* species. It should be noted that the nutritional value of algae for herbivorous zooplankton is dependent on the culture conditions of the algae (Sterner *et al.*, 1993; van Donk & Hessen, 1993), and hence comparisons between studies should be made with caution. Furthermore, as a result of the presence of *Daphnia*, algae may change their morphology, leading to a decrease in digestibility of the algal cells for the daphnids (Hessen & van Donk, 1993). However, as we replaced the algal media daily, changes in morphology are unlikely to have been important in this study.

A number of other papers have considered the effects of two different food algae fed in combination to zooplankton (Jamieson, 1980; Lewis & Maki, 1981; Hanazato & Yasuno, 1984; Roman, 1984; Cowgill *et al.*, 1985; Rothhaupt, 1991; Lundstedt & Brett, 1991; Vanni & Lampert, 1992). Hanazato & Yasuno (1984) fed a mixture of *Chlorella sp.* and *Microcystis aeruginosa* to *Moina macrocopa*. Although *Microcystis* is usually not considered to be assimilated by zooplankton or even reported to be toxic (DeMott *et al.*, 1991), Hanazato & Yasuno found that growth of animals reared on a mixture of *Chlorella* and *Microcystis* in a ratio of 9:1 was higher than the growth of animals reared on pure *Chlorella* media. Rothhaupt (1991) described a synergistic effect of *Anabaena sp.* and *Monoraphidium minutum* on the population growth rate of the rotifer *Brachionus rubens*. Lundstedt & Brett (1991) found a marked synergistic effect of *Stephanodiscus hantzschii* and *Rhodomonas minuta* on growth and reproduction of *Bosmina longispina*. Vanni & Lampert (1992) also reported that *Daphnia galeata* individuals grown on a mixture of *Oocystis lacustris* and *Scenedesmus acutus* tended to have higher numbers of offspring compared to those grown in the monoalgal media. Here, we found a definite synergistic effect of *Chlamydomonas globosa* and *Scenedesmus obliquus* on growth and reproduction of *D. galeata*.

We observed a significant positive relationship between the values of SCC and the rate of population increase (r). This means that measuring carbon content under field conditions would not only give information on the condition of single animals, but could also provide some insight into the reproductive potential, as was expressed in the laboratory derived values for r. Furthermore, SCC data from the field could provide us with an alternative method for the estimation of death rates (d) under field conditions. If we assume that the relation between SCC and r is the same in the field as in the laboratory, it would be possible to compute the difference in r found in the field with the r derived from the SCC values of the animals in the field. This difference would reflect the higher mortality in the field, and hence could provide insight into the death rates caused by predation, since non-predation mortality is already included in the value for r derived from the SCC values. This is of course assuming that the r-values from the laboratory and the field are derived at similar temperatures, and that the intrinsic mortality in the field and laboratory are equal.

In conclusion, a synergistic effect of two different algal species seems to be the rule rather than the exception, unless one of the species is hardly taken up at all, as is the case with most toxic or filamentous blue-greens (Gliwicz & Lampert, 1990). In this study, the culture conditions of the two algal species were the same. Hence it is unlikely that the two species were limited by different factors, especially since the C/N ratios of the algae were similar (Boersma, unpubl. data). It is, however, known that the fatty acid composition of *Scenedesmus* and *Chlamydomonas* are not the same (Ahlgren *et al.*, 1990), and that some of these fatty acids are essential (Müller-Navarra, 1993). This leads to the conclusion that the two algal species were probably not simply substitutable resources for the daphnids, but rather that the two different food species provided complementary resources, each containing different proportions of two or more nutritional essential elements (Tilman, 1982). Although poorly studied in zooplankton, the phenomenon of complementary resources is likely to be important in a wide variety of consumers (see references in Pennings *et al.*, 1993). This implies that not only the algal densities, but also the composition of the algal community, even if it were comprised of equal-sized edible species, might be important in the regulation of aquatic herbivore densities.

Acknowledgements

This study was supported by the Life Science Foundation (SLW), which is subsidized by the Netherlands Organization for Scientific Research (NWO). We thank Aafje Landman, Theo Frank and Diana Tischer for their assistance in the laboratory, and Ramesh Gulati, Joop Ringelberg, Onno van Tongeren and Karen Wiltshire for their comments on the manuscript.

References

Ahlgren, G., L. Lundstedt, M. T. Brett & C. Forsberg, 1990. Lipid composition and food quality of some freshwater phytoplankton for cladoceran zooplankters. J. Plankton Res. 12: 809–818.

Boersma, M. & J. Vijverberg, 1994a. Possible toxic effects on *Daphnia* resulting from the green alga *Scenedesmus obliquus*. Hydrobiologia 294: 99–103.

Boersma, M. & J. Vijverberg, 1994b. The effect of preservation methods on the Carbon content of *Daphnia*. Arch. Hydrobiol. 130: 241–247.

Boersma, M. & J. Vijverberg, 1994c. Seasonal variations in the condition of two *Daphnia* species and their hybrid in a eutrophic lake: evidence for food limitation. J. Plankton Res. 16: 1793–1809.

Cowgill, U. M., K. I. Keating & I. T. Takahashi, 1985. Fecundity and longevity of *Ceriodaphnia dubia/affinis* in relation to diet at two different temperatures. J. Crust. Biol. 5: 420–429.

DeMott, W. R., Q. Z. Zhang & W. W. Carmichael, 1991. Effects of toxic cyanobacteria and purified toxins on the survival and feeding of a copepod and 3 species of *Daphnia*. Limnol. Oceanogr. 36: 1346–1357.

Duncan, A., W. Lampert & O. Rocha, 1985. Carbon weight on length regressions of *Daphnia* spp. grown at threshold food concentrations. Verh. int. Ver. Limnol. 22: 3109–3115.

Gliwicz, Z. M. & W. Lampert, 1990. Food thresholds in *Daphnia* species in the absence and presence of blue-green filaments. Ecology 71: 691–702.

Hanazato, T. & M. Yasuno, 1984. Growth, reproduction and assimilation of *Moina macrocopa* fed on *Microcystis* and/or *Chlorella*. Jpn. J. Ecol. 34: 195–202.

Hartmann, H. J. & D. D. Kunkel, 1991. Mechanisms of food selection in *Daphnia*. Hydrobiologia 225 (Dev. Hydrobiol. 71): 129–154.

Hessen, D. O. & E. van Donk, 1993. Morphological changes in *Scenedesmus* induced by substances released from *Daphnia*. Arch. Hydrobiol. 127: 129–140.

Jamieson, C. D., 1980. Observations on the effect of diet and temperature on rate of development of *Mesocyclops leuckarti* (Claus) (Copepoda, Cyclopoida). Crustaceana 38: 145–154.

Kerfoot, W. C. & K. L. Kirk, 1991. Degree of taste discrimination among suspension-feeding cladocerans and copepods – Implications for detritivory and herbivory. Limnol. Oceanogr. 36: 1107–1123.

Lampert, W., 1977. Studies on the carbon balance of *Daphnia pulex* de Geer as related to environmental conditions. II. The dependence of carbon assimilation on animal size, temperature, food concentration and diet species. Arch. Hydrobiol. Suppl. 48: 310–335.

Lewis, M. A. & A. W. Maki, 1981. Effects of water hardness and diet on productivity of *Daphnia magna* Straus in laboratory culture. Hydrobiologia 85: 175–179.

Lundstedt, L. & M. T. Brett, 1991. Differential growth rates of 3 cladoceran species in response to mono-algal and mixed-algal cultures. Limnol. Oceanogr. 36: 159–165.

Meyer, J. S., C. G. Ingersoll, L. L. McDonald & M. S. Boyce, 1986. Estimating uncertainty in population growth rates: Jacknife *vs* Bootstrap techniques. Ecology 67: 1156–1166.

Müller-Navarra, D., 1993. Quantifizierung von Nahrungsqualität für herbivores Zooplankton. Thesis, University of Kiel, 137 pp.

Pennings, S. C., M. T. Nadeau & V. J. Paul, 1993. Selectivity and growth of the generalist herbivore *Dolabella auricularia* feeding upon complementary resources. Ecology 74: 879–890.

Roman, M. R., 1984. Utilization of detritus by the copepod, *Acartia tonsa*. Limnol. Oceanogr. 29: 949–959.

Rothhaupt, K. O., 1991. The influence of toxic and filamentous blue-green algae on feeding and population growth of the rotifer *Brachionus rubens*. Int. Rev. ges. Hydrobiol. 76: 67–72.

Salonen, K., 1979. A versatile method for the rapid and accurate determination of carbon by high temperature combustion. Limnol. Oceanogr. 24: 177–183.

Sterner, R. W., D. D. Hagemeier & W. L. Smith, 1993. Phytoplankton nutrient limitation and food quality for *Daphnia*. Limnol. Oceanogr. 38: 857–871.

Taub F. B. & A. M. Dollar, 1968. The nutritional inadequacy of *Chlorella* and *Chlamydomonas* as food for *Daphnia pulex*. Limnol. Oceanogr. 13: 607–617.

Tessier, A. J. & C. E. Goulden, 1982. Estimating food limitation in cladoceran populations. Limnol. Oceanogr. 27: 707–717.

Tilman, D., 1982. Resource competition and community structure. Princeton University Press, Princeton, NJ.

van Donk, E. & D. O. Hessen, 1993. Grazing resistance in nutrient-stressed phytoplankton. Oecologia 93: 508–511.

Vanni, M. J. & W. Lampert, 1992. Food quality effects on life history traits and fitness in the generalist herbivore *Daphnia*. Oecologia 92: 48–57.

Vijverberg, J., 1989. Culture techniques for studies on the growth, development and reproduction of copepods and cladocerans under laboratory and in situ conditions: a review. Freshwat. Biol. 21: 317–373.

Hydrobiologia **307**: 117–126, 1995.
P. Larsson & L.J. Weider (eds), *Cladocera as Model Organisms in Biology.*

Effects of *Microcystis aeruginosa* on interference competition between *Daphnia pulex* and *Keratella cochlearis*

Alison Sartonov[1]
Department of Biological Sciences, Dartmouth College, Hanover, NH 03755-3576, USA
[1]*Current address: USAID, American Embassy – Moscow, Department of State, Washington, DC 20521-5430, USA*

Key words: Microcystis aeruginosa, competition, population growth, *Daphnia pulex, Keratella cochlearis*

Abstract

Laboratory experiments tested the hypothesis that a toxic strain of *Microcystis aeruginosa* decreases the ability of *Daphnia pulex* to interfere with *Keratella cochlearis*. To test a variety of conditions, juvenile and adult *Daphnia* were exposed to the cyanobacterium for different time periods prior to, and during the experiments. Adult *Daphnia* not only suppressed rotifers over successive two-day intervals, but also had a significant impact within a 24-hour period. However, the presence of *Microcystis* (5×10^5 cells ml^{-1}) decreased the *Daphnia* effect in both experiments. Although juvenile *Daphnia* also significantly suppressed *Keratella* population growth, the presence of *Microcystis* (10^5 and 5×10^5 cells ml^{-1}) caused a significant reduction in daphniid body size and decreased the ability of both nonacclimated and acclimated daphniids to suppress rotifers. *Keratella* inhalation and mortality are positively correlated with filtering rates and body size of *Daphnia*. Therefore, the feeding rates and size structure of a *Daphnia* population will determine its potential to interfere with vulnerable rotifers. In all experiments the presence of *Microcystis* significantly decreased the ability of *Daphnia* to interfere with this rotifer despite the fact that *Keratella* was also inhibited. In the field this effect might be augmented if *Microcystis* colonies are more easily ingested by cladocerans than by the rotifers.

Introduction

Selective predation by fish (Brooks & Dodson, 1965) and invertebrates (Dodson, 1974) has long been recognized as a major force in structuring zooplankton communities. In the absence of significant predation, interspecific competition for limiting resources is thought to become increasingly important (Romanovsky & Feniova, 1985). However, predation and competition often interact to determine species composition, particularly on a seasonal basis (Kerfoot & Pastorok, 1978; Cooper & Smith, 1982). Within a community, relative competitive abilities also vary with changing environmental conditions (see review in Bengtsson, 1987). Recent studies have revealed additional factors which can change competitive relations among zooplankton. Suspended sediments (Kirk & Gilbert, 1990) and non-toxic filamentous cyanobacteria (Gliwicz & Lampert, 1990), both of which may interfere with feeding behavior, can affect threshold food concentrations and alter competitive dominance hierarchies. Toxic factors, such as chemicals or toxic cyanobacteria, may also affect competitive ability, but studies have concentrated on relative susceptibilities (Smith & Gilbert, 1995) and little is known of their effects on species interactions.

Blooms of cyanobacteria are often accompanied by changes in zooplankton community structure where small cladocerans and rotifers dominate (Edmondson & Litt, 1982; Infante & Riehl, 1984; Orcutt & Pace, 1984; Fulton & Paerl, 1987a, 1988b; Gilbert, 1990; Fulton & Jones, 1991). Laboratory studies have focused on differential susceptibilities, assuming this aspect to be the sole determinant of observed patterns in the field (Smith & Gilbert, 1995). A number of mechanisms have been proposed to explain the apparently greater susceptibilities of larger species. Cyanobacterial morphology (filament length or colony size) may interfere with feeding in large cladocerans (Gliwicz, 1977; Webster & Peters, 1978; Gilbert & Durand,

1990), while making the cyanobacteria inaccessible to smaller zooplankton (Fulton & Paerl, 1987a,b, 1988a; Gilbert & Durand, 1990). Cyanobacterial endotoxins may affect feeding on nutritious food by depressing filtering rates (Lampert, 1981a,b, 1982), particularly in large cladocerans (Lampert, 1982). Finally, the presence of toxic cyanobacteria can differentially reduce survivorship and population growth rates (*r*) of zooplankton species (Lampert, 1981a,b; Fulton & Paerl, 1987a, 1988a,b; Gilbert, 1990; DeMott *et al.*, 1991; Henning *et al.*, 1991). However, with the exception of some cladocerans (Gilbert, 1990), no clear pattern relating to body size has been documented.

Cladocerans and rotifers are common components of freshwater ecosystems where they often compete for the same food resources. Large cladocerans (>1.2 mm body length), such as *Daphnia pulex*, are superior competitors, capable of suppressing or causing extinction of rotifers through both exploitative and interference competition (see review in Gilbert, 1988b). Any factor that inhibits large cladocerans to a greater degree than rotifers should alter zooplankton community structure by favoring rotifers. *Microcystis aeruginosa*, which differentially affects *Keratella cochlearis* and *Daphnia pulex*, may be such a factor.

Microcystis aeruginosa is a common, bloom-forming, unicellular cyanobacterium which often grows in colonies. Some strains produce microcystins, cyclic heptapeptides (Krishnamurthy *et al.*, 1986), which are toxic to mammals (mouse bioassay). Although experiments with *Microcystis* cell extracts (Jungmann *et al.*, 1991) and with purified microcystins (DeMott *et al.*, 1991) indicate that these toxins are also inhibitory or lethal to some zooplankton, they may not be the sole source of the effects observed in the laboratory and in nature (DeMott *et al.*, 1991; Jungmann, 1992). An earlier study (Smith & Gilbert, 1995) showed that the rotifer, *Keratella cochlearis*, is somewhat less susceptible to single cells of *Microcystis* strain UTEX 2061 than the cladoceran, *Daphnia pulex*. *Keratella* is a small rotifer, and one of the most vulnerable to interference competition from large *Daphnia* (Gilbert & Stemberger, 1985; Burns & Gilbert, 1986a,b; Gilbert, 1988a; Gilbert & MacIsaac, 1989; MacIsaac & Gilbert, 1989). The susceptibility of *D. pulex* to some *Microcystis* strains, evidenced by reductions in population growth, adult body size (Smith & Gilbert, 1995) and filtering rates (Lampert, 1981a,b, 1982) could reduce the extent to which this cladoceran can suppress the rotifer.

Studies of differential effects of cyanobacteria on potentially competing zooplankton have typically taken two approaches, both of which have limitations. Broad, easily defined classes (e.g. adults or juveniles) may be studied as even-aged cohorts without carefully considering the relevance or significance of such divisions. Changes in relative competitive ability may then be inferred from reductions in reproduction and survivorship. In contrast, competition experiments, which follow populations through several generations, do incorporate age structure, but in an uncontrolled manner. In addition, the outcome of such experiments often depends on the initial conditions. Experiments with even-aged cohorts may fail to detect sensitivities inherent in untested age classes, while experiments with entire populations may obscure age-specific effects.

Although food thresholds are better determinants of competitive ability than measures of *r*, interference competition between cladocerans and rotifers can be significant even when food is abundant (Gilbert & Stemberger, 1985; MacIsaac & Gilbert, 1989). *Microcystis* suppresses population growth in both *Daphnia* and *Keratella* when food is not limiting, but inhibition of *Daphnia* appears to be greater (Smith & Gilbert, 1995). The purpose of this set of experiments is to test the hypothesis that *Microcystis* will also favor *K. cochlearis* by decreasing the strength of interference competition from *Daphnia pulex*.

To study Daphnia interactions with *Keratella* in the presence of *Microcystis*, I did experiments with juvenile and adult *Daphnia* which were exposed to the cyanobacterium for different time periods. Lampert (1982) found that *Microcystis*-induced reductions in filtering rates were quickly reversed when daphniids were transferred to pure food suspensions. However, with prolonged exposure, irreversible effects such as decreased growth (Lampert, 1981a,b; Smith & Gilbert, 1995), reproduction and longevity (Fulton & Paerl, 1987a, 1988b; DeMott *et al.*,1991; Henning *et al.*, 1991; Smith & Gilbert, 1995) are also likely to occur. The severity of these permanent effects depends on the age of the animals as well as on the duration of exposure (Lampert, 1981b; DeMott *et al.*, 1991; Smith & Gilbert, 1995). Exposure time itself may depend on the spatial distributions of *Microcystis* (Reynolds, 1978) and zooplankton in the water column, particularly when diel vertical migration occurs (Haney, 1987; Forsyth *et al.*, 1990). Thus, this experimental approach allowed me to explore the potential impact of *Microcystis* on daphniid-rotifer competitive interactions under a range of conditions likely to be met in the field.

Methods

General

Daphnia pulex and *Keratella cochlearis* were collected from Pond View (Lyme, New Hampshire) and Star Lake (Norwich, Vermont), respectively, and maintained in the laboratory for many generations prior to the start of this investigation. Neither lake is known to have *Microcystis* blooms. A clone of each species was cultured in glass-fiber (Whatman 934-AH) filtered lake water (Norford Lake, Vermont) supplemented (9:1) with modified MBL medium (Stemberger, 1981) and fed *Cryptomonas* sp. isolated from Pleasant Lake, New Hampshire (Gilbert, 1985).

Microcystis aeruginosa strain UTEX 2061 was provided by W. Carmichael. Although nontoxic to mammals (W. Carmichael, pers. comm.), this strain was previously shown to be toxic to *D. pulex* under current laboratory conditions (Smith & Gilbert 1995). Xenic cultures were maintained as single cells in BG-11 medium (Stanier *et al.*, 1971). Cells were harvested by centrifugation and resuspended in the filtered lake water medium. All concentrations used were within the range of natural bloom conditions (Infante & Riehl, 1984; R. Fulton, pers. comm.). All cultures used were approximately 60-days old (see Smith & Gilbert, 1995).

Experiments were conducted in a walk-in incubator at 20 °C with dim lighting (L:D 16:8). Cell densities for *Cryptomonas* and *Microcystis* suspensions were determined by direct hemacytometer count, after which appropriate dilutions were prepared. Replicate ($n = 10$ per treatment) *Keratella* populations (20 individuals) were initiated in 80-ml screw-cap glass vials (sealed with parafilm to prevent air bubbles) and maintained on a plankton wheel (rotating 90 sec every 10 min). All experiments were designed to assess the effects of *Microcystis* on interference competition between *Daphnia pulex* and *Keratella cochlearis* in the presence of adequate food (10^4 cells ml^{-1} *Cryptomonas* sp.). *Cryptomonas* cells were always abundant at the end of the two-day intervals. In order to avoid damaging the animals, handling was gentle and minimal. At the conclusion of each experiment, the lengths (excluding tailspine) of surviving daphniids were measured (at 24×) using a dissecting microscope and ocular micrometer.

One-day interference experiment with adult Daphnia

This experiment was designed to assess the short-term effect of *Microcystis* (5×10^5 cells ml^{-1}) on the interaction between *Daphnia* and *Keratella.* Neither species was exposed to *Microcystis* prior to the start of the experiment. Twenty non-ovigerous *Keratella* were selected from batch cultures (fed 10^4 cells ml^{-1} *Cryptomonas* sp.) to initiate each population. One adult *Daphnia pulex*, randomly chosen from batch cultures (fed 10^4 cells ml^{-1} *Cryptomonas* sp.) was added to each bottle. After 24 hours the rotifers in each of the replicate populations ($n = 10$ per treatment) were counted. The mean length (±SD) of the *Daphnia* was 1.83 ± 0.14 mm ($n = 20$).

Interference experiment with nonacclimated adult Daphnia

The purpose of this experiment was to evaluate the long-term effect of *Microcystis* (5×10^5 cells ml^{-1}) on *Keratella* population growth in the presence and absence of *Daphnia.* Neither species was exposed to *Microcystis* prior to the start of the experiment. In this, and all successive experiments, 20 *Keratella* were randomly selected from batch cultures (fed 10^4 cells ml^{-1} *Cryptomonas* sp.) without regard to reproductive status to initiate each population ($n = 10$ per treatment). Individual adult *Daphnia pulex* were also randomly chosen from batch cultures (fed 10^4 cells ml^{-1} *Cryptomonas* sp.). *Keratella* populations were counted and transferred to fresh media every 2 days (8 days total). At these times, the *Daphnia* were replaced (one per bottle). Offspring produced by the *Daphnia* were discarded. Replicates in which the daphniid died during the two-day interval were excluded from the analysis. Care was taken to avoid removing any rotifers with the *Daphnia.* Since exposure was limited to a maximum of 48 hours, the *Microcystis* effect on *Daphnia* is limited and considered to be short-term. The average size (± SD) of *Daphnia* used in this experiment was 1.91 ± 0.22 mm ($n = 47$). Population growth (r) for rotifer populations is commonly determined using the following equation, where N_0 is the initial population size and N_t is the population size at time t:

$$r = (\ln N_t - \ln N_0)/t$$

However, the accuracy of the estimate for each population was enhanced by using the data collected every two days over a six-day interval to calculate a regression line, the slope of which is *r*. Although the total length of the interval varied among experiments, the initial 2-day acclimation period was always excluded from the analysis. *Keratella* population growth (*r*) was compared among treatments using ANOVA and Tukey's HSD for multiple comparisons (SAS Institute 1986).

Interference experiment with nonacclimated juvenile Daphnia

This experiment also determined the long-term effect of *Microcystis* (5×10^5 cells ml^{-1}) on *Keratella* population growth in the presence and absence of *Daphnia*. Twenty *Keratella* were randomly selected from batch cultures (fed 10^4 cells ml^{-1} *Cryptomonas* sp.) to initiate each population ($n = 10$ per treatment). However, in this case individual, 5-day old *D. pulex* were randomly chosen from even-aged batch cultures (fed 10^4 cells ml^{-1} *Cryptomonas* sp.). Four-day old *D. pulex* are large enough (1.22 ± 0.16 mm, $n = 28$, measurements of nonexperimental animals) to interfere with rotifers (Burns & Gilbert, 1986a). *Keratella* were counted and transferred to fresh media every 2 days along with the daphniid (one per bottle). Any offspring produced by the *Daphnia* were discarded. *Keratella* in each replicate were counted only as long as the *Daphnia* remained alive. *Keratella* populations which persisted for less than 4 days after the acclimation period were excluded from the analysis. The changes in rotifer number during a 10-day interval (maximum) were used to calculate the growth rate (*r*) for each replicate population using the method described above. At this time all rotifers in the *Daphnia* treatment were extinct.

Interference experiment with acclimated juvenile Daphnia

This experiment was designed to evaluate the long-term effect of *Microcystis* (10^5 cells ml^{-1}) on *Keratella* population growth in the presence and absence of *Daphnia*. However, it also provides data on the long-term effect of *Microcystis* on *D. pulex*. Here, a lower *Microcystis* concentration was used to ensure adequate survival of *Daphnia* juveniles. These individuals had been exposed to 10^5 cells ml^{-1} *Microcystis* from birth to the start of the experiment at age 4 days (= acclimated).

Twenty *Keratella* were randomly selected from batch cultures (fed 10^4 cells ml^{-1} *Cryptomonas* sp.) to initiate each population ($n = 10$ per treatment). *Keratella* populations were counted and transferred to fresh media every 2 days along with the daphniid (one per bottle). Any offspring produced by the *Daphnia* were counted and discarded. *Keratella* in each replicate were counted only as long as the *Daphnia* remained alive. *Keratella* populations which persisted for less than 4 days after the acclimation period were excluded from the analysis. The changes in rotifer number during an 8-day interval (maximum) were used to calculate the growth rate (*r*) for each replicate population as previously described. At this time all rotifers in the *Daphnia* treatment were extinct. To compare *Daphnia* population growth between treatments, bootstrap methods (Smith & Gilbert, 1995) were used to generate replicate *r*-values from the life table data. However, since a maximum of only three broods was produced by the controls a more appropriate value for comparisons is $\sum l_x m_x$ (see Gilbert, 1990; Kirk & Gilbert, 1990).

Table 1. Effects of *Microcystis aeruginosa* (5×10^5 cells ml^{-1}) on *Daphnia pulex* interference with *Keratella cochlearis* during a one day interval. At the start of the experiment, each replicate population consisted of 20 non-ovigerous *Keratella*. Values shown are means ± se, $n = 10$. Results of *t* tests compare each treatment with the control.

Treatment	Number of *Keratella* remaining	t_{18}	*P*
Control	19.5 ± 0.2	–	–
Microcystis	19.1 ± 0.4	0.857	>0.4
Microcystis + *Daphnia*	19.9 ± 0.1	1.661	>0.1
Daphnia	12.0 ± 1.0	7.031	<0.001

Results

24-hr Interference experiment with adult Daphnia

Microcystis significantly reduced the ability of *Daphnia* to interfere with *Keratella* during a 24-hour period. *Keratella* population numbers were reduced 40 percent by *Daphnia* alone ($P<0.001$). However, *Daphnia* had no effect on *Keratella* in the presence of *Microcystis*. In addition, *Microcystis* did not affect *Keratella* populations during this time period (Table 1).

Table 2. Effects of *Microcystis aeruginosa* (5 × 10^5 cells ml^{-1}) on *Daphnia pulex* (nonacclimated adults) interference with *Keratella cochlearis* population growth. ANOVA table and comparison of slopes (r) for Fig. 1. Means (± se) followed by the same letter are not significantly different at the 0.05 level (Tukey's HSD).

Source	df	SS	MS	F	P
Model	3	3.2217	1.0739	159.56	0.0001
Error	29	0.1951	0.0067		

Treatment	n	r
Control	10	0.272 ± 0.008 A
Microcystis	10	0.149 ± 0.010 B
Microcystis + *Daphnia*	8	−0.076 ± 0.044 C
+ *Daphnia*	5	−0.665 ± 0.059 D

Table 3. Effects of *Microcystis aeruginosa* (5 × 10^5 cells ml^{-1}) on *Daphnia pulex* (nonacclimated juveniles) interference with *Keratella cochlearis* population growth. ANOVA table and comparison of slopes (r) for Fig. 2. Means (± se) followed by the same letter are not significantly different at the 0.05 level (Tukey's HSD).

Source	df	SS	MS	F	P
Model	3	2.1112	0.7037	37.89	0.0001
Error	32	0.5943	0.0186		

Treatment	n	r
Control	10	0.251 ± 0.008 A
Microcystis	10	0.008 ± 0.017 B
Microcystis + *Daphnia*	10	−0.117 ± 0.071 B
+ *Daphnia*	6	−0.484 ± 0.058 C

Table 4. Effects of *Microcystis* (10^5 cells ml^{-1}) on *Daphnia pulex* (acclimated juveniles) interference with *Keratella cochlearis* population growth. ANOVA table and comparison of slopes (r) for Fig. 3. Means (± se) followed by the same letter are not significantly different at the 0.05 level (Tukey's HSD).

Source	df	SS	MS	F	P
Model	3	3.7334	1.2445	52.47	0.0001
Error	32	0.7589	0.0237		

Treatment	n	r
Control	10	0.164 ± 0.012 A
Microcystis	10	0.112 ± 0.011 A
Microcystis + *Daphnia*	9	−0.388 ± 0.081 B
+ *Daphnia*	7	−0.621 ± 0.078 C

Table 5. Effects of *Microcystis aeruginosa* (10^5 cells ml^{-1}) on size of survivors and population parameters for *Daphnia pulex*. Values shown are means ± SD. Results of t tests are given (NS P>0.05, ** P<0.01, *** P<0.001).

	Control	*Microcystis*	t
Size (mm)	2.3±0.1	1.9±0.1	t_{13}=7.407 ***
r	0.175±0.017	0.145±0.018	t_{17}=3.704 **
$\sum l_x m_x$	8.2±1.8	5.6±1.1	t_{17}=3.939 **
Generation time (days)	12.3±0.6	11.9±0.5	t_{17}=1.584 NS
Number reproducing (%)	10 (100)	9 (100)	
Number surviving to end of experiment (%)	9 (90)	6 (67)	

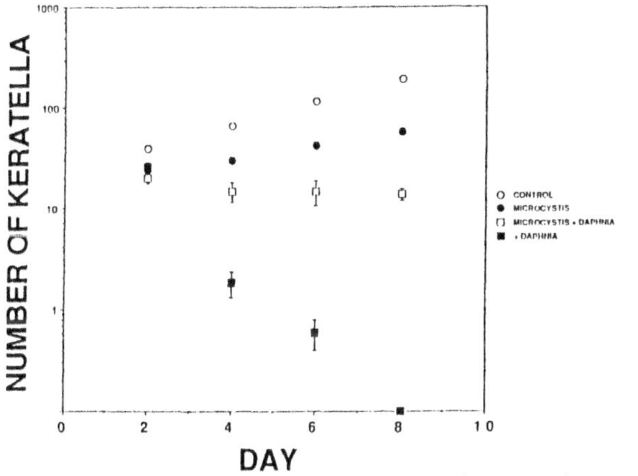

Fig. 1. Effects of *Microcystis aeruginosa* (5 × 10^5 cells ml^{-1}) on *Daphnia* (nonacclimated adults) interference with *Keratella* population growth. At the start of the experiment, each replicate population (n = 10) consisted of 20 randomly chosen *Keratella*. Values shown are means ± se. In some cases symbols obscure error bars. The slope of each line = r. *Daphnia* (adults not previously exposed to *Microcystis*) were replaced every two days.

Interference experiment with nonacclimated adult Daphnia

Microcystis significantly reduced the ability of nonacclimated adult *Daphnia* to interfere with *Keratella*. Over a six-day interval, the presence of *Microcystis* caused a significant reduction in *Keratella* population growth compared to the controls (Fig. 1 & Table 2). The combined effects of *Daphnia* and *Microcystis* caused a further significant reduction resulting in a negative r-value. However, the most dramatic effect occurred when *Daphnia* was present without *Microcystis*. In this case all rotifer populations were extinct by the 8th day of the experiment.

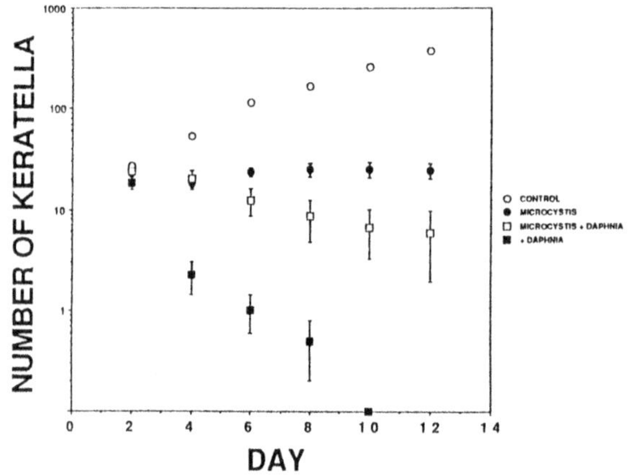

Fig. 2. Effects of *Microcystis aeruginosa* (5×10^5 cells ml^{-1}) on *Daphnia* (nonacclimated juveniles) interference with *Keratella* population growth. At the start of the experiment, each replicate population ($n = 10$) consisted of 20 randomly chosen *Keratella*. Values shown are means ± se. In some cases symbols obscure error bars. The slope of each line = *r*. *Daphnia* (5 days old, not previously exposed to *Microcystis*) remained until death or the end of the experiment.

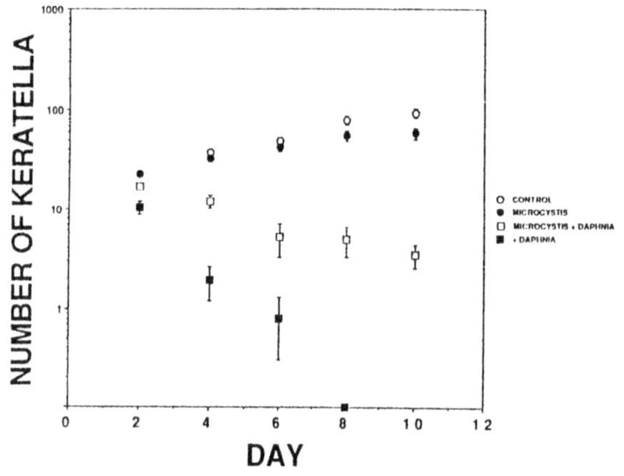

Fig. 3. Effects of *Microcystis aeruginosa* (10^5 cells ml^{-1}) on *Daphnia* (acclimated juveniles) interference with *Keratella* population growth. At the start of the experiment, each replicate population ($n = 10$) consisted of 20 randomly chosen *Keratella*. Values shown are means ± se. In some cases symbols obscure error bars. The slope of each line = *r*. *Daphnia* (5 days old, exposed to 10^5 cells ml^{-1} *Microcystis* from birth) remained until death or the end of the experiment.

Interference experiment with nonacclimated juvenile Daphnia

Microcystis significantly reduced the ability of nonacclimated juvenile *Daphnia* to interfere with *Keratella*. Although the presence of *Microcystis* significantly reduced rotifer population growth (to near zero), the additional suppression due to the presence of *Daphnia* was not significant (Fig. 2 & Table 3). However, once again, all *Keratella* populations went rapidly extinct (by day 10) when *Daphnia* was present without *Microcystis*. Juvenile *Daphnia* were exposed to *Microcystis* for 12 days (from age 5–17 days). Most had reached reproductive maturity by age 13 days. Despite this limited exposure to *Microcystis*, survivors were significantly smaller than those in the controls [1.77 ± 0.04 mm ($n = 4$) compared to 2.33 ± 0.07 mm ($n = 7$) $t_9 = 14.4937$, $P<0.001$].

Interference experiment with acclimated juvenile Daphnia

Microcystis significantly reduced the ability of acclimated juvenile *Daphnia* to interfere with *Keratella*. At the lower concentration (10^5 cells ml^{-1}) *Microcystis* did not suppress *Keratella* population growth (Fig. 3 & Table 4). Therefore, the significant suppression (negative *r*) evident in the *Microcystis* + *Daphnia* treatment must be attributed to the *Daphnia* effect. However, the suppression was significantly greater when *Daphnia* was present without *Microcystis* (extinction by day 8).

Since neonates (0–24 h old) were exposed to *Microcystis* from birth to the end of the experiment (a total of 14 days), the effect of *Microcystis* on *Daphnia* can also be evaluated. Compared to controls, surviving *Daphnia* were smaller at maturity after a lifelong exposure to *Microcystis* ($P<0.001$) (Table 5). Although generation times were similar, reproduction ($\sum l_x m_x$) was significantly suppressed ($P<0.01$). All of the daphniids survived to reproductive maturity, but mortality was 10 percent and 33 percent for the control and *Microcystis* treatments respectively (Table 5).

Discussion

The potential effects of *Microcystis aeruginosa* on zooplankton community structure are far more complex than previously believed. Earlier experiments demonstrated that *Daphnia pulex* was somewhat more sensitive to *Microcystis* than *Keratella cochlearis* when population growth rates were compared (Smith & Gilbert, 1995). However, *Microcystis* can also alter competitive interactions between the species, further favoring the rotifer. Although the magnitude of the effect varied with experimental conditions, in all cases the presence of *Microcystis* significantly decreased the ability of *Daphnia* to interfere with *Keratella*.

In addition to population growth suppression of *D. pulex*, other factors, such as reductions in filtering rates and adult daphniid body size can affect interactions between the two species. The ability of *Daphnia* to mechanically interfere with rotifers is well documented (Gilbert & Stemberger, 1985; Burns & Gilbert, 1986a,b; Stemberger & Gilbert, 1987; Gilbert, 1988a,b). Since body size and filtering rate of *Daphnia* are positively correlated with *Keratella* inhalation and mortality (Burns & Gilbert, 1986a), the feeding rates and size structure of a *Daphnia* population will determine not only its competitive interactions (exploitative) with other cladocerans (Neill, 1975; Lynch, 1979; Romanovsky & Feniova, 1985), but also its potential effects (interference competition) on vulnerable rotifers. Therefore, anything that influences body size, filtering rate, or both should directly affect daphniid-rotifer interactions.

Adult *Daphnia* rapidly reduced *Keratella* population numbers in two experiments. Individual *Daphnia* not only suppressed rotifers over successive two-day intervals (Fig. 1 & Table 2), but also had a significant impact within a 24-hour period (Table 1). However, the presence of *Microcystis* moderated the *Daphnia* effect in both experiments. With *Microcystis*, suppression was not as strong over the two-day intervals, and *Daphnia* was unable to reduce *Keratella* population numbers in a 24-hour period. In experiments with natural communities, (Gilbert (1989) found that *Keratella* mortality due to interference was less than expected and only significant after a two-day period. He suggested that the presence of *Anabaena affinis* filaments in the lake water may have reduced the potential for *Daphnia* to interfere with *Keratella*.

The mortality rate imposed by *Daphnia* on *Keratella* in treatments without *Microcystis* ($0.4\ d^{-1}$) was higher than predicted from earlier studies (Burns & Gilbert, 1986a; Gilbert, 1989). The predicted death rate ($d_D d^{-1}$), using the experimental density (N) of $12.5\ L^{-1}$ and mean daphniid body length (L) of 1.78 mm, was 0.3 (Stemberger, unpubl., cited in Gilbert, 1989). Since the replicate populations in the 24-hour experiment (Table 1) were initiated with nonovigerous *Keratella*, the discrepancy between the observed and predicted values may be partially due to the absence of reproduction in the 24-hr period. Non-ovigerous rotifers are no more vulnerable to interference than those carrying eggs (Burns & Gilbert, 1986b). However, the chance of an individual surviving inhalation and rejection increases with age (Gilbert & Stemberger, 1985) and may be a factor in these experiments.

Juvenile *Daphnia* also significantly suppressed *Keratella* population growth (Fig. 2 & Table 3, Fig. 3 & Table 4). However, the presence of *Microcystis* caused a significant reduction in daphniid body size and decreased the ability of *Daphnia* to suppress rotifers whether acclimated to *Microcystis* (Fig. 3 & Table 4) or not (Fig. 2 & Table 3). In addition, reproduction ($\sum l_x m_x$) of acclimated *Daphnia* was significantly suppressed (Table 5). Reductions in both adult body size and population growth rates have been documented in earlier experiments (Smith & Gilbert, 1995).

Numerous studies have demonstrated the ability of *Microcystis* to reduce cladoceran filtering rates (Lampert, 1981a,b, 1982; Nizan *et al.*, 1986; Fulton & Paerl, 1987a; Fulton, 1988; Fulton & Paerl, 1988a; Henning *et al.*, 1991; Jungmann *et al.*, 1991). For example, Lampert, 1981a) found that very low concentrations of *Microcystis* (1.4×10^3 cells ml^{-1}) in the normal food reduced filtering rates of *Daphnia pulicaria* by 17 percent. Filtering rates were further depressed with an increase in *Microcystis* concentration. However, the mechanism for this inhibition is still poorly understood and may involve more than one factor.

In all of the current experiments the *Cryptomonas* concentration was slightly below the incipient limiting level of 1.5×10^4 cells ml^{-1} (cited as unpublished results in Burns & Gilbert, 1986a). Theoretically, the addition of *Microcystis* could have exceeded this level, causing a decrease in filtering rate regardless of any inherent inhibitory potential (Rigler, 1961). However, this possibility has been investigated for a number of cladoceran species. These studies concluded that *Microcystis* decreased filtering rates regardless of total cell concentration (Lampert, 1981a,b; Lampert, 1982; Fulton & Paerl, 1987a).

The possibility that two separate inhibitory mechanisms could be operating simultaneously is an important consideration when evaluating the relative toxic effects of *Microcystis* on daphniids and other zooplankton. Experiments with purified microcystin (DeMott *et al.*, 1991) have demonstrated that this toxin can be inhibitory or lethal to some zooplankton species. However, as these researchers cautioned, microcystin may not be the sole source of the effects observed in the laboratory and nature. There is increasing evidence that microcystin is not the cause of filtering rate reductions in cladocerans. Although Lampert (1981b) reported that the chemical responsible for filtering inhibition was an endotoxin, this conclusion seems implausible given the speed of the response (<< 10 min). In a later study, Lampert (1982) found that the effect was readily reversible when *Microcystis* was removed and suggested that a chemosensory mechanism, such as taste, was involved.

Jungmann *et al.* (1991) compared the effects of intact *Microcystis* cells and various treated *Microcystis* cells on the filtering rate of *Daphnia pulicaria*. Invariably the intact cells were more inhibitory than extracted cells. Lipophilic extraction of the *Microcystis* particles, however, completely removed the inhibitory factor from the cells. Demott *et al.* (1991) obtained similar results comparing intact cells with purified microcystin-LR and nodularin. Their observation that *Daphnia* filtering rates were depressed only by intact cells, and not by the purified toxins, together with the results of Jungmann *et al.* (1991) suggest that some physical or chemical phenomenon on the cell surface (Gerritson & Porter, 1982) may be responsible, rather than a toxin contained in the cells.

The complexity of *Microcystis* effects on zooplankton is similar to that now recognized for some chemical toxins. In the past, acute toxicity tests with zooplankton were used almost exclusively. However, realization that chronic exposure to toxicants could have important population level effects led to more extensive experimentation. Studies of the long-term effects of radiation (Marshall, 1962, 1966), cadmium (Marshall, 1978) and chromium (Coniglio & Baudo, 1989) documented reductions in individual fecundity and survival which contributed to an overall decrease in population growth. In addition, others found sublethal effects in *Daphnia* species which, although not investigated, have the potential to affect interactions with rotifers vulnerable to interference competition. The herbicide atrazine decreased adult body size in *Daphnia pulex* (Schober & Lampert, 1977) and depressed feeding rates in *D. pulicaria* (Lampert *et al.*, 1989) while the insecticide lindane decreased filtering rates in *D. pulex* (Gliwicz & Sieniawska, 1986). Toxicity studies concerned only with lethal effects or relative susceptibilities ignore the aspect of species' interactions within complex ecosystems (Winner & Farrell, 1976).

The results of these experiments illustrate the profound effect of *Microcystis* on *Daphnia* interactions with *Keratella*. In all cases the presence of *Microcystis* significantly decreased the ability of *Daphnia* to interfere with this vulnerable rotifer, despite the fact that *Keratella* was also inhibited. In nature, this effect might be much more pronounced. *Microcystis aeruginosa* is largely a colonial species, although high numbers of single cells and small colonies have been observed (Benndorf & Henning, 1989). Large colonies (>60 mm) may be inaccessible to zooplankton (Reynolds *et al.*, 1982), but small colonies are often eaten by cladocerans (Infante & Riehl, 1984; Schoenberg & Carlson, 1984; Reynolds *et al.*, 1982). Since colonies may be too large for rotifers to ingest, the inhibitory effect of *Microcystis* on *Keratella* would be lessened. In addition, over a longer time period, *Daphnia* population level responses (Smith & Gilbert, 1995) could further diminish the impact of *Daphnia* on rotifers, dramatically favoring the latter in natural communities.

Acknowledgements

I'd like to thank R. Stemberger and S. Wickham for statistical advice and M. Bean for technical assistance in the laboratory. Comments by J. Gilbert, J. Haney and W. DeMott improved the manuscript and are greatly appreciated. This work was supported in part by a Dartmouth College Cramer Fellowship to the author and NSF grant BSR-8705233 to JJG.

References

Bengtsson, J., 1987. Competitive dominance among Cladocera: Are single-factor explanations enough? An examination of the experimental evidence. Hydrobiologia 145 (Dev. Hydrobiol. 35): 245–257.

Benndorf, J. & M. Henning, 1989. *Daphnia* and toxic blooms of *Microcystis aeruginosa* in Bautzen Reservoir (GDR). Int. Revue ges. Hydrobiol. 74: 233–248.

Brooks, J.L. & S.I. Dodson, 1965. Predation, body-size, and composition of plankton. Science 150: 28–35.

Burns, C.W. & J.J. Gilbert, 1986a. Effects of daphnid size and density on interference between *Daphnia* and *Keratella cochlearis*. Limnol. Oceanogr. 31: 848–858.

Burns, C.W. & J.J. Gilbert, 1986b. Direct observations of the mechanisms of interference between *Daphnia* and *Keratella cochlearis*. Limnol. Oceanogr. 31: 859–866.

Coniglio, L. & R. Baudo, 1989. Life-tables of *Daphnia obtusa* (Kurz) surviving exposure to toxic concentrations of chromium. Hydrobiologia 188/189 (Dev. Hydrobiol. 54): 407–410.

Cooper, S.D. & D.W. Smith, 1982. Competition, predation and the relative abundances of two species of *Daphnia*. J. Plankton Res. 4: 859–879.

DeMott, W. R., Q. Zhang & W. Carmichael, 1991. Effects of toxic cyanobacteria and purified toxins on the survival and feeding of a copepod and three species of *Daphnia*. Limnol. Oceanogr. 36: 1346–1357.

Dodson, S. I., 1974. Zooplankton competition and predation: An experimental test of the size efficiency hypothesis. Ecology 55: 605–613.

Edmondson, W. T. & A. H. Litt, 1982. *Daphnia* in Lake Washington. Limnol. Oceanogr. 27: 272–293.

Forsyth, D. J., M. R. James & M. Cryer, 1990. Alteration of seasonal and diel patterns in vertical migration of zooplankton by *Anabaena* and planktivorous fish. Arch. Hydrobiol. 117: 385–404.

Fulton, R. S.,III, 1988. Resistance to blue-green algal toxins by *Bosmina longirostris*. J. Plankton Res. 10: 771–778.

Fulton, R. S., III & R. C. Jones, 1991. Growth and reproductive responses of *Daphnia* to cyanobacterial blooms on the Potomac River. Int. Revue ges. Hydrobiol. 76: 5–19.

Fulton, R. S., III & H. W. Paerl, 1987a. Toxic and inhibitory effects of the blue-green alga *Microcystis aeruginosa* on herbivorous zooplankton. J. Plankton Res. 9: 837–855.

Fulton, R. S., III & H. W. Paerl, 1987b. Effects of colonial morphology on zooplankton utilization of algal resources during blue-green algal (*Microcystis aeruginosa*) blooms. Limnol. Oceanogr. 32: 634–644.

Fulton, R. S., III & H. W. Paerl, 1988a. Zooplankton feeding selectivity for unicellular and colonial *Microcystis aeruginosa*. Bull. mar. Sci. 43: 500–508.

Fulton, R. S., III & H. W. Paerl, 1988b. Effects of the blue-green alga *Microcystis aeruginosa* on zooplankton competitive relations. Oecologia 76: 383–389.

Gerritson, J. & K. G. Porter, 1982. The role of surface chemistry in filter feeding by zooplankton. Science 216: 1225–1227.

Gilbert, J. J., 1985. Competition between rotifers and *Daphnia*. Ecology 66: 1943–1950.

Gilbert, J. J., 1988a. Susceptibilities of ten rotifer species to interference from *Daphnia pulex*. Ecology 69: 1826–1838.

Gilbert, J. J., 1988b. Suppression of rotifer populations by *Daphnia*: A review of the evidence, the mechanisms, and the effects on zooplankton community structure. Limnol. Oceanogr. 33: 1286–1303.

Gilbert, J. J., 1989. The effect of *Daphnia* interference on a natural rotifer and ciliate community: Short-term bottle experiments. Limnol. Oceanogr. 34: 606–617.

Gilbert, J. J., 1990. Differential effects of *Anabaena affinis* on cladocerans and rotifers: Mechanisms and implications. Ecology 71: 1727–1740.

Gilbert, J. J. & M. W. Durand, 1990. Effect of *Anabaena flos-aquae* on the abilities of *Daphnia* and *Keratella* to feed and reproduce on unicellular algae. Freshwat. Biol. 24: 577–596.

Gilbert, J. J. & H. J. MacIsaac, 1989. The susceptibility of *Keratella cochlearis* to interference from small cladocerans. Freshwat. Biol. 22: 333–339.

Gilbert, J. J. & R. S. Stemberger, 1985. Control of *Keratella* populations by interference competition from *Daphnia*. Limnol. Oceanogr. 30: 180–188.

Gliwicz, Z. M., 1977. Food size selection and seasonal succession of filter feeding zooplankton in an eutrophic lake. Ekol. Pol. 25: 179–225.

Gliwicz, Z. M. & W. Lampert, 1990. Food thresholds in *Daphnia* species in the absence and presence of blue-green filaments. Ecology 71: 691–702.

Gliwicz, M. Z. & A. Sieniawska, 1986. Filtering activity of *Daphnia* in low concentrations of a pesticide. Limnol. Oceanogr. 31: 1132–1138.

Haney, J. F., 1987. Field studies on zooplankton-cyanobacteria interactions. N. Z. J. mar. freshw. Res. 21: 467–475.

Henning, M., H. Hertel, H. Wall & J.-G. Kohl, 1991. Strain-specific influence of *Microcystis aeruginosa* on food ingestion and assimilation of some cladocerans and copepods. Int. Revue ges. Hydrobiol. 76: 37–45.

Infante, A. & W. Riehl, 1984. The effect of cyanophyta upon zooplankton in a eutrophic tropical lake (Lake Valencia, Venezuela). Hydrobiologia 113 (Dev. Hydrobiol. 23): 293–298.

Jungmann, D., 1992. Toxic compounds isolated from *Microcystis* PCC7806 that are more active against *Daphnia* than two microcystins. Limnol. Oceanogr. 37: 1777–1783.

Jungmann, D., M. Henning & F. Juttner, 1991. Are the same compounds in *Microcystis* responsible for toxicity to *Daphnia* and inhibition of its filtering rate? Int. Revue ges. Hydrobiol. 76: 47–56.

Kerfoot, W. C. & R. A. Pastorok, 1978. Survival versus competition: Evolutionary compromises and diversity in zooplankton. Verh. int. Ver. Limnol. 20: 362–374.

Kirk, K. L. & J. J. Gilbert, 1990. Suspended clay and the population dynamics of planktonic rotifers and cladocerans. Ecology 71: 1741–1755.

Krishnamurthy, T., W. W. Carmichael & E. W. Sarver, 1986. Toxic peptides from freshwater cyanobacteria (blue-green algae). I. Isolation, purification and characterization of peptides from *Microcystis aeruginosa* and *Anabaena flos-aquae*. Toxicon 24: 865–873.

Lampert, W., 1981a. Inhibitory and toxic effects of blue-green algae on *Daphnia*. Int. Revue ges. Hydrobiol. 66: 285–298.

Lampert, W., 1981b. Toxicity of the blue-green *Microcystis aeruginosa*: Effective defence mechanism against grazing pressure by *Daphnia*. Verh. int. Ver. Limnol. 21: 1436–1440.

Lampert, W., 1982. Further studies on the inhibitory effect of the toxic blue-green *Microcystis aeruginosa* on the filtering rate of zooplankton. Arch. Hydrobiol. 95: 207–220.

Lampert, W., W. Fleckner, E. Pott, U. Schober & K-U Storkel, 1989. Herbicide effects on planktonic systems of different complexity. Hydrobiologia 188/189 (Dev. Hydrobiol. 54): 415–424.

Lynch, M., 1979. Predation, competition, and zooplankton community structure: An experimental study. Limnol. Oceanogr. 24: 253–272.

MacIsaac, H. J. & J. J. Gilbert, 1989. Competition between rotifers and cladocerans of different body sizes. Oecologia 81: 295–301.

Marshall, J. S., 1962. The effects of continuous gamma radiation on the intrinsic rate of natural increase of *Daphnia pulex*. Ecology 43: 598–607.

Marshall, J. S., 1966. Population dynamics of *Daphnia pulex* as modified by chronic radiation stress. Ecology 47: 561–571.

Marshall, J. S., 1978. Population dynamics of *Daphnia galeata mendotae* as modified by chronic cadmium stress. J. Fish. Res. Bd Can. 35: 461–469.

Neill, W. E., 1975. Experimental studies of microcrustacean competition, community composition and efficiency of resource utilization. Ecology 56: 809–826.

Nizan, S., C. Dimentman & M. Shilo, 1986. Acute toxic effects of the cyanobacterium *Microcystis aeruginosa* on *Daphnia magna*. Limnol. Oceanogr. 31: 497–502.

Orcutt, J. D. & M. L. Pace, 1984. Seasonal dynamics of rotifer and crustacean zooplankton populations in a eutrophic, monomictic lake with a note on rotifer sampling techniques. Hydrobiologia 119: 73–80.

Reynolds, C. S., 1978. IX. Ecology of Aquatic Organisms. 2. Algae and other Plants: Stratification in natural populations of bloom-forming blue-green algae. Verh. int. Ver. Limnol. 20: 2285–2292.

Reynolds, C. S., J. M. Thompson, A. J. D. Ferguson & S. W. Wiseman, 1982. Loss processes in the population dynamics of phytoplankton maintained in closed systems. J. Plankton Res. 4: 561–600.

Rigler, F. H., 1961. The relation between concentration of food and feeding rate of *Daphnia magna* Straus. Can. J. Zool. 39: 857–868.

Romanovsky, Y. E. & I. Y. Feniova, 1985. Competition among Cladocera: Effect of different levels of food supply. Oikos 44: 243–252.

SAS Institute Inc., 1986. SAS system for linear models. SAS Institute, Cary (N.C.), 210 pp.

Schober, U. & W. Lampert, 1977. Effects of sublethal concentrations of the herbicide Atrazine on growth and reproduction of *Daphnia pulex*. Bull. envir. Contam. Toxicol. 17(3): 269–277.

Schoenberg, S. A. & R. E. Carlson, 1984. Direct and indirect effects of zooplankton grazing on phytoplankton in a hypereutrophic lake. Oikos 42: 291–302.

Smith, A. & J. J. Gilbert, 1995. Relative susceptibilities of rotifers and cladocerans to *Microcystis aeruginosa*. Arch. Hydrobiol. in press.

Stanier, R. Y., R. Kunisawa, M. Mandel & G. Cohenbaz, 1971. Purification and properties of unicellular blue-green algae (order Chroococcales). Bact. Rev. 35: 171–205.

Stemberger, R. S., 1981. A general approach to the culture of planktonic rotifers. Can. J. Fish. aquat. Sci. 38: 721–724.

Stemberger, R. S. & J. J. Gilbert, 1987. Multiple-species induction of morphological defenses in the rotifer *Keratella testudo*. Ecology 68: 370–378.

Webster, K. E. & R. H. Peters, 1978. Some size-dependent inhibitions of larger cladoceran filterers in filamentous suspensions. Limnol. Oceanogr. 23: 1238–1245.

Winner, R. W. & M. P. Farrell, 1976. Acute and chronic toxicity of copper to four species of *Daphnia*. J. Fish. Res. Bd Can. 33: 1685–1691.

Hydrobiologia **307**: 127–138, 1995.
P. Larsson & L.J. Weider (eds), Cladocera as Model Organisms in Biology.

The influence of prey hardness on *Daphnia*'s selectivity for large prey

William R. DeMott
Dept. of Biology and Crooked Lake Biological Station, Indiana-Purdue University, Fort Wayne, IN 46805, USA

Key words: Daphnia, food selection, grazing, prey hardness, size-selection, zooplankton

Abstract

Two comparable methods were used to study the feeding of four species of *Daphnia* on large spherical particles which differed in size and hardness. The first method used gut analysis to estimate the selectivities of daphnids feeding in a broad size range of a single particle type, including polystyrene beads (4–60 μm diameter) in the laboratory and *Eudorina* colonies (10–90 μm) in the field. In the second method, *Daphnia* of different sizes fed in a mixture of 6.5 μm *Chlamydomonas* and one of eight test particles. Smaller daphnids were less effective in feeding on large test particles. Nonlinear regression was therefore used to estimate the *Daphnia* body size at which the clearance rate on a test particle was reduced to 50% of that for *Chlamydomonas*. The results of both methods show that prey size and hardness are both very important in determining daphnid feeding selectivity. For a given particle size, 'soft' algae (naked and gelatinous flagellates) are more readily ingested than 'hard' algae (diatoms and dinoflagellates), and 'hard' algae are more readily ingested than polystyrene beads. *Daphnia* can feed effectively on algae that are 2–5 times larger than the largest ingestible bead.

Introduction

In general, larger predators can capture and ingest larger prey. Such a pattern was shown in a classical paper by Burns (1968), who found a linear relationship between carapace length and the maximum-sized polystyrene bead ingested by individuals of six species of daphnids and *Bosmina longirostris*. Although polystyrene beads are convenient for studying particle size selection, they may not always be representative of natural foods of similar size and shape. DeMott (1986, 1988) demonstrated that daphnids, unlike copepods, *Bosmina*, and certain rotifers, did not discriminate between flavored and unflavored beads or between algae and beads of the same size. However, these experiments were conducted with particles in a small, readily ingested size range (6–12 μm diameter). Bern (1990a) found that *Daphnia cucullata* fed nonselectively on small beads and algae but also observed discrimination against beads near the upper size limit for ingestion.

Since algae are softer than polystyrene beads and can be broken or compressed during ingestion, a large alga should be easier to swallow than a bead of comparable size. The same reasoning suggests that soft algae could be more readily ingested than hard algae. Furthermore, Kerfoot & Kirk (1991) showed weak, but statistically significant discrimination between flavored and unflavored beads near the upper size for ingestion by *Daphnia ambigua*. Their results suggest that taste could also be involved in the selection of large particles by daphnids.

There are many reports of Cladocera feeding effectively on natural prey considerably larger than predicted by the Burns' (1968) regression. The biggest discrepancies are reported for *Bosmina*, which often feeds effectively on large algae, including *Pandorina* (20 μm; Knoechel & Holtby, 1986), *Cosmarium* (28 × 16 × 12 μm; Bleiwas & Stokes, 1985), and *Microcystis* (60–100 μm diameter; Jarvis *et al.*, 1987) as well as large ciliates (40–120 μm length; Jack & Gilbert, 1993). Burns' (1968) bead regression predicts that the largest particle ingested by a cladoceran of *Bosmina*'s size (*c.* 0.4 mm) is only about 10 μm. Daphnids also seem to feed more effectively than predicted on large, soft prey. For example, Schoenberg & Carlson (1984) reported that 'even the smallest *Daphnia* (0.68 mm) caused a substantial decline in *Microcys-*

tis despite an initial colony diameter (36 μm) nearly twice the maximum ingestible particle size predicted by Burns' (1968) regression.' *Daphnia pulex* as small as 1.25 mm fed effectively on a 60–100 μm size fraction of *Microcystis* (Jarvis *et al.*, 1987). Daphnids also feed more effectively than expected on large ciliates (Jack & Gilbert, 1993; Wickham & Gilbert, 1993).

The goal of this study was to test and quantify the effects of particle hardness on *Daphnia*'s selectivity for large particles. Two kinds of feeding experiments were conducted with polystyrene spheres and a variety of 'hard' and 'soft' algae. In the first method, daphnids ranging from neonates to large adults were allowed to feed on a broad size range of particles. Selection was then determined by comparing the particle size distributions in guts with size distributions in the feeding suspension. This approach was used in the laboratory with *Daphnia pulicaria* feeding on polystyrene beads and with *Daphnia pulicaria* and *Daphnia galeata* feeding in Crooked Lake on the spherical colonial green flagellate *Eudorina*. The second approach used radioisotope labeling to examine selection between a readily ingested reference particle and a variety of large test particles of differing hardness.

Methods

Experimental organisms

Most experiments used a clone of *Daphnia pulicaria* obtained from Crooked Lake, near Columbia City, Indiana. Some experiments also included *Daphnia galeata* hybrids from Crooked Lake (Taylor & Hebert, 1992), *Daphnia ambigua* from adjacent Little Crooked Lake or *Daphnia pulex* from a nearby temporary pond. *Daphnia* were cultured in filtered water (GF/C glass fibre) from Crooked Lake in 3.8 litre bottles with *Chlamydomonas reinhardii* as food.

Algae were grown in batch cultures in Erlenmeyer flasks with a modified MBL medium (Stemberger, 1981). Table 1 lists the sources and characteristics of the algae and beads used in the pair-wise laboratory experiments. All algae are spherical or ovoid in shape. Algal dimensions were estimated by measuring at least 50 cells or colonies of each taxon with a microscope and calibrated micrometer. 'Hardness' was not directly measured. Instead, algae were placed in categories based on how easily cells or colonies might be broken, compressed, or distorted during ingestion. Algae with hard, brittle frustules or cell walls, including the diatom *Cyclotella* and the armored dinoflagellate *Peridinium* are considered 'hard algae,' whereas the naked (*Chlamydomonas* spp. and *Carteria*) and gelatinous (*Pandorina*) flagellates are considered 'soft algae.' *Eudorina* cultured in the laboratory was considered 'very soft' because its gelatinous sheath appeared less robust and more flexible than the sheaths of *Pandorina* from laboratory cultures or *Eudorina* from Crooked Lake. Moreover, when concentrated by pouring through a screen, nearly all laboratory *Eudorina* from 50–60 μm in diameter passed through a 48-μm screen, whereas *Pandorina* > 50 μm were retained by the same screen. Since solid polystyrene beads are not broken by zooplankton and should retain their shape during ingestion, beads are considered 'very hard.'

Particle selection experiments

The gut analysis experiments used broad size distributions of single particle types. A broad size distribution of polystyrene beads was created by combining beads from vials with nominal mean diameters of 6, 12, 25, 45, and 90 μm. Very large beads were eliminated by sieving through a 100-μm screen, and by settling. Differential settling was used to make a relatively smooth size distribution in the 30–60 μm size range. An effort was made to include enough beads in the upper size range so that selectivity for larger particles could be accurately assessed (cf. McCauley & Downing, 1985). Measurement of 530 randomly chosen beads, using a microscope (600×), revealed a distribution with 35% of the total number of beads < 10 μm, 26% from 10–15.5 μm, and a broad tail from 15.5–60 μm comprising 39%.

Daphnia pulicaria were first placed in 300 ml bottles containing filtered lake water and 10^4 cells ml^{-1} of *Chlamydomonas reinhardii* and allowed to acclimate for about 1 h. Beads were then added at a concentration of 600 or 1000 ml^{-1}. Bottles were inverted several times at the beginning of each feeding trial and after five or six min to maintain an evenly dispersed mixture of beads. After a 12 min feeding period, the animals were anesthetized with carbonated water and preserved with formalin-sucrose. Bead ingestion was analyzed by placing animals on microscope slides, clearing body tissues with a tissue solubilizer and counting and measuring beads at 600× magnification. Since very large beads are often rejected at the mouth (Bern, 1990b), my data analysis only includes beads found within the gut. In some trials all ingested beads were counted and measured whereas in others I only measured the

Table 1. Sources, mean spherical diameters, and characteristics of the particles used in laboratory feeding experiments (see text for further explanation)

Taxon or type	Source	Diam. (μm)	Description
Carteria olivieri	CBS	26	Naked green, soft
Chlamydomonas reinhardii	UT90	6.5	Naked green, soft
Chlamydomonas capensis	UT1753	25	Naked green, soft
Cyclotella meneghiniana	LCC	23	Diatom, hard
Eudorina sp.[1]	CBS	76	Gelatinous green, very soft
Pandorina morum[2]	CBS	38	Gelatinous green, soft
Peridinium sp	CBS	42	Dinoflagellate, hard
Polystyrene beads	DS	12	Very hard
Polystyrene beads	DS	25	Very hard

[1]Collected on a 48 μm screen.
[2]Poured through a 48 μm screen and collected on a 30 μm screen.
Particles were obtained from Carolina Biological Supply (CBS), University of Texas Culture Collection (UT), Loras College Freshwater Diatom Culture Collection (LCC) and Duke Scientific (DS).

largest bead ingested by each individual daphnid. The total length of each animal was measured from the tip of the head to the base of the tail spine.

Comparable gut content data were obtained for *Daphnia pulicaria* and *Daphnia galeata* feeding on *Eudorina* in Crooked Lake in May during the 'clear water phase' (Secchi depth 7 m; DeMott, unpubl.). Animals were collected in a vertical plankton net tow from 6 m depth to the surface. A single daphnid was placed on a microscope slide and the partially dissected gut was squashed under a microscope cover slip (Kerfoot *et al.*, 1985). Whole, apparently healthy spherical colonies of *Eudorina* were common in guts, whereas colony fragments were not observed and digested cells were rare. Thus, *Eudorina* colonies were ingested whole and seemed to experience little or no damage during gut passage. Each ingested *Eudorina* colony was measured at 400×. Data on the size distribution of *Eudorina* in the plankton were obtained for 300 randomly chosen colonies from water samples collected over the same 0–6 m depth range, preserved in Lugol's solution, and measured using the Utermöhl technique at 400×.

To facilitate comparisons among radiotracer experiments, a readily ingested reference algae, *Chlamydomonas reinhardii* (6.5 μm diameter), was paired with a variety of test particles (Table 1). In each experiment, animals were first acclimated to unlabeled algae for a minimum of 1 h before a short (10 min) feeding trial with radioactive food at the same low concentration ($0.25\ mm^3\ l^{-1}$, about $50\ \mu g\ C\ l^{-1}$ total). Animals ranging from neonates to large adults were used in each experiment. Each measurement of feeding rates and selectivity was typically based on 5–10 juveniles within a 0.1 mm length size class or 2–5 adults within a 0.2 mm length size class. Experiments were run in beakers with 300 ml of feeding suspension. At least three replicate beakers were used for each test particle and some experiments were repeated on different days.

The 'zooplankton sandwich technique' (DeMott, 1988) was used to compare selection between the reference particle (*Chlamydomonas reinhardii)* and a narrow size range of polystyrene beads. Animals were fed a mixture of radioactively labelled *Chlamydomonas* and beads. After the feeding trial, animals were anesthetized with carbonated water and animals within a size class were placed on a microscope cover slip and covered with several drops of tissue solubilizer and a second coverslip. After about 1 h at room temperature, body tissues were dissolved and beads were clearly visible in the guts. After the beads were counted, each sandwich was grasped with a pair of forceps and placed in a scintillation vial in preparation for scintillation counting. Thus, feeding rates and selection between *Chlamydomonas* and beads were measured for the same individuals.

Selectivity and feeding rates on *Chlamydomonas reinhardii* and a test alga were quantified using a dual label technique in which one food type was

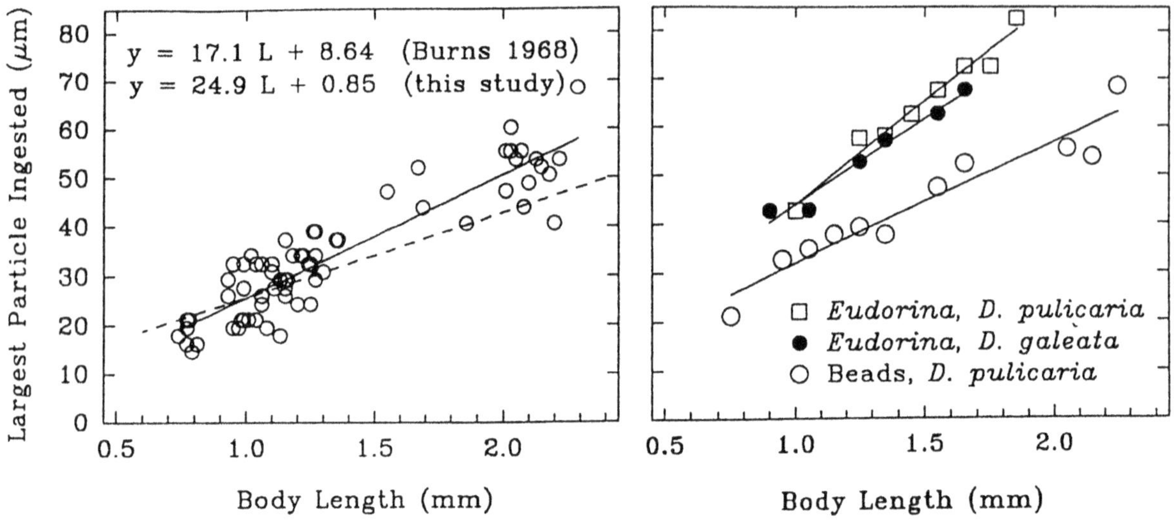

Fig. 1. Linear regressions for body length (L, mm) versus the largest particle ingested (y, μm) for individuals of *Daphnia pulicaria* feeding on beads (left) and for daphnids grouped in 0.1 mm size classes feeding on beads in the laboratory or *Eudorina* in Crooked Lake (right; see text for further explanation).

labelled with ^{14}C-bicarbonate and the other with ^{32}P-orthophosphate (Lampert & Taylor, 1985; DeMott, 1989; DeMott & Moxter, 1991). In experiments with *Pandorina* and *Eudorina*, narrower size distributions were obtained by pouring the algae through sieves after labelling, just before the beginning of feeding trials (Table 1). As in the sandwich experiments, animals were anesthetized with carbonated water, sorted by size and immediately placed in scintillation vials. In both the sandwich and dual label experiments three or six trials were run for each pair of food particles. Methods for preparing labelled algae, scintillation counting and calculation of clearance rates followed Lampert & Taylor (1985) and DeMott & Moxter (1991).

Data analysis

Daphnia typically show nonselective, maximal clearance rates over a broad intermediate particle size range and reduced clearance rates at the upper and lower extremes (Lampert, 1987). Selectivity is expressed here as the clearance ratio, $W_i' = C_i/C_{pref}$, where C_i and C_{pref} are, respectively, the clearance rates on prey i and the preferred prey (Vanderploeg & Scavia, 1979). W' is just the clearance rate (also termed 'filtering rate') normalized to that of the preferred prey. In experiments based on measuring a wide range of particles in daphnid guts, I assumed that the smaller beads or smaller *Eudorina* would be 'preferred' and normalized relative clearance rates to the mean of the two smallest size classes. In dual label experiments, the clearance rates are normalized to the small, readily ingested reference particle, *Chlamydomonas reinhardii*. Thus, a value of 0.0 indicates no feeding on the larger test particle, whereas a W' value of 1.0 indicates nonselective feeding. Although experimental variability around W' values of 1.0 was expected, values significantly > 1.0, which would indicate preferences for large test particles, were not expected.

As mentioned above, many experiments examined selectivity between a reference particle, *C. reinhardii*, and a test particle, using a broad range of animals sizes. One might expect that a moderately large test particle would be too big for small animals but readily ingested by large animals. When plotting selectivity versus body size, this would result in an 'S-shaped' logistic curve, with very low selectivity for small animals, nonselective feeding by large animals and a transition between the two. Therefore, the selectivity (W') versus body length data for each *Daphnia* species and reference particle/test particle combination were fit to a four-parameter, least squares, logistic equation, using an iterative process and the Marquardt-Levenberg algorithm (Sigma-Plot 5.0, Jandel Scientific, 1992). For comparisons between experiments, the resulting logistic equations were solved iteratively to estimate the body size at which the clearance rate on the test particle was reduced by 50% relative to *Chlamydomonas*. This body size where W' = 0.50 is termed the 'L_{50}.'

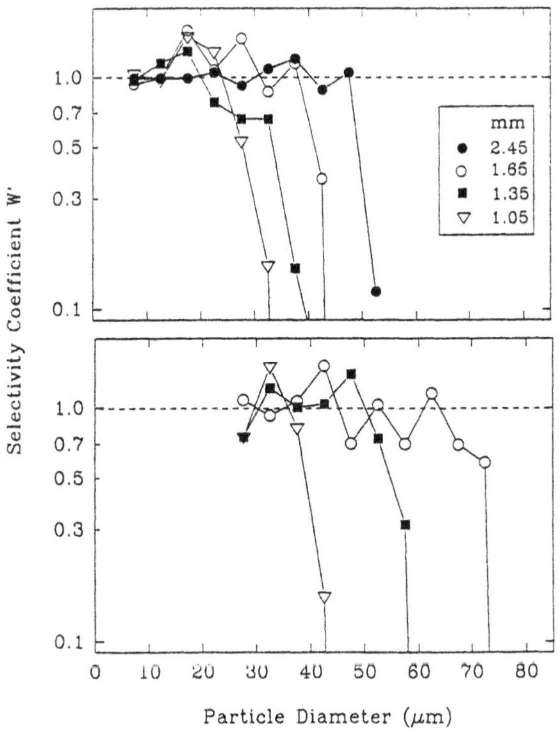

Fig. 2. Selectivity curves for *Daphnia pulicaria* feeding on beads (above) and for *Daphnia pulicaria* and *Daphnia galeata* feeding on *Eudorina* (below). Selectivity values < 0.1, mostly zero values, are not shown on these log scale plots (see text for further explanation).

Results

Gut contents experiments

A linear relation was found between the largest bead ingested by individual *Daphnia pulicaria* and body length (Fig. 1, left, solid line and symbols; $r^2 = 0.83$, $p < 0.001$). Since the regression of Burns (1968) was based on carapace length, not total length, I converted her regression equation to total length, using a linear regression for total length (L) and carapace length (CL) for *Daphnia pulicaria* from Crooked Lake ($L = 1.27 \times CL + 0.094$, $r^2 = 0.99$). My regression curve (solid line in Fig. 1, left) for the largest bead ingested has a higher slope than that obtained by Burns (1968; dotted line in Fig. 1, left), but the results are similar.

Whereas *Daphnia* in the laboratory experiments ingested 10 to 400 beads per individual, with most values > 20, the guts of *Daphnia* from Crooked contained only 0 to 13 *Eudorina* per individual. Since smaller animals ingested fewer *Eudorina*, an analysis based on individuals would provide biased results for *Eudorina*. Therefore, the maximum-sized *Eudorina* ingested by *Daphnia* in Crooked Lake was analyzed for 0.1 mm length classes of *Daphnia* representing a minimum of 20 *Eudorina* ingested per class. A linear relation was found between the largest *Eudorina* ingested (y, μm) and body length (L, mm) for both *D. pulicaria* and *D. galeata* (Fig. 1, right; *D. pulicaria*, $y = 43.2 \times L + 0.383$, $r^2 = 0.97$, $p < 0.001$; *D. galeata*, $y = 35.4 \times L + 8.46$, $r^2 = 0.97$, $p < 0.001$). The lower values for *D. galeata* may reflect its larger helmet. The laboratory bead data were also grouped by 0.1 mm size classes of *Daphnia* to allow direct comparison with the *Eudorina* data. The largest *Eudorina* ingested was substantially larger than the largest bead ingested (Fig. 1, right, grouped bead data; $y = 24.9 \times L + 7.01$, $r^2 = 0.92$, $p < 0.001$).

Figure 2 shows selectivity plots based on measurements of all of the beads or *Eudorina* ingested by 0.1 mm size classes of *Daphnia*. Three size classes were chosen for comparison because each was represented by a minimum of 90 particles ingested for both beads and *Eudorina*. Larger sample sizes were obtained by pooling data for both *D. pulicaria* and *D. galeata* feeding on *Eudorina*. A fourth '2.45 mm' class is based on measurements of 393 beads ingested by a single *D. pulicaria*. The log scale for W' de-emphasizes presumably random fluctuations around 1.0 and emphasizes the upper size limit for each daphnid size class. The results indicate relatively nonselective feeding over a broad size range with distinct upper particle size limits for each animal size class. Thus, these results offer support for using the largest particle ingested as a measure of selectivity, if the sample size is sufficiently large. The selectivity curves also show differences in size selection for beads and for *Eudorina*, especially for larger animals. The curves show that *Daphnia* readily ingests *Eudorina* that are larger than the largest bead ingested.

Radiotracer experiments

The radiotracer experiments provide extensive data on relations between body size and both clearance rates and selectivity. The clearance rate versus body size data were fit to power functions of the form $CR = aL^b$, where CR is the clearance rate (ml $ind^{-1}h^{-1}$), L is the body length (mm), and a and b constants (Table 2). Regressions for the reference particle, *C. reinhardii*, exhibited only modest variability. The mean coefficient b (± S.E.), the slope in a log-log regression, was 2.72 ± 0.10 for 8 experiments with *D. pulicaria* and 2.84 ± 0.12 for 4 experiments with *D. ambigua*. The mean

Table 2. Parameters for log-log regressions of clearance rate versus body size for dual label experiments with *Chlamydomonas reinhardii* and various test particles (Diam., diameter in μm), where clearance rate (ml ind^{-1} h^{-1}) = aL^b, L is body length (mm), a and b are constants, and n is the number of data points for both the test particle and *Chlamydomonas*. Four species of *Daphnia* used in the experiments (*D. ambigua*, Da; *D. pulicaria*, Dpa; *D. galeata*, Dg; and *D. pulex*, Dpx)

				Test particle			*Chlamydomonas*		
Test particle	Diam. (μm)	Sp.	n	b	a	r^2	b	a	r^2
Beads	12	Da	18	5.31	0.487	0.78	2.93	0.621	0.87
	12	Dpa	14	2.93	0.478	0.95	2.32	0.659	0.96
	25	Dpa	30	7.49	0.038	0.94	3.05	0.499	0.94
Cyclotella	23	Da	13	4.59	0.294	0.91	2.85	0.400	0.92
	23	Dpa	37	3.99	0.190	0.92	2.71	0.393	0.94
	23	Dg	16	4.85	0.192	0.91	2.84	0.406	0.96
C. capensis	25	Da	15	3.02	0.480	0.96	3.06	0.470	0.96
	25	Dpa	19	2.73	0.472	0.98	2.78	0.450	0.98
Carteria	26	Dpa	21	3.85	0.343	0.78	3.11	0.517	0.83
Pandorina	38	Dpa	38	4.11	0.320	0.86	2.45	0.640	0.94
Peridinium	42	Dpa	19	4.80	0.084	0.89	2.59	0.406	0.81
Eudorina	75	Da	16	4.79	0.191	0.86	2.51	0.369	0.84
	75	Dpa	16	3.75	0.244	0.93	2.72	0.422	0.96
	75	Dpx	13	3.81	0.317	0.97	2.92	0.499	0.98

a coefficients were 0.50 ± 0.04 and 0.47 ± 0.06 for the same data. Neither the slope (b) nor the intercept (a) coefficients differed between *D. pulicaria* and *D. ambigua* for the reference particle regressions (t-tests, N.S.).

In contrast to the reference particle regressions, the test particle regressions showed highly variable coefficients (Table 2). Since these experiments involved large particles, large slope (b) coefficients indicate a test particle which is more readily eaten by large animals than by small animals.

Figure 3 shows results for the 'bead sandwich' experiments, including clearance rate versus body length data (above) and selectivity (W') versus body length data (below). The selectivity plots illustrate how the logistic regressions were used to estimate the L_{50} body length. Estimates of L_{50} for 12 μm beads were similar for *D. ambigua* and *D. pulicaria* (lower left; 0.74 and 0.78 mm, respectively). As expected, however, the L_{50} for *D. pulicaria* feeding on 25 μm beads was considerably larger (lower right, 1.36 mm). Except for small daphnids, which ingested few beads, the number of beads collected but not ingested was generally < 10% of the number of beads in the gut.

The dual label experiments pairing the reference particle with various test algae provide a variety of results, the extremes of which are shown in Fig. 4. An experiment with 42 μm *Peridinium* illustrates a sigmoidal relation between selectivity and body length (Fig. 4, lower left). Small animals exhibited low selectivity for *Peridinium* whereas large animals fed nonselectively in the same mixture of *Peridinium* and *C. reinhardii*. In contrast, even the smallest daphnids fed nonselectively on mixtures of *Chlamydomonas reinhardii* and 25 μm *Chlamydomonas capensis* (Fig. 4, lower right).

Regression parameters for the selectivity data are listed in Table 3. Most of the data could be fit to a four-parameter logistic equation of the form

$$W' = \frac{a-d}{1+(\frac{L}{c})^b} + d. \quad (1)$$

In this equation, *a* and *d* are the asymptotic maximum and minimum values, respectively, *b* is the slope parameter and *c* is the value of L (length, mm) at the inflection point. Thus, values of *a* approaching 1.0 would show nonselective feeding by larger animals and values of *d* approaching 0 would indicate no feeding on the test particle by smaller animals. The best fits were obtained in cases where the data suggest distinct upper and lower asymptotes, such as the results for 25 μm beads and for *Peridinium*. Experiments with less dis-

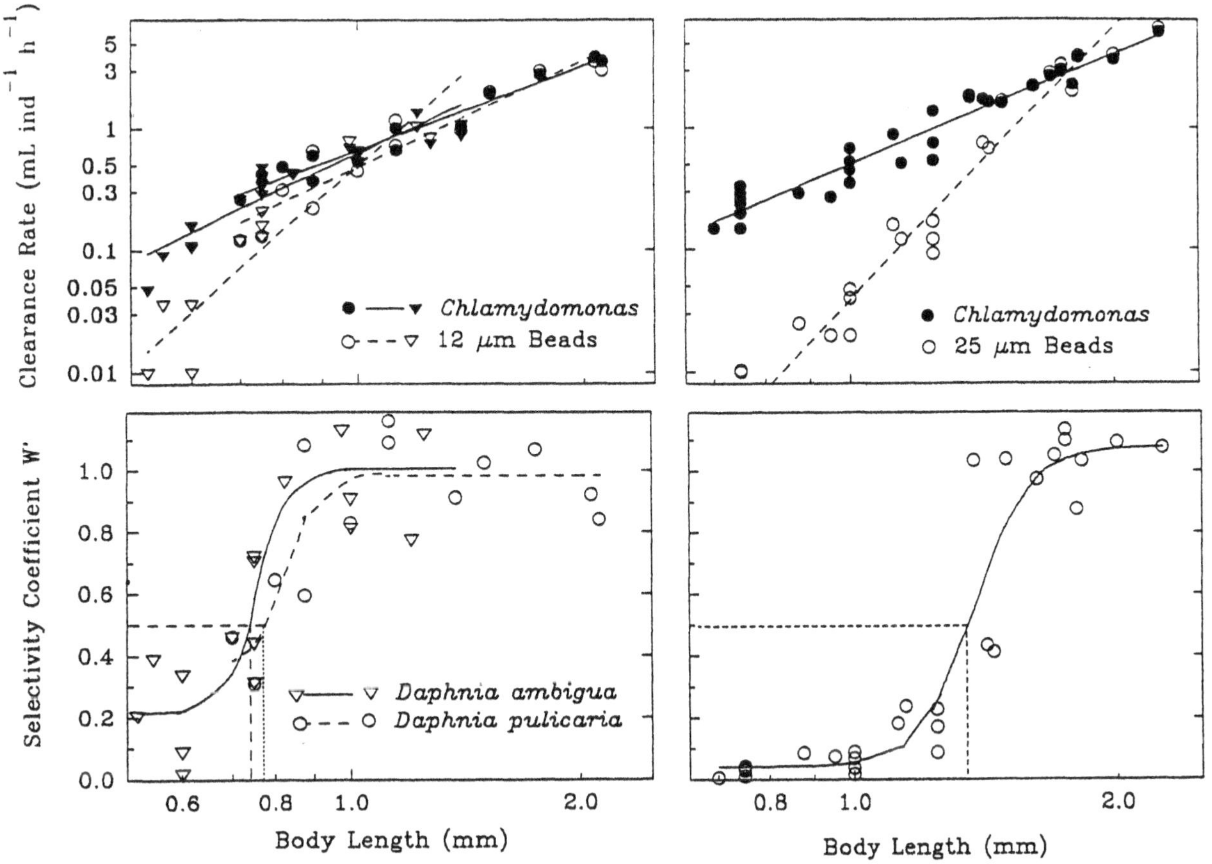

Fig. 3. Clearance rate and selectivity curves for daphnids feeding on mixtures of *C. reinhardii* and either 12 μm beads (left) or 25 μm beads (right). Upper graphs show log-log linear regressions. The selectivity curves (below) are nonlinear logistic regressions. Dashed lines beginning at W' = 0.50 provide graphical estimates of L_{50}'s (see text for further explanation).

tinct upper and lower asymptotes could have been fit by simpler equations, but the logistic equation was used where possible for consistency. In general, most upper asymptote (*a*) values were near 1.0 and most lower asymptote (*d*) values were near 0.0 (Table 3). However, the smallest *Daphnia* species, *D. ambigua,* was evidently too small to show distinct upper asymptotes in experiments with *Cyclotella* and *Eudorina* (*a* values of 1.56 and 1.42, respectively). Similarly, the larger *Daphnia* species were too large to exhibit distinct lower asymptotes in experiments with 12 μm beads, *Carteria*, and *Eudorina*.

Data from two experiments did not give satisfactory fits to the four-parameter logistic equation. A three-parameter model was used for *Carteria* because the lack of low values for W' precluded estimation of the lower asymptote, d (Table 3). The selectivity data for *Chlamydomonas capensis* were fit to linear regressions since there was no relation between selectivity and body size (N.S. slope for both *D. pulicaria* and *D. ambigua*).

Figure 5 shows the selectivity versus body size regressions, each plotted over the experimental range of body lengths. To avoid clutter, the data points were not plotted. Except for the experiments with *Chlamydomonas capensis*, selectivity values increased with increasing *Daphnia* body size. Curves that are shifted to the right indicate particles that are more difficult to ingest. The plots show that particles of similar size and shape can have very different selectivity curves whereas other particles that differ in size can have similar selectivity curves. Moreover, the positions of the curves for particles of similar size correspond with particle hardness. Thus, in the grouping of particles with mean diameters from 23–26 μm (Fig. 5, upper), the 'soft' flagellates *Chlamydomonas capensis* and *Carteria* are more readily ingested than the 'hard' diatom *Cyclotella*, which is more readily ingested than beads. Results for three species of *Daphnia* feeding on

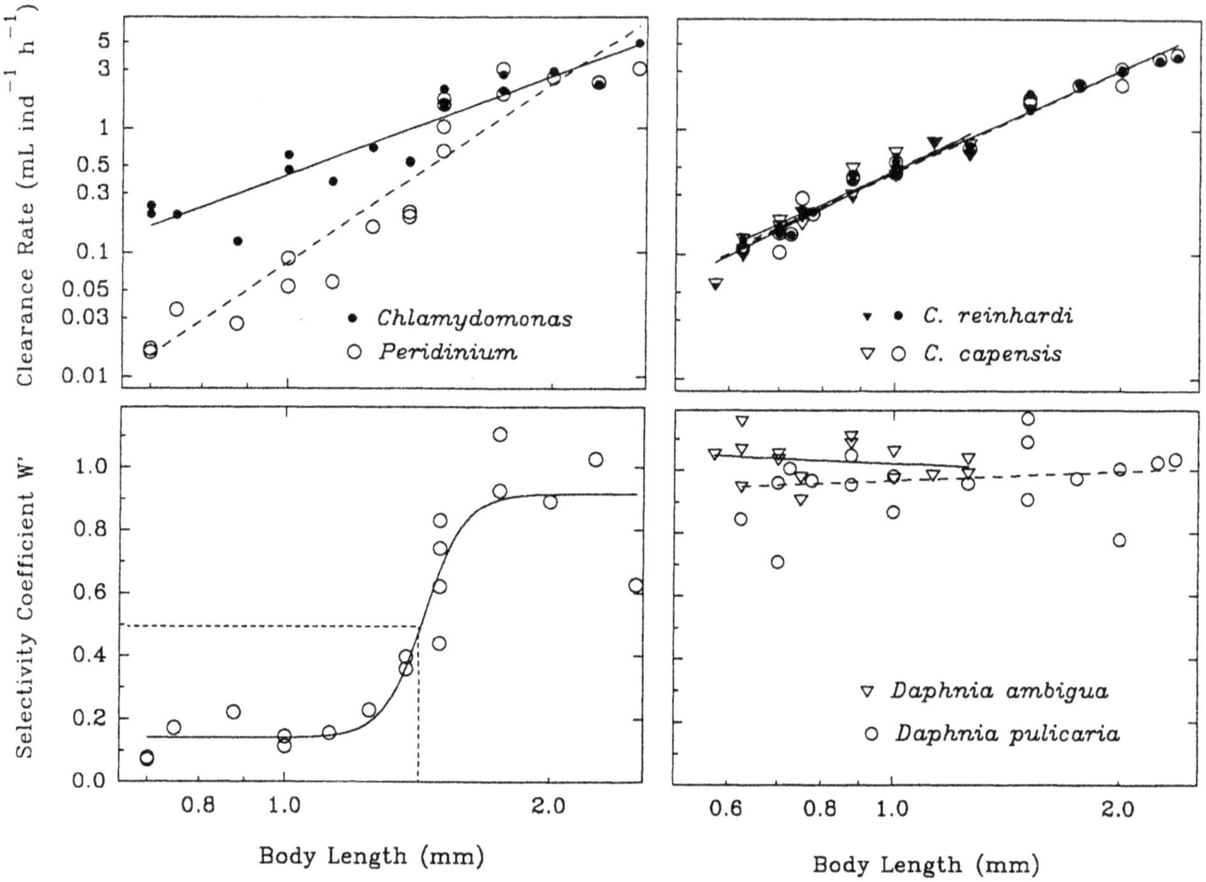

Fig. 4. Clearance rate and selectivity curves for daphnids feeding on mixtures of *C. reinhardii* and either 42 μm *Peridinium* (left) or 25 μm *Chlamydomonas capensis* (right). The selectivity curve for *Peridinium* is a nonlinear logistic regression (lower left), whereas the selectivity curves for *C. capensis* are linear regressions (lower right).

Cyclotella, however, are very similar. Very soft (*Eudorina*) and soft (*Pandorina*) gelatinous greens that differ by a factor of two in size produced very similar selectivity curves (Fig. 5, below). Both of these colonial flagellates were more readily ingested than the 'hard' dinoflagellate, *Peridinium.*

As described above, each *Daphnia* species/test particle combination that was fit to a logistic curve provided a single estimate of L_{50}, the body length where the clearance rate on the test particle is reduced to 50% that of the reference particle (Table 3). For gut contents studies with *Eudorina* from Crooked Lake, the particle diameter where W' = 0.50 was estimated for 0.1 mm size classes of *Daphnia pulicaria* and *D. galeata*. Plots of particle size versus L_{50} provide an overview of the experimental results (Fig. 6). A series of lines were obtained when data for particles of similar hardness were connected by linear regression lines. Although the data points are too few for statistical testing, the five regressions shown in Fig. 6 appear to be more or less parallel. The slopes range from 21.2 for the 'bead sandwiches' to 40.5 for *D. pulicaria* feeding on *Eudorina* in the field, giving a mean ($\pm$ S.E.) of 31.9 $\pm$ 3.2. The slopes for the four algal regressions ranges from 31.1 to 40.5 with a mean of 34.6 $\pm$ 2.2.

The L_{50} plots emphasize that for particles of similar size, beads are eaten less readily than 'hard' algae (*Cyclotella* and *Peridinium*) which are eaten less readily than 'soft' algae (*Carteria* and *Pandorina* in lab experiments; *Eudorina* from the field), which are eaten less readily than 'very soft' algae (*Eudorina* from laboratory cultures). A vertical slice through Fig. 6 can also be used to estimate the effects of particle hardness on the selectivity of animals of a given size. For example, for a body length of 1.0 mm, the same W' = 0.50 value is predicted for an 18 μm bead, a 30 μm 'hard alga,' a 43 μm 'soft alga,' and a 'very soft' alga > 75 μm.

Table 3. Regression coefficients for four-parameter logistic curves comparing selectivity between *C. reinhardii* and various test particles versus *Daphnia* body length. Experiments were run with four species of *Daphnia (D. ambigua,* Da; *D. pulicaria,* Dpa; *D. galeata,* Dg; and *D. pulex,* Dpx). Equations were solved to find the body length (L_{50}, mm) at which the clearance rate on a given test particle was reduced by 50% relative to the reference particle. Asterisks (*) indicate that no estimate was obtained (see text for further explanation)

Test particle	μm	Sp.	a	b	c	d	r^2	L_{50}
Beads	12	Da	1.01	−20.9	0.756	0.217	0.81	0.74
		Dpa	0.98	−20.9	0.825	0.365	0.71	0.78
	25	Dpa	1.08	−13.6	1.38	0.041	0.91	1.36
Cyclotella	23	Da	1.56	−2.60	1.09	0.069	0.88	0.77
		Dpa	0.97	−4.96	0.853	0.00	0.82	0.86
		Dg	0.95	−7.13	0.887	0.098	0.81	0.87
C. capensis[1]	25	Da	1.02	−0.010	*	*	0.17	*
		Dpa	0.97	0.095	*	*	0.15	*
Carteria[2]	26	Dpa	1.10	−1.87	0.547	*	0.52	0.50
Pandorina	38	Dpa	1.13	−3.87	0.813	0.084	0.75	0.73
Peridinium	42	Dpa	0.916	−22.1	1.43	0.142	0.92	1.42
Eudorina	75	Da	1.42	−3.45	1.11	0.010	0.68	0.92
		Dpa	1.01	−11.9	1.02	0.372	0.88	0.91
		Dpx	1.00	−16.9	0.903	0.386	0.83	0.83

[1] Fit to a linear regression of the form y = bx + a.
[2] A three-parameter logistic equation was used due to the lack of low values for W'.

Discussion

In a review on *Daphnia* feeding and nutrition, Lampert (1987) concluded that food selection is primarily determined by particle diameter and shape. Shape is especially important for thin, elongate particles; *Daphnia* can readily ingest filaments ≥ 1 mm length. My study provides support for the importance of size constraints in food selection, but also gives evidence for the importance of another parameter, particle hardness. Even for the spherical particles used in this study, the selectivity spectrum (*sensu* McCauley & Downing, 1985) is not a simple two dimensional particle size versus selectivity curve, but rather, a family of curves for particles which differ in both size and hardness.

In most of my experiments, selectivity varied markedly with *Daphnia* body size. In each of the radiotracer experiments, large *Daphnia* fed nonselectively on mixtures of *Chlamydomonas* and large particles, whereas smaller animals fed selectively on the small reference particle. These results strongly support the view that selectivity of smaller *Daphnia* was largely due to size constraints. Although other factors, such as taste, motility, and particle charge have not been rigorously excluded, size and hardness can account for the observed patterns of selectivity.

Algae and protozoa, unlike polystyrene beads, can be broken, compressed, or distorted in shape during ingestion. Flagellates are usually destroyed during ingestion and are, therefore, often difficult to recognize in gut analysis studies (Porter, 1975; Kerfoot *et al.*, 1985). Thus, a weak cell wall and breakage during ingestion may explain the efficient ingestion of the flagellate *Chlamydomonas capensis* by even the smallest *D. ambigua* (0.55 mm length). Infante & Edmondson (1985) noted that colonial diatoms, such *Fragilaria*, *Synedra*, and *Tabellaria* as well as the large centric, *Stephanodiscus niagarae* (60 μm), could be eaten by small (< 0.9 mm) *Daphnia* and were usually broken during ingestion. My results show that 23 μm *Cyclotella* are more difficult for *Daphnia* to ingest than the 25 μm flagellate *Chlamydomonas capensis* or the 26 μm flagellate *Carteria olivieri*. This suggests that cracking open the diatoms may be more difficult and less consistent than for flagellates.

Breakage was not a factor with *Eudorina* from Crooked lake, as *Daphnia* ingested whole colonies which appeared to pass through the gut undigested.

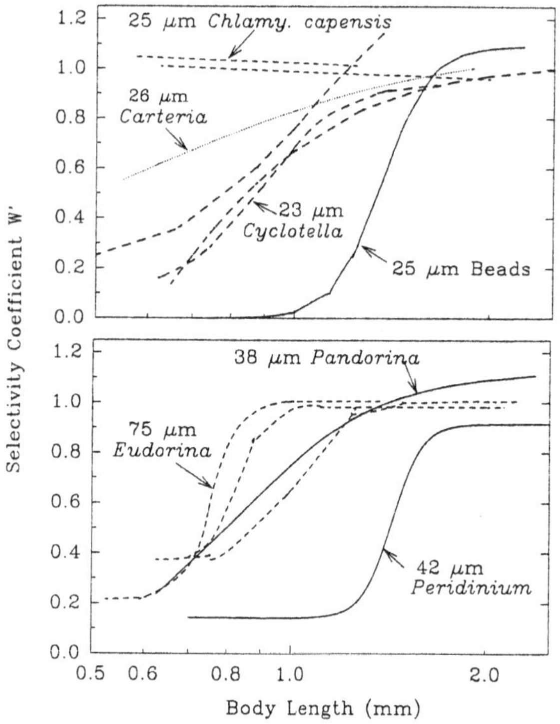

Fig. 5. Selectivity regression curves for experiments pairing a reference particle (*C. reinhardi*) and various larger test particles (< 30 μm diameter above; > 30 μm below). Replicate curves represent regressions for different *Daphnia* species (see Table 3 for regression coefficients).

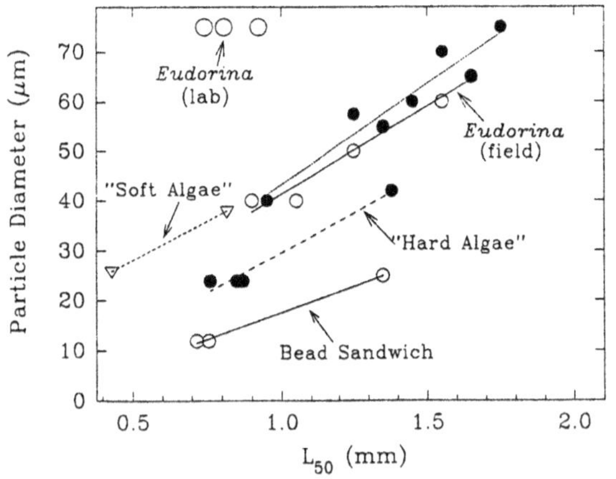

Fig. 6. Relation between particle diameter and L_{50}. Particles are grouped by hardness, including beads, 'hard algae' (*Cyclotella* and *Peridinium*), 'soft algae' (*Carteria*, and *Pandorina*) and 'very soft algae' (*Eudorina* from laboratory; see text for further explanation).

Gelatinous algae such as *Eudorina* are, however, easily flattened and distended under a coverslip, like a balloon filled with water (DeMott, pers. obs.). Perhaps gelatinous algae are distorted and distended during ingestion. Several feeding studies have shown that daphnids, *Bosmina*, and *Brachionus* are effective in ingesting colonies of *Microcystis* (Schoenberg & Carlson, 1984; Fulton & Paerl, 1987; Jarvis *et al.*, 1987). Since *Microcystis* colonies are soft and amorphous, their ingestion might involve both distortion and breakage. Presumably, ciliates and other protozoa could also be broken or distorted during ingestion.

Contrary to my results, Ferguson *et al.* (1982) reported that only small (< 50 μm) colonies of *Eudorina elegans* were commonly observed in the guts of *Daphnia hyalina* in Blelham Tarn. Since they did not present quantitative data, it is not possible to determine whether the difference is real, or is due to observations on smaller *Daphnia*, 'harder' *Eudorina*, or, perhaps, a lack of quantitative analysis. My results show that 'very soft' *Eudorina* from laboratory cultures are more readily ingested than the 'soft' *Eudorina* from Crooked Lake. Thus factors affecting the rigidity of the gelatinous sheath, including the algal species, and, perhaps, the growth conditions, can probably influence ingestibility. As in Crooked Lake (DeMott, unpub.), Reynolds *et al.* (1982) and Reynolds & Rogers (1983) noted that *Eudorina* was abundant in Blelham Tarn during a period of high *Daphnia* abundance in late spring. Although they attributed the success of *Eudorina* to a size refuge from grazing for colonies > 50 μm diameter, my results suggest that digestion-resistance is also important.

In agreement with Burns (1968), my results suggest that the ability to ingest larger prey increases linearly with *Daphnia* body size. The regressions in Fig. 6 further suggest that the slope of the food size/body size relation is similar among prey which differ in hardness. Such a consistency of slope would simplify estimating and modeling selectivity. On the other hand, both my bead data and those of Burns (1968) exhibit lower slopes than the algal data. This may indicate a relative advantage of larger body size in cracking open or slurping down large algae.

My experiments were all run at low food concentrations where clearance rates are expected to be maximal and rejection of particles is expected to be minimal. Although daphnids have limited abilities to handle or reject individual particles, there is evidence for behavioral mechanisms which could modify size selectivity (Gliwicz & Siedlar, 1980; Meise *et al.*, 1985; Hartmann & Kunkel, 1991; Kerfoot & Kirk, 1991). Rejection of low quality or interfering particles is most likely when the total food concentration is high (DeMott,

1993). Thus, particle size selection could also be influenced by food concentration and alternative foods.

The results presented here show how experiments with beads can lead researchers to underestimate the abilities of zooplankton to feed on large natural prey. Although experiments with beads can be used to show seasonal and zooplankton species and body size trends in food selection (e.g. Gliwicz, 1977), such experiments are misleading when extrapolated to algae of similar size. Gut analysis studies provide useful information on patterns and mechanisms of selection but also must be interpreted with caution. The size of particles in the gut cannot always be equated with the size of captured particles. For example, DeMott & Moxter (1991) observed that copepods feeding on *Oscillatoria* filaments 1 mm in length had guts filled with 10 μm fragments. Moreover, the absence of a prey in the gut could be due to rapid digestion as well as the lack of ingestion. Thus, one can question whether the frequently reported absence of cyanobacteria such as *Microcystis* in zooplankton diets (reviewed by Haney, 1987) is real or an artifact. Kobayashi (1991) noted species differences in the sizes of particles in the guts of lake zooplankton but also observed many dates when there was no relation between body size and the largest particle ingested. As noted by McCauley & Downing (1985), such results are probably due to a scarcity of large particles in the water column, which precludes the measurement of size differences. Finally, although gut content studies provide useful information about differences in diet among zooplankton species (e.g. Infante & Edmondson, 1985; Kerfoot *et al.*, 1985), they rarely provide quantitative data on selectivity.

McCauley & Downing (1985) and Bern (1994) both point out that there have been very few studies of food selection by Cladocera over a broad range of food sizes. Even though daphnids are accurately described as 'bulk-processing suspension-feeders,' carefully designed experiments with natural prey are needed to understand their interactions with food resources.

Acknowledgements

This study was supported by NSF grant BSR9006770. I thank for Karen McLellan for measuring and counting *Eudorina* in the guts of daphnids from Crooked Lake and Christine Barlow for analyzing beads in the guts of daphnids in laboratory experiments. Carolyn Burns, Martina Teschner, Karl-Otto Rothhaupt and an anonymous reviewer provided helpful comments on the manuscript. I also wish to thank Petter Larsson and his colleagues at the University of Bergen for organizing this symposium. My participation in the symposium was supported by the Max-Planck-Gesellschaft.

References

Bern, L., 1990a. Size-related discrimination of nutritive and inert particles by freshwater zooplankton. J. Plankton Res. 12: 1059–1067.

Bern, L., 1990b. Post-capture particle size selection by *Daphnia cucullata* (Cladocera). Limnol. Oceanogr. 35: 923–926.

Bern, L., 1994. Particle selection over a broad size range by crustacean zooplankton. Freshwat. Biol. 32: 105–112.

Bleiwas, A. H. & P. M. Stokes, 1985. Collection of large and small particles by *Bosmina*. Limnol. Oceanogr. 30: 1090–1092.

Burns, C. W., 1968. The relationship between body size of filter-feeding Cladocera and the maximum size of particle ingested. Limnol. Oceanogr. 13: 675–678.

DeMott, W. R., 1986. The role of taste in food selection by freshwater zooplankton. Oecologia 69: 334–340.

DeMott, W. R., 1988. Discrimination between algae and artificial particles by freshwater and marine copepods. Limnol. Oceanogr. 33: 397–408.

DeMott, W. R., 1989. Optimal foraging theory as a predictor of chemically-mediated food selection by suspension-feeding copepods. Limnol. Oceanogr. 34: 140–154.

DeMott, W. R. & F. Moxter, 1991. Foraging on cyanobacteria by copepods: Responses to chemical defenses and resource abundance. Ecology 72: 1820–1834.

DeMott, W. R., 1993. Hunger-dependent diet selection in suspension-feeding zooplankton. In Hughes, R. N. (ed.), Diet selection: An interdisciplinary approach to foraging behavior, Blackwell Scientific, Oxford: 102–123.

Ferguson, A. J. D., J. M. Thompson & C. S. Reynolds, 1982. Structure and dynamics of zooplankton communities maintained in closed systems with special reference to algal food supply. J. Plankton Res. 4: 523–543.

Fulton, R. S. III & H. W. Paerl, 1987. Effects of colonial morphology on zooplankton utilization of algal resources during blue-green algal (*Microcystis aeruginosa*) blooms. Limnol. Oceanogr. 32: 634–644.

Gliwicz, Z. M., 1977. Food size selection and seasonal succession of filter feeding zooplankton in an eutrophic lake. Ekol. Pol. 25: 179–225.

Gliwicz, Z. M. & E. Siedlar, 1980. Food size limitation and algae interfering with food collection in *Daphnia*. Arch. Hydrobiol. 88: 155–177.

Haney, J. F., 1987. Field studies on zooplankton-Cyanobacteria interactions. N. Z. J. mar. Freshwat. Res. 21: 467–475.

Hartmann, H. J. & D. D. Kunkel, 1991. Mechanisms of food selection in *Daphnia*. Hydrobiologia 225 (Dev. Hydrobiol. 71): 129–154.

Infante, A. & W. T. Edmondson, 1985. Edible phytoplankton and herbivorous zooplankton in Lake Washington. Arch. Hydrobiol. Beih. Ergenb. Limnol. 21: 161–171.

Jack, J. D. & J. J. Gilbert, 1993. Susceptibilities of different-sized ciliates to direct suppression by small and large cladocerans. Freshwat. Biol. 29: 19–29.

Jandel Scientific, 1992. SigmaPlot Scientific Graph System: Transforms and Curve Fitting Reference.

Jarvis, A. C., R. C. Hart & S. Combrink, 1987. Zooplankton feeding on size fractionated *Microcystis* colonies and *Chlorella* in a hypertrophic lake (Hartbeespoort Dam, South Africa): implications to resource utilization and zooplankton succession. J. Plankton Res. 9: 1231–1249.

Kerfoot, W. C., W. R. DeMott & D. L. DeAngelis, 1985. Interactions among cladocerans: food limitation and exploitative competition. Arch. Hydrobiol. Beih. Ergebn. Limnol. 21: 431–451.

Kerfoot, W. C. & K. L. Kirk, 1991. Degree of taste discrimination among suspension-feeding cladocerans and copepods: Implications for detritivory and herbivory. Limnol. Oceanogr. 36: 1107–1123.

Knoechel, R. & L. B. Holtby, 1986. Cladoceran filtering rate: body length relationships for bacterial and large algal particles. Limnol. Oceanogr. 31: 195–200.

Kobayashi, T., 1991. Body lengths and maximum gut food-particles sizes of the dominant cladocerans and calanoid copepods in Wallerawang Reservoir, New South Wales. Aust. J. mar. Freshwat. Res. 42: 399–408.

Lampert, W., 1987. Feeding and nutrition in *Daphnia*. In Peters, R.H. & R. de Bernardi (eds), *Daphnia*. Mem. Ist. ital. Idrobiol. 45: 143–192.

Lampert, W. & B. E. Taylor, 1985. Zooplankton grazing in a eutrophic lake: implications of diel vertical migration. Ecology 66: 68–82.

McCauley, E. & J. A. Downing, 1985. The prediction of cladoceran grazing rate spectra. Limnol. Oceanogr. 30: 202–212.

Meise, C. J., W. R. Munns, Jr. & N. G. Hairston, Jr., 1985. An analysis of the feeding behavior of *Daphnia pulex*. Limnol. Oceanogr. 30: 862–870.

Porter, K. G., 1975. Viable gut passage of gelatinous green algae ingested by Daphnia. Verh. int. Ver. Limnol. 19: 2840–2850.

Reynolds, C. S. & M. W. Rogers, 1983. Cell- and colony-division in *Eudorina* (Chlorophyta: Volvocales) and some ecological implications. Br. phycol. J. 18: 111–119.

Reynolds, C. S., J. M. Thompson, A. J. D. Ferguson & S. W. Wiseman, 1982. Loss processes in the population dynamics of phytoplankton maintained in closed systems. J. Plankton Res. 4: 561–600.

Schoenberg, S. A. & R. E. Carlson, 1984. Direct and indirect effects of zooplankton grazing on phytoplankton in a hypereutrophic lake. Oikos 42: 291–302.

Stemberger, R. S., 1981. A general approach to the culture of planktonic rotifers. Can. J. Fish. aquat. Sci. 38: 721–724.

Taylor, D. J.& P. D. N. Hebert, 1992. *Daphnia galeata mendotae* as a cryptic species complex with interspecific hybrids. Limnol. Oceanogr. 37: 658–665.

Vanderploeg, H. A. & D. Scavia, 1979. Two electivity indices for feeding with special reference to zooplankton grazing. J. Fish. Res. Bd Can. 36: 362–365.

Wickham, S. A. & J. J. Gilbert, 1993. The comparative importance of competition and predation by *Daphnia* on ciliated protists. Arch. Hydrobiol. 126: 289–313.

Hydrobiologia **307**: 139–145, 1995.
P. Larsson & L.J. Weider (eds), Cladocera as Model Organisms in Biology.

Egg size and egg mass of *Daphnia magna*: response to food availability

Irina Trubetskova[1] & Winfried Lampert*
Max Planck Institute of Limnology, Postfach 165, 24302 Plön, Germany (author for correspondence)*
[1]*Present address: Oismae 96–53, Tallinn-EE0035, Estonia*

Key words: Daphnia, dry egg weight, egg size, egg size-mass relationship, food availability, specific gravity of eggs

Abstract

The influence of different food availability on egg size and egg mass in *Daphnia magna* Straus was studied in long-term experiments using a flow-through system. *Daphnia* were either kept at constant high or low food levels or subjected to alternating periods of high food and starvation. Some animals were starved continuously after they had deposited their first clutch of eggs. Eggs were measured and weighed and their density (dry mass per volume) was determined. The results support the model of Glazier (1992), which defines a region of 'reproductive constraint' at very low food concentrations and a region of 'adaptive response' as food concentrations increase. Egg sizes were largest under continuously low food concentrations (0.1 mg C l^{-1}), which indicates that the maximum of Glazier's non-linear response curve is at very low food levels. Eggs produced during starvation were small, probably as a result of reproductive constraints. Egg density was about 0.37 mg dry weight mm^{-3} and did not differ between treatments.

Introduction

Egg size and neonate size in *Daphnia* are positively correlated (Glazier, 1992; Lampert, 1993). Therefore, the life-history consequences of varying egg size in *Daphnia* (Lynch, 1989) have become a topic of interest in the recent literature. Tradeoffs between offspring size and number (Ebert, 1993), the optimum amount of energy allocated per egg under varying food conditions (Glazier, 1992; Gliwicz & Guisande, 1992) and in the presence of predators (Stibor, 1992), and the determination of maturation size by maternal effects via egg size (Lampert, 1993) are especially interesting. The two most important factors in determining *Daphnia* egg size are probably size of the mother (Glazier, 1992; Lampert, 1993) and food supply (Guisande & Gliwicz, 1992). Ebert (1993) presented a very clear graph of the combined effect of clutch number and food level on neonate size – a close correlate of egg size – for *D. magna*.

The effect of food availability on egg size and neonate size has been particularly well-studied but with contradictory results. A number of studies found a positive correlation between egg size and food supply (Green, 1956; Brambilla, 1980; Urabe, 1988; Lynch, 1989; Tessier & Consolatti, 1991), others detected a negative relationship (Baichorov, 1992; Glazier, 1992; Guisande & Gliwicz, 1992; Taylor, 1985) or found no correlation (Glazier, 1992; Semenchenko & Semenjuk, 1988; Taylor, 1985). In a critical analysis of available data, Glazier (1992) proposed two main reasons for the conflicting results: the use of different ration levels and food conditions and the lack of consideration of maternal effects. There is also a genetic component in neonate size variability (Tessier & Consolatti, 1991; Ebert, 1993), i.e, one must expect differences in egg size responses to changing food availability between species and clones.

Glazier (1992) proposed a hypothetical model of the response of *Daphnia* egg size to food availability, which implies a maximum egg size at a relatively low food level. With decreasing food availability, egg size increases as larger neonates are more starvation resistant (Gliwicz & Guisande, 1992). However, when food availability decreases further, eggs will become smaller again due to energetic constraints. This model has not been tested quantitatively. It is possible that daphnids rarely reach the low-food region of the model when the food supply is low but constant. On the other hand, energy constraints may be effective if well-fed females are starved and use their limited amount of energy reserves to still produce some eggs. These eggs should then be small. One aim of this study is to test this hypothesis.

Direct measurements of egg dry mass are a rather delicate and time-consuming procedure, hence, egg mass is usually estimated from egg size. These estimates are based on microscopic measurements of egg size and calculation of egg volume. Egg volume is converted to wet egg mass assuming a density of 1 mg mm^{-3} (Edmondson & Winberg, 1971). Then wet mass must be converted to dry mass. There is a surprisingly large variation in the applied conversion factors. They range from 0.1 (De Bernardi *et al.*, 1978) to 0.4 (Ivanova, 1985). Although some of the variability may be due to measuring errors, another source of variation may be that the conversion factor is influenced by internal or external factors. For example, the ratio dry mass:volume of the eggs may depend on food availability or egg size. As this is an important factor in energetic consideration for constraints and tradeoffs, we use our data to calculate the dry mass:volume ratio of eggs in a clone of *D. magna* and compared them with results from the literature.

Methods

Experiments were carried out with a clone of *D. magna* that had been maintained in the laboratory for many years. Experimental animals were kept at 20 °C with unlimited food (*Scenedesmus acutus* Meyen) under dim continuous light. Freshly born neonates were placed randomly in the chambers of a flow-through system (Lampert *et al.*, 1988) and grown at constantly high (1.5 mg carbon l^{-1}) or constantly low (0.1 mg C l^{-1}) concentrations of *Scenedesmus* in membrane-filtered (0.45 μm) lake water (10 animals per vessel; flow rate 1.5 l d^{-1}). The food suspension was freshly prepared every day. After the daphnids had produced their first clutch of eggs, they were transferred to clean vessels and were subsequently subjected to different feeding protocols. They were either fed continuously (1.5 or 0.1 mg C l^{-1}) or exposed to alternating periods of feeding and starvation until they had their eighth clutch. Only primiparae grown at the high food level were exposed to alternating food conditions. The alternating food regime started with a period of starvation. Daphnids were transferred to different food conditions (to starvation or 1.5 mg C l^{-1}) whenever they produced a new clutch of eggs. Thus, clutches 1, 3, 5, and 7 were produced after a period of feeding while clutches 2, 4, 6, and 8 were produced after a period of starvation. The maximum time between egg deposition and transfer to new food conditions was 12 hours. As this applied to transfers in both directions, we considered the effect of the delay on egg size negligible. Some primiparae from both the high-food and the low-food treatments were isolated and starved to death. During starvation, *Daphnia* remained in the flow-through system but they received only membrane-filtered water.

Egg measurements were taken within a few hours of the deposition of a new clutch of eggs into the brood pouch (i.e., egg stage 1, Threlkeld, 1979). Eggs were carefully removed from the brood pouch with fine needles. Their diameter was measured to the nearest 0.01 mm under a dissecting microscope equipped with a semi-automatic measuring device (Wild MMS 235). Repeated measurement of individual eggs at different angles showed very little variability. Assuming the shape to be spherical, the egg volume was calculated from the diameter. Individual clutches were transferred to small aluminum containers, dried for 24 hours at 60 °C, cooled in a desiccator, and weighed to the nearest 0.1 μg on an electronic balance. Mean egg mass was calculated as dry weight of the clutch divided by the number of eggs. The density of an egg d (mg mm^{-3}) was calculated from egg dry mass W (mg) and volume V (mm^{-3}):

$$d = W/V. \qquad (1)$$

Assuming a density of 1, d can be used for the conversion factor wet mass: dry mass.

Results

Experimental treatments started after the deposition of the first clutch, therefore the first eggs available for analysis came from clutch two. These eggs are smaller than those of the third and subsequent clutches but eggs of the third clutch differ only slightly from eggs of later clutches (Glazier, 1992; Ebert, 1993; Lampert, 1993). In order to eliminate the maternal effect (size of the mother) we analyzed two separate groups, the pooled eggs from clutches 3 to 8 and the eggs of the second clutch. Continuously starved *Daphnia* produced no more than two additional clutches, thus they

Table 1. Effects of different food levels on egg diameter (D), egg mass (W) and specific density of eggs (d) in pooled clutches number 3 to 8 of *D. magna* (mean ± 1 S.D.). Values connected by a horizontal line are not significantly different from each other (ANOVA, $p > 0.05$)

Food level	High	Fluctuating		Low
		Fed	Starved	
(mg C l^{-1})	(1.5)	(1.5)	(0.0)	(0.1)
D (mm)	0.35 ± 0.01	0.34 ± 0.01	0.37 ± 0.03	0.39 ± 0.01
W (μg)	7.95 ± 0.37	6.96 ± 0.46	10.27 ± 2.02	12.20 ± 1.79
d (mg mm^{-3})	0.37 ± 0.02	0.35 ± 0.01	0.39 ± 0.11	0.39 ± 0.05

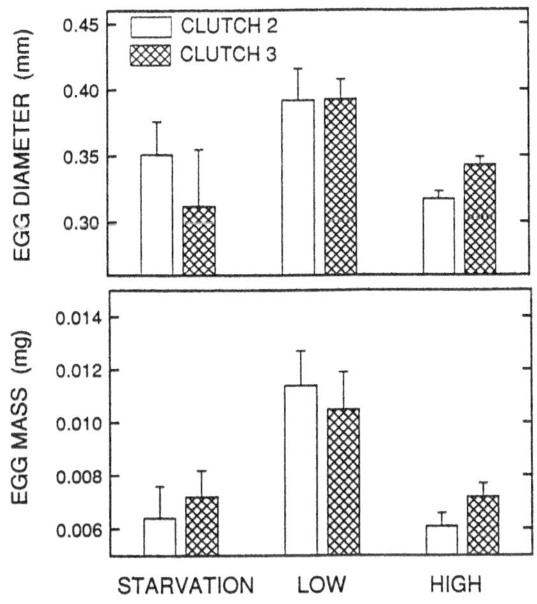

Fig. 1. Egg size and egg mass (mean ± 1 S.D.) of the second and third clutches of *D. magna* in response to different feeding regimes.

could not be included in the analysis of clutches 3 to 8.

In pooled clutches 3 to 8, egg sizes differed significantly (ANOVA, $p < 0.05$) between high and low food levels; eggs were larger at the low food level (Table 1). Egg sizes also differed between periods of feeding and starvation in the fluctuating food regime. Eggs deposited after a starvation period were larger. No difference was found between egg sizes at continuously high food and at the end of each feeding period, and between egg sizes at continuously low food and at the end of each starvation period.

The same trend is found in individual egg mass (Table 1), although eggs from periodic feeding were slightly lighter than eggs from continuous feeding at high food. Overall, eggs were heavier during low food availability. However, the dry mass per volume (d) did not differ between high and low food treatments (ANOVA, $p = 0.638$). The mean (± S.D.) conversion factor was 0.378 ± 0.05. The variability of egg mass between clutches was higher for low than for high food availability (Table 1). This may be a consequence of a lower number of eggs per clutch and, thus, a higher error in weighing the eggs.

Feeding prehistory determined the reproduction of daphnids under continuous starvation. Starving primiparae grown at low food concentrations (0.1 mg C l^{-1}) stopped reproduction immediately, while animals raised at high food levels (1.5 mg C l^{-1}) had enough material in reserve to produce a second and in some cases a third clutch. Time until death was about 10 days, thus the food concentration during starvation in the flow-through vessels was well below the threshold for maintenance metabolism, although some bacteria may have been available to the daphnids.

Egg sizes of clutches 2 and 3 were compared to the sizes in the respective clutches under the continuous feeding regime (Fig. 1). There were no significant differences between the second and the third clutches in egg size and egg mass in all treatments (ANOVA, $p > 0.05$). As in the later clutches, eggs from the low food treatment were significantly ($p < 0.05$) larger than eggs from the high food treatment. However, under starvation, eggs produced from energy reserves were smaller than eggs from the low food treatment. They were as small as eggs from the high food treatment. The effect was particularly clear for egg masses (Fig. 1).

The conversion factor (d) did not differ between treatments and was similar to the factor obtained for later clutches. Thus the density of eggs is not affected by food conditions.

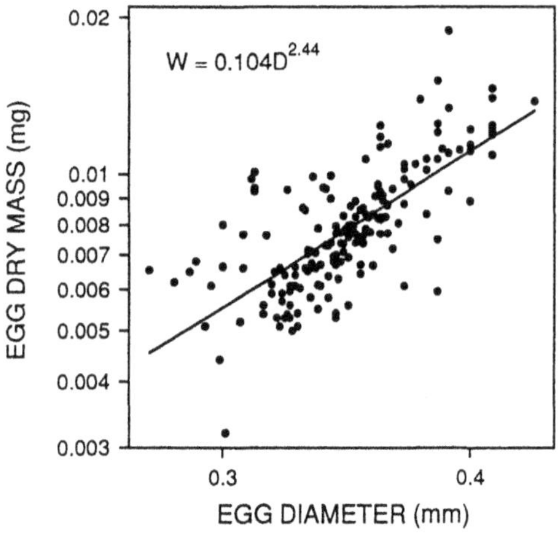

Fig. 2. Log-log regression of egg mass on egg size. Pooled data from all treatments.

Measurements from all treatments were pooled to calculate a general log-log regression of egg mass on egg diameter (Fig. 2). The resulting relationship is

$$W = 0.104 \times D^{2.44} \quad (2)$$

$$(r^2 = 0.53; n = 181; p < 0.0001).$$

The exponent 2.44 is lower than the expected 3.0 for a sphere. This seems to be caused by the relatively high mass of a few very small eggs. Exclusion of eggs with diameters of less then 0.312 mm (lowest mean value in Fig. 1) results in the relationship

$$W = 0.190 \times D^{3.03} \quad (3)$$

$$(r^2 = 0.64; n = 164; p < 0.0001).$$

The conversion factor (*d*) calculated from this relationship is about 0.37. The exclusion of the very small eggs is probably justified as those were often not viable.

Discussion

When Glazier (1992) developed his qualitative model on dependence of offspring size on food quantity, it was not possible to make quantitative predictions as the available data came from different clones and species, food conditions, and experimental protocols. Thus the assignment of different data sets to his optimum curve is somewhat arbitrary. A rigorous test of the model would require quantitative data on the response of egg size to food availability. The low-food treatment used by Glazier (0.3 mg C l^{-1}) was not low enough to fall on the decreasing branch of the response curve left of the optimum (the 'reproductive constraint' range). Thus he could not quantitatively determine the food concentration for maximum egg size.

Although, with only two concentrations tested, it is possible that the two points are on different sides of the maximum, even our low-food treatment (0.1 mg C l^{-1}) was probably not low enough to detect any reproductive constraint. As egg mass decreased from 0.1 to 1.5 mg C l^{-1}, this range seems to fall on the decreasing branch above the optimum (the 'adaptive response'). This negative relationship between food concentration and egg (neonate) size has been found by several authors (for review see Guisande & Gliwicz, 1992; Glazier, 1992). Table 2 compares egg masses which have been obtained with the same experimental set up and with various concentrations of the same food species for various species of *Daphnia*. With one exception (Taylor, 1985; *D. pulicaria*) egg mass decreases in the range between 0.1 and 1.5 mg C l^{-1}. This suggests that the optimum, if it exists, is below 0.1 mg C l^{-1}. There is, however, only one measurement below 0.1 mg C l^{-1}. Guisande & Gliwicz (1992) performed an experiment at 0.04 mg C l^{-1}, which is close to the threshold concentration for growth (Gliwicz, 1990). In fact, they found egg mass decreased at very low concentrations (although not significantly) and they came to the same conclusion as Glazier (1992), i.e., that egg size was constrained by insufficient energy availability. With a different food source (*Chlamydomonas*), Urabe (1988) observed a decrease in egg size of *D. galeata* at 0.05 mg C l^{-1}. Both studies suggest that, under constant food conditions, the 'reproductive constraint' is only effective when the food concentration comes close to the threshold for growth.

Energetic constraint should be expected if *Daphnia* are starved and produce eggs from body reserves only. Such a situation is likely in nature. It may occur if a *Daphnia* population overshoots the carrying capacity and creates a clear-water phase. *D. magna* cultured at low-food conditions did not have enough reserves to produce any eggs during starvation, but those cultured at high food produced two more clutches, though egg numbers were reduced. This is consistent with the idea that total egg formation needs two instars and the number of eggs is determined during each instar (Ebert & Yampolsky, 1992).

Eggs produced during starvation were in fact smaller than eggs produced at 0.1 mg C l^{-1}, but had the same

Table 2. Available data on food dependence of egg mass (mean ± S.D.) in *Daphnia* obtained by different authors with the same experimental design (flow-through system, *Scenedesmus acutus* Meyen as food, 20 °C). (F) indicates fluctuating food conditions)

Species	Clutch number	Food level (mg C l^{-1})	Egg mass (μg)	Source
D. pulex	1–10	0.1	2.80 ± 0.22	Taylor (1985)
	1–10	1.0	2.40 ± 0.25	
D. pulicaria	1–10	0.1	2.80 ± 0.38	
	1–10	1.0	2.80 ± 0.27	
D. hyalina	1–3	0.2	3.55 ± 0.17	Guisande & Gliwicz (1992)
	1–3	1.0	2.04 ± 0.15	
D. pulicaria	1–2	0.04	2.70 ± 0.20	
	1–3	0.2	2.96 ± 0.18	
	1–3	1.0	2.20 ± 0.20	
D. magna	2	0.0	6.39 ± 1.19	This study
	2	0.1	11.14 ± 1.34	
	2	1.5	6.10 ± 0.54	
	3	0.0	7.16 ± 0.99	
	3	0.1	10.50 ± 1.38	
	3	1.5	7.21 ± 0.46	
	3–8	0.1	12.20 ± 1.79	
	3–8	1.5	7.95 ± 0.37	
	4, 6, 8	0.0(F)	10.27 ± 2.02	
	3, 5, 7	1.5(F)	6.96 ± 0.46	

Table 3. Density (*d*) of *Daphnia* eggs (mg dry mass mm^{-3}) under varying food conditions. Sources: (1) Green (1956); (2) Baichorov (1992); (3) this study; (4) Dumont *et al.* (1975); (5) Mityanina (1978); (6) Urabe (1988), recalculated from carbon mass assuming 50 % carbon in dry mass; (7) Stibor (1991); (8) Khmeljova & Golubev (1984); (9) Berberovic (1990). (F) indicates fluctuating food conditions)

Species	Food	Food level (mg C l^{-1})	*d* (mg mm^{-3})	Source
D. magna	*Chlorella vulgaris*	> 1.0	0.33	(1)
		1.6	0.30	(2)
	Scenedesmus acutus	0.0	0.38	(3)
		0.1	0.38	(3)
		1.5	0.38	(3)
		0.0(F)	0.36	(3)
		1.5(F)	0.37	(3)
	Field samples			
	Belgium		0.30	(4)
	Belarus		0.26	(5)
	Karelia		0.29	(5)
D. galeata	*Chlamydomonas reinhardi*	0.05–2.50	0.30	(6)
D. hyalina	*Scenedesmus acutus*	1.0	0.47	(7)
D. longispina	Field samples (Belarus)		0.51	(8)
D. galeata	Field s. (L. Constance)		0.32	(9)
D. hyalina	Field s. (L. Constance)		0.33	(9)

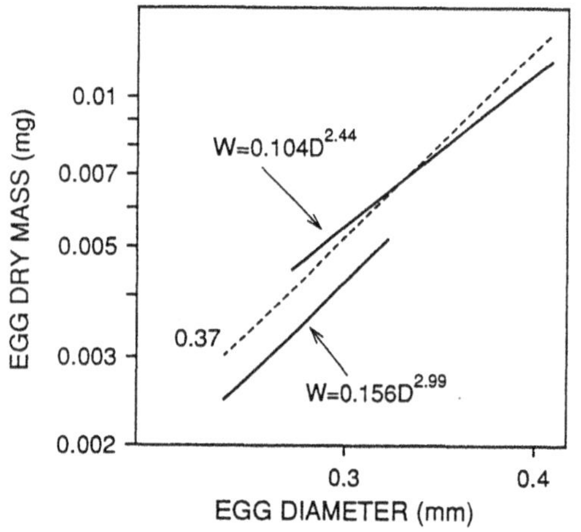

Fig. 3. Comparison of the relationship of egg mass to egg size (Fig. 2; upper solid line) with the relationship given by Baichorov (1992, lower solid line). The dashed line represents the theoretical relation assuming a perfect sphere, a density of 1, and a conversion factor (*d*) dry mass:wet mass of 0.37.

size as eggs produced at 1.5 mg C l^{-1}. The size reduction is, however, probably not due to the fact that these animals had been raised at high food before they were starved. The 'memory effect' is unlikely to have lasted to the third clutch that produced eggs of the same size as the second one. Thus, the sequence of egg masses in Fig. 1 seems to reflect the 'reproductive constraint' and the 'adaptive response' of Glazier's (1992) model.

Although we cannot exclude the possibility that small eggs have a slightly higher dry mass per volume, we are probably justified in assuming a mean density (*d*) of 0.37 mg mm^{-3}. Urabe (1982) found a similar effect. The smallest eggs measured had a slightly higher carbon:volume ratio, but there was no significant correlation with egg volume. Most values reported in the literature are slightly lower than our measurements (Table 3). The only values which are considerably higher (about 0.5) are the values for the closely related *D. longispina* and *D. hyalina*. As we did not find differences in egg density between treatments, the variability to be seen in Table 3 may be due to clonal differences. Clonal variability in egg size and neonate size has been well-documented (Tessier & Consolatti, 1991; Glazier, 1992; Ebert, 1993), but information on egg density is lacking. It is, therefore, interesting to compare our regression of egg mass on egg size with the one provided by Baichorov (1992) (Fig. 3).

The two regressions cover different size ranges of eggs. Baichorov's regression is truncated at the higher end, as all eggs were produced at high food levels (1.6 mg C l^{-1}). Although the different size distributions of eggs are probably not clone specific but can be explained by the experimental set up, the two regressions produce different values for *d* (Table 3), which may be an indication of clonal variation. Consequently, microscopic measurements of egg size may yield reasonable estimates of egg mass in studies of reproductive investment within a single clone, but *d* should be determined for each clone or species until we know more about its variability.

Acknowledgements

I. Trubetskova was supported by a research fellowship from the Max Planck Society and – during the preparation of the manuscript – by her German friends Hildegard and Heinz Pack to whom she is heartily grateful. I. Trubetskova also thanks all members of the Max Planck Institute of Limnology at Plön for their permanent interest, concern and support. She is especially thankful to Heike Mumm, Gisela Stolpe, Joanna Pijanowska and William DeMott for help with processing the data.

References

Baichorov, B. M., 1992. Relations between fertility, egg size, and reproductive effort in *Daphnia magna* under various density and temperature regimes. Zhurn. Obsh. Biol. 53: 830–839. (In Russian).

Berberovic, R., 1990. Biomass composition of two sympatric *Daphnia* species: Impact of environmental factors and life history strategies. Ph.D. Thesis, University of Constance, Germany, 174 pp.

Brambilla, D. J., 1980. Seasonal change in size in maturity in small pond *Daphnia*. In Kerfoot, W. C. (ed.), Evolution and Ecology of Zooplankton Communities. The University Press of New England, Hanover (N.H.), London: 438–455.

De Bernardi, R., P. Lacqua & E. Soldavini, 1978. Effects of temperature and food on developmental time and growth in *Daphnia obtusa* Kurz and *Simocephalus vetulus* (O. F. Müller) (Crustacea, Cladocera). Mem. Ist. ital. Idrobiol. 36: 171–191.

Dumont, H. J., I. van de Velde & S. Dumont, 1975. The dry weight estimate of biomass in a selection of Cladocera, Copepoda and Rotifera from the plankton, periphyton and benthos of continental waters. Oecologia 19: 75–97.

Ebert, D., 1993. The trade-off between offspring size and number in *Daphnia magna*: The influence of genetic, environmental and maternal effects. Arch. Hydrobiol. Suppl. 90: 453–473.

Ebert, D. & L. Yu. Yampolsky, 1992. Family planning in Daphnia: When is clutch size determined? Russ. J. Aquat. Ecol. 1: 143–147.

Edmondson, W. T. & G. G. Winberg, 1971. A manual on methods for the assessment of secondary productivity in fresh waters. IBP Handbook No. 17. Oxford, Blackwell, 358 pp.

Glazier, D. S., 1992. Effects of food, genotype, and maternal size and age on offspring investment in *Daphnia magna*. Ecology 73: 910–926.

Gliwicz, Z. M., 1990. Food threshold and body size in cladocerans. Nature 343: 638–640.

Gliwicz, Z. M. & C. Guisande, 1992. Family planning in *Daphnia*: resistance to starvation in offspring born to mothers grown at different food levels. Oecologia (Berl.) 91: 463–467.

Green, J., 1956. Growth,size and reproduction of *Daphnia* (Crustacea: Cladocera). Proc. zool. Soc. Lond. 126: 173–204.

Guisande, C. & Z. M. Gliwicz, 1992. Egg size and clutch size in two *Daphnia* species grown at different food levels. J. Plankton Res. 14: 997–1007.

Ivanova, M. B., 1985. Production of planktonic crustaceans in fresh waters. Leningrad, Zool. Inst. Acad. Sci. USSR, 222 pp. (In Russian).

Khmeljova, N. N. & A. P. Golubev, 1984. Production of fodder and commercial crustaceans. Minsk, Nauka i Tekhnika, 216 pp. (In Russian).

Lampert, W., 1993. Phenotypic plasticity of the size at first reproduction in *Daphnia*: the importance of maternal size. Ecology 74: 1455–1466.

Lampert, W., R. D. Schmitt & P. Muck, 1988. Vertical migration of freshwater zooplankton: a *test* of some hypotheses predicting a metabolic advantage. Bull. Mar. Sci. 43: 620–640.

Lynch, M., 1989. The life history consequences of resource depression in *Daphnia pulex*. Ecology 70: 246–256.

Mityanina, I. F., 1978. Changes in weight and calorific value of *Daphnia magna* during embryogenesis. Vesti ANBSSR. Ser. bial. navuk. 5: 85–88. (In Russian).

Semenchenko, V. P. & G. A. Semenjuk, 1988. Production of *Moina macrocopa* (Straus) at different temperature and trophic conditions. Dokl. Akad. Nauk BSSR 32: 856–858. (In Russian).

Stibor, H., 1991. Größenvariabilität von *Daphnia* spp. bei der ersten Reproduktion. Diploma thesis, University of Kiel, Germany, 65 pp.

Stibor, H., 1992. Predator-induced life-history shifts in a freshwater cladoceran. Oecologia 162–165.

Taylor, B. E. 1975. Effects of food limitation on growth and reproduction of *Daphnia*. Arch. Hydrobiol. Beih. Ergebn. Limnol. 21: 285–296.

Tessier, A. J. & N. L. Consolatti, 1991. Resource quantity and offspring quality in *Daphnia*. Ecology 72: 468–478.

Threlkeld, S. T., 1979. Estimating cladoceran birth rates: the importance of egg mortality and egg age distribution. Limnol. Oceanogr. 24: 601–612.

Urabe, J., 1988. Effect of food concentration on the net production of *Daphnia galeata*: Separate assessment of growth and reproduction. Bull. Plankton Sci. Jpn. 35: 159–174.

Hydrobiologia **307**: 147–151, 1995.
P. Larsson & L.J. Weider (eds), Cladocera as Model Organisms in Biology.

Reduced digestibility of UV-B stressed and nutrient-limited algae by Daphnia magna

Ellen van Donk[1] & Dag O. Hessen[2]
[1]*Agricultural University Wageningen, Dept. of Water Quality Management and Aquatic Ecology, P.O. Box 8080, 6700 DD Wageningen, The Netherlands*
[2]*University of Oslo, Biological Institute, Div. Limnology, P.O. Box 1027, Blindern, 0316 Oslo, Norway*

Key words: Daphnia, digestibility, grazing, nutrient limitation, phytoplankton, UV-B

Abstract

Daphnia magna was fed the green alga *Selenastrum capricornutum*, cultured under four different growth conditions: (1) phosphorus limitation, (2) nitrogen limitation, (3) UV-B irradiation, and (4) no nutrient limitation, no UV-B irradiation. Contrary to non-limited algal cells, nutrient-limited cells were not efficiently assimilated. Especially, P-limited cells passed through the gut mostly intact, while N-limited cells were partly assimilated. Also, algae exposed to moderate doses of UV-B radiation (0.3 mW cm^{-2} of UV_{312}) were less efficiently assimilated after being grazed. Digestibility of the algae decreased with increased UV-B exposure time. Nutrient-limited and UV-B stressed algal cells increased in volume and became granular in appearance. These changes in the algal cells, combined with changed cell wall properties, most probably reduced their digestibility.

Introduction

There are numerous studies on the effects of nutrient-limitation (e.g. Hecky & Kilham, 1988) and UV-B radiation (e.g. Steemann-Nielsen, 1964; Lorenzen, 1979; Worrest, 1982; Cullen & Lesser, 1991) on phytoplankton growth and productivity. Both nutrient limitation and UV-B radiation may induce a delay in algal cell division. Furthermore, current levels of solar UV-B radiation are known to inhibit other metabolic processes in algal cells such as nutrient uptake and flagellate motility (Döhler, 1985; Häder & Häder, 1988). Nutrient limitation and UV-B stress may also change the cell morphology and structure of the algae. An increase in average cell size has been observed both for UV-B irradiated algal cells (Karentz *et al.*, 1991; Veen, 1991) and for nutrient limited cells (Mitchell *et al.*, 1992; Van Donk & Hessen, 1993). These changes in stressed algal cells may also influence their digestibility for zooplankton. Recently Van Donk & Hessen (1993) demonstrated that P-limited green algae (*Scenedesmus subspicatus* and *Selenastrum capricornutum*) passed largely intact through the gut of *Daphnia* and were thus protected from heavy grazing losses. The reduced digestibility of P-limited cells was attributed to structural and morphological changes in the P-limited cells. Other studies have demonstrated decreased growth and reproduction in *Daphnia* if fed P-limited algae even when food was abundant (Hessen, 1990; Watanabe, 1990; Sommer, 1992; Sterner, 1993). So, apparently P-limitation of the phytoplankton can alter trophic interactions, reducing the transfer of energy to zooplankton.

In order to analyse whether reduced digestibility is also present among algae subject to other stress factors than P-limitation, we conducted a series of experiments with *D. magna* feeding on N-limited and UV-B irradiated cells of the green alga *Selenastrum capricornutum*. Also, a replicate of the grazing experiment with P-limited cells, as described in Van Donk & Hessen (1993), was carried out.

Materials and methods

The green alga *Selenastrum capricornutum* NIVA CHL 10, used in the experiments, was obtained from the culture collection of the Norwegian Institute for Water Research (Skulberg and Skulberg, 1990). An inoculum of the axenic algal culture was incubated in the inorganic nutrient medium Z8 20% at 20 °C (Skulberg & Skulberg, 1990). Illumination was provided by cool-white fluorescent tubes at 70 μE m^{-2} s^{-1} using a 14:10 h LD cycle.

To obtain P- and N-limited algal cells, exponentially growing cells were inoculated into flasks containing a phosphorus-free and nitrogen-free medium, respectively. The cells entered stationary state after five days (P-limitation) and three days (N-limitation).

To obtain UV-stressed algae, cells harvested during exponential growth, were incubated in complete medium in flat-bottom beakers and exposed to UV_{312}-radiation. A Vilber-Lourmat VL-115M, 15 Watts Lamp, with peak irradiance at 312 nm was used as the UV source. Additional white light illumination was given as described above. A gradient of UV-B doses was achieved by exposing the algae to UV_{312} during various periods (2, 4, 8, 12, 20, 25 and 48 h). We used a standard intensity of 0.3 mW cm^{-2}, or 0.018 J cm^{-2} min^{-1}, corresponding to maximum, mid-summer irradiance as measured near Oslo (Hessen & Van Donk, 1994). These exposure periods represented doses of 2.2, 4.3, 8.6, 13.0, 21.6, 27.0 and 51.8 J cm^{-2} UV_{312} respectively. Light intensity was measured with a Vilber-Lourmat VLX-3W Radiometer, with peak sensitivity at 312 nm. The UV_{312}-dose recorded by this instrument is somewhat higher than those measured with the commonly used LI-COR spectro-radiometer, since the former integrates some radiation on both sides of 312 nm.

The grazing experiments were carried out with the zooplankter *Daphnia magna*, obtained from a laboratory culture grown on *S. capricornutum* for more than two years. In the first two experiments, feeding by *D. magna* on the green alga *S. capricornutum* cultured under P- and N-limited conditions, was studied. In the third experiment *D. magna* was fed UV-B stressed algae. A control experiment was carried out with non-limited, non-irradiated algal cells.

All experiments were conducted in the dark in 0.5 l Erlenmeyer flasks and six daphnids (length 2.8–3.3 mm, mean dry weight 160 μg ind^{-1}) were incubated in 400 ml algal suspensions. For each treatment, algae without zooplankters were also incubated in the dark to control whether algal numbers changed due to factors other than grazing.

Table 1. Disappearance rates of the green alga *Selenastrum capricornutum* due to grazing by *Daphnia magna* (15 ind l^{-1}) grazing on P-limited (Exp. I), N-limited algae (Exp. II) and UV-B stressed algae (Exp. III) (h = hours of exposure to 0.018 J cm^{-2} min^{-1} UV_{312}). C:N:P ratios and cell volumes of the cells are measured at the start of the experiment. In the controls, algal cells were non-limited and not stressed by UV-B. (Standard errors are given in parentheses). An asterisk (*) indicates that the values differ significantly ($p<0.05$) from the controls

Treatment	C:N:P in algae (atomic ratios)	Cell volume (μm^3)	Disappearance rate (day^{-1})
Control	67:9:1	31.5 (±1.8)	1.35 (±.08)
Exp. I			
P-lim	714:71:1	58.6 (±2.7)*	0.22 (±.03)*
Exp. II			
N-lim	55:2:1	39.4 (±4.3)*	0.94 (±.18)*
Exp. III			
UV_{312}			
2h		33.6 (±1.3)	1.33 (±.04)
4h		26.7 (±2.3)*	1.48 (±.22)
8h		36.6 (±3.4)*	1.25 (±.14)
12h		42.2 (±1.6)*	1.06 (±.21)*
20h		53.7 (±2.2)*	0.62 (±.03)*
25h		56.2 (±2.1)*	0.47 (±.08)*
48h		64.2 (±3.5)*	0.32 (±.03)*

The initial algal cell concentration in all experiments was *c.* 2 × 10^5 cells ml^{-1}. The mean volume of the algal cells was measured at the start of the grazing experiments by an electronic particle counter (Coulter Counter). Volumes of stressed and non-stressed algal cells were statistically compared by applying the non-parametric Mann-Whitney U test (Fowler & Cohen, 1993). The critical probability value was set at $p<0.05$. The grazing experiments lasted for 2–3 days. Flasks were shaken manually four times a day. At the start of Exp. I and II the C, N and P contents of both non- and nutrient-limited algae were analyzed. C and N contents of algae, filtered onto acid-washed pre-ignited GF/F-filters, were measured on a Carlo-Erba CHN 1106 elemental analyzer, and P content of the total samples was measured after peroxo-disulfate digestion.

Algae were counted at intervals of four to eight hours during the three days of incubation (*Daphnia* introduced at t = 0). Samples were preserved with Lugol's solution. Algae were counted microscopically

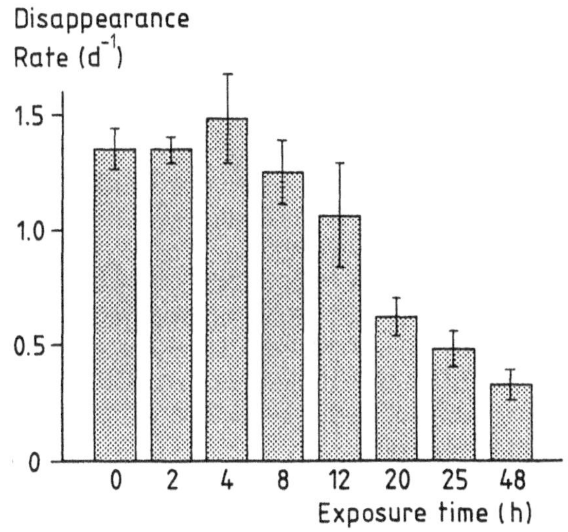

Fig. 1. Mean disappearance rates (day^{-1}) of the green alga *Selenastrum capricornutum* caused by grazing of *Daphnia magna*. The algae were first exposed to UV_{312}-radiation (0.018 J cm^{-2} min^{-1}) during different periods. Standard error bars are given.

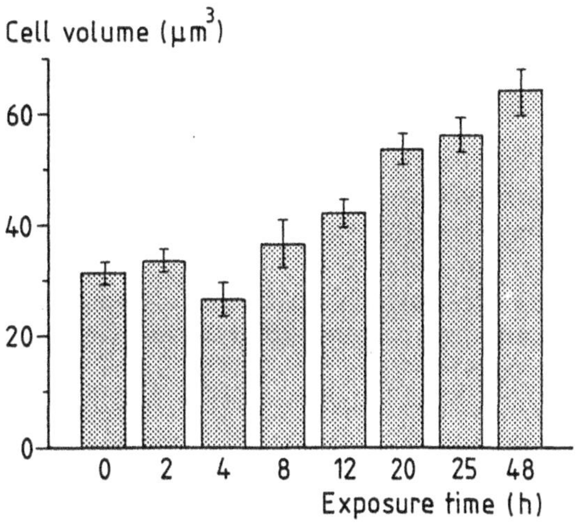

Fig. 2. Mean cell volumes of the green alga *Selenastrum capricornutum* after exposure to UV_{312}-radiation (0.018 J cm^{-2} min^{-1}) during different periods. Standard error bars are given.

and disappearance rates of the algae were calculated from the slope of the regression between the decrease in cell numbers (ln(cells/ml)) and time (days). Because measurements for replicate flasks were similar compared with other sources of variation, the replicates were pooled. The calculated disappearance rates of the algae were not true clearance rates because a large proportion of the algae in the nutrient-deficient and UV-irradiated treatments were grazed, but not digested. The disappearance rates measured in Exp. I, II and III were compared statistically with the rate in the controls by applying a t-test (critical probability value of $p<0.05$) on the regression lines (Fowler & Cohen, 1993).

The egestion products of the daphnids were analysed during each experiment. Animals were placed individually on slides under cover slips. After defecation, rectum contents of *Daphnia* were dispersed and examined microscopically (Porter, 1975).

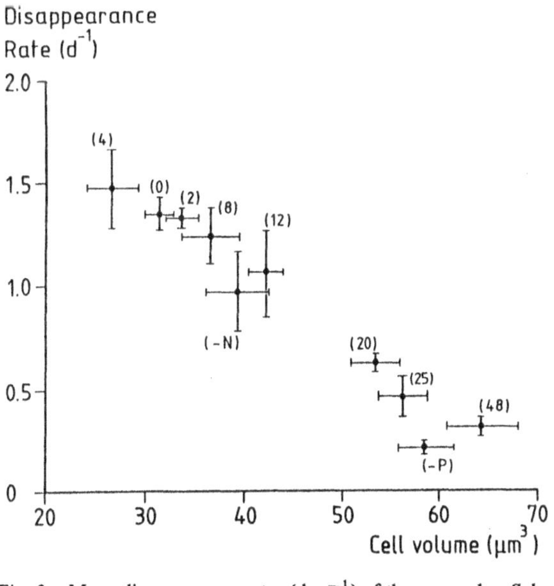

Fig. 3. Mean disappearance rates (day^{-1}) of the green alga *Selenastrum capricornutum* grazed by *Daphnia magna* as a function of mean algal cell volumes, measured during nutrient-limitation (−P = P-limitation, −N = N limitation) and UV-B stress (2–48 h = hours of exposure to UV_{312}-radiation of 0.018 J cm^{-2} min^{-1}). Standard error bars are given.

Results

The algal disappearance rates determined for *D. magna* feeding on P- and N-limited algal cells in experiments I and II and UV-B irradiated cells in experiment III are given (Table 1). Further, the C:N:P ratios and mean volumes of the algal cells are given, as well as disappearance rate for non-limited and nonirradiated algae in the control experiment. Algae incubated in the dark without *Daphnia* demonstrated no significant changes in cell numbers during the three days of incubation.

A significant lowered algal disappearance rate (80–85% reduction relative to the control) was observed in experiment I when *D. magna* was feeding on P-limited

S. capricornutum. *Daphnia* feeding on N-limited algae in experiment II demonstrated a significant reduction in disappearance rate of approximately 30% (Table 1).

In experiment III when *Daphnia* were grazing on UV-B stressed algae, disappearance rates were found to be significantly reduced only for algae irradiated for 12 h or more (>13.0 J cm^{-2} UV_{312}). Algae stressed by UV-B for 48 h (51.8 J cm^{-2} UV_{312}) demonstrated a reduction of approximately 75% relative to the control (Table 1 and Fig. 1).

Both nutrient limited and UV-B stressed algae increased their cell volumes. Under the light microscope, the large cells had a granular appearance with internal, spherically-shaped clear areas. The largest cell volumes were found for P-limited algae and 48 h UV-B irradiated algae (approximately twice the normal cell volume; Table 1). A significant increase in cell volume was found for all algae irradiated for 8 h or more (>8.6 J cm^{-2} UV_{312}). A significant decrease in cell volume was measured only for algae irradiated during 4 h (Table 1; Fig. 2). A negative relationship was found between algal cell volumes and disappearance rates (Fig. 3).

Fecal analyses revealed that the algal cells with a larger volume were normally grazed, but a high number of these cells passed undamaged through the daphnid gut, probably because of reduced digestibility. In contrast, nonlimited and nonirradiated cells were assimilated efficiently.

Discussion

Cell volume of both nutrient-limited and UV-B stressed cells increased and under the light microscope, these cells had a granular appearance. Such changes in algal cells have also been observed in other studies where algae were cultured under nutrient limitation (Mitchell *et al.*, 1992; Sterner *et al.*, 1993; Van Donk & Hessen, 1993) or UV-B stress (Karentz *et al.*, 1991; Veen, 1991). Karentz *et al.* (1991) and Veen (1991) stated that an increase in cell dry weight of UV-B stressed algae reflects a storage of proteins and carbohydrates due to delayed cell division. Myklestad (1977) and Søndergaard & Schierup (1982) observed the release of photosynthetic products by algae growing under nutrient-stress.

Our results indicate that digestibility of algae by *Daphnia* may be severely influenced by the physiological state of the algae. Contrary to nonlimited and nonirradiated algal cells, nutrient limited and UV-B irradiated cells were assimilated with very low efficiency. Specifically, P-limited and UV-B stressed algae passed largely intact through the gut.

Van Donk & Hessen (1993) found a doubling of cell volume for P-limited cells of *Selenastrum capricornutum* and a considerable reduction in the assmilation rate by *D. pulex* (approximately 75%) and *D. magna* (approximately 50%). In these experiments digestibility of the P-limited cells decreased while feeding rates were not influenced. A decrease in clearance and feeding rates was measured by Sterner *et al.* (1993) for *D. pulex* grazing on severely P- and N-limited green alga *Scenedesmus acutus*. To our knowledge no previous grazing experiments with UV-B stressed algae have been carried out.

Preliminary results (Van Donk *et al.*, in prep.) from our experiments with the green alga *Chlamydomonas reinhardtii*, cultured under P- and N-limited conditions, support the results presented here. Transmission electron microscopic photographs of nutrient-limited *Chlamydomonas* revealed intracellular storage of photosynthetic products, considerable thickening of the cell wall mainly consisting of glycoproteins, and accumulation of mucous carbon compounds around the cell wall. The same grazing experiments, executed with a mutant clone of *Chlamydomonas reinhardtii* lacking a cell wall, showed, however, efficient assimilation of non-limited as well as nutrient-limited cells. Although the structure of the nutrient-limited cells changed due to storage of photosythetic products, the volumes of the mutant cells did not increase significantly. These results indicate that changes in and around the cell wall are most likely responsible for the decrease in digestibility. One might speculate whether UV-stress inhibits P-uptake in algae, giving rise to the same physiological properties as those observed under real P-deficiency. Döhler & Biermann (1987) measured a reduced ^{15}N-nitrate uptake for two marine diatoms under UV-B irradiation.

Summarizing, we conclude that UV-B stressed and nutrient-limited phytoplankton strongly increased their resistance to digestibility, most probably due to both mucous secretion and changes in the cell wall structure. Thus, at times when phytoplankton growth rates are reduced by these stress factors, also grazing pressure on these phytoplankters would be reduced. These grazing experiments were only conducted with stressed green algae. Studies on other edible algae are necessary to assess if these effects form a common phenomenon among phytoplankton populations. Several studies report differential sensitivity of freshwater assemblages to inhibition by UV-B radiation (e.g. Gala & Giesy, 1991; Döhler, 1985). Reduced digestibility of stressed algal cells, if also present among natural populations, may significantly alter trophic interactions and reduce the transfer of energy between primary producers and consumers in aquatic ecosystems.

Acknowledgements

We would like to thank the staff at the Norwegian Institute for Water Research, Oslo, for the use of algal cultures and laboratory facilities and Dr R. D. Gulati and Prof Dr W. J. Wolff for their constructive comments on earlier versions of the manuscript.

References

Cullen, J. J. & M. P. Lesser, 1991. Inhibition of photosynthesis by ultraviolet radiation as a function of dose and dosage rate: results for a marine diatom. Mar. Biol. 111: 183–190.

Döhler, G., 1985. Effect of UV-B radiation (290–320 nm) on the nitrogen metabolism of several diatoms. J. Plant Physiol. 118: 391–400.

Döhler, G. & J. Biermann, 1987. Effect of U.V.-B irradiation on the response of ^{15}N-nitrate uptake of *Lauderia annulata* and *Synedra planctonica*. J. Plankton Res.: 881–890.

Fowler, J. & L. Cohen, 1993. Practical statistics for field Biology. John Wiley & Sons Ltd, England.

Gala, W. R. & J. P. Giesy, 1991. Effects of ultraviolet radiation on the primary production of natural phytoplankton assemblages in Lake Michigan. Ecotoxicology and Environmental Safety 22: 345–361.

Hecky, R. E. & P. Kilham, 1988. Nutrient limitation of phytoplankton in freshwater and marine environments: A review of recent evidence on the effects of enrichment. Limnol. Oceanogr. 33: 796–822.

Hessen, D. O., 1990. Carbon, nitrogen and phosphorus status in *Daphnia* at varying food conditions. J. Plankton Res. 12: 1239–1249.

Hessen, D. O. & E. van Donk, 1994. Effects of UV-radiation of humic water on primary and secondary production. Water, Air and Soil Pollution 75: 325–338.

Häder, D. P. & M. A. Häder, 1988. Inhibition of motility and phototaxis in the green flagellate, *Euglena gracilis*, by UV-B radiation. Arch. Microbiol. 150: 20–25.

Karentz, D., J. E. Cleaver & D. L. Mitchell, 1991. Cell survival characteristics and molecular responses of antartic phytoplankton to ultraviolet-B radiation. J. Phycol. 27: 326–341.

Lorenzen, C. J., 1979. Ultraviolet radiation and phytoplankton photosynthesis. Limnol. Oceanogr. 24: 1117–1120.

Mitchell, S. F., F. R. Trainor, P. H. Rich & C. E. Goulden, 1992. Growth of *Daphnia magna* in the laboratory in relation to the nutritional state of its food species, *Chlamydomonas reinhardtii*. J. Plankton Res. 14: 379–391.

Myklestad, S, 1977. Production of carbohydrates by marine planktonic diatoms. II Influence of the N/P ratio in the growth medium on the assimilation ratio, growth rate, and production of cellular and extracellular carbohydrates by *Chaetoceros affinis* var. *willei* (Gran) Hustedt and *Skeletonema costatum* (Grev.) Cleve. J. exp. mar. Biol. Ecol. 29: 161–179.

Porter, K. G., 1975. Viable gut passage of gelatinous green algae ingested by Daphnia. Verh. int. Ver. Limnol. 19: 2840–2850.

Skulberg, O. M. & R. Skulberg, 1990. Research with algal cultures-NIVA's Culture Collection of Algae. NIVA-Report ISBN 82–551743–6.

Sommer, U, 1992. Phosphorus-limited *Daphnia*: Intraspecific facilitation instead of competition. Limnol. Oceanogr. 37: 966–973.

Søndergaard, M. & H. H. Schierup, 1982. Release of extracellular organic carbon during a diatom bloom in Lake Mossø: molecular weight fractionation. Freshwat. Biol. 12: 313–320.

Steemann-Nielsen, E., 1964. On a complication in marine productivity work due to the influence of ultraviolet light. J. Cons. inst. Explor. Mer. 29: 130–135.

Sterner, R. W., D. D. Hagemeier, W. L. Smith & R. F. Smith, 1993. Phytoplankton nutrient limitation and food quality for *Daphnia*. Limnol. Oceanogr. 38: 858–871.

Sterner, R. W., 1993. *Daphnia* growth on varying quality of *Scenedesmus:* mineral limitation of zooplankton. Ecology 74: 2351–2360.

Van Donk, E. & D. O. Hessen, 1993. Grazing resistance in nutrient stressed phytoplankton. Oecologia 93: 508–511.

Veen, A., 1991. Experimental analysis of phytoplankton production as effected by enhanced UV-B radiation. Annual Scientific Report Institute of Public Health and Environmental Protection, Bilthoven ISBN 90–6960–012–9.

Watanabe, Y., 1990. C:N:P ratios of size-fractionated seston and planktonic organisms in various trophic levels. Verh. int. Ver. Limnol. 24: 195–199.

Worrest, R. C., 1982. Review of literature concerning the impact of UV-B radiation upon marine organisms. In Calkins, J. (ed.), The Role of Solar Ultraviolet in Marine Ecosystems. Plenum, New York: 309–346.

Hydrobiologia **307**: 153–159, 1995.
P. Larsson & L.J. Weider (eds), *Cladocera as Model Organisms in Biology.*

UV-B-tolerance of alpine and arctic *Daphnia*

Iris D. Zellmer
Zoological Institute, Ludwig-Maximilians-University, Munich, Germany; Present address: Wilhelm-Buschstr. 14, 06118 Halle, Germany

Key words: Daphnia, UV-B, survival, melanism, reproduction, behavior

Abstract

Survival of transparent and pigmented *Daphnia* from alpine as well as arctic habitats, exposed to solar radiation with total or reduced UV-B, was measured. In the alpine pond, survivorship and reproduction of transparent juvenile as well as adult *Daphnia* were significantly reduced due to UV-B. Transparent adult arctic *Daphnia* also showed significant reduction in survivorship when exposed to total, when compared with controlled solar radiation. In contrast, heavily melanized adult arctic *Daphnia* showed no increase in mortality when exposed to natural solar radiation. There appeared to be differences both in tolerance and behavior in the two arctic forms when exposed to solar radiation.

Introduction

During the last few years, the reduction of stratospheric ozone has resulted in an increase in the UV-portion of solar radiation measured at terrestrial stations (Blumthaler & Ambach, 1991; Smith *et al.*, 1992). However, for many years UV-radiation was thought to be of no importance to aquatic organisms, since water and especially humic acids and dissolved particles absorb UV readily. In contrast to this, it has been demonstrated in the field (Hunter *et al.*, 1981; Siebeck, 1983; Hebert & Emery, 1990) as well as in laboratory experiments (Karanas *et al.*, 1981; Dey *et al.*, 1987), that UV-radiation does penetrate water to an extent that aquatic organisms can be harmed. Melanin pigmentation in aquatic organisms has recently been discussed as a photo-protective mechanism that reduces the detrimental effects of harmful near UV- and UV-radiation (Luecke & O'Brien, 1983; Hebert & Emery, 1990; Hobaeck & Wolf, 1991).

Alpine and arctic habitats have many characteristics in common. For example, high radiation input especially in the short wavelength range are typical for both environments. Therefore, in alpine and arctic habitats, photoprotection especially against the most harmful short wavelengths of UV radiation may be of vital importance to aquatic organisms. This is to be expected, especially in shallow water bodies with high radiation transmittance.

In this paper, data from field experiments will be presented for members of the *Daphnia pulex* group, which were sampled in alpine and arctic habitats under natural solar radiation. I ask questions about the specific UV-B sensitivity of these species with regard to age, pigmentation, water depth, as well as behavioural aspects.

Methods

In July and August of 1991, and June–August of 1992, experiments with transparent (clear) alpine *Daphnia pulex* (Linné emend. Leydig) were conducted in their natural habitat, a shallow eutrophic alpine pond (1240 m a.s.l., mean depth = 16 cm) at the 'Kampenwand' in the German Alps. *Daphnia* were caught with a dip net and selected by size (females significantly smaller than the smallest egg-bearing female were categorized as juveniles). Thirty *Daphnia* (either juveniles or adults) were placed in each experimental container (400 ml volume) filled with filtered pond water. Water depth was either 4 or 8 cm in the experiments with juveniles, and 4 cm in the experiments with adults. Screened windows (mesh size = 180 μm) in the walls of the containers allowed easy water exchange (including algae) between the containers and the pond. The containers were placed in floats within the pond and left for seven days; they were exposed to either total solar radiation or to reduced solar radiation (> 50% of UV-B being screened out via a foil, Folex, Folanorm, Germany, Fig. 1). The foil-covered containers served as controls. Each experiment consisted of 3–5 replicates (90–150 animals in total). Pond algae served as food (via diffusion through the windows) with species composition and density being identical to that of the pond. After seven days, the number of adult *Daphnia*, the number of egg-bearing *Daphnia* as well as the number of juvenile *Daphnia* were recorded and a new set of experiments with new *Daphnia* was started. Differences between experimental and control treatments were tested with the Rank-Wilcoxon-test (Sokal & Rohlf, 1981).

Corresponding experiments with arctic *Daphnia* of the *Daphnia pulex* group were conducted at Igloolik (N.W.T., Canada) with deeply melanized as well as with transparent (clear) *Daphnia* from several shallow ponds (maximum depths ranged from 40–150 cm). *Daphnia* were selected by age and pigmentation. Ten adult melanized and ten adult clear Daphnia were placed in open plastic containers (60 ml volume) and filled with (paper) filtered original pond water to a depth of 4 cm. The experiments with transparent and melanic *Daphnia* were conducted simultaneously with the experimental containers being arranged randomly in a tray for radiation exposure. *Daphnia* were fed daily with a mixture of green algae, which was frozen until the day of use.

For six days, containers were either exposed to total solar radiation or to solar radiation with the major part of UV-B being excluded by the same foil type used in the alpine experiments (control). The number of live individuals was recorded each day. Significant differences in percent survivorship between experimental and control treatments were tested using the binomial test (Sokal & Rohlf, 1981).

On several days, the number of *Daphnia* in the upper two-thirds of the water column and the number in the lower third were recorded hourly from sunrise to sunset. The numbers were then transformed to percentages of live animals in each container. Percentages of all replicates of each treatment (with and without foil, transparent and melanic *Daphnia*) were analyzed statistically (Student's t-test; Sokal & Rohlf, 1981).

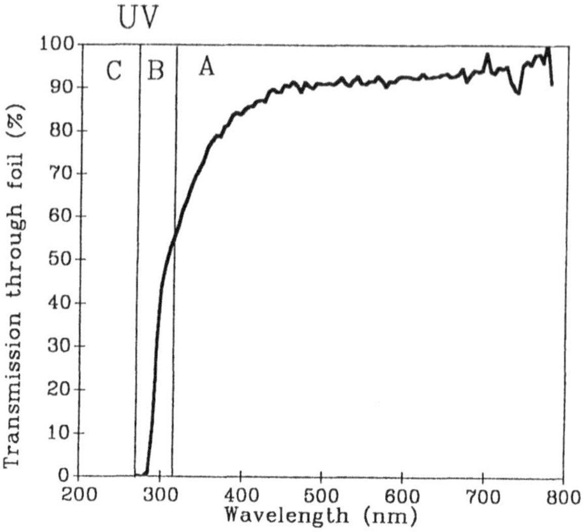

Fig. 1. Transmittable spectrum of solar radiation (%) through UV-B absorbing foil.

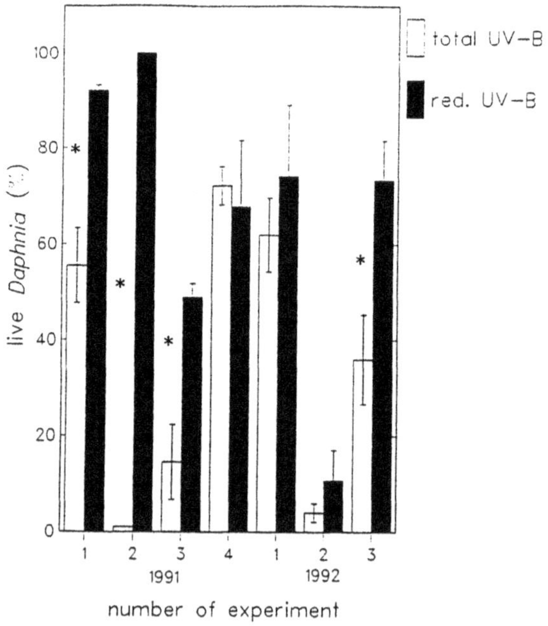

Fig. 2. Average survivorship (%) of at inoculation juvenile transparent alpine *Daphnia* after 7 days of exposure to solar radiation with total or reduced UV-B. Experiments were conducted in the original pond in containers with a 4 cm deep water column of natural pond water with constant water exchange (n = 90–150 *Daphnia* mean and S.E., * = differences are significant).

Results

In their natural alpine habitat, exposure to total solar radiation of transparent juvenile *D. pulex* resulted in significant increases in mortality for both water depths

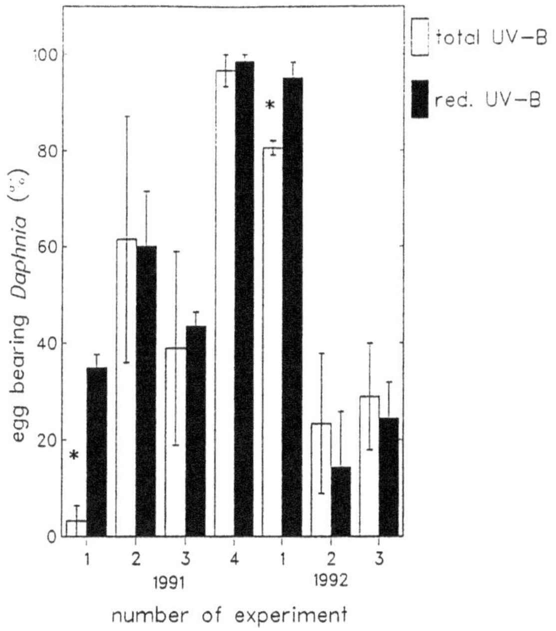

Fig. 3. Mean number of egg-bearing *Daphnia* (% of surviving *Daphnia* in the same experiments as in Fig. 2 (n = 90–150 *Daphnia*, mean and S.E., * = differences are significant).

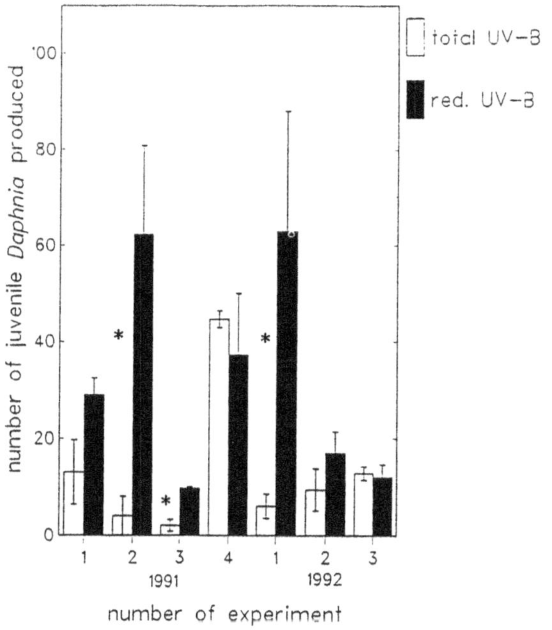

Fig. 4. Mean number of juveniles produced at inoculation by juvenile *Daphnia* (30 *Daphnia* container^{-1}) in the same experiments as in Fig. 2 (n = 90–150 *Daphnia*, mean and S.E., * = differences are significant).

(Fig. 2). Mean survivorship differences between UV-B protected and UV-B exposed *Daphnia* were 31.7% ($p < 0.02$) for the 4 cm deep treatment and 11.4% ($p < 0.05$) for the 8 cm deep treatment. Similarily, the mean survivorship difference of adult *Daphnia* in the 4 cm water column was 17.7% between UV-B protected and UV-B exposed treatments ($p < 0.05$, data not shown).

The reduction in juvenile survivorship was greater for the 4 cm deep treatment, when compared to 8 cm deep treatment. In addition, UV-induced mortality in juvenile *Daphnia* exceeded that of adult mortality under identical experimental conditions.

When juvenile *Daphnia* were inoculated, the proportion of egg-bearing *Daphnia* was significantly reduced in UV-B exposed containers for both the 4 and 8 cm depth treatments ($p < 0.05$, Fig. 3). The mean difference in the proportion of egg-bearing animals between UV-B protected animals and animals exposed to total solar radiation was 5.4% in the 4 cm water column, and 7.1% in the 8 cm water column. When adult *Daphnia* were inoculated into the 4 cm water column, the proportion of egg-bearing animals was also significantly reduced ($p < 0.05$) after exposure to total solar radiation (mean difference of 4.9%). No major differences were observed between the two water depths as well as between the two age classes with respect to the effect of UV-B on the percentage of egg-bearing *Daphnia*.

The effect of UV-B on the percentage of egg-bearing females was considerably smaller than the detrimental effect UV-B had on the number of juveniles produced in all three experimental setups (Fig. 4). When juvenile *Daphnia* were inoculated into the 4 cm water column, UV-B protected animals produced significantly ($p < 0.05$) more (on average 19.8) offspring. For the 8 cm water column, the mean difference between UV-B protected and UV-B exposed treatments was 19.9 offspring produced ($p < 0.02$). In most cases no significant differences were found between the two tested water depths (4 cm and 8 cm) with respect to the UV-B effect on offspring production. When adult *Daphnia* were inoculated, the mean difference in the number of juveniles produced in UV-B protected and the UV-B exposed treatments was 50.4 offspring ($p < 0.05$, data not shown) in the 4 cm water column.

In the high Arctic, exposure of deeply melanized adult *Daphnia* to total solar radiation over a six day period resulted in no significant increase in mortality compared to UV-B reduced controls (Fig. 5A). In contrast, transparent *Daphnia* showed a significant

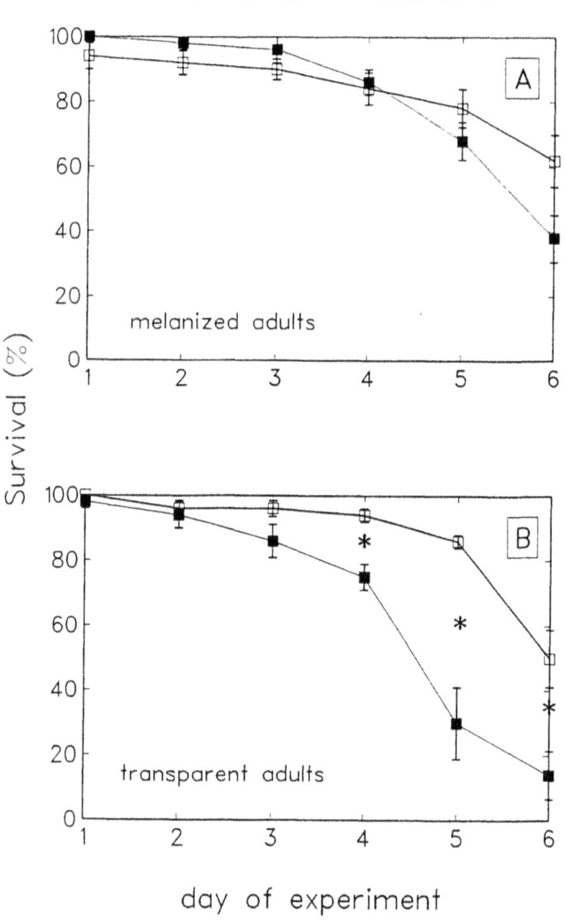

Fig. 5. Survival (%) of melanic (A) and transparent (B) adult arctic *Daphnia* exposed to solar radiation with total or reduced UV-B in a 4 cm deep water column (n = 50, * = differences are significant, experiment was performed betwen 24–30 August 1992).

increase in mortality after 4 days of exposure to total solar radiation compared with the UV-B reduced controls (Fig. 5B).

In addition, there appeared to be differences with respect to the behaviour between melanic and transparent adult *Daphnia* when exposed to total solar radiation. Melanic *Daphnia* swam actively during daylight hours, while transparent *Daphnia* remained nearly motionless at the bottom of the experimental containers for most of the daylight hours. Also, on average, melanic *Daphnia* were more evenly distributed throughout the water column, when compared to clear Daphnia (Figs 6 & 7). This was true on the overcast day with little direct solar radiation (Figs 6 & 8A) as well as on the clear day with bright sunshine (Figs 7 & 8B). On the overcast day, however, melanic *Daphnia* under reduced UV-B treatment were even more abundant in the upper water layers compared to those

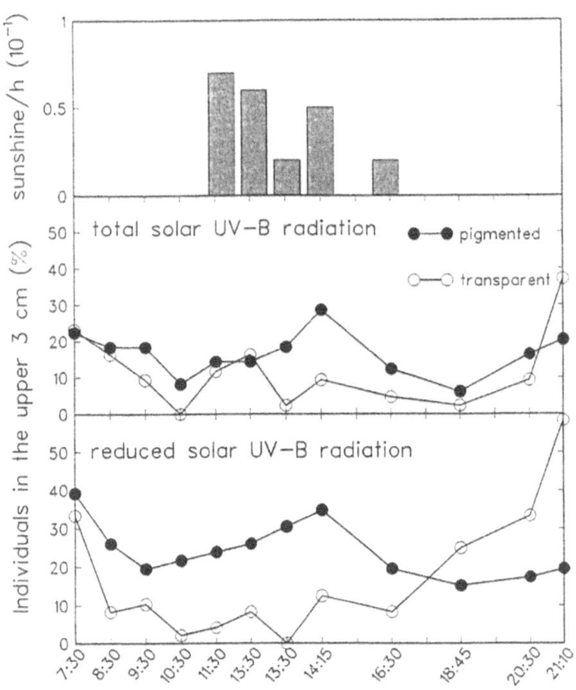

Fig. 6. Percentage of adult transparent or pigmented arctic *Daphnia* in the upper 3 cm of the water column under exposure to solar radiation with total or reduced UV-B in relation to daylight hours on an overcast day (n = 50, 27 August 1992).

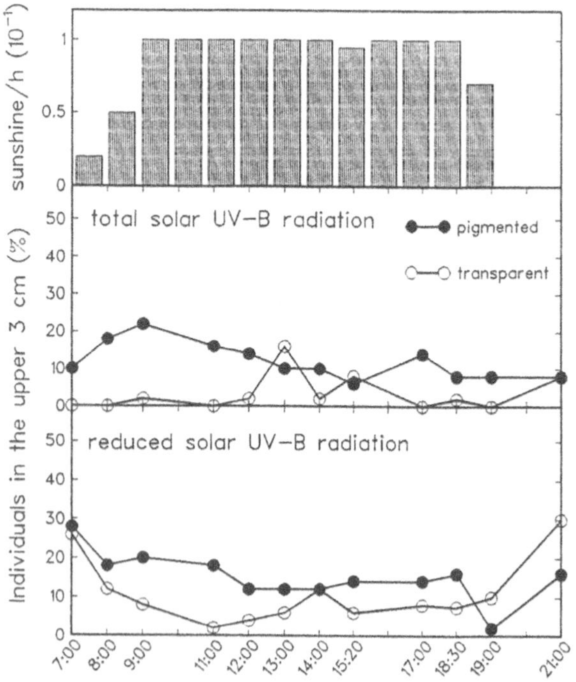

Fig. 7. Percentage of adult transparent or pigmented arctic *Daphnia* in the upper 3 cm of the water column under exposure to solar radiation with total or reduced UV-B in relation to daylight hours on a bright sunny day (n = 50, 28 August 1992).

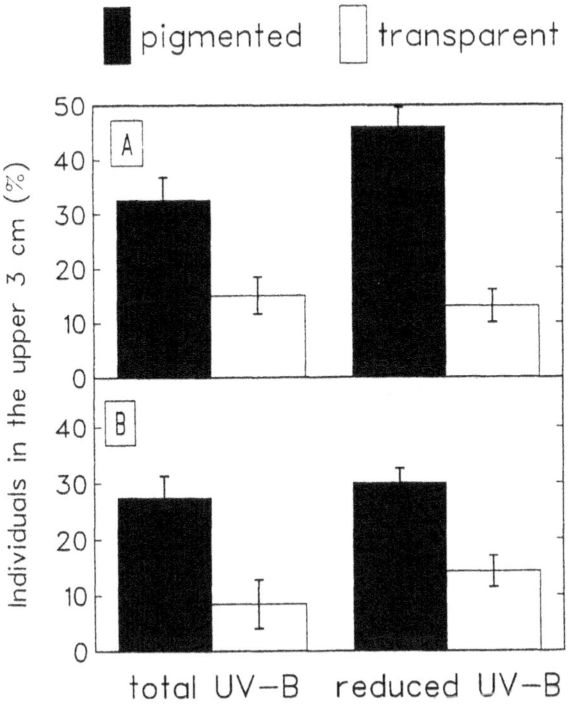

Fig. 8. Average percentage of adult transparent or pigmented arctic *Daphnia* (with S.E.) in the upper 3 cm of the water column with total or reduced UV-B exposure on an overcast (A) and a sunny (B) day (n = 50). Data were collected from 8.30 to 16.30 h on the overcast day, and from 8.00 to 19.00 h on the sunny day.

exposed to unshielded solar radiation (Fig. 8A). This was not found when bright sunshine was recorded during the entire day (Fig. 8B).

With the onset of sunset, transparent *Daphnia* began to swim more actively and moved into the upper water layers in greater numbers than did melanic *Daphnia* (Figs 6 & 7). The onset of this behavioural change occurred earlier (Fig. 6, onset at about 16.30 h) in UV-B protected containers than in exposed containers (Fig. 6, onset at about 19.00 h). This behaviour was delayed in UV-B protected containers on days of bright sunshine (Fig. 7, onset at about 19.00 h) when compared to overcast days (Fig. 6, onset at about 16.30 h).

Discussion

In comparison with sea-level habitats, alpine as well as arctic habitats receive increased doses of UV radiation due to altitude (high UV-intensity) and latitude (lower UV intensity but longer duration), respectively. Therefore, damaging effects of UV radiation are to be expected. However, the alpine pond tested here was characterized by high phytoplankton densities (1 – 6 × 10^6 cells ml^{-1}) and a considerable amount of dissolved humics (deep brown color). Both factors are known to reduce UV transmittance drastically (Wetzel, 1983).

The question arose, whether there exist any measurable UV effects on aquatic organisms in such eutrophic, humic water bodies. Due to the experimental set up, it was possible to test the selective impact of the most harmful range of solar UV radiation, defined as UV-B (290 nm to 315 nm), on these aquatic organisms. All longer wavelengths of natural solar emission were not affected by the UV-B absorbing foil. Due to the remote location of the pond, permanent radiation measurements were not possible. Short-term radiation measurements were of no value, since weather conditions in the pond area (especially cloud cover) changed dramatically within minutes and on almost all days. So no reasonable estimates could have been gained from such data. Therefore, the experiments were conducted in a way that conclusions on the effect of UV-B could be drawn from qualitative rather than quantitative comparisons of experimental and control treatments.

The results show that even in the highly eutrophic, humic alpine pond, juvenile as well as adult mortality increased significantly in the unprotected containers compared to UV-B shielded controls (Fig. 2). This is in agreement with the results of Hunter *et al.* (1981), who also found reduced survivorship of anchovy larvae (*Engraulis mordax*) when exposed to total solar radiation in comparison to UV-reduced controls within 17 cm water column.

The great variability in survivorship and reproductive parameters within the UV-B protected controls of the different experiments in both years reflect the differences in experimental conditions. One factor which varied, was the size (age) of the juveniles at inoculation. Since the alpine *Daphnia* population was highly synchronized (Zellmer, 1993), the age of the majority of juveniles used in the experiments varied from week to week. As a result of such variability, differences in UV-B sensitivity are to be expected. In addition, weather conditions in the Alps changed dramatically during the course of the experiments. Periods of mainly overcast and rainy weather caused high sediment input into the pond. The following days of bright sunshine and high radiation input caused algal blooms accompanied by changes in the physico-chemistry of the pond (temperature, oxygen content, pH). Thus, the physiological condition of the *Daphnia* was affected by all these factors. This was true, however, for both the experimental and control containers. The only difference between the two sets of containers was exposure to solar UV-B.

Mortality is a rather crude parameter of UV-B impact, since different biochemical and physiological effects may occur before death. Therefore, in the alpine experiments, reproductive parameters were chosen as presumably more sensitive parameters with respect to UV-B damage. In the alpine environment, significant reductions in the percentage of egg-bearing *Daphnia* (Fig. 3) as well as in the number of offspring (Fig. 4) were seen as a consequence of exposure to total solar radiation, in comparison to UV-B reduced controls at the tested water depths. Similarily, reduced survivorship and reproduction under artificial UV-irradiation were reported by Karanas *et al.* (1979) for *Acartia clausii* (Copepoda) and by Damkaer *et al.* (1980) for marine zooplankton. It is interesting to note, that there was on average, no difference between the two water depths in the UV-B induced reduction of the percentage of egg-bearing *Daphnia*, as well as in the number of juveniles produced. In contrast, juvenile survivorship was greater in the 8 cm deep containers compared to the 4 cm deep ones.

In the alpine pond, juvenile *Daphnia* showed a greater UV-B sensitivity with respect to survivorship and reproductive parameters compared to adults. This is in agreement with the study of Jüttner (1989) on age dependence of UV damage in *Daphnia pulex*. She found decreased UV-B sensitivity with increasing age in *D. pulex* under standardized laboratory conditions. These results may be caused by several factors, which vary between juvenile and adult *Daphnia*. One factor is the greater susceptibility in juvenile animals towards mutagens such as UV-B, since juveniles show higher cell division rates compared to adults. In addition, the photo-repair system may not be as efficient in juvenile *Daphnia* as that found in adults. Furthermore, the amount of energy stored by an individual daphnid of a given age, may play an important role in this regard, since the costly photo-repair processes certainly will be affected negatively by limitations in energy supply.

The results of the alpine pond *Daphnia* experiments demonstrate that UV-damage repair mechanisms and photoprotection in the tested *Daphnia* were not sufficient to balance the UV-B induced damage. Thus, a natural population may not only be reduced due to mortality after intense UV-exposure, but may be altered due to reduced reproductive rates.

The effects of UV-B radiation on transparent alpine *Daphnia* were similar to those found in transparent arctic *Daphnia*. Here again, survivorship of adult *Daphnia* was significantly reduced upon exposure to total solar radiation compared to UV-B shielded controls (Fig. 5B). However, water transparency is much greater in non-humic, oligotrophic arctic ponds compared to most alpine ponds. The algal food suspension sedimented rather quickly thus providing no shielding against radiation.

In contrast to the significant increase in mortality in transparent adult arctic *Daphnia*, melanic adult arctic *Daphnia* showed no significant increase in mortality after exposure to full solar radiation (Fig. 5A). Similar results have been reported by Luecke & O'Brien (1983) and Hebert & Emery (1990). However, this is the first report, where only the UV-B portion of the spectrum was excluded. This allows one to distinguish between the effect of total solar radiation from any UV-B related phenomena.

The differences in mortality between unshielded and shielded containers can most likely be attributed to differences in UV-B impact, since there were no differences in food quantity and physico-chemical factors between the shielded and the fully exposed containers.

The observed differences in survivorship of melanic and transparent *Daphnia* could be attributed to the photoprotection of the melanin shield, which covers most of the dorsal portion of the carapace and antennae of melanic *Daphnia* as observed by Hebert & Emery (1990). They reported that melanin has a high UV absorption. However, melanic *Daphnia* also showed an increased mortality after several days of exposure (Fig. 5A). This was greater in the unprotected than in the UV-B protected containers. This could reflect the higher sensitivity of these *Daphnia* to UV radiation during the molting stage. During molting, *Daphnia* shed the entire carapace and lose (at least temporarily) their melanin protection until melanin is resynthesized.

Melanic and transparent *Daphnia* do co-occur in some ponds in the Arctic. This observation poses the question why melanic and clear *Daphnia*, being significantly different in their ability to endure UV-B radiation, can coexist within the same water body under the same radiation conditions? One answer may be seen in the results of the behavioural observations. While melanic *Daphnia* swam actively during daylight hours and showed a more even distribution throughout the water column, the transparent animals rested at the bottom of the containers during daylight and only moved into the upper water strata after sunset (Figs 6

& 7). Thus during the day, transparent *Daphnia* avoided the upper water strata where radiation intensity was greater than in the lower strata. At sunset, transparent *Daphnia* moved upwards and were found at an even greater percentage in the upper water strata than the pigmented forms. In the UV-B protected containers, the upward movement of the transparent *Daphnia* was initiated much earlier (more than 2 hrs earlier) than in the unprotected containers. Therefore UV-B could have a key function in these behavioural differences in transparent *Daphnia*.

These observations agree well with the findings of Hebert & Emery (1990). They reported that melanic arctic *Daphnia* do occur in the upper water strata during daylight hours while transparent *Daphnia* do not. After sunset, the situation was reversed.

The reason why *Daphnia* invest so much energy in building up the costly photo-protective melanin shield is not clear. Melanic *Daphnia* do survive better in the upper water strata than clear *Daphnia*, and they probably use more time for active food collection in the upper water layers without being severely affected by UV-B radiation.

Acknowledgements

I thank P. D. N. Hebert for his support of the arctic research and O. Siebeck for his support of the alpine experiments. I also wish to thank two anonymous referees for valuable comments.

References

Blumthaler, M. & W. Ambach, 1991. Indication of increasing solar ultraviolet-B radiation flux in Alpine regions. Science 248: 206.

Dey, D. B., D. M. Damkaer & G. A. Heron, 1987. UV-B dose/dose-rate responses of seasonally abundant copepods of Pudget Sound. Contrib. #40 School of Oceanography, University of Washington, Seattle, USA.

Hebert, P. D. N. & C. J. Emery, 1990. The adaptive significance of cuticular pigmentation in *Daphnia*. Funct. Ecol. 4: 703–710.

Hobaek, A. & H. G. Wolf, 1991. Ecological genetics of Norwegian *Daphnia* II. Distribution of *Daphnia longispina* genotypes in relation to short-wave radiation and water colour. Hydrobiologia 225 (Dev. Hydrobiol. 71): 229–243.

Hunter, J. R., S. E. Kaupp & J. H. Taylor, 1981. Effects of solar and artificial ultraviolet-B radiation on larval northern anchovy, *Engraulis mordax*. Photochem. Photobiol. 34: 477–486.

Jüttner, I., 1989. Untersuchungen zur UV-Toleranz von *Daphnia pulex obtusa* unter besonderer Berücksichtigung verschiedener Populationsparameter. Diplomarbeit, Ludwig-Maximilians-University, Department of Biology, Munich, Germany.

Karanas, J. J., R. C. Worrest & H. Van Dyke, 1981. Impact of UV-B radiation on fecundity. Mar. Biol. 65: 125–133.

Luecke, C.& J. W. O'Brien, 1983. Photoprotective pigments in a pond morph of *Daphnia middendorffiana*. Arctic 36: 365–368.

Siebeck, O., 1983. UV als Umweltfaktor in aquatischen Biotopen. Verh. Gesell. Ökologie. 10: 583–589.

Smith, R. C., B. B. Prezelin, K. S. Baker, R. R. Bidigare, N. P. Boucher, T. Coley, D. Karentz, S. McIntyre, H. A. Matlick, D. Menzies, M. Ondrusek, Z. Wam & K. J. Waters, 1992. Ozone depletion: ultraviolet radiation and phytoplankton biology in Antarctic waters. Science 255 (5047): 952–959.

Sokal, R. R., Rohlf F. J., 1981. Biometry. W. H. Freeman and Company, New York.

Zellmer, I. D., 1993. Der Einfluss des Futters auf die UV Toleranz von *Daphnia pulex* unter Labor- und Freiland-Bedingungen. Ph.D. Thesis, Ludwig-Maximilians-Universität Munich.

Hydrobiologia **307**: 161–165, 1995.
P. Larsson & L.J. Weider (eds), Cladocera as Model Organisms in Biology.

An account of a preliminary mechanistic model of swimming behaviour in *Daphnia*: its use in understanding diel vertical migration

J. Ringelberg
Department of Aquatic Ecology, University of Amsterdam, Kruislaan 320, 1098 SM Amsterdam, The Netherlands

Key words: Daphnia, diel vertical migration, model, phototaxis

Abstract

This preliminary mechanistic model of normal swimming and phototactic behaviour in individual *Daphnia* was constructed using data and assumptions based on experiments and observations. Swimming under constant light intensity is characterized by short periods of upward movements alternating with equal periods of downward movements. Two oscillators are proposed that generate these phases in swimming. Unexpected shifts in depth, as observed in *D. magna* and *D. 'longispina'*, are also present in the swimming of the computer daphnid and thus seem to be inherent to the underlying mechanism. As in real daphnids, during relative decreases in light intensity of low velocity, positive phototactic upward swimming is stepwise. With increasing velocity in the change in light, these steps disappear. When the model is triggered by a natural increase in light at dawn, a small downward movement results. Migration distance can be increased to commonly found depths of migrating *Daphnia* by the introduction of a 'fish exudate factor' into the model, which enhances the phototactic response. Since attenuation of light in the water affects the phototactic swimming response, it also influences migration distance. The results of model calculations agree quite well with an empirical relationship between Secchi disc depth and amplitude of diel vertical migration in a number of lakes.

Introduction

The causal (proximate) explanation of diel vertical migration (DVM) in *Daphnia* is based on phototactic behaviour elicited by changes in light intensity around sunrise and sunset. A large and consistent body of knowledge about photo-behaviour in *Daphnia* is available (Ringelberg, 1987). This knowledge is largely based on laboratory experimentation in tubes, and, consequently, reactive phototactic swimming is over short distances and of short durations, as compared to vertical migrations in lakes. This confinement is often thought of as to be a disadvantage, which diminishes the relevance of the results. It has, however, no bearing on the explanation of DVM. Reaction parameters like stimulus thresholds and swimming velocities have to be used in the mechanistic explanation of migration. A photoresponse occurs as long as the stimulus, a relative change in light intensity, is of supra-threshold strength. In nature, swimming continues as long as changes in light intensity around sunrise and sunset are of magnitudes that surpass the threshold. These supra-threshold changes last for about two hours and start about 90 min. before sunrise or 30 min. before sunset, respectively. Early morning descent and late evening ascent coincide indeed with these periods as was shown for *D. galeata* × *hyalina* in Lake Maarsseveen (Ringelberg *et al.*, 1991). So, the next step in explaining DVM in terms of phototactic behaviour is to build a model of swimming *Daphnia* based (as far as possible) on experimental evidence and to trigger the computer daphnid by sunrise and sunset changes in light intensity. It is expected that the simulated migrations show features (e.g. timing, amplitude) comparable to those of vertical migrations in lakes.

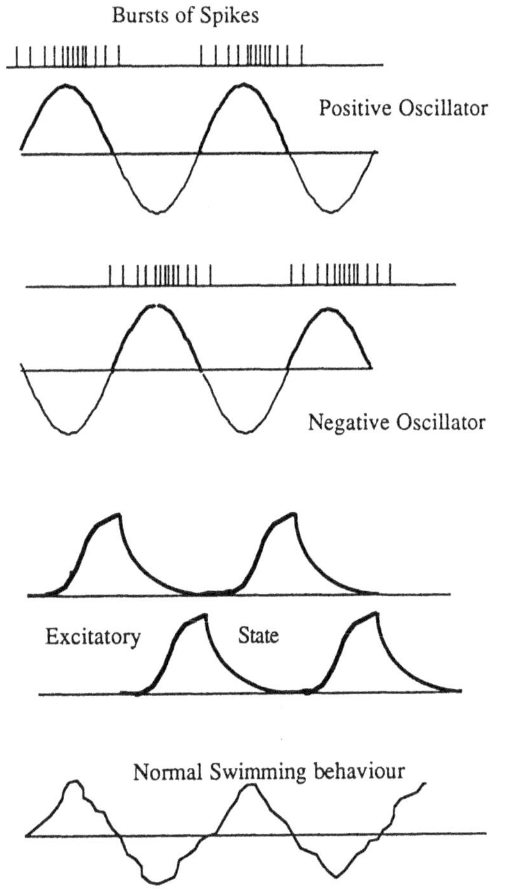

Fig. 1. Assumed physiological mechanism as the basis for normal swimming behaviour of *Daphnia* as incorporated into the model. See text.

Construction of the model

Modelling software (Stella, High Performance Systems, Inc.) was used to simulate, firstly, normal swimming behaviour and, secondly, phototactic swimming as caused by relative changes in light intensity. The integration of these two aspects makes up the essential part of the model from which conclusions about DVM can be inferred.

'Hop-and-sink' swimming in *Daphnia* consists of a stroke of the second antennae, resulting in a small upward movement, followed by a pause during which the animal sinks. When light intensity is constant, a series of strong strokes alternate with weak strokes, and result in an oscillation around a more-or-less constant depth. However, these alternating phases are irregular and sudden vertical displacements over larger distances occur. Nevertheless, autocorrelation analysis of 2–3 hour recordings in *D. magna* revealed a periodicity of about two minutes (Ringelberg, unpublished). The average duration of upward and downward phases in *D. 'longispina'* is 60.4 sec (s.d. = 16.54, n = 98) and, thus, a similar periodicity of about two minutes also seems to be present in this species. These data were obtained in experimental tubes under constant light and temperature following the method of Ringelberg (1964). (Probably, this *D. 'longispina'* from Lake Maarsseveen was actually *D. galeata* or *D. galeata* × *hyalina*. Allozyme analyses, which were performed by Spaak and Hoekstra (1993), have revealed that the daphnids in this lake consist of the aforesaid taxa, which can change in relative dominance from year to year and also within a given season).

Based on this periodicity, two sinusoidal functions were assumed with default period lengths of two minutes and which were shifted in phase by 180° to model normal swimming (Fig. 1). One can imagine that in the central nervous system, two (groups of) neurons fire with a periodicity of two minutes, one group stimulating the vigour of antennal beats, the other inhibiting this activity. For reasons of simulation, spike frequency was changed into amplitude, and an excitatory state (Ringelberg, 1964) is formed in proportion to this amplitude. Excitation in sensory and nervous systems disintegrate with time, therefore, an exponential decrease was assumed. The value of the disintegration constant was calculated from the stimulus strength-stimulus duration response curve (rheobase-chronaxie curve), which was obtained from phototactic reactions (i.e. caused by relative decreases in light intensity, Ringelberg, 1993) using *D. 'longispina'*. Both positive and negative excitatory states were summed and the sum determined whether the computer *Daphnia simplicitas* swims upwards or downwards with a proportional velocity. The default amplitude of the positive and negative oscillator was calibrated in order to achieve the average normal swimming distance of 2.9 cm (s.d. = 1.39, n = 98).

It cannot be expected that the period and the amplitude of the oscillators are constant since also durations and velocities of upward and downward swimming phases vary and are normally distributed. Therefore, in the model, at randomly chosen times, the values of the period and the amplitude were replaced by values chosen randomly from normal distributions with default values as means and standard deviations of 10% about these means.

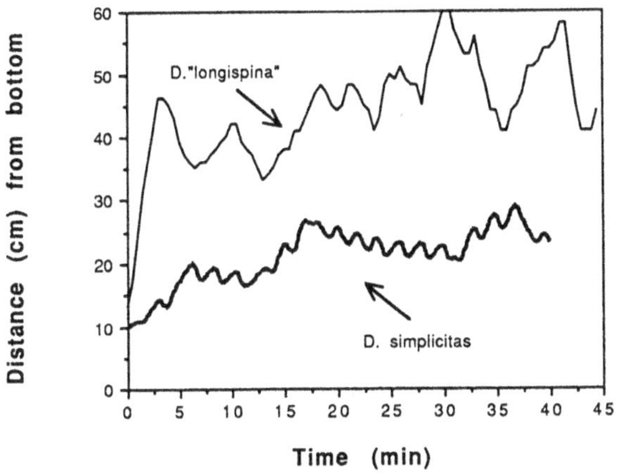

Fig. 2. An example of a swimming track of *D. 'longispina'* compared to a simulation run of *D. simplicitas*.

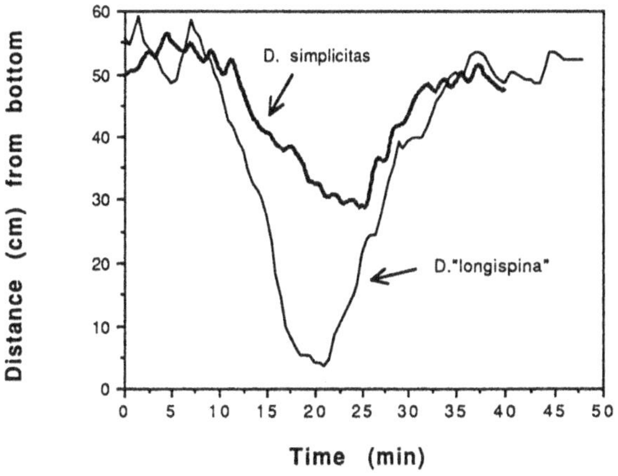

Fig. 3. Illustration showing how normal swimming may be interrupted by sudden changes in depth position without any concomitant changes in environmental conditions. Note that this occurs in both real and simulated *Daphnia*.

The second step was to incorporate phototactic swimming into the model. This was done by introducing two additional excitatory states; the first one increased proportionally to the velocity of relative decreases, while the second one was proportional to the velocity of relative increases in light intensity. Whether a phototactic swimming response does or does not occur will depend on the combination of velocity and duration of the light change. At a certain low velocity of the relative change in light intensity, an infinite duration of this light stimulus is needed because the velocity of increase in excitation equals the disintegration velocity of this excitation (see Ringelberg, 1964, 1987 for theoretical treatment). A disintegration constant used in the model, was calculated from experimental values (Ringelberg, 1993). Since it was demonstrated (Ringelberg, 1964) that the effect of relative changes in light intensity are superimposed on normal swimming, the four excitatory states were summed to obtain a final value of excitation that determined swimming direction and proportional swimming velocity.

Still, another aspect of photo-reactive swimming must be considered. If a *Daphnia* swims up or down, it experiences an increase or decrease in light intensity, respectively. The extent of this change is a function of swimming velocity and the turbidity of the water. This consequence of phototactic upward and downward swimming was also incorporated into the model by means of an extinction coefficient.

Finally, the model was run with a natural light intensity increase measured on June 18, 1989 at 0.30 m below the water surface in Lake Maarsseveen. Absolute light intensity and relative increases, from 90 min. before until 90 min. after sunrise, were used as input. The effect of turbidity on the amplitude of migration was studied by varying the extinction coefficient in the model. The results were compared with the correlation between amplitude and Secchi-disc depths, as compiled for a variety of lakes, by Dodson (1990). Extinction coefficients were transformed into Secchi-disc depths using a formula given by Talling in George (1983).

Complete information on the functional relations and parameter values used in the model can be obtained from the author upon request.

Results and discussion

Normal swimming of *Daphnia simplicitas* and either *D. 'longispina'* or *D. magna* compares reasonably well (Fig. 2). Tracks of the latter species were smoothed for comparison because of irregularities due to 'hop-and-sink' swimming, and time lags of the observer, which were absent in the computer runs. Of special interest are sudden vertical displacements, performed by the computer *Daphnia* as well as the real species (Fig. 3). Previously, these interruptions of the regular oscillations were thought to be caused by unknown internal and/or external influences. However, it cannot be excluded that they are inherent to the basic mechanism

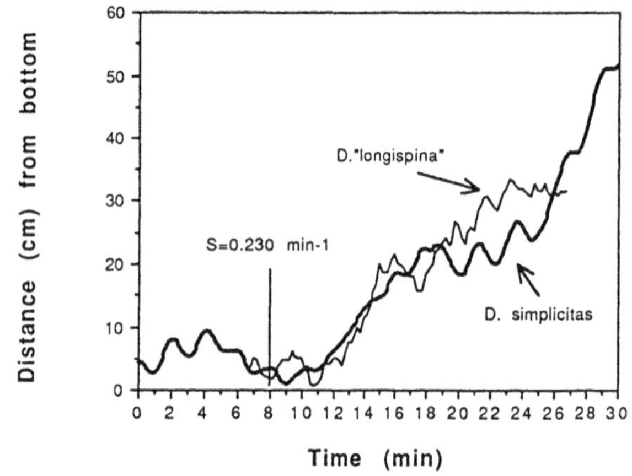

Fig. 4. An example of a positively phototactic reaction evoked by a relative decrease in light intensity ($S = 0.230 \text{ min}^{-1}$) at which step-wise upward swimming is still present in both a real and simulated *Daphnia*.

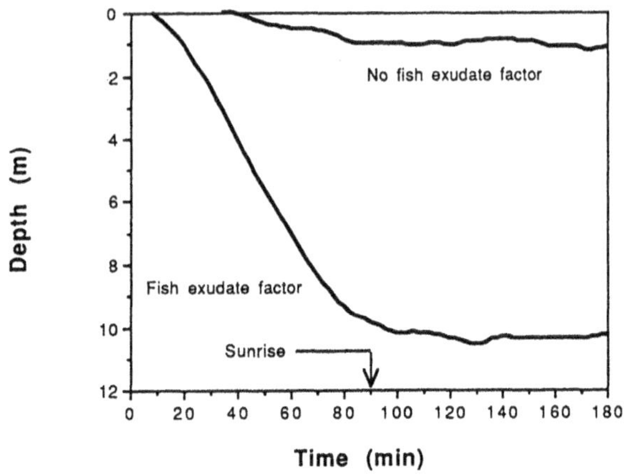

Fig. 6. A natural increase in light intensity at dawn was used to trigger the model. Without evaluation ('no fish exudate') of the excitatory state, which results from the increase in light, the amplitude of migration is small; realistic amplitudes are achieved by the introduction of an appropriate value for the excitatory state.

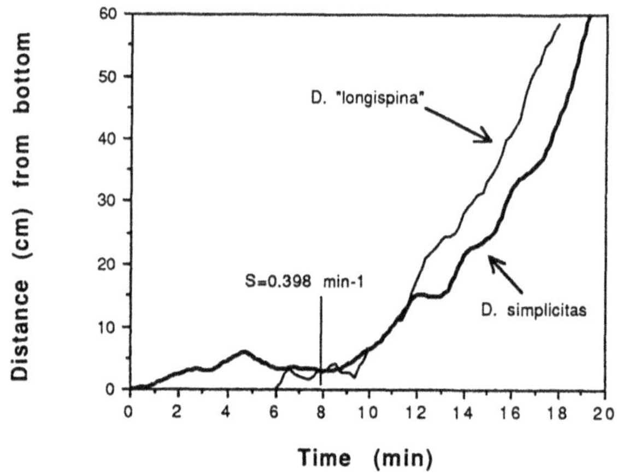

Fig. 5. As in Fig. 4 but at a higher rate of relative decrease in light intensity ($S = 0.398 \text{ min}^{-1}$). Note that the steps in upward swimming (as depicted in Fig. 4) disappear.

of normal swimming. In the model, they are the result of the stochastic succession of high amplitudes and/or long periods of either oscillator, which were accompanied by low values of the other oscillator. Of course, this interesting similarity in behaviour of the computer daphnid and *D. magna* or *D. 'longispina'* does not prove that oscillators are responsible for swimming behaviour in the real species.

Positively phototactic reactions compare well (Figs 4 and 5). As in *D. magna* and *D. 'longispina'*, at low values of relative decreases in light intensity, upward swimming is stepwise. The upward phase of normal swimming is prolonged by the light decrease, while the influence of the negative oscillator is decreased. The periods between upward gain become shorter with increasing stimulus strength, and result in an approximated continuous upward swimming motion (Fig. 5).

The data on phototactic reactions, used to build the model, were obtained at a time when the enhancing effect of juvenile perch (*Perca fluviatilis*) on these reactions was unknown (Ringelberg, 1991). At the time of the experiments, an incidental observation (June 26–27, 1979) in Lake Maarsseveen revealed a very weak DVM with a difference in mean population daytime and night-time vertical distribution of only 1.98 m (Ringelberg, 1993). A comparable small amplitude was obtained by triggering the model with a natural light increase around sunrise (Fig. 6).

Although the mechanism of phototaxis is always functioning in the absence of fish exudates, reactivity is poor and it does not lead to extensive vertical migrations. To achieve such migrations, the excitatory states (which were elicited by relative changes in light intensity) were given greater weight than those of the oscillators. Simulated migration amplitudes were modelled as presently found in Lake Maarsseveen (Ringelberg *et al.*, 1991).

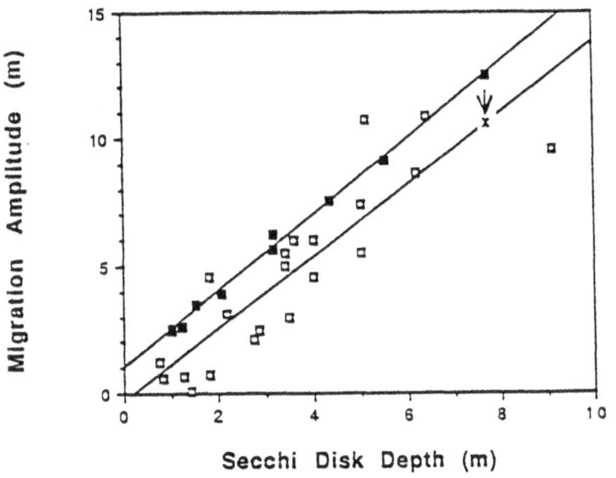

Fig. 7. Amplitude of DVM as a function of turbidity of the water as expressed by Secchi-disc depth. Open squares and the lower regression line (y = −0.299 + 1.4092 x) are based on empirical data from different lakes as compiled from the literature by Dodson (1990). The black squares are generated by the computer model triggered by a natural increase in light intensity and different extinction coefficients. The latter were recalculated as Secchi-disc depths for comparison. The regression function is y = 1.024 + 1.4887x. The arrow indicates the result (x) of a computer run with an adjusted evaluation factor of excitatory states.

Finally, the model was run using different extinction coefficients. After Secchi-disc depths were recalculated, the amplitudes of migration became a function of turbidity, and this function was strikingly similar to the one published by Dodson (1990), which was based on migrations in a number of lakes (Fig. 7). The slope of the regression is important; the vertical position of the calculated regression line, which is too high in Fig. 7 as compared to the empirical one, can be adjusted quite easily to the level of the empirical one by using a slightly smaller value for the constant, evaluating excitatory states (this was done for one turbidity, see cross and arrow, Fig. 7). Evidently, the influence of water turbidity on day depth in lakes can be explained by light intensity decreases, which result from downward swimming, and thus lead to a diminished influence of relative increases in light intensity at dawn. In clear water, these decreases are smaller, as compared to more turbid water and phototactic downward swimming is less affected. Since the depth at which visual predators are able to forage is inversely related to light extinction, the adaptive significance of this property of phototaxis is evident.

Amplitudes of DVM collected for various lakes by Dodson (1990), show considerable scatter around the regression line (see Fig. 7). Since Loose (1993) demonstrated that vertical migration amplitude depends on the concentration of fish exudates, part of the variance may be explained by this factor. In the model, amplitude of migration is sensitive to the factor evaluating excitatory states, which was introduced to simulate fish exudates. Experiments on phototaxis using various concentrations of exudates, have to be conducted before the functional significance of this mechanism can be firmly established.

This preliminary mechanistic model simulates quite well, the normal swimming and photo-behaviour in individual *Daphnia*. When triggered by light intensity changes at dawn, the model simulates diel vertical migrations with several realistic properties. Therefore, it is worthwhile to continue experiments designed to evaluate the underlying physiological mechanism(s) and to refine the model accordingly.

References

Dodson, S., 1990. Predicting diel vertical migration of zooplankton. Limnol. Oceanogr. 35: 1195–1201.

George, D. G., 1983. Interrelations between the vertical distribution of *Daphnia* and chlorophyll-a in two large limnetic enclosures. J. Plankton Res. 5: 457–475.

Loose, C. J., 1993. *Daphnia* diel vertical migration behavior: response to vertebrate predator abundance. Arch. Hydrobiol. Ergebn. Limnol. 39: 29–36.

Ringelberg, J., 1964. The positively phototactic reaction of *Daphnia magna* Straus: A contribution to the understanding of diurnal vertical migration. Neth. J. Sea Res. 2: 319–406.

Ringelberg, J., 1987. Light induced behaviour in *Daphnia*. In Peters, R. H. & R. de Bernardi (eds), *Daphnia*. Mem. Ist. Idrobiol. 45: 285–323.

Ringelberg, J., 1993. Phototaxis as a behavioural component of diel vertical migration in a pelagic *Daphnia*. Arch. Hydrobiol. Ergebn. Limnol. 39: 45–55.

Ringelberg, J., B. J. G. Flik, D. Lindenaar & K. Royackers, 1991. Diel vertical migration of *Daphnia hyalina* (*sensu latiori*) in Lake Maarsseveen: Part 1. Aspects of seasonal and daily timing. Arch. Hydrobiol. 121: 129–145.

Spaak, P. & J. R. Hoekstra, 1993. Clonal structure of the *Daphnia* population in Lake Maarsseveen: its implications for diel vertical migration. Arch. Hydrobiol. Beih. Ergebn. Limnol. 39: 157–165.

Hydrobiologia **307**: 167–175, 1995.
P. Larsson & L.J. Weider (eds), Cladocera as Model Organisms in Biology.

Life history characteristics of *Daphnia magna* clones differing in phototactic behaviour

L. De Meester
Laboratory of Animal Ecology, University of Ghent, K.L. Ledeganckstraat 35, B-9000 Gent, Belgium

Key words: Clone, *Daphnia*, life history, offspring size, phototaxis, vertical migration

Abstract

In an analysis of a life table experiment involving positively, intermediately and negatively phototactic *Daphnia magna* clones, life history traits such as the average duration of the adult instar, neonate and adult body size were found to be correlated with phototactic behaviour. The size of the eggs and neonates was positively correlated with adult body size, and with egg development time. Adult body size was positively correlated with the size of the second and subsequent clutches. I argue that the intrinsic positive correlation between offspring size and egg development time is a key factor structuring the differences in life history patterns observed between the positively and intermediately phototactic *Daphnia* genotypes, and that the two life history patterns are to be considered alternatives suited for different environmental conditions (e.g. habitats with and without strong predation pressure on adults).

Introduction

In a life table experiment using *Daphnia magna* clones with different phototactic behaviour (De Meester, 1994a), it was found that differences in several life history characteristics were associated with habitat preference (daytime vertical distribution). Positively phototactic clones achieve high intrinsic rates of increase at smaller adult body sizes through rapid development, at a cost of producing small neonates. Negatively and intermediately phototactic clones have larger adult body sizes, and produce larger neonates that are more starvation resistant than those of positively phototactic clones. Intermediately phototactic clones achieve high intrinsic rates of increase through the production of large clutches. These results suggest that positively phototactic clones are efficient exploiters of optimal conditions, whereas intermediately and negatively phototactic clones are more stress-resistant. These differences in life histories of *D. magna* clones with different day depth preference are similar to those observed by Stich & Lampert (1984) between two different *Daphnia* species that coexist in Lake Constance and which differ in vertical migration behaviour.

The striking difference in life histories between *D. magna* clones differing in phototactic behaviour offers an opportunity to get some insight into questions such as: how and to what extent (different) life histories are structured by (the same) trade-offs? Using the data obtained in the life table experiment described in De Meester (1994a), here I explore the relationships between various life history traits in an effort to provide more insight into the mechanisms underlying the observed differences in life history strategies.

Material and methods

The origin of the clones used and the protocol of the life table experiment is described in detail in De Meester (1994a). I worked with nine clones, three of which (the positively phototactic clone C_1, the intermediately phototactic clone C_4, and the negatively phototactic clone C_3) were isolated from a fishless city pond in Ghent (Citadelpark). The six other clones are two groups of sister clones obtained by two subsequent generations of selfing a clone (P_1) which was isolated from a pond near the river Schelde (Driehoekvijver, Heusden; with variable but low predation by fish). Clones $P_1 12$ and $P_1 32,48$ are negatively phototactic, clones $P_1 32$, $P_1 81$ and $P_1 32,70$ are intermediately phototactic, and clone $P_1 32,85$ is positively phototactic. Due to their similar genetic background, the use of sister clones with different phototactic behaviour reduces the risk that a linkage between phototactic behaviour and life history characteristics is due to chance effects.

From each clone, five to six animals were cultured individually in 1 l jars filled with dechlorinated tap water. Female neonates less than 24 h old were inoculated and provided with *Scenedesmus acutus* at a concentration of 2×10^5 cells ml^{-1} for the first three days, after which the food concentration was restored daily to 6.5–7 $\times 10^5$ cells ml^{-1}. The medium was renewed every day (see De Meester, 1994a). Only second clutch neonates were used in the experiment; their mothers and grandmothers had been cultured under identical conditions as provided in the life table experiment, except for the number of animals per liter being four to eight instead of one.

The following life history traits were measured: age at maturity (AGEMAT, in days); age at release of first and subsequent broods [AGE(x), in days]; time between broods, averaged over the first five broods [INTER(5)] and over the first 10 broods [INTER(10), in days]; life span (LONGEVITY, in days); number of offspring in the first and subsequent clutches [BR(x)]; number of clutches (BRNR); total number of offspring produced (BRTOT, including male offspring); percentage of males in each of the clutches (%MALES); the intrinsic rate of increase (*r*, estimated by solving Euler's equation iteratively); size of the first adult instar and of subsequent instars [SIZE(x), with SIZE(1) = SIZEMAT; in mm]; size of the eggs (length of the longest axis) of the first and second clutch [SIZEGG(x); average of 10 eggs, with 3–4 eggs measured in three mothers per clone]; and size of the neonates of the first and second clutch [SIZENEO(x); average of 10 neonates, from three mothers per clone].

Correlations (Kendall's τ; Sokal & Rohlf, 1981) between several life history characteristics, and correlations between these traits and phototactic behaviour were calculated at three levels. In addition to phenotypic correlations using measurements on all individuals ($n = 52$, except for missing values), I calculated correlations of clonal averages ($n = 9$) and phenotypic correlations using individual measurements, but grouping individuals according to phototactic behaviour ($n = 12$ for positively phototactic animals, $n = 23$ for intermediately phototactic animals, and $n = 17$ for negatively phototactic animals, except for missing values). Overall phenotypic correlations provide information on the general picture of associations between traits, and may reveal trade-offs between life history traits. However, patterns may be obscured due to genotype-dependent acquisition of resources and genotype-dependent allocation patterns (Van Noordwijk & De Jong, 1986). Correlations of clonal means provide some insight into 'acquired' relationships between traits (Bell, 1984), and may reveal intrinsic trade-offs through which the life histories of genotypes are structured. Caution in interpreting the resulting patterns is warranted, as clonal means also contain an environmental component. In the present experiment, the overall patterns observed can be considered reliable, however, since the between-genotype component of variation for most life history traits was large compared to the within genotype component (see De Meester, 1994a). The number of individuals per clone was too low to calculate reliable within clone phenotypic correlations (to measure 'variable' relationships sensu Bell, 1984). However, since most of the between-genotype variability is associated with differences in phototactic behaviour between the clones used in the experiments (De Meester, 1994a), phenotypic correlations grouping clones with different phototactic behaviour was considered a reasonable alternative. The comparison of the correlation structure among groups of clones differing in phototactic behaviour (positively, intermediately, and negatively phototactic clones) is central to the present study, as it may indicate the mechanism by which the different life history patterns have been structured.

All statistical analyses were done using the SYSTAT program (Wilkinson, 1990). Not all types of correlations could be calculated for all the character pairs. For all analyses involving phototactic behaviour or the size of the eggs and neonates, only correlations of clonal means could be obtained, because no data on individuals were available. Phototactic behaviour was determined for the different clones (see De Meester, 1991), but was not quantified for the individuals used in the life history experiments. For the size of eggs and neonates, observations on offspring from several individuals were pooled in the measurements.

Table 1. Summary table of life history characteristics of nine *D. magna* clones. Means ± S.D. of six (five in clones $P_1 12$ and $P_1 32$) individuals are given. Clones are grouped according to their phototactic behaviour (−, x, +: negatively, intermediately, and positively phototactic, respectively). AGE(1): age (in days) at release of first clutch (time from neonate to neonate); LONGEVITY: age at death (in days); BRNR: number of broods produced; BRTOT: total number of (female + male) offspring produced; MALE: percentage of male neonates among offspring (averaged over all broods). For more detailed information on analysis of these life history characteristics, see De Meester (1994a).

Clone	AGE(1)	LONGEVITY	BRNR	BRTOT	MALE
(−)					
C_3	8.33 ± 0.52	37.83 ± 16.86	9.50 ± 4.76	423.3 ± 262.4	0.07 ± 0.16
$P_1 12$	9.80 ± 0.45	49.00 ± 13.17	11.20 ± 2.95	318.2 ± 98.7	35.48 ± 17.64
$P_1 32,48$	10.00 ± 0.00	38.50 ± 15.04	8.67 ± 4.08	289.7 ± 164.7	19.13 ± 24.67
(x)					
C_4	8.67 ± 0.52	23.00 ± 13.54	5.67 ± 4.68	283.3 ± 299.7	0.00 ± 0.00
$P_1 32$	8.40 ± 0.55	43.20 ± 16.21	11.60 ± 5.23	786.6 ± 395.0	0.00 ± 0.00
$P_1 81$	8.67 ± 0.52	47.17 ± 10.07	12.83 ± 2.86	869.5 ± 137.4	0.00 ± 0.00
$P_1 32,70$	9.00 ± 0.00	41.67 ± 11.95	9.67 ± 2.88	574.2 ± 137.4	0.00 ± 0.00
(+)					
C_1	8.67 ± 0.52	28.83 ± 14.89	7.50 ± 5.21	362.3 ± 292.9	0.00 ± 0.00
$P_1 32,85$	8.50 ± 0.55	29.50 ± 10.80	7.83 ± 4.22	292.2 ± 171.4	0.00 ± 0.00

Results

General features of the life history of Daphnia magna

It took *D. magna* individuals under the given conditions five to eight days to mature, and eight to 10 days before the first clutch was released from the brood pouch (time from neonate to neonate; Table 1). Individual females were found to survive for up to 61 days, but most animals (71.2%) lived for less than 50 days (Table 1). Though there is a decline in fertility towards the end of life, with the exception of three individuals, all animals produced broods right until they died. The largest clutches were produced between the fifth and twelfth adult instars, with the number of offspring produced in one single clutch being in some cases larger than 100. Overall, the size of the neonates produced in the first clutch was smaller (average 0.743 mm) than that of the subsequent clutches (average of second clutch: 0.809 mm; eighth clutch: 0.880 mm). The size at maturity varied from 2.43 to 3.33 mm. The largest size attained by any individual was 5.76 mm.

Offspring sex

Under the environmental conditions provided in the experiment (1 individual culture^{-1}, high food concentration, long day photoperiod), male offspring were only observed in the three negatively phototactic clones. The incidence of male offspring was low in clone C_3, but quite high in clones $P_1 12$ and $P_1 32,48$ (Table 1). For the latter two clones, 70.8% of the combined clutches (n = 106) were all-female, 12.3% were single sex male, and 17.0% were mixed. The incidence of mixed broods was particularly high in clone $P_1 12$, with more than 60% of the clutches with male neonates also having some female neonates.

Phototaxis and life history traits

There is an overall tendency for negative correlations between phototactic behaviour and age- and size-related life history characteristics, whereas reproductive life history traits tend to be positively correlated with phototactic behaviour (Table 2; Fig. 1). All correlations between phototaxis and body size are negative: positively phototactic clones are on average smaller, and produce smaller offspring than negatively and intermediately phototactic clones (Fig. 1). This ten-

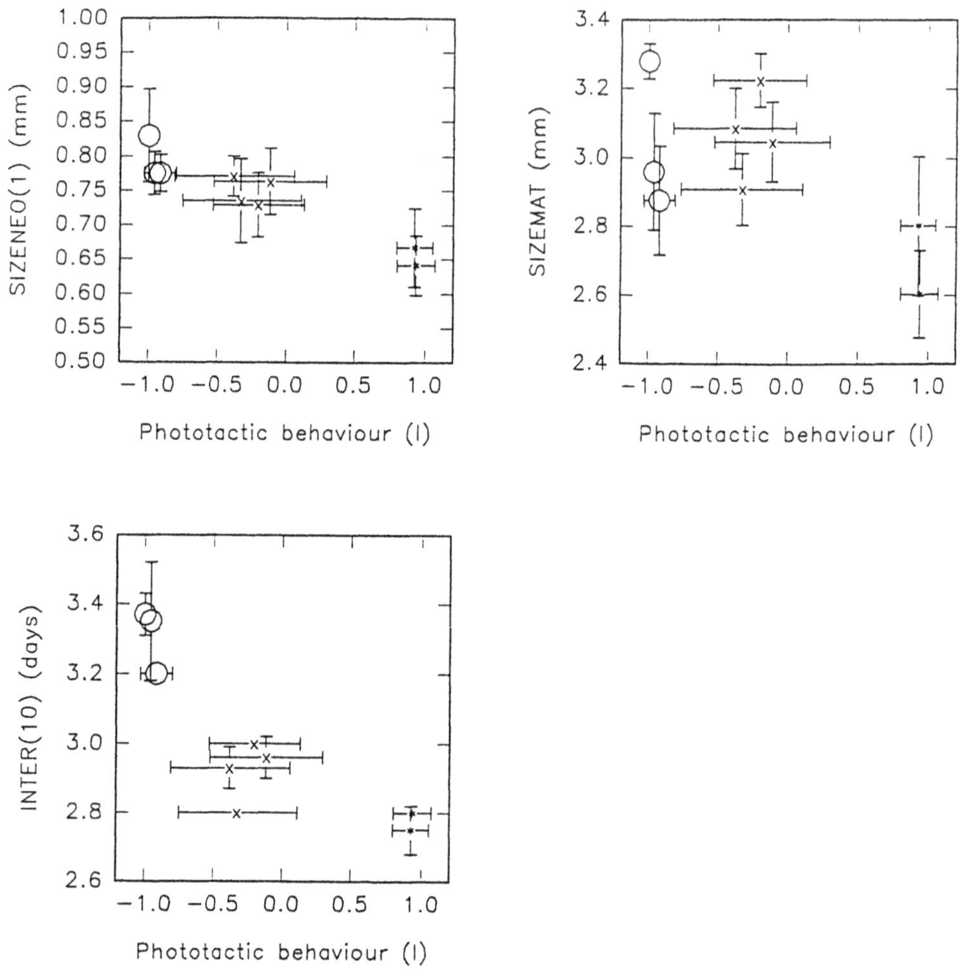

Fig. 1. Scatterplots showing correlations between clonal means of phototactic behaviour (phototactic index I, see De Meester, 1991) and size of the neonates of the first clutch [SIZENEO(1)], size at maturity (SIZEMAT) and the adult instar duration [averaged over the first 10 adult instars: INTER(10)]. All error bars: 1 S.D.. o – negatively phototactic, x – intermediately phototactic, * – positively phototactic.

Table 2. Correlations (Kendall's τ) between clonal means of phototaxis, size of the eggs and size of the neonates with clonal means of various life history traits of nine *Daphnia magna* clones with different phototactic behaviour (n = 9; critical value for $p = 0.05$: $|\tau| = 0.556$). All $|\tau|$ values > 0.25 are reported; nsl: not significant and 'low', i.e. $|\tau| > 0.25$. See Material and methods for life history trait designations.

	INTER (10)	BR (1)	BR (2)	BR (3)	SIZE MAT	SIZE (3)	SIZE EGG (1)	SIZE EGG (2)	SIZE NEO (1)	SIZE NE4O (2)
Photo	−0.83	+0.52	+0.26	+0.33	−0.39	−0.39	−0.40	−0.65	−0.86	−0.59
SIZEEGG(1)	+0.57	−0.48	nsl	nsl	+0.48	+0.48				
SIZEEGG(2)	+0.54	−0.67	nsl	nsl	+0.39	+0.44	+0.37			
SIZENEO(1)	+0.69	−0.65	nsl	−0.25	+0.42	+0.42	+0.37	+0.76		
SIZENEO(2)	+0.65	−0.61	−0.28	−0.28	+0.33	+0.28	+0.48	+0.61	+0.65	

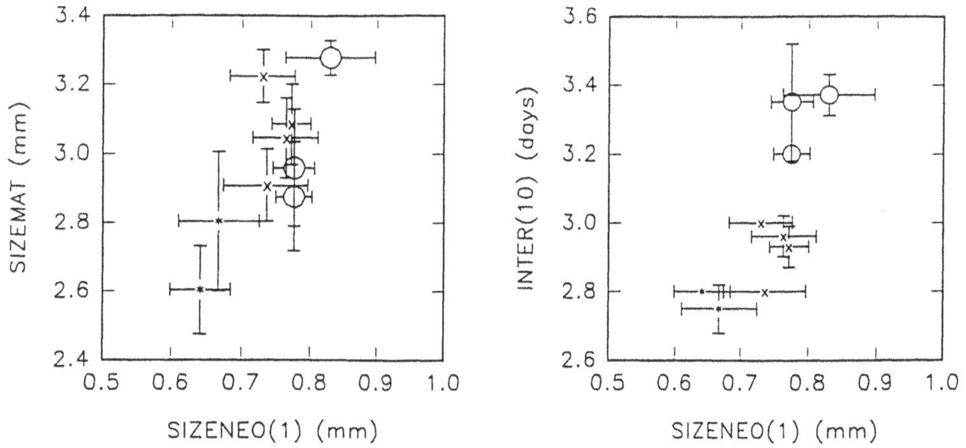

Fig. 2. Scatterplots showing correlations between clonal means of the size of the neonates of the first clutch [SIZENEO(1)] and size at maturity (SIZEMAT), and the adult instar duration [averaged over the first 10 adult instars: INTER(10)]. All error bars: 1 S.D.. o – negatively phototactic, x – intermediately phototactic, * – positively phototactic.

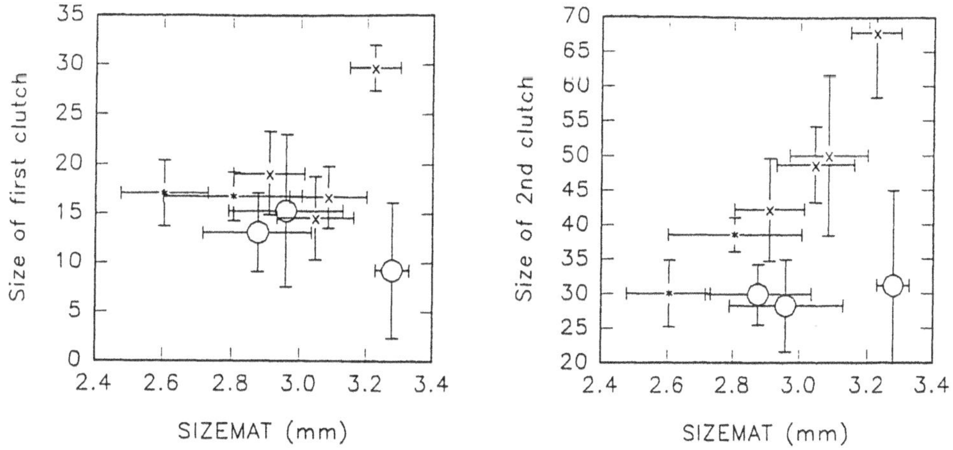

Fig. 3. Scatterplots showing correlations between clonal means of the size at maturity (SIZEMAT) and the size of the first and second clutch. All error bars: 1 S.D.. o – negatively phototactic, x – intermediately phototactic, * – positively phototactic.

dency is strongest for the size of neonates (of the first and second clutch). Correlations between phototactic behaviour and reproductive life history traits are relatively weak (Table 2) because the intermediately phototactic clones tend to have the largest clutches (see Fig. 3).

Size and age-related traits

The sizes of the eggs and neonates are positively correlated, the correlation being significant for SIZEGG(2) with SIZENEO(1) and SIZENEO(2), and for SIZENEO(1) with SIZENEO(2) (Table 2). In addition, the size of the eggs and neonates of the first as well as the second clutch show a consistent positive relationship with adult body size (Table 2; Fig. 2). There is a strong tendency for positive correlations between size of the eggs and neonates and the average duration of the adult instar (Table 2; Fig. 2).

Size and reproductive traits

Adult body size is positively correlated with the sizes of both the second and the third clutch (Table 3; Fig. 3). This relationship holds for all phenotypic correlations as well as for clonal means, though it is stronger for intermediately and positively phototactic clones than for negatively phototactic ones. Correlations between adult body size and the size of the first clutch are weak (Table 3; Fig. 3).

The pattern is totally different for the size of eggs and neonates, as these traits show a negative correla-

Table 3. Correlations (Kendall's τ) between important age, reproductive and size related life history traits of nine *Daphnia magna* clones with different phototactic behaviours. Ph: overall phenotypic correlations pairing values for each individual (n = 52 except for missing values); G: correlations of clonal (genotypic) means (n = 9; critical value for $p = 0.05$: $|\tau| = 0.556$); −, x, +: phenotypic correlations grouping observations for individuals of negatively (n = 17; critical value for $p = 0.05$: $|\tau| = 0.368$), intermediately (n = 23; critical value for $p = 0.05$: $|\tau| = 0.296$) and positively phototactic (n = 12; critical value for $p = 0.05$: $|\tau| = 0.455$) clones, respectively. For phenotypic correlations, all significant correlations are reported; for correlations of clonal means, all $|\tau|$ values >0.25 are reported; ns: not significant at $p = 0.05$; nsl: not significant and 'low', i.e. $|\tau| < 0.25$. See Material and methods for life history trait designations.

BR (1)	BR (2)	BR (3)	*r*	SIZE MAT	SIZE (3)		
ns	ns	ns	−0.36	+0.36	+0.33	Ph	INTER(10)
−0.48	nsl	nsl	−0.29	+0.48	+0.42	G	
ns	ns	ns	−0.65	ns	ns	-	
+0.41	+0.52	+0.70	+0.45	+0.56	+0.67	x	
ns	ns	ns	ns	ns	ns	+[1]	
	+0.35	+0.34	+0.35	ns	ns	Ph	BR(1)
	+0.33	nsl	+0.42	nsl	nsl	G	
	ns	ns	ns	ns	−0.39	-	
	+0.44	+0.42	ns	+0.48	+0.33	x	
	ns	ns	ns	ns	ns	+	
		+0.79	+0.61	+0.43	+0.43	Ph	BR(2)
		+0.89	+0.76	+0.39	+0.44	G	
		+0.75	ns	+0.47	+0.48	-	
		+0.72	+0.58	+0.69	+0.74	x	
		+0.70	+0.68	+0.77	+0.73	+	
			+0.59	+0.45	+0.45	Ph	BR(3)
			+0.65	+0.28	+0.33	G	
			ns	+0.39	+0.43	-	
			+0.51	+0.75	+0.78	x	
			ns	+0.85	+0.84	+	
				ns	ns	Ph	*r*
				+0.25	+0.31	G	
				+0.48	−0.51	-	
				+0.49	+0.59	x	
				ns	+0.50	+	
					+0.66	Ph	SIZEMAT
					+0.83	G	
					+0.39	-	
					+0.70	x	
					+0.86	+	

[1] n= 5 due to missing values (critical value, $p = 0.05$: $|\tau| = 1$).

tion with the size of the first clutch, and no consistent relationship with the size of subsequent clutches (Table 2).

Age and reproductive traits

Correlations between the size of the first, second and third clutches and the average duration of the adult instar are weak, with the exception of the recurrent positive phenotypic correlations between age-related traits and clutch sizes observed for intermediately phototactic clones (Table 3; no such pattern was observed for the positively or negatively phototactic clones).

Correlations among the sizes of different clutches tend to be positive, though the pattern is weak for those relationships involving the size of the first clutch (Table 3).

The intrinsic rate of increase tends to be negatively correlated with the average duration of the adult instar, with the exception, however, of a clearly positive phenotypic correlation of *r* with INTER(10) observed for animals of the intermediately phototactic clones (Table 3). Overall, the intrinsic rate of increase is positively correlated with clutch sizes. However, only a weak relationship is apparent with the size of the first clutch, and animals of negatively phototactic clones do not conform to the overall pattern, as their *r* values tend to be not related with clutch sizes (Table 3). *r* tends to be positively correlated with adult body size, though the relation is most consistent for the intermediately phototactic animals (Table 3).

Discussion

As discussed in De Meester (1994a), the present life table experiment reveals a strong association between phototactic behaviour and life history traits in the *Daphnia* clones studied: positively phototactic animals have a relatively short development time, are relatively small and produce small eggs and offspring, whereas intermediately and negatively phototactic animals mature more slowly and have a longer egg development time, but grow larger and produce larger eggs and offspring.

As expected, egg size is positively correlated with size of neonates, and in accordance with other studies (e.g. Ebert, 1991; Lampert, 1993), both are positively correlated with adult body size and with adult instar duration. Though one cannot infer causality from correlations, my results strongly suggest that the production of large offspring entails a longer egg development time (as measured by adult instar duration; see Fig. 2). Ebert (1993) reported a positive correlation between neonate size and the duration of the instar during which vitellogenesis takes place.

Whereas there is a positive correlation of the clonal means of adult body size with the size of the second and subsequent clutches, at least for intermediately and positively phototactic clones (see Fig. 3), no such pattern was found for the first clutch. However, there is a consistent negative correlation between the size of the eggs and neonates and the size of the first clutch (a relationship which does not hold for the second or third clutch, Table 2). This pattern can be explained by strong selection for a large size of the first clutch. Whereas negatively and intermediately phototactic clones are constrained by the need to produce high quality offspring, positively phototactic clones are able to produce a relatively large first clutch at a small adult body size, by producing smaller eggs and consequently, smaller offspring. They can do so because their habitat preference for the near-surface water (De Meester, 1993) entails a lower risk of starvation. The offspring size of the positively phototactic clones may lie close to the physiological minimum size for viable offspring in *Daphnia magna* (Lampert, 1993).

There are two ways by which an iteroparous organism can achieve a higher intrinsic rate of increase: it can start to reproduce at an earlier age (having a shorter development time), or it can produce more offspring per day during the reproductive period. The latter can be achieved by producing larger clutches, or by reducing the time between two subsequent clutches. Reducing the time between two subsequent clutches can be mediated through a shorter egg development time, and can then be viewed as a component of the option to shorten the development time. In the present life table experiment, positively and intermediately phototactic *Daphnia magna* clones have a similarly high intrinsic rate of increase, though their life history characteristics are strikingly different. My results indicate that these differences can be understood by considering the following arguments: (1) both intermediately and positively phototactic clones are *r*-selected, i.e. there is strong selection for a short development time and for the production of a large number of offspring in the first clutch; (2) intermediately phototactic animals are more constrained than the positively phototactic ones with respect to the minimal size of the offspring, because they have to produce starvation-resistant offspring; (3) there is an intrinsic positive relationship between offspring size and egg development time. In order to combine the production of large offspring with the production of a high number of eggs in the first clutch, intermediately phototactic animals mature at a relatively large size, a strategy which may be dictated by an energy or a size (content of brood pouch) constraint. The relatively long egg development time associated with the large eggs is offset by the capacity to produce a large number of offspring during subsequent clutches. It is noteworthy that the average duration of the adult instar in intermediately phototactic animals shows a positive correlation with the intrinsic rate of increase (see Table 3). This is counter-intuitive, but can be explained by the strong positive correlation between the average duration of the adult instar and clutch size in intermediates (see Table 3; in positively and negatively phototactic animals the correlation is non-significant or negative). The correlation of r with adult body size is indeed strongest (and positive) when only intermediately phototactic animals are considered.

So far, I have explained why the intermediately phototactic animals mature at a relatively large size. Given that positively phototactic animals can afford to produce smaller neonates, why do they not grow as large as intermediates and produce more offspring in the first and subsequent clutches? It may be the result of selection for a short development time: if positively phototactic animals would mature at a large size and produce small neonates, then their post-embryonic development time, and hence their total development time, would be inflated. The positive correlation between neonate and adult body size reported here and in several other studies (e.g. Ebert, 1991; Lampert, 1993) may indeed reflect selection for a short total (egg-to-egg) development time. It is noteworthy that, as this relationship between neonate and adult body size seems to be generally valid (for both intra- and interclonal variability; Ebert, 1991), the production of starvation-resistant offspring (see Tessier & Consolatti, 1989; Gliwicz & Guisande, 1992) in *Daphnia* entails a relatively large adult body size.

Though I have attempted to explain the small adult body size in positively phototactic animals in terms of an optimization of total development time, the small size of the animals may also be adaptive in relation to the habitat preference of the animals, as it may partially compensate for the higher visibility of positively phototactic animals residing in near-surface water during the daytime by making them less vulnerable to visual predation.

The difference between positively and intermediately phototactic clones with respect to the size of the neonates has in the present experiment been quantified under high food conditions. Under low food conditions, a shift to the production of larger eggs and neonates is to be expected in both types of animals (e.g. Gliwicz & Guisande, 1992).

The results of the present life table experiment suggest that the two alternative life history strategies described above are associated with differences in habitat preference (see De Meester, 1993). It is likely, however, that a similar association can be found at the interpopulational level. Indeed, the short development time strategy is expected to be dominant in populations facing high predation pressure on adults, whereas in populations from relatively "safe" habitats (including habitats with refuges) the large-clutch strategy may be more important, especially if these habitats (refuges) are associated with less optimal conditions.

Negatively phototactic clones do not conform to the above pattern. As discussed in De Meester (1994a), negatively phototactic animals probably have special adaptations for a life in the marginal near-benthic habitat they occur in, as they seem able to coexist with positively and intermediately phototactic clones, irrespective of the large differences in intrinsic rate of increase (De Meester, 1994b). They have a low intrinsic rate of increase, but produce offspring of high quality.

In addition, all three negatively phototactic clones produced males in the life table experiment, whereas none of the positively or intermediately phototactic animals did. This indicates that negatives engage in sexual reproduction more readily than the other clones, which is consistent with the frequent observation of ephippial females in well-fed and apparently healthy cultures (pers. obs.). Negatively phototactic clones seem to be 'fugitive' genotypes, occurring in a relatively safe but marginal habitat, and they reduce their risk of extinction by frequent production of sexual offspring. It is noteworthy that two out of three negatively phototactic clones also showed a relatively high proportion of mixed broods (17.0%; Barker & Hebert (1986) reported 7.6% of the broods being of mixed sex in natural populations of *Daphnia magna* during a sexual period).

The differences in neonate size between positively, intermediately and negatively phototactic *Daphnia* may shed light on the question: why is there an increase in neonate size with consecutive clutches (Glazier, 1992; Ebert, 1993)? Irrespective of the important differences in size of the neonates between different clones in my experiment, they all show an increase in neonate size with increasing clutch number. Glazier (1992) pointed out that the small size of the neonates of the first clutch may be the result of selection for increased fecundity being stronger in younger than in older females. Ebert (1993) disagrees with this explanation, and he argues that a reduction in offspring size has essentially the same negative effect on fitness as reducing clutch size. However, a *Daphnia* which is able to produce several clutches probably occupies a habitat with relatively low adult mortality, i.e. a habitat with no or very low predation pressure, or with predation on juveniles (e.g. by invertebrates). Under both conditions, the production of large neonates is advantageous either because the offspring are likely to grow up under crowded conditions, or because larger offspring are less vulnerable, or vulnerable during a shorter period, to (invertebrate) predators (e.g. Lynch, 1980). Since the production of several clutches is intrinsically linked with conditions in which larger offspring have a relatively large fitness advantage, selection for a large clutch size is more strongly counter-acted by selection for large offspring in later than in earlier clutches.

In conclusion, my results indicate that it is likely that the differences in life histories of the positively, intermediately and negatively phototactic *Daphnia* genotypes studied here are structured by the intrinsic relationship between offspring size, egg development time and adult body size. As such, the present results suggest that offspring size is likely to be a key factor in our overall efforts to understand the evolution of *Daphnia* life histories.

Acknowledgements

The author is a postdoctoral researcher with the National Fund of Scientific Research. I thank P. Larsson, A. Hobaek, O. Kleiven and other colleagues from the Institute of Zoology, University of Bergen, for a splendid organization of the Symposium on Cladocera. The National Fund for Scientific Research, Belgium is gratefully acknowledged for travel support. The University of Ghent (Faculty of Science) covered part of the registration and accomodation expenses. Prof. H. J. Dumont critically read the text. I thank J. Pijanowska and L. Weider for many helpful suggestions on earlier versions of the manuscript.

References

Barker, D. M. & P. D. N. Hebert, 1986. Secondary sex ratio of the cyclic parthenogen *Daphnia magna* (Crustacea: Cladocera) in the Canadian Arctic. Can. J. Zool. 64: 1137–1143.

Bell, G., 1984. Measuring the cost of reproduction. I. The correlation structure of the life table of a plankton rotifer. Evolution 38: 300–313.

De Meester, L., 1991. An analysis of the phototactic behaviour of *Daphnia magna* clones and their sexual descendants. Hydrobiologia 225 (Dev. Hydrobiol. 71): 217–227.

De Meester, L., 1993. The vertical distribution of *Daphnia magna* genotypes selected for different phototactic behaviour: Outdoor experiments. Arch. Hydrobiol. Beih. Ergebn. Limnol. 39: 137–155.

De Meester, L., 1994a. Life histories and habitat selection in *Daphnia*: Divergent life histories of *D. magna* clones differing in phototactic behaviour. Oecologia 97: 333–341.

De Meester, L., 1994b. Habitat partitioning in *Daphnia*: Coexistence of *Daphnia magna* clones differing in phototactic behaviour. In Beaumont, A. (ed.), Genetics and Evolution of Aquatic Organisms. Chapman and Hall, London: 323–335.

Ebert, D., 1991. The effect of size at birth, maturation threshold and genetic differences on the life-history of *Daphnia magna*. Oecologia 86: 243–250.

Ebert, D., 1993. The trade-off between offspring size and number in *Daphnia magna*: The influence of genetic, environmental and maternal effects. Arch. Hydrobiol. Suppl. 90: 453–473.

Glazier, D. S., 1992. Effects of food, genotype, and maternal size and age on offspring investment in *Daphnia magna*. Ecology 73: 910–926.

Gliwicz, Z. M. & C. Guisande, 1992. Family planning in *Daphnia*: Resistance to starvation in offspring born to mothers grown at different food levels. Oecologia 91: 463–467.

Lampert, W., 1993. Phenotypic plasticity of the size at first reproduction in *Daphnia*: The importance of maternal size. Ecology 74: 1455–1466.

Lynch, M., 1980. The evolution of cladoceran life histories. Q. Rev. Biol. 55: 23–41.

Sokal, R. R. & F. J. Rohlf, 1981. Biometry. W.H. Freeman and Company, San Francisco.

Stich, H. B. & W. Lampert, 1984. Growth and reproduction of migrating and non-migrating *Daphnia* species under simulated food and temperature conditions of diurnal vertical migration. Oecologia 61: 192–196

Tessier, A. J. & N. L. Consolatti, 1989. Variation in offspring size in *Daphnia* and consequences for individual fitness. Oikos 56: 269–276.

Van Noordwijk, A. J. & G. De Jong, 1986. Acquisition and allocation of resources: Their influence on variation in life-history tactics. Am. Nat. 128: 137–142.

Wilkinson, L., 1990. SYSTAT: The system for statistics. Systat Inc, Evanston.

Hydrobiologia **307**: 177–184, 1995.
P. Larsson & L.J. Weider (eds), Cladocera as Model Organisms in Biology.

Diel horizontal migration and swarm formation in *Daphnia* in response to *Chaoborus*

Olaug Vetti Kvam & Ole T. Kleiven
Institute of Zoology, Department of Ecology, University of Bergen, Allégaten 41, N-5007 Bergen, Norway

Key words: Chaoborus flavicans, chemical communication, *Daphnia longispina*, *Daphnia pulex*, horizontal migration, swarming

Abstract

Dense swarms of *Daphnia longispina* (up to 4000 animals l^{-1}) were recorded along the littoral zone in a lake where *Chaoborus flavicans* is considered the main predator. *D. longispina* coexisted with *D. pulex*, but there were no *D. pulex* in the littoral swarms. Swarms were less dense at night (about 1/10 the density), and *D. longispina* exhibited diel horizontal migrations by aggregating in the littoral during the day and spreading out at night. Laboratory experiments showed that *Chaoborus* capture efficiency on juvenile daphnids was higher in the light compared to darkness, and that *Daphnia* exhibited a behavioural response to water that had previously contained *Chaoborus*. We conclude that predation from *Chaoborus* can be an important factor affecting the distribution patterns of *Daphnia* observed in this lake. The behavioural experiments indicated that this influence might be partly mediated by chemical agents.

Introduction

Flocking is a well-known behaviour to avoid predation in terrestrial vertebrates (Humpries & Driver, 1967; Hamilton, 1971; Pulliam, 1973; Treisman, 1975a, b; and Caraco *et al.*, 1980). The same mechanisms operate among fish and in fish-zooplankton interactions (Neill & Cullen, 1974; Milinski & Heller, 1978; Heller & Milinski, 1979; Milinski, 1979; Jakobsen & Johnsen, 1987, 1988a). The efficiency of swarming to avoid predation is due to two effects: a) confusion, used against visual predators, and b) dilution, i.e. reduction in the chances of a given individual being eaten. The main cost of swarming is considered to be a rapid reduction in the concentration of food within the swarm, and it has been shown that zooplankton conspecifics tend to be more aggregated under excess food compared to low food conditions (Jakobsen & Johnsen, 1988b).

The impetus of our work was based on observations of dense swarms of the waterflea *Daphnia longispina* O.F. Müller (Crustacea: Cladocera) within and along the littoral zone of a mesotrophic lake: Myravatn, Bergen, Norway (Giske, 1986). In Myravatn, *D. longispina* coexists with *Daphnia pulex* Leydig, and it is assumed that there is habitat segregation between the two species. There are no planktivorous fish in the lake. According to Giske (1986), the phantom midge *Chaoborus flavicans* Meigen is the main factor regulating the *Daphnia* population in Myravatn. Because it is subjected to low predation pressure in Myravatn, *Chaoborus* is able to go through two generations per year, in contrast to one generation in lakes in the same area inhibited by planktivorous fish. The *Chaoborus* in Myravatn perform diel vertical migration, but to a lesser extent than would be expected in a lake with planktivorous fish (pers. comm., Jan Wivegh, University of Bergen). *Chaoborus* feed on juvenile daphnids, but since *Daphnia* release offspring approximately every third day (at 20 °C), one might assume that also adult egg-bearing females distribute themselves in response to the predator, in order to protect their offspring just after birth.

Davies (1985) reported diel horizontal migration in *Daphnia hyalina lacustris* with aggregations of > 1000 animals l^{-1} in the littoral. He considered these dense concentrations of daphnids in the littoral to be an escape response from fish predators in the pelagial. De Meester *et al.* (1993) also described a similar type of diel horizontal migration in *Scapholebris mucronata*, with > 1000 individuals l^{-1} swarming in the littoral zone during day and dispersing at night. These distribution patterns are also considered to be the result of active avoidance from fish predators. Whether *D. longispina* exhibit similar mechanisms to avoid predation from *Chaoborus*, depends partly on the question of modes of predation. *Chaoborus* is generally considered to be a tactile sit-and-wait predator (Pastorok, 1978), and Swift and Forward (1981) did not find that *Chaoborus* had a higher strike efficiency in the light compared to dark conditions when feeding on copepods. It should be noted however, that the 'confusion effect' of swarming can apply even if olfactory or tactile senses are used to detect the prey item, and that the dilution effect will be the same.

During the last decade, several authors have demonstrated that chemical signals can be important factors regulating aquatic environments, especially concerning predator-prey interactions. A review of chemical communication in planktonic animals is given by Larsson and Dodson (1993). Dodson (1988) demonstrated experimentally that some *Daphnia* species responded to chemical signals from *Chaoborus americanus* by changing their vertical migration pattern. We therefore assumed that the *Daphnia* distributions in Myravatn could be affected by the detection of chemical substances released by *Chaoborus*.

Our aim in this study was to: a) describe the distribution patterns and the swarming phenomenon in Myravatn; b) conduct experiments to test whether *Chaoborus* had higher predation success on daphnids in the light rather than in darkness; and c) conduct experiments focusing on the role of chemical substances from *Chaoborus*, which affect the horizontal migration of *Daphnia*.

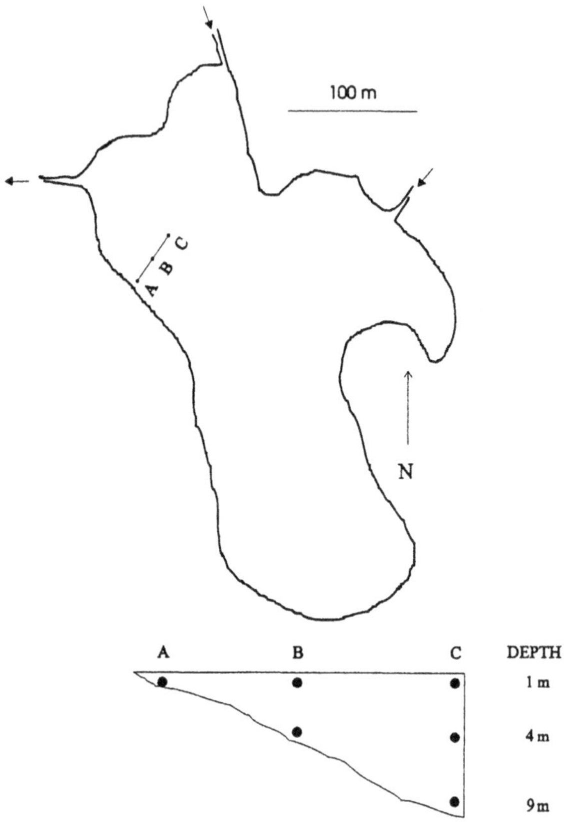

Fig. 1. Map of Myravatn with the sampling stations A, B and C, and a cross-section showing the six locations where the day and night samples were collected.

Materials and methods

Field

The lake studied, Myravatn, is located 8 km south of Bergen, Norway. It is a small (0.55 km^2) temperate mesotrophic lake without planktivorous fish. The littoral vegetation is well developed, and consists of a continuous vegetation belt around most of the lake, composed mainly of *Equisetum fluviatile*, *Nymphaea alba*, *Nuphar lutuem*, and *Potamogeton natans*.

The horizontal and vertical distribution of daphnids in Myravatn was analysed by two kinds of samples. Firstly, day and night series of plankton samples were taken at the end of every month from May to September. We collected at three stations at the northern end of the lake (Fig. 1): station A was at the edge of the vegetation zone; station B was 15 m offshore; and station C was 30 m offshore (at the deepest point of the basin). At each station, samples were taken at each of the following depths: 1m at station A, 1 and 4 m at station B, and 1, 4 and 9 m at station C. Twenty samples were taken at each location. Secondly, cross-sectional samples from station A to C were taken in the middle of the summer to ensure that there was no area where aggregations or swarms were not accounted for in the monthly samples. In the cross-sections, a total of 121 samples were collected in a transect from the shore to the deepest point of the basin; samples were taken

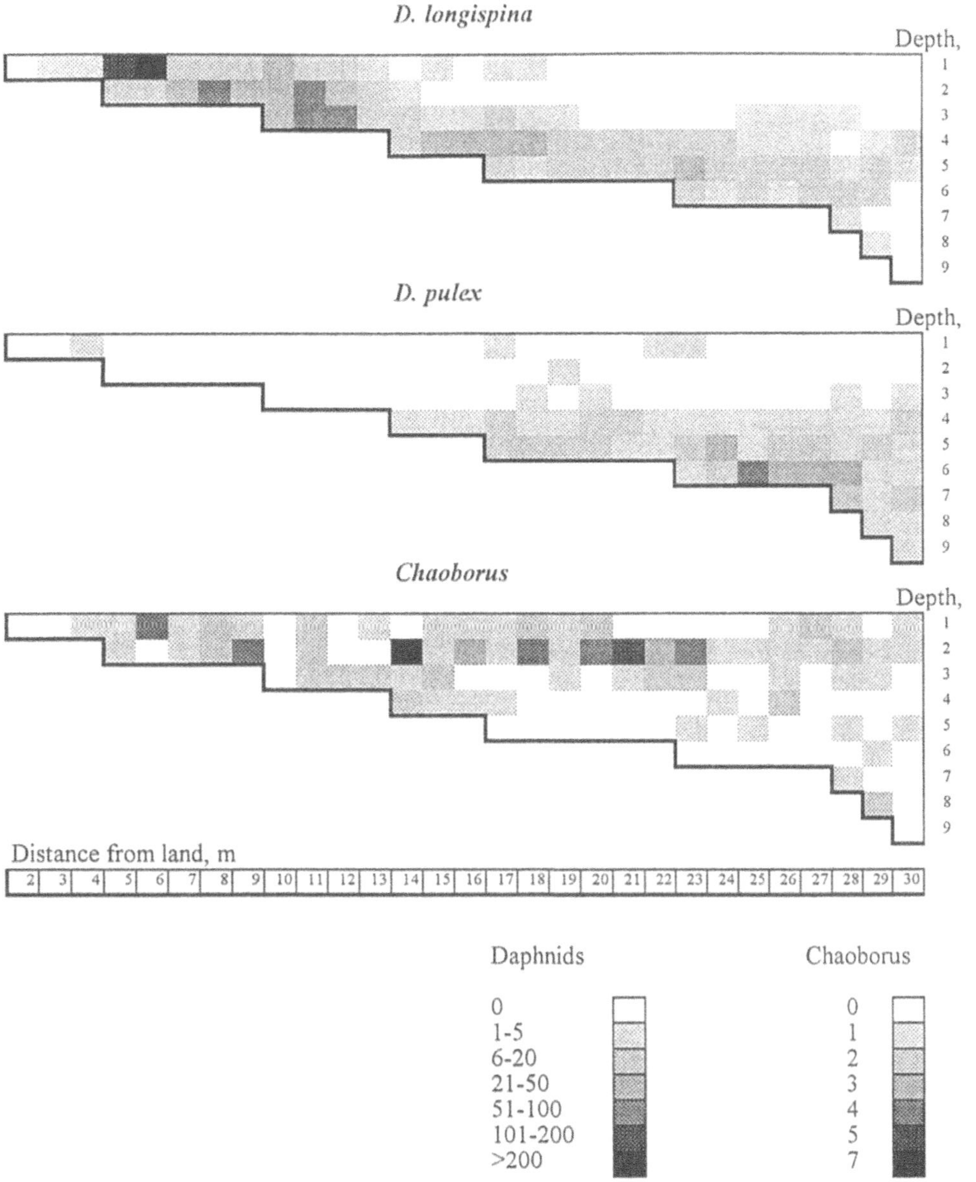

Fig. 2. Depiction of the distribution patterns of *D. longispina*, *D. pulex* and *Chaoborus* in Myravatn on 28 July 1990. The level of shading indicates the number of animals l^{-1}.

at 1 m intervals, both horizontally and vertically (Fig. 2).

All samples were collected with a 1-liter Ruttner water sampler, filtered through a plankton net, and preserved in 96% ethanol. Because of very clear distribution patterns, the daphnids were only counted in five randomly chosen samples from the total of 20 parallels for each depth/station in the day and night samples, whereas all samples were screened for *Chaoborus*. The Ruttner sampler is not a very good device for quantifying *Chaoborus*, because it will underestimate the total density. Still we decided to include the results, because they indicated a trend in the proportions between the different locations. Because only the large larvae are a threat to the daphnids, only third and fourth instars of *Chaoborus* were counted. The cross-sectional sampling in the middle of the summer was tested by a Kolmogorov-Smirnov two-sample test to determine if the populations of *D. pulex*, *D. longispina*, and *Chaoborus* had divergent distributions. The numbers

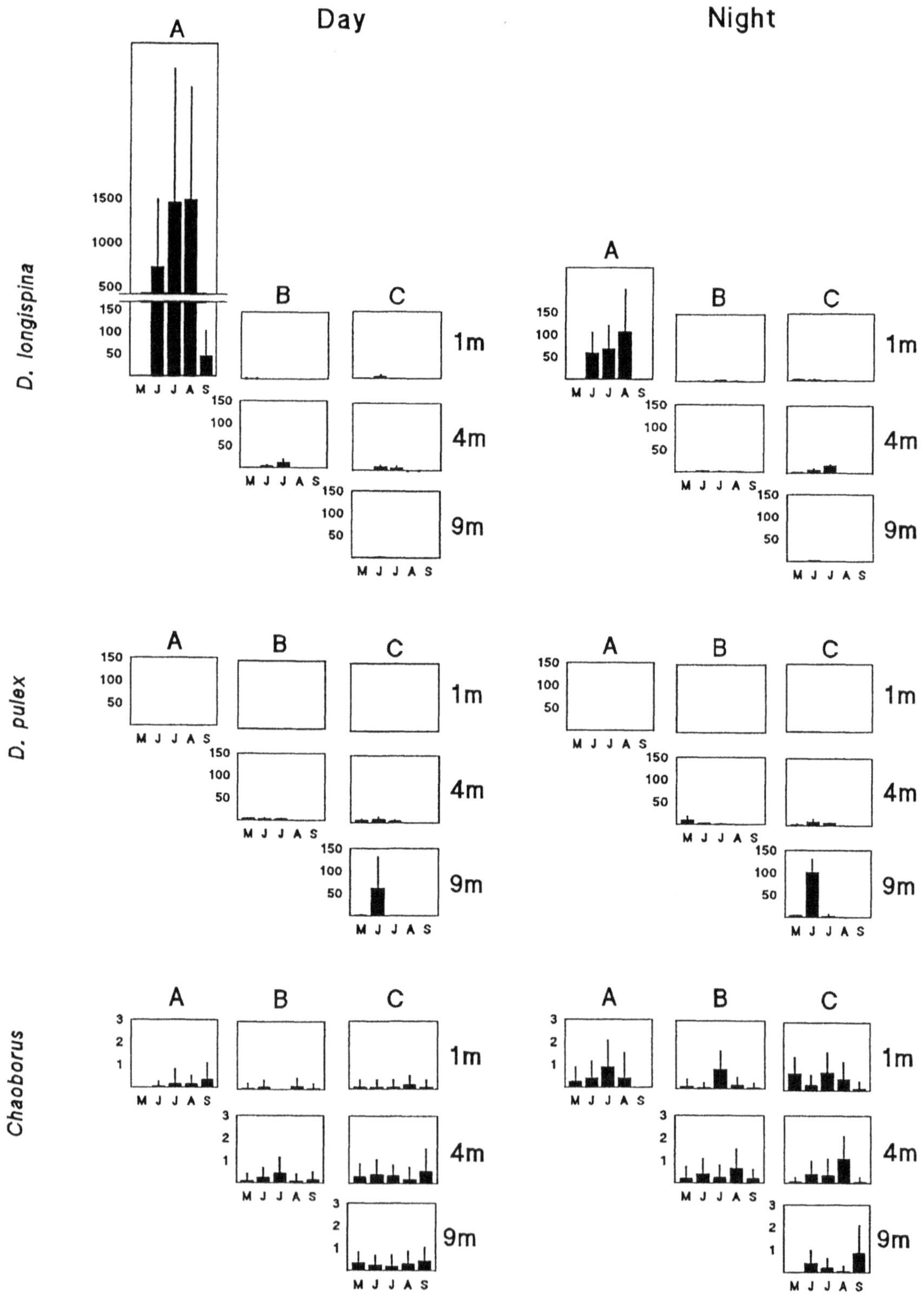

Fig. 3. The distributions of *D. longispina, D. pulex* and *Chaoborus* in day and night routine samples at stations A, B and C, at depths 1, 4 and 9 m. Each histogram shows the number of animals l^{-1} in samples taken at the end of May, June, July, August and September.

of the different species in each sample were tested in pairwise comparisons. Because of clear results (Fig. 3), tests for swarming and segregation between the two *Daphnia* species were not conducted for the monthly samples.

Experiment 1

To test whether *Chaoborus* had a higher predation success on daphnids in the light compared to darkness, a series of simple laboratory experiments was set up. One fourth instar *Chaoborus* larva was added to 1 litre jars containing 10 new-born *D. longispina* and kept there for four hours. The daphnids were given surplus food (1 mg dry weight of *Scenedesmus cf. acutus* l^{-1}), while the chaoborid was starved for 24–48 hours prior to each experiment. Each experiment consisted of two jars, one of the glasses was placed in the light, whereas the other was kept in total darkness. After four hours, the remaining daphnids in each jar were counted. A total of 13 replicates was run. To control for mortality, which was not related to predation, and to ensure that none of the small neonates were able to avoid our detection, jars without *Chaoborus* were also set up.

The basic medium for all the laboratory experiments was 'Standard water', deionized water to which irradiated and filtered seasalt and $CaCO_3$ were added to a conductivity of about 400 mS cm-1 (Hobæk & Larsson, 1990). All experiments were conducted with *Daphnia*-clones from Myravatn that had been cultured in the laboratory for more than two years. The *Chaoborus* were collected directly from the lake and conditioned for some days prior to the experiments.

Experiment 2

To test possible behavioural reactions of *D. longispina* and *D. pulex* to chemicals released from *Chaoborus*, experiments were conducted in an experimental set-up described by Kleiven *et al.* (1994). Five flow-through chambers (plastic cell culture flasks 135 × 11 × 4.5 cm) were glued together and interconnected with an opening (1.5 × 3 cm) large enough for adult daphnids to traverse. Ten adult daphnids with eggs were placed in each chamber, and water that had previously contained *Chaoborus* was added in chamber #1. The other chambers received an input of Standard water. In the controls, Standard water was added to all chambers. The flow-through system created a gradient of experimental water from chamber #1 to chamber #5, which was tested by adding coloured water to chamber #1. The experiments lasted for four hours, and during this time, experimental water from chamber #1 was able to mix gradually with the water in chambers #2 and #3. Small amounts of experimental water were detected in chamber #4, but none of this water was detected in chamber #5.

To avoid microbial degradation of the predator substance (Loose *et al.*, 1993), we used fresh medium. *Chaoborus* were allowed to feed on *D. longispina* and *D. pulex* juveniles. After four hours of feeding, *Chaoborus* were pipetted into a tank with Standard water at a density of 10 Chaoborus l^{-1}. The '*Chaoborus* water' was pumped from the tank directly into the chamber. To prevent *Chaoborus* from getting into the chambers, a plankton net was used at the inlet of the peristaltic pump. All experiments were conducted with excess food (1 mg dry weight of *Scenedesmus* cf. *acutus* l^{-1}) and a flow-through rate of 4 ml min^{-1}. The experiments were conducted both under light and dark conditions. In each replicate, one row of chambers containing *D. longispina* and one row containing *D. pulex* were run simultaneously.

To test whether there was any escape response from the experimental water, a regression analysis between the numbers of daphnids in the five chambers was calculated for each replicate. The mean regression line of all replicates was then tested against a deviation from a mean value of zero (no horizontal displacement) using a t-test.

Results

Field

The day and night routine samples revealed that the swarms of *D. longispina* were about ten times denser during the day than during night (Fig. 3). *Chaoborus* performed a weak diel vertical migration, while *D. pulex* was found mainly in the deeper water strata. The dense aggregations of *D. longispina* along the littoral persisted from June until the end of August, and visual observations made it clear that the aggregations were formed as a 1–2 m broad belt of swarms all the way around the lake. The animals were not evenly distributed within this belt, but occurred in local swarms. The maximum number of adult *D. longispina* found in a 1-l sample was 4038 animals, while the maximum number of juveniles was 4392 animals. The juvenile *Daphnia* distribution was similar to the adult *D. longispina* distribution, but was not included in the figures because

Table 1. Chaoborus predation efficiency on juvenile *D. longispina*

Exp. #	Light Proportion captured	Dark Proportion captured
1	0.7	0.2
2	0.6	0.2
3	0.6	0.2
4	0.6	0.3
5	0.4	0.4
6	0.7	0.2
8	0.4	0.3
9	0.2	0.2
10	0.4	0.1
11	0.3	0.5
12	0.4	0
13	0.4	0.2

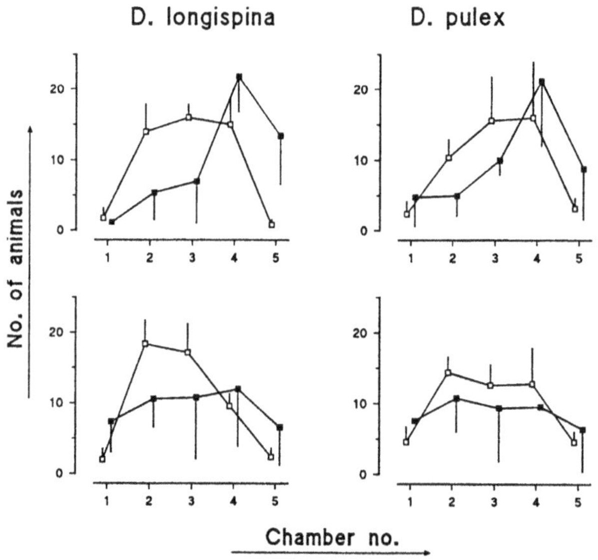

Fig. 4. Horizontal distributions of *D. longispina* and *D. pulex* in experiments conducted under light (open squares) and dark conditions (filled squares). The squares represent the average ± (S.D.) number of animals in each chamber. Upper figures are experiments with input of '*Chaoborus*-water' into chamber #1, lower figures are control experiments with Standard water input into all chambers.

of difficulties in determining with certainty the species identity for juveniles.

The cross-sectional sampling series from the littoral to the deepest point in the lake (Fig. 2) gives the trends of the distribution patterns in Myravatn. *D. longispina* exhibits dense swarms within and along the littoral, while *D. pulex* are found at greater depths. *Chaoborus* are more evenly distributed in the water column than would be expected in a lake with planktivorous fish. The Kolmogorov-Smirnov two-sample test revealed that the three populations had divergent distributions ($p < 0.05$). From the cross-sectional samples, it seems that the most important aspects of the distribution patterns were detected in the day and night routine samples.

Experiment 1

Chaoborus capture efficiency on juvenile *D. longispina* under light and dark conditions are given (Table 1). After four hours, *Chaoborus* demonstrated a significantly higher success rate in detecting and eating juvenile daphnids in the light compared to dark conditions (t-test, $p < 0.005$). In the control jars, all the animals were found, and no mortality was detected.

Experiment 2

The experiments, which tested the response of *D. longispina* and *D. pulex* to water that had previously contained *Chaoborus*, revealed that the daphnids were able to distribute themselves horizontally according to chemical cues from *Chaoborus* (Fig. 4). The experiments also indicated a difference in behavioural response between light and dark conditions. In darkness, both *D. pulex* and *D. longispina* demonstrated a significant escape response ($p < 0.05$) from the chamber which received water from *Chaoborus*, i.e. they increased their distance from the input of '*Chaoborus* water'. In the light, both *D. longispina* and *D. pulex* responded by aggregating in the middle chambers, but showed no escape response (Fig. 4). The same tendency towards aggregating in the middle was also found in the control experiments.

Discussion

Swarming along the littoral vegetation in Myravatn is contrary to the horizontal distribution patterns observed in many other lakes. There are several examples in the literature where *Daphnia* avoid the littoral zone. This is commonly referred to as shore-avoidance. Siebeck (1980) suggested that optical orientation towards the land shadow was a mechanism that could lead to different distributions among different species. Pennak (1973) discussed the effects of repellents from macrophytes on *Daphnia*, and Gliwicz and Rykowska (1992) considered predation from littoral fish, as the ultimate factor responsible for shore-avoidance in *Daphnia*.

The fact that the swarms in Myravatn are less dense at night (about 1/10 of the density during daytime), indicates that *D. longispina* exhibits a diel horizontal migration similar to the migration observed by Davies (1985) and De Meester *et al.* (1993). Our sampling does not show where the animals go when the swarms disintegrate, but it seems likely that the daphnids spread out into the pelagial, as described in the studies mentioned above. If the dense swarms spread into the much larger volume of water in the pelagial during the night, then one would not be able to detect an increase in numbers anywhere in the pelagial, which would correspond to the reduction in density in the littoral.

The dispersion of swarms at night indicates an avoidance behaviour in response to a visual or day-active predator. The feeding-experiments show that *Chaoborus* eats significantly more daphnids in the light, and this illustrates the importance of visual prey detection. This is in contrast to the results of Swift & Forward (1981), who did not find differences in *Chaoborus* capture efficiency on copepods under light and dark conditions. This contradiction might be attributed to the differences in criteria for measuring *Chaoborus* predation and/or the different modes of swimming in copepods compared to daphnids. If *Chaoborus* is a more efficient predator in the light, it will induce a higher degree of swarming during the day, when compared to the night. *Chaoborus* in Myravatn exhibit some diel vertical migration (Fig. 3), but to a lesser extent than in lakes with planktivorous fish (Jan Wivegh, pers. comm.). Hence, a larger number of *Chaoborus* is present in the upper layers of the lake during the day than in lakes with planktivorous fish, and the potential swarm-inducing effect from *Chaoborus* on daphnids during the day will be more pronounced than in other lakes.

Gradually, a reduction in food concentration within a swarm will force the animals to disperse. The aggregations of *D. longispina* in Myravatn are very dense (up to 4000 animals l^{-1}), and it is not likely that phytoplankton production is great enough to support continuous swarming. Therefore, the animals have to disperse at certain time intervals. Since phytoplankton production is reduced at night, food is also a factor that will enhance swarming during the day and dispersing during the night.

Both the cross-sectional samples and the day and night routine samples showed that *D. longispina* and *D. pulex* had divergent distributions, and that *D. longispina* exhibited denser swarms than *D. pulex*. These differences in distribution patterns between the two species of *Daphnia* might be caused by different modes of predation defense. Morphological defenses like neckteeth formation are found in both *D. pulex* and *D. longispina* in Myravatn (Giske, 1986), but they are commonly more pronounced in *D. pulex*. We can not say unequivocally that *D. pulex* use more morphological defenses, while *D. longispina* rely more on behavioural predation avoidance. Still, if the smaller size of juvenile *D. longispina* makes them vulnerable to *Chaoborus* predation for a longer time period than the larger *D. pulex*, and if the larger *D. pulex* has higher food requirements, there would be a difference in the cost/benefit ratio of swarming between the two species. Therefore, swarming behaviour might be favoured in *D. longispina*, while morphological defenses might be favoured in *D. pulex*.

The distribution of daphnids in Myravatn might also be influenced by other predators, but we did not conduct experiments with predators other than *Chaoborus*. In addition to *Chaoborus*, there are also pike (*Esox lucius*), water mites, Notonectidae and Corixidae in the lake. Pike are not considered planktivores, and zooplankton have not been found in the stomach contents of pike from this lake (Nilsen, 1980). Water mites can be a dominant predator in lakes (Butler & Burns, 1993), and Notonectidae can exhibit high predation rates as well (Murdoch *et al.*, 1984). However, since *Chaoborus* is by far the most abundant invertebrate predator in Myravatn, we still believe that the distribution patterns of daphnids are influenced primarily by this predator.

Our behavioural experiments revealed differences in escape response between light and dark conditions. In darkness, both species responded with a significant escape response away from the '*Chaoborus* water'. This indicates that the daphnids distribute themselves horizontally according to chemical signals from *Chaoborus*. In the light, however, the daphnids responded by aggregating in the middle chamber even when there were chemical substances that indicated the presence of a predator (Fig. 4). Aggregations in the light and in the presence of surplus food are in accordance with the findings of Jakobsen and Johnsen (1988b). A stronger tendency to swarm in the light compared to darkness might also explain the aggregation in the middle chamber, instead of an escape response when chemical substances are present.

The results of both the field work and the experiments indicate that *Chaoborus* is an important factor in explaining the atypical *Daphnia* distribution patterns found in Myravatn. The behavioural experiments showed that daphnids are able to respond adequately to chemical substances emitted from their main predator in Myravatn. Therefore, we suggest strongly that the *Daphnia* distributions observed in Myravatn are partly mediated by chemical substances from *Chaoborus*.

Acknowledgements

We would like to thank Inger Haugsbø for her excellent work in the laboratory, and two anonymous reviewers for their constructive criticisms. We also want to thank Petter Larsson, Elisabeth Müller Lysebo, Harald Sægrov, Kurt Urdal and Larry Weider for helpful comments on earlier versions of the manuscript, and Anders Hobæk for useful discussions.

References

Butler, M. I. & C. Burns, 1993. Water mite predation on planktonic Cladocera: parallel curve analysis of functional responses. Oikos 66: 5–16.

Caraco, T. S., S. Martindale & H. R. Pulliam, 1980. Avian flocking in the presence of a predator. Nature 285: 400–401.

Davies, J., 1985. Evidence for a diurnal horizontal migration in *Daphnia hyalina lacustris* Sars. Hydrobiologia 120: 103–105.

De Meester, L., S. Maas, K. Dierckens & H. J. Dumont, 1993. Habitat selection and patchiness in *Scapholebris*: horizontal distribution and migration of S. mucronata in a small pond. J. Plankton Res. 15: 1129–1139.

Dodson, S., 1988. The ecological role of chemical stimuli for the zooplankton: Predator-avoidance behavior in *Daphnia*. Limnol. Oceanogr. 33: 1431–1439.

Giske, J., 1986. (In Norwegian). Populasjonsregulerende faktorer hos to arter *Daphnia* i et vann uten planktivor fisk. Master thesis, Zoological Museum, University of Bergen.

Gliwicz, Z. M. & A. Rykowska, 1992. 'Shore avoidance' in zooplankton: a predator-induced behavior or predator-induced mortality? J. Plankton Res. 14: 1331–1342.

Hamilton, W. D., 1971. Geometry for the selfish herd. J. Theor. Biol. 31: 295–311.

Heller, R. & M. Milinski, 1979. Optimal foraging of sticklebacks on swarming prey. Anim. Behav. 27: 1127–1141.

Hobæk, A. & P. Larsson, 1990. Sex determination in *Daphnia magna*. Ecology 71: 2255–2268.

Humpries, D. A & P. M. Driver, 1967. Erratic display as a defence against predators. Science 156: 1767–1768.

Jakobsen, P. J. & G. H. Johnsen, 1987. The influence of predation on horizontal distribution of zooplankton species. Freshwat. Biol. 17:501–508.

Jakobsen, P. J. & G. H. Johnsen, 1988a. Differences in individual protection against predation from fish in swarming *Bosmina longispina*. Anim. Behav. 36: 986–990.

Jakobsen, P. J. & G. H. Johnsen, 1988b. The influence of food limitation on swarming behaviour in the waterflea *Bosmina longispina*. Anim. Behav. 36: 991–995.

Kleiven, O.T., P. Larsson & A. Hobæk, 1995. *Daphnia* in the dark: the role of chemical signals in spatial distribution. (unpublished manuscript).

Larsson, P. & S. Dodson, 1993. Chemical communication in planktonic animals. Arch. Hydrobiol. 129: 129–155.

Loose, C., E. Elert & P. Dawidowicz, 1993. Chemically-induced diel vertical migration in *Daphnia*: a new bioassay for kairomones exuded by fish. Arch. Hydrobiol. 126: 329–337.

Milinski, M., 1979. Can an experienced predator overcome the confusion of swarming prey more easily? Anim. Behav. 27: 1122–1126.

Milinski, M. & R. Heller, 1978. Influence of a predator on the optimal foraging behaviour of sticklebacks (*Gastrerosteus aculeatus* L.). Nature 275: 642–644.

Murdoch, W. W., M. A. Scott & P. Ebsworth, 1984. Effects of the general predator, *Notonecta* (Hemiptera) upon a freshwater community. J. Anim. Ecol. 53: 791–808.

Neill, S. R. St. J. & J. M. Cullen, 1974. Experiments on whether schooling by their prey affects the hunting behaviour of cephalopods and fish predators. J. Zool. 172: 549–569.

Nilsen, M., 1980. Gjedda, *Esox lucius* L., i Myravatn. Populasjon, ernæring og reproduksjon. Master thesis, Zoological Museum, University of Bergen.

Pastorok, R. A., 1978. Predation by *Chaoborus* larvae and its impact on the zooplankton community. Ph. D. Dissertation, Univ. Washington, 238 pp.

Pennak, R. W., 1973. Some evidence for Aquatic macrophytes as repellents for a limnetic species of *Daphnia*. Int. Revue ges. Hydrobiol. 58: 569–576.

Pullian, H. R., 1973. The advantages of flocking. J. Theor. Biol. 38: 419–422.

Siebeck, O., 1980. Optical orientation of pelagic crustaceans and its consequence in the pelagic and littoral zones. In Kerfoot, W. C. (ed.), Evolution of Zooplankton Communities. University Press of New England, Hanover, NH: 28–38 pp.

Swift, M. C. & R. B. Forward, 1981. *Chaoborus* prey capture efficiency in the light and dark. Limnol. Oceanogr. 26: 461–466.

Treisman, M., 1975a. Predation and the evolution of gregariousness. I. Models for concealment and evasion. Anim. Behav. 23: 779–800.

Treisman, M., 1975b. Predation and the evolution of gregariousness. II. An economic model for predator-prey interaction. Anim. Behav. 23: 801–825.

Hydrobiologia **307**: 185–191, 1995.
P. Larsson & L.J. Weider (eds), Cladocera as Model Organisms in Biology.

Behavioural and morphological changes in polymorphic *Daphnia* related to different predation regimes

Elisabeth Müller Lysebo
Institute of Zoology, University of Bergen, Allégaten 41, N-5007 Bergen, Norway

Key words: Cyclomorphosis, *Daphnia*, habitat preference, invertebrate predation

Abstract

Morphology and behaviour of *Daphnia galeata* in Lake Lille Lungegårdsvann, Bergen, Norway, were studied in two consecutive years. A portion of the population showed helmet cyclomorphosis in 1992, but no helmets developed in 1993. The non-helmeted individuals in 1992 were found in the littoral habitat during summer, whereas the helmeted morph stayed mainly in the pelagial. These differences in morphology and behaviour within and between years are likely to be caused by changes in predation regimes. Larvae of the predacious phantom midge, *Chaoborus flavicans*, were observed only in 1992, whereas the three-spined stickleback, *Gasterosteus aculeatus*, was present both years.

Introduction

Antipredator defensive mechanisms exhibited by zooplankton in a given lake, are related to the predation regime. Defenses can be morphological, behavioural or life-history features, and many of these are reported to be induced by the presence of a predator (Havel, 1987; Harvell, 1990; Dawidowicz *et al.*, 1990) or by chemicals from the predator (reviewed by Larsson & Dodson, 1993).

Cyclomorphosis, a possible morphological defensive mechanism in *Daphnia*, is an annual change in morphs over time, which can involve changes in neckteeth, tailspine, and head shape or length. Water temperature, turbulence, food and light have been reported as cues that can induce cladoceran cyclomorphosis (Jacobs, 1987). Mort (1989) showed that the period of highest helmet formation in a field population of *Daphnia* also coincided with peak densities of invertebrate predators like *Chaoborus* and *Leptodora*, and other studies have shown that helmet size in daphnids increases with increasing invertebrate predation pressure (Pijanowska, 1990). Others (Krueger & Dodson, 1981; Hebert & Grewe, 1985; Dodson, 1988b, 1989a; Tollrian, 1990; Hanazato, 1991) have found that invertebrates such as *Chaoborus*, induce cyclomorphic features in daphnids, structures that might reduce vulnerability to *Chaoborus* predation (Krueger & Dodson, 1981; Havel & Dodson, 1984).

Other inducible defensive mechanisms in aquatic organisms are behavioural. Changes in distribution, either horizontal, vertical or swarm formation, have been shown to serve as antipredator defenses (Milinski, 1979; Davies, 1985; Jakobsen & Johnsen, 1987, 1988; Dodson, 1988a; Dawidowicz *et al.*, 1990; Gliwicz & Rykowska, 1992; De Meester *et al.*, 1994; Kvam & Kleiven, 1995).

Many defenses are believed to be costly. Costly inducible defenses should be reduced in the absence of the predator. Hence, in this study I have looked for morphological and behavioural changes in *Daphnia galeata* within and between two consecutive years, 1992 and 1993, during which the predator regimes switched from combined invertebrate and fish predation to solely fish predation.

Study site and methods

Eutrophic Lake Lille Lungegårdsvann is a former marine lagoon in the center of Bergen, Norway. Surface area is approximately 26 000 m^2, mean depth is

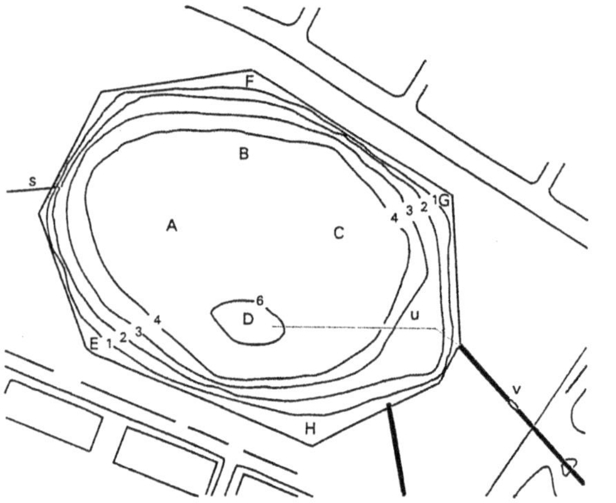

Fig. 1. Map of Lake Lille Lungegårdsvann. A–D represent the pelagic sampling stations, E–H the littoral stations. Numbers indicate depth in meters, s and b are water inlets. The outlet, U, is an Olzewsky tube with a one-way valve (v) (Modified from Nordeide & Oftedal, 1985 and the Municipality of Bergen).

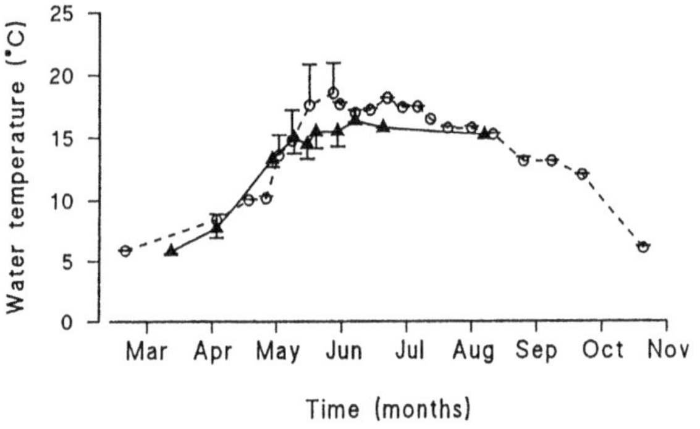

Fig. 2. Seasonal change in mean temperature in Lake Lille Lungegårdsvann. o = 1992, ▲ = 1993. Bars represent standard errors (S. E.).

4 m, and maximum depth is 6 m. The lake has been biomanipulated and monitored since 1985 (Larsson & Kambestad, in press).

The sampling periods were from March to November 1992 and from April to August 1993. Samples were taken before noon. Night samples were taken on July 17, 1992.

Plankton samples were taken with a Patalas-Schindler 25-l trap in the pelagial, (stations A-D) and in the littoral (E-H) (Fig. 1). Samples were taken at every meter. At station D, physico-chemical data also were obtained. Oxygen was measured in the field using a YSI Oxymeter and conductivity (μS/cm) was measured at room temperature using a CDM80 conductivity meter. From March to the end of June 1992, only samples at station D were taken, but when a belt of *Daphnia* was observed in the littoral as in 1991, systematic sampling of both the pelagial and the littoral was started. Preliminary samples were taken at one littoral station on June 4 and 15. In 1993, vertical and horizontal hauls also were taken on some of the dates, using a plankton net with mesh size of 90 μm and a diameter of 1 m to sample for *Chaoborus flavicans* and *Bythotrephes longimanus*. *Gasterosteus aculeatus* was caught using plexiglass traps placed on the bottom (Breder, 1960).

Zooplankton were preserved in a 5% sucrose-formaldehyde mixture (Haney & Hall, 1972). A subsample of at least 30 daphnids from each sample were scored in five classes of different degrees of helmet development (1992) and three classes of tailspine development. When the entire sample contained less than 30 daphnids, all daphnids were scored. When looking for *Chaoborus* larvae or exuviae of pupa, the entire sample was counted.

To test whether the two *Daphnia* morphs had different distributions, the density-independent G-test was used (Sokal & Rohlf, 1981). The critical probability value was set at $p < 0.001$ due to the large number of animals tested.

Results

Abiotic and biotic conditions

Conductivity ranged between 100 and 300 μS cm^{-1} during 1992, but increased in November 1992 to more than 1100 μS cm^{-1}. Due to sea water inflow during winter 1992/93, the bottom water had a conductivity of around 6000 μS cm^{-1} until the middle of June 1993, when it started to decline. Throughout the whole sampling period, conductivity was approximately 1000 μS cm^{-1} down to a depth of 5 m. The average pH-value rose from 7.93 ± 0.66 in 1992 to 8.90 ± 1.21 in 1993. Anoxic conditions occurred between 5 and 6 m throughout all of 1993, whereas this period only lasted from mid-May to late July in 1992. In August 1993, there was less than 50% oxygen saturation at all depths. Water temperature was higher in June and July 1992 compared to 1993 (Fig. 2).

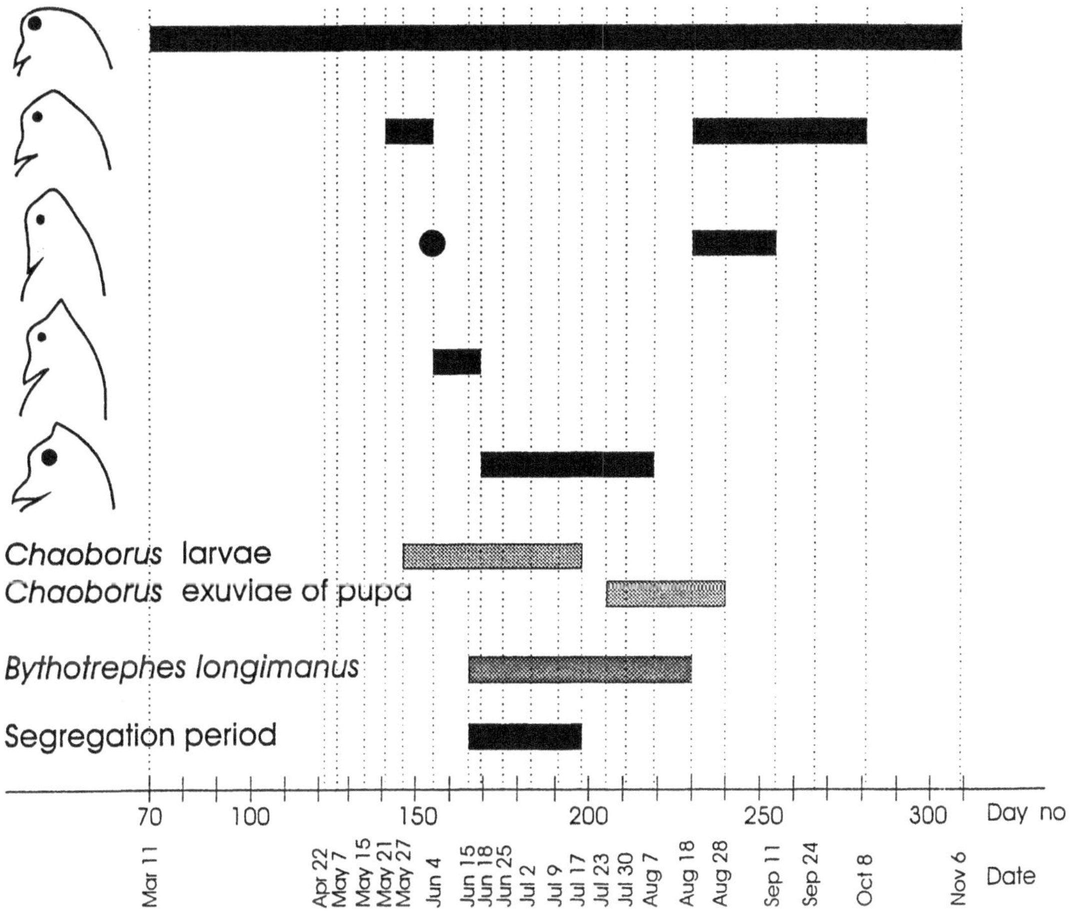

Fig. 3. Cyclomorphosis and the period of habitat segregation between the non-helmeted and helmeted morphs of *Daphnia* related to the presence of *Chaoborus* larvae and exuviae of pupa and *Bythothrephes* in 1992. The degree of helmet formation was scored into five classes.

Daphnia species present in 1992 were *D. galeata* and *D. longispina*, whereas only *D. galeata* was found in 1993. Potential predators in 1992 were larvae of the phantom midge, *C. flavicans*, three-spined stickleback (*G. aculeatus*), and *B. longimanus*. *Chaoborus* larvae were found from late May until mid-July, but were not detected in high numbers due to sampling limitations. The larvae found in July were fourth instar. After July 23, only exuviae of pupa were found (Fig. 3). *B. longimanus* was observed from mid-June to mid-August (Fig. 3), and reached average maximum densities of approximately 0.4 ind l^{-1} by mid-July. All *Chaoborus* larvae and most of the *Bythotrephes* were caught in the pelagial. *Chaoborus* was absent during winter, probably because of the unexpected increase in salinity (Lysebo, 1993). A continuous high conductivity level throughout 1993 seemed to prevent *Chaoborus* from reestablishing itself in the lake. *Gasterosteus aculeatus* was present in both years, but at considerably lower densities than when biomanipulation started. The number of fish $\times$ trap^{-1} $\times$ hr^{-1} in 1984 was on average 200, whereas from 1991 onwards, fish density was reduced to less than five fish $\times$ trap^{-1} $\times$ hr^{-1} (Larsson & Kambestad, in press).

Morphology

D. galeata in L. Lille Lungegårdsvann showed helmet cyclomorphosis in 1992 (Fig. 3), but not in 1993. Helmet development started in late May, and reached maximum expression in June and July, and then declined in the last half of August. By early October, the helmets had disappeared. In all body size classes except for the round-headed neonates, there were

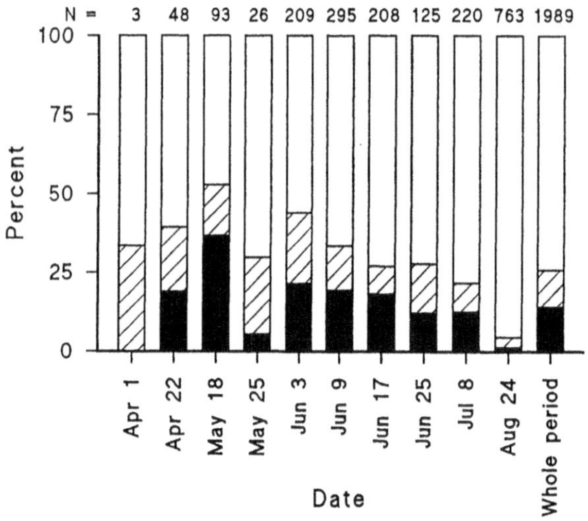

Fig. 4. Seasonal change in the proportion of *D. galeata* with long tailspines (open bars), with reduced spines (hatched bars), and without spines (filled bars) in 1993. The last histogram shows the average proportion for the whole sampling period. Numbers above each bar represent total number of animals counted.

Table 1. Results from the G-tests testing differences in distributions of the helmeted and non-helmeted daphnids in each of the two habitats (littoral and pelagic zones).

Date	Number of daphnids	$G_{adj.}$	*P*-value
Jun 4	328	12.28	< 0.001
Jun 15	401	209.43	< 0.001
Jul 2	136	91.62	< 0.001
Jul 9	508	58.76	< 0.001
Jul 17	354	17.67	< 0.001
Jul 17N	741	33.79	< 0.001
Jul 23	304	8.56	N. S.
Jul 30	149	0.81	N. S.
Aug 7	373	11.88	< 0.001
Aug 18	291	1.25	N. S.
Aug 28	567	37.59	< 0.001
Sep 11	775	8.09	N. S.
Sep 24	715	1.36	N. S.
Oct 8	746	2.77	N. S.
1992	6388	20.72	< 0.001

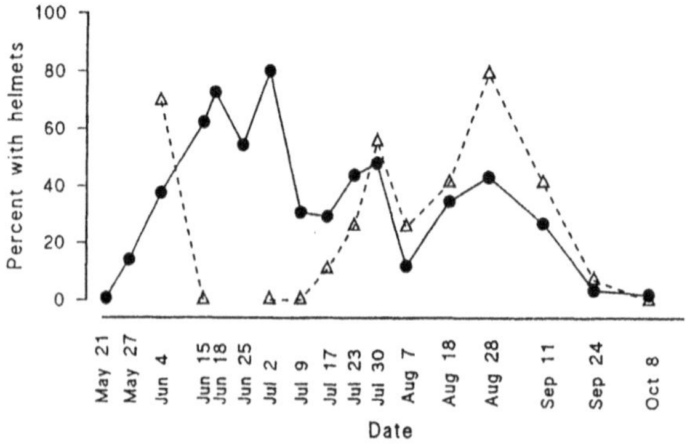

Fig. 5. Proportion of helmeted *Daphnia galeata* from each habitat in 1992. The habitats are the littoral (Δ) and the pelagic (•) zones. Before May 21 and after Oct 8, none of the daphnids had helmets (N > 32 for each habitat).

both smooth and helmeted individuals. *D. longispina* did not develop helmets. Reduced or absent tailspines were observed in adult *D. longispina* and both *D. galeata* morphs, but the proportion of short-spined daphnids in 1992 never exceeded 10%. The proportion of individuals with reduced tailspines was higher in 1993 than in 1992 (Fig. 4); it increased from April and reached a maximum in mid-May. The proportion declined in late May, increased again in early June, and thereafter, declined steadily until sampling was ended.

Behaviour

Figure 5 shows the proportion of helmeted individuals from the pelagial and the littoral during day. On July 17, when additional night samples were taken, the distributions were about the same for the day and night samples.

From the beginning to the last half of July 1992, less than 30% of the individuals found in the littoral had helmets. In the littoral sample taken on June 15, there were no helmeted individuals, and only 0.02% of the individuals sampled in the littoral in the first half of July had helmets. The proportion of helmeted individuals in the pelagial increased from late May, when the first helmeted individuals were found, and reached maximum levels during the last half of June and early July. From mid-June to mid-July, the distribution between the two morphs was significantly different (G-test, $p < 0.001$) with the helmeted individuals staying in the pelagial and the non-helmeted animals occupying the littoral (Fig. 5, Table 1). This coincided with the occurrence of fourth instar larvae of *Chaoborus* (Fig. 3). On June 4, there was a significantly higher proportion of helmeted individuals in the littoral compared to the pelagial. The proportion of helmeted daphnids in the pelagial dropped markedly on July 9. From late July, when exuviae of *Chaoborus* pupae were found, there were no significant differences in the distribution between the two morphs – with the exception of two days in August when the highest proportion of helmeted daphnids was found in the littoral (Fig. 5, Table 1).

Discussion

Morphology

Helmet cyclomorphosis as observed in the 1992 population of *D. galeata* is well-known in this and other *Daphnia* species (Jensen, 1985; Mort, 1989; Pijanowska, 1990). In Lake Lille Lungegårdsvann, helmet cyclomorphosis appears to be induced by invertebrate predators, particularly *Chaoborus*. The occurrence of fourth instar larvae of *Chaoborus* and *Bythotrephes* when *D. galeata* has the most developed helmets, supports this supposition. This is consistent with Mort (1989) who reported maximum helmet development in *D. galeata mendotae* when densities of the invertebrate predators, *Chaoborus* and *Leptodora*, reached maximum levels. Further, laboratory experiments have shown that invertebrate predators can induce helmet formation (Hanazato, 1991).

There is a problem with this scenario; the neonates and the smallest juveniles, which are most vulnerable to invertebrate predators, are not producing helmets. *Chaoborus* is known to feed selectively on smaller zooplankton (Dodson, 1974; Havel & Dodson, 1984; Brett, 1992) although it prefers the intermediate-sized individuals within its feeding range (Pastorok, 1981; Pijanowska, 1990). It has been reported both from field observations (Mort, 1989) and laboratory experiments (Dodson, 1988b; Tollrian, 1990; Hanazato, 1991) that neonates and juveniles develop helmets, although there is some discrepancy concerning which size classes develop the most marked helmets. One could speculate that a high helmet might protect a medium or large-sized daphnid against predation by gape-limited, small planktivores like juvenile *G. aculeatus*. In L. Lille Lungegårdsvann, the reproductive period for *G. aculeatus* extends from late April to mid-July (Nordeide, 1986). This results in newly hatched juveniles present during the same period as helmet formation. However, fish are probably not inducing helmets in *D. galeata* in L. Lille Lungegårdsvann because the animals did not develop helmets in 1993, even though *G. aculeatus* was present at approximately the same densitiy as in the previous year. It should be noted, however, that in 1992 the daphnids in L. Lille Lungegårdsvann were exposed to two categories of predators who fed selectively on either large or small prey. If an individual daphnid appears small and therefore less profitable, by not developing a helmet as a neonate or small juvenile, this might also serve as a defensive mechanism against predation by *G. aculeatus*.

Cyclomorphosis has often been correlated with abiotic factors such as temperature (Mort, 1989), but in L. Lille Lungegårdsvann, temperature alone cannot explain helmet formation since the temperatures in the two years were not significantly different from each other during the sampling season. Although the temperature in June and July 1993 was lower than the same period in 1992, which was also the period of the highest helmets, the temperature was above the helmet extension threshold of 10–15 °C (Wetzel, 1983).

Helmet development did not take place in all the adult *D. galeata*. This might imply that there are genetically distinct clones in the lake, and that one or more of these clones do not possess the ability to develop helmets. Spitze (1992) found significant differences in neckteeth expression between different clones of *D. pulex*. Thus, I cannot rule out that a clonal replacement was responsible for the lack of helmet formation in 1993.

Adult *D. galeata* with reduced or no tailspines were observed in both years. The spines were not broken mechanically. One might interpret this as a defense against fish predation. Without a tailspine, an animal might look smaller and thereby appear less profitable to the fish. Spine-reduction occurred in a larger proportion of the daphnids in 1993 than in 1992; these lower proportions in 1992 may be due to conflicting demands presented to *Daphnia* by the two categories of size-selective predators. In 1993 the greatest proportions of 'well-spined' animals were observed from late April to July. This coincided with the reproductive period in *G. aculeatus* (Nordeide, 1986) when stickleback densities had increased.

Neonates and juveniles always had long tailspines, which might be interpreted as a defense against predation by *Chaoborus* and *Bythotrephes*. Pijanowska (1990) suggests that the 'tailspine is required mostly in the youngest individuals, which need to be better protected than older ones'. She proposed that the spine serves a similar function to a lizard's tail. When the spine has been broken by a predator, the animal may continue to swim and chances for a second attack are probably very low.

One additional factor, the influence of salinity increase on the morphology in *D. galeata*, is unknown. It is possible that the energy expenditure connected with the new osmotic situation in 1993 was too high for the daphnids to be able to allocate energy to helmet development. Further, it might also influence the tailspine development.

Behaviour

There are indications of a morph-specific, but not a species-specific, habitat segregation in 1992 in L. Lille Lungegårdsvann from mid-June until mid-July with the non-helmeted daphnids staying in the littoral and the helmeted individuals occupying the pelagial.

Habitat segregation has been reported from other lakes. Kvam & Kleiven (this issue) report that two coexisting species, *D. pulex* and *D. longispina*, segregate in a lake where *Chaoborus* is the main predator. The smaller species, *D. longispina*, aggregated in the littoral, whereas *D. pulex* stayed closer to the bottom. They suggest that predation might have lead to these distribution patterns. In L. Lille Lungegårdsvann, the habitat segregation might have been caused by *Chaoborus*, although competition between the morphs cannot be excluded as an influencing factor. Habitat segregation coincided with observations of predatory fourth instar larvae of *Chaoborus* (Fig. 3). Helmeted daphnids might have a better ability to survive in the habitat with the highest densities of invertebrate predators, whereas the non-helmeted daphnids must escape into the littoral zone, which could serve as a refuge from invertebrate predators. This refuge might result from the addition of rainbow trout (*Onchorhynchus mykiss*) and brown trout (*Salmo trutta*) into the lake, which resulted in a reduction in the pelagic stickleback population. These top predators might have forced the sticklebacks into the littoral, thus providing *Chaoborus* with an opportunity to establish in the pelagial (P. Larsson, personal communication).

The observed habitat segregation of daphnids is consistent with the hypothesis that helmets defend against invertebrate predation (Dodson, 1989b; Pijanowska, 1990). These observations are also in accordance with Davies (1985) and De Meester *et al.* (1994), who explained high concentrations of zooplankton in the littoral as an escape from predation in the pelagial. Daphnids might use chemical cues released by *Chaoborus* to shift their habitat use, as has been shown experimentally by Larsson *et al.* (submitted).

Thus, the shift in *Daphnia* morphology and behaviour accompanied by a shift in the predation regime in consecutive years, indicates that helmet development and habitat segregation are anti-predator defenses induced by the presence of the predator.

Acknowledgements

I would like to thank Petter Larsson, Olaug Vetti Kvam, Per J. Jakobsen, Ole T. Kleiven, and Kurt Urdal for their comments on the manuscript and suggestions for improvement. I am also grateful to two anonymous reviewers for their valuable criticisms. Jarle Lysebo helped with one of the figures. This work has received financial support from the Norwegian Research Council (NFR) div. Norwegian Research Council for Science and Humanities (NAVF), NAVF as a part of the research programme 'Dynamics and Enhancement of Freshwater Fish Populations' (FFT), University of Bergen, and the Miss Lina Asbjørnsens scholarship fund.

References

Breder, C. M., 1960. Design for a fry trap. Zoologica, N. Y. 45: 155–159.

Brett, M. T., 1992. *Chaoborus* and fish-mediated influences on *Daphnia longispina* population structure, dynamics and life history strategies. Oecologia 89: 69–77.

Davies, J., 1985. Evidence for a diurnal horizontal migration in *Daphnia hyalina lacustris* Sars. Hydrobiologia 120: 103–105.

Dawidowicz, P., J. Pijanowska & K. Ciechomski. 1990. Vertical migration of *Chaoborus* larvae is induced by the presence of fish. Limnol. Oceanogr. 35: 1631–1637.

De Meester, L., S. Maas, K. Dierckens & H. J. Dumont, 1994. Habitat selection and patchiness in *Scapholeberis*: horizontal distribution and migration of *S. mucronata* in a small pond. J. Plankton Res. 15: 1129–1139.

Dodson, S. I., 1974. Adaptive change in plankton morphology in response to size-selection predation: a new hypothesis of cyclomorphosis. Limnol. Oceanogr. 19: 721–729.

Dodson, S. I., 1988a. The ecological role of chemical stimuli for the zooplankton: Predator-avoidance behavior in *Daphnia*. Limnol. Oceanogr. 33: 1431–1439.

Dodson, S. I., 1988b. Cyclomorphosis in *Daphnia galeata mendotae* Birge and *D. retrocurva* Forbes as a predator-induced response. Freshwat. Biol. 19: 109–114.

Dodson, S. I., 1989a. The role of chemical stimuli for the zooplankton: Predator-induced morphology in *Daphnia*. Oecologia 78: 361–367.

Dodson, S. I., 1989b. Predator-induced reaction norms. Cyclic changes in shape and size can be protective. BioScience 39: 447–452.

Gliwicz, M. Z. & A. Rykowska, 1992. 'Shore avoidance' in zooplankton: a predator-induced behavior or predator-induced mortality? J. Plankton Res. 14: 1331–1342.

Hanazato, T., 1991. Induction of development of high helmets by a *Chaoborus*-released chemical in *Daphnia galeata*. Arch. Hydrobiol. 122: 167–175.

Haney, J. F. & D. J. Hall, 1972. Sugar-coated *Daphnia*. A new preservation technique for Cladocera. Limnol. Oceanogr. 18: 331–333.

Harvell, C. D., 1990. The ecology and evolution of inducible defenses. Quart. Rev. Biol. 65: 323–340.

Havel, J. E., 1987. Predator-induced defenses: A review. In Kerfoot, W. C & A. Sih (eds), Predation. University Press of New England: 263–278.

Havel, J. E. & S. I. Dodson, 1984. *Chaoborus* predation on typical and spined morphs of *Daphnia pulex*: Behavioral observations. Limnol. Ocenogr. 29: 487–494.

Hebert, P. D. N. & P. M. Grewe, 1985. *Chaoborus*-induced shifts in the morphology of *Daphnia ambigua*. Limnol. Ocenogr. 30: 1291–1297.

Jacobs, J., 1987. Cyclomorphosis in *Daphnia*. In Peters, R. H. & R. de Bernardi (eds), '*Daphnia*'. Mem. Ist. Ital. Idrobiol. 45: 325–252.

Jakobsen, P. J. & G. H. Johnsen, 1987. The influence of predation on horizontal distribution of zooplankton species. Freshwat. Biol. 17: 501–508.

Jakobsen, P. J. & G. H. Johnsen, 1988. Size-specific protection against predation by fish in swarming waterfleas, *Bosmina longispina*. Anim. Behav. 36: 986–990.

Jensen, J. W., 1985. The morphology and ecology of coexisting *Daphnia galeata galeata* Sars and *Daphnia longispina caudata* Sars. Verh. Int. Ver. Limnol. 22: 3051–3057.

Krueger, D. A. & S. I. Dodson, 1981. Embryological induction and predation ecology in *Daphnia pulex*. Limnol. Oceanogr. 26: 219–223.

Kvam, O. V. & O. T. Kleiven, 1995. Diel horizontal migration and swarm formation in *Daphnia* in response to *Chaoborus*. In Larsson, P. & L. J. Weider (eds), Cladocera as Model Organisms in Biology. Developments in Hydrobiology 107. Kluwer Academic Publishers, Dordrecht: 177–184. Reprinted from Hydrobiologia 307.

Larsson, P. & S. I. Dodson, 1993. Chemical communication in planktonic animals. Arch. Hydrobiol. 129: 129–155.

Lysebo, E. M., 1993. Salinity tolerance in larvae of the phantom midge *Chaoborus flavicans* Meigen. Master of Science thesis. University of Bergen, Norway.

Milinski, M. Z., 1979. Can an experienced predator overcome the confusion of swarming prey more easily? Anim. Behav. 27: 1122–1126.

Mort, M. A., 1989. Cyclomorphosis in *Daphnia galeata mendotae*: Variation and stability in phenotypic cycles. Hydrobiologia 171: 159–170.

Nordeide, J. T., 1986. Life history strategy of a dense population of the three-spined stickleback (*Gasterosteus aculeatus* L.). Master of Science thesis. University of Bergen, Norway.

Nordeide, J. T. & S. Oftedal, 1985. Prosjekt Lille Lungegårdsvann (in Norwegian). Zoologisk museum; Fylkesmannen i Hordaland, Miljøvernavdelingen; Institutt for mikrobiologi og plantefysiologi; Institutt for marinbiologi.

Pastorok, R. A., 1981. Prey vulnerability and size selection by *Chaoborus* larvae. Ecology 62: 1311–1324.

Pijanowska, J., 1990. Cyclomorphosis in *Daphnia*: an adaptation to avoid invertebrate predation. Hydrobiologia (Dev. Hydrobiol. 60): 198: 41–50.

Sokal, R. R. & F. J. Rohlf, 1981. Biometry. 2nd edn. W. H. Freeman & Co, New York.

Spitze, K., 1992. Predator-mediated plasticity of prey life history and morphology: *Chaoborus americanus* predation on *Daphnia pulex*. Am. Nat. 139: 229–247.

Tollrian, R., 1990. Predator-induced helmet formation in *Daphnia cucullata* (Sars). Arch. Hydrobiol. 119: 191–196.

Wetzel, R. G., 1983. Limnology, 2nd edn. Saunders College Publishing, Orlando, Florida.

Hydrobiologia **307**: 193–201, 1995.
P. Larsson & L.J. Weider (eds), Cladocera as Model Organisms in Biology.

Seasonal occurrence and diel locomotor activity in littoral Cladocera in a mesohumic lake in Norway

Jan Ivar Koksvik
University of Trondheim, The Museum, Erling Skakkesgate 47, N-7004 Trondheim, Norway

Key words: Littoral Cladocera, seasonal distribution, gamogenesis, diel activity

Abstract

In situ studies were carried out using activity traps during the course of one year. Of 35 species recorded, only one, *Alonella nana*, was found to be perennial. The majority of species hatched in May/June, at temperatures of 4–15 °C and disappeared in October/November at temperatures of 1–6 °C. Males were found in 23 species, and always at the end of the active period. No species were found to have more than one period of gamogenesis. Day activity, which was most intense in the afternoon, seemed to be the common pattern. The activity pattern in some species seemed to be influenced by the presence of males.

Introduction

Compared to planktonic Cladocera, little work has been done on the ecology, life history and diel activity of littoral species. The main reason for this has probably been the lack of adequate sampling methods. This was pointed out by Pennak (1962) and emphasized in the IBP-Handbook No 17 (Edmondson & Winberg, 1971) as well. The few contributions on these topics in recent decades indicate that sampling methodology is still a problem. The present study was based on a trapping method that catches animals moving on the substrate or swimming close to it. It gives relatively clean samples and seems effective for studying a number of aspects of cladoceran life.

In the last 15–20 years, littoral Cladocera have been collected from a large variety of lakes in Norway. Much of the work has been carried out as part of general freshwater investigations in connection with water power development and water-course conservancy plans. These projects have amassed a great deal of zoogeographical data on this group, and have generated numerous reports, but there have been few opportunities for penetrating more deeply into problems of the ecology of these organisms. The aim of this study is to expand our understanding of the life histories and behavioural ecology of littoral Cladocera.

The lake

Målsjøn is a lake situated 25 km south of the city of Trondheim, Norway, at 63 °14′N, 10 °26′E. It is 165 m above sea level and its surface area is 0.25 km. The maximum depth is 13 m and 67% of the lake is less than 3 m deep. The lake is surrounded by coniferous forest, mires and 0.25 km of cultivated farmland. The bedrock geology of the 2.5 km^2 catchment area is dominated by calcareous greywacke and shale.

Normally the lake is ice covered from early November to early May. Maximum ice thickness is usually 70–80 cm, in March–April. Two full circulation periods occur, one immediately after the ice disappears and one in October. The lake is well stratified both in summer and winter. Maximum surface temperatures of 18-22 °C occur in July–August. Summer temperatures in the littoral vary up to 2–3 °C within a given 24 hour period. Daily maxima were normally recorded in the afternoon.

The lake has a well-developed littoral zone with *Carex rostrata*, *Phragmites communis*, *Potamogeton* spp., *Equisetum fluviatile*, *Nymphaea candida*, *Nuphar luteum* and *Chara globularis* dominating. The littoral zoobenthos is dominated by Corixidae, Ephemeroptera, Coleoptera, Gastropoda and Trichoptera, in addition to Cladocera and Copepoda. The fish populations of the lake consist of burbot (*Lota lota*

L.) arctic char (*Salvelinus alpinus* L.) and trout (*Salmo trutta* L.)

Målsjøn can be characterized as a mesohumic lake. $KMnO_4$ consumption ranged between 40–60 mg l^{-1}, Secchi disc transparancy was between 1.4–3.5 m and the water colour ranged from brown to yellowish brown. Light transmission (1%) varied from 1.8 to 3.8 m. Normal summer values for total hardness were 1.5–2.0 °dH in the surface layers, conductivity ranged from 50–70μS cm^{-1} (K_{18}) and pH varied between 7.0–7.3.

Methods

Littoral Cladocera were collected throughout one year (1972) using plexiglass traps designed by Breder (1960), and described in Ricker (1968). The traps were slightly modified to aid emptying by having five holes of approximately 1 cm diameter covered by a metal net with effective mesh size 0.28–0.30 mm placed on the back wall. The width of the entrance aperture was set to approximately 1 cm. The traps were placed on the bottom in two groups, four at a depth of 0.2 m and four at a depth of 1 m. The traps were manually emptied every two hours throughout a 24-hour period. The investigation consisted of 40 separate sampling periods, which covered a full year. In winter, the traps were operated through holes in the ice, while in summer, the traps were operated from a boat at 1 m depth and a quay at 0.2 m depth.

Trapped animals were collected on a 90 μm plankton screen and backwashed into a vial with 70% ethanol for later counting and species determination. The nomenclature used follows Flössner (1972), except for *Ophryoxus gracilis* Sars, which has a northern distribution and was not mentioned by Flössner (1972). The original description of the species is given in Sars (1993).

Results

Seasonal occurrence

A total of 35 species of cladocerans were recorded during the investigation. Seven of these are traditionally considered as planktonic. These are *Diaphanosoma brachyurum*, *Holopedium gibberum*, *Daphnia longispina*, *Daphnia galeata*, *Ceriodaphnia quadrangula*, *Bosmina longispina* and *Bythotrephes longimanus*. Because some of these (*B. longispina*, *D. brachyurum* and *C. quadrangula*) were abundant in the littoral, it was natural to include them in this study.

The occurrence of the different species throughout the year is shown in Fig. 1. *Chydorus* spp. were found to be *C. sphaericus*, *C. latus* and *C. piger*, but since it was not always possible to differentiate them, they are listed simply as *Chydorus* spp.

Very few species were trapped in the winter months of December to April. Only one, *Alonella nana*, was recorded in every month and can therefore be definitely denoted as perennial.

Quite a few species seem to have hatched immediately after ice out. One week later, on the 12–13 May, nine species were found, compared to only two species during ice out on the 4–5 May. The water temperature in the littoral had risen from 4 to 10 °C during this period. Many species hatched concurrently with the increasing temperatures through May. At the end of May, 16 species that had been absent from at least February, reappeared in the littoral. Eight additional ones were caught in June. The highest numbers of species were found in July and August, totally 31 in each month. Between 24 and 29 species were found during each trapping period. Gradually decreasing numbers were then found from September to the end of the year, with only four species detected in December. When ice initially formed on 1 November, 15 species were still present. The temperature at the bottom in the littoral was then 1 °C.

Most of the species that hatched in May disappeared during October and November, having been present in catches for six to seven months. Species caught from June onwards disappeared from September to November, having been found for a total of four to six months. Only four species, in addition to *Chydorus* spp., were caught in seven or more months. These were *B. longispina*, *Acroperus harpae*, *Alona quadrangularis* and *Simocephalus vetulus*. Four species were detected in four or fewer months: *Alona rectangula*, *Alonella exigua*, *Alona intermedia* and *Sida crystallina*. They were all caught for the first time as late as July or August.

The species could be grouped into two categories based on how abundant they were through the season: (1) Species having their highest abundance peak in autumn (September–October), typified by *Eurycercus lamellatus* (Fig. 2); others were *Ophryoxus gracilis*, *S. vetulus* and *Acroperus elongatus*. (2) Species having their highest abundance peak in summer (July to early

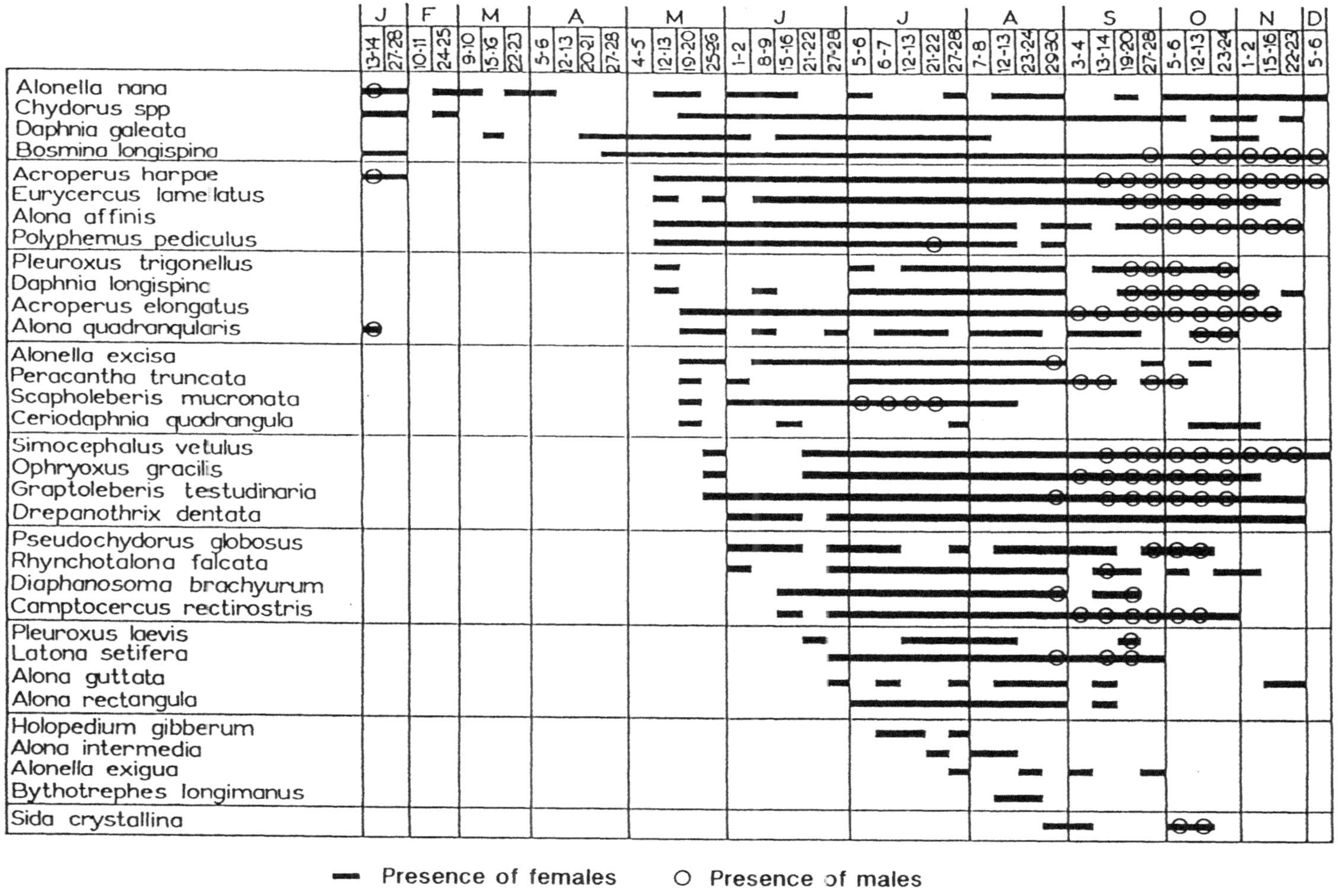

Fig. 1. Time periods for the presence of different cladoceran species in the catches.

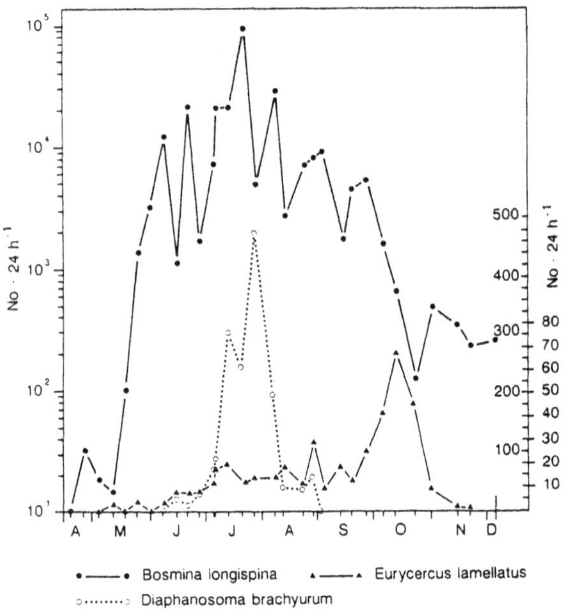

Fig. 2. Patterns of seasonal development. Ordinates correspond from left to right to *Bosmina longispina, Diaphanosoma brachyurum* and *Eurycercus lamellatus.*

August). This category may be subdivided into two groups: (a) species having a long active period, with a gradual increase to a peak, and a similar decrease, typified by *B. longispina* (Fig. 2); others were *Drepanothrix dentata, Alona affinis* and *Rhynchotalona falcata*; (b) species having a short active period, with a rapid development to a pronounced peak followed by a similarly rapid decrease in numbers. *D. brachyurum* (Fig. 2) showed this pattern, as did *Alonella excisa, Peracantha truncata, Scapholeberis mucronata* and *Polyphemus pediculus.* Species not mentioned above either did not show a distinct pattern of development, or too few were caught to reveal a pattern.

Sexual periods

Males were found for 23 species, usually at the end of the active period for the species (Fig. 1). Only one sexual period was recorded in all species. Females with ephippia were generally found within the period when males were present. When ephippial females were present after males had disappeared, they always occurred in very low numbers. For the majority of species, the most intense period of gamogenesis was September and October. The duration of the sexual period varied.

Males were found for a period that ranged from less than a month up to four months in the case of *A. harpae*, which was represented by males in all sampling periods from mid-September to mid-January. Male *S. vetulus, A. elongatus* and *B. longispina* were found for 2–2.5 months, male *Graptoleberis testudinaria, O. gracilis* and *A. affinis* for 1.5–2 months, and male *P. truncata, Camptocercus rectirostris, Pleuroxus trigonellus, D. longispina* and *E. lamellatus* for 1–1.5 months.

Two species, *S. mucronata* and *P. pediculus*, had much earlier sexual periods than the rest. Males were found in July. Females were also found somewhat later, but both species disappeared before the end of August. Only three species showed the first appearance of males later than September. These were *S. crystallina, A. quadrangularis* and *A. nana.*

The relative number of males varied greatly between the species. *O. gracilis* had the highest proportion (90%) of males during two sampling periods in October. A male proportion of 50–90% during the most intense period of gamogenesis was found in several of the more abundant species, e.g. *A. elongatus* (50%), *A. harpae* (66%), *S. vetulus* (88%) and *E. lamellatus* (84%). In some species, the proportion of males was always low. In *B. longispina* for example, males never exceeded 10% of the total number of individuals.

Diel activity

Activity periods were defined as parts of the day or night when the number of captured individuals of a certain species exceeded the mean for the 2-h trapping intervals. Due to limited numbers, activity patterns could not be detected for all species. In species that showed a stable activity pattern, day activity was most prominent, and the highest activity was most often recorded in the afternoon.

A typical representative of this pattern is *A. harpae* (Fig. 3). The highest activity was always recorded under conditions with light intensity above 5 lux. During the summer, the highest activity was recorded during the afternoon and evening; in autumn, the activity period gradually shortened and was concentrated during the lightest period around midday. An equivalent pattern of daytime activity was found in *P. truncata, D. brachyurum* and *P. pediculus* for the shorter period during the summer months when they were present in the catches.

A. elongatus had regular daytime activity during June and July, but from August onwards its locomotor activity seemed to have been arrhythmic (Fig. 4).

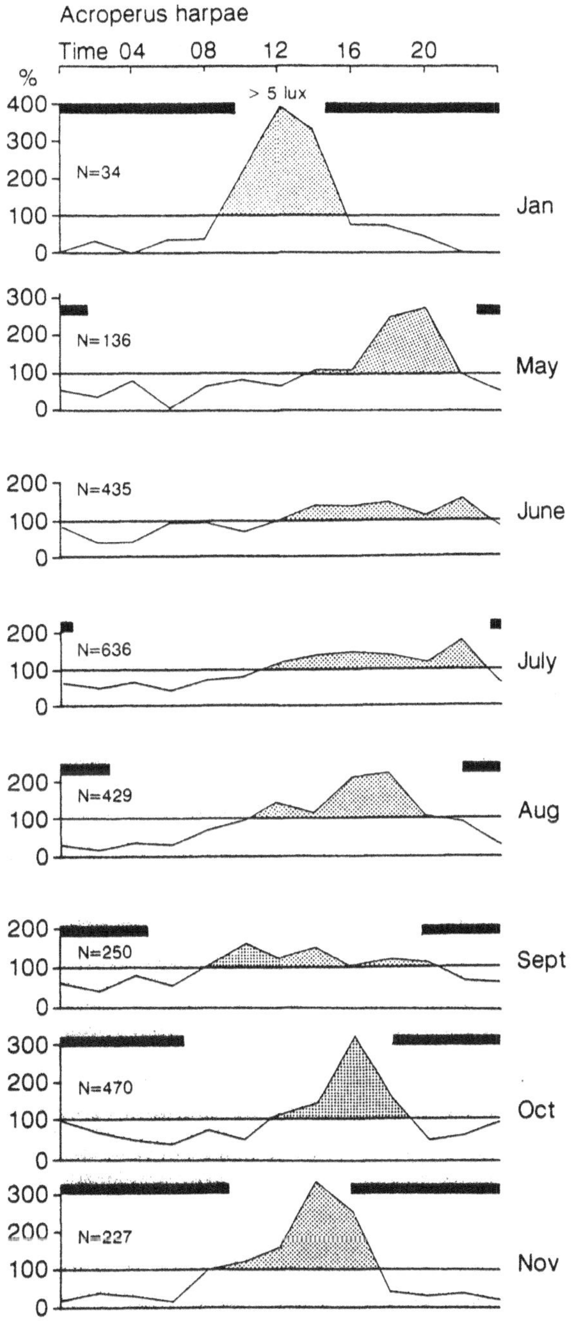

Fig. 3. Diel activity in *Acroperus harpae*. The curves show the numbers of individuals caught at 2-h intervals as percentages of the mean. Shaded areas indicate higher captures than the mean and are defined as activity periods. Horizontal thick lines show periods when light intensity was less than 5 lux. N = total numbers captured each month.

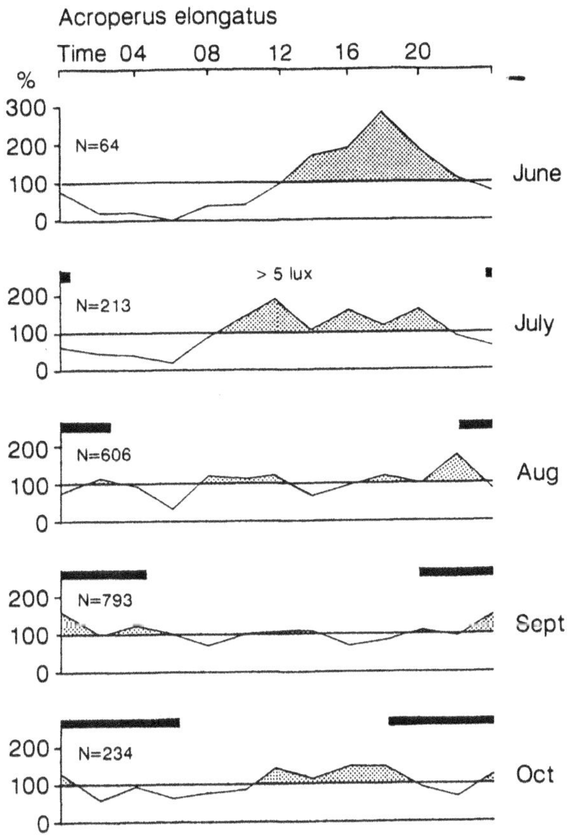

Fig. 4. Diel activity in *Acroperus elongatus*. Explanations in Fig. 3.

Males were present from the beginning of September, often in equal numbers to females. Both sexes lacked a rhythmic pattern.

B. longispina differed from the other species by having different activity patterns at depths of 0.2 and 1 m. At 0.2 m (Fig. 5), the species was clearly day active throughout the season, whereas it mostly showed a nocturnal activity pattern at 1 m depth (Fig. 6). A two-tailed t-test of differences in size of *B. longispina* individuals from the two depths showed that those caught at 0.2 m were significantly smaller ($p<0.001$). This was the case for all five periods when measurements were carried out from June to October, both in day and night catches. At 0.2 m, young individuals were numerically dominant, and egg/neonate-bearing females were rarely observed. At 1 m, the proportion of egg/neonate-bearing females varied, but was always much higher than at 0.2 m.

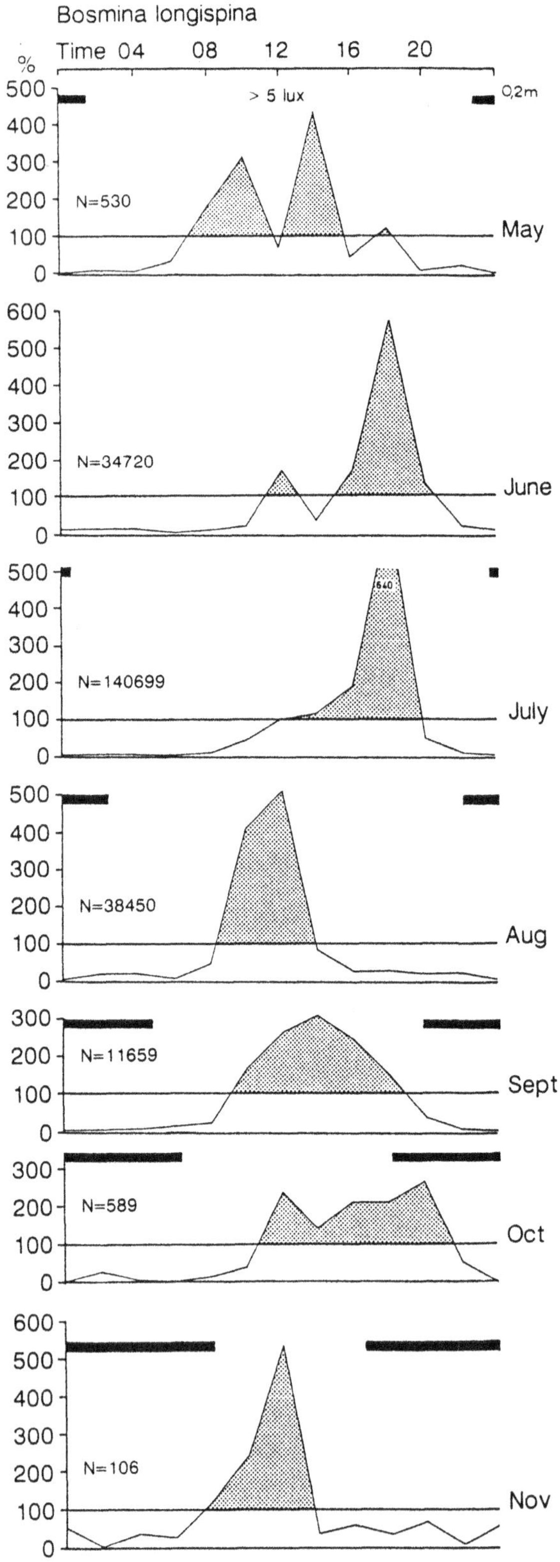

Fig. 5. Diel activity in *Bosmina longispina* at 0.2 m depth. Explanations in Fig. 3.

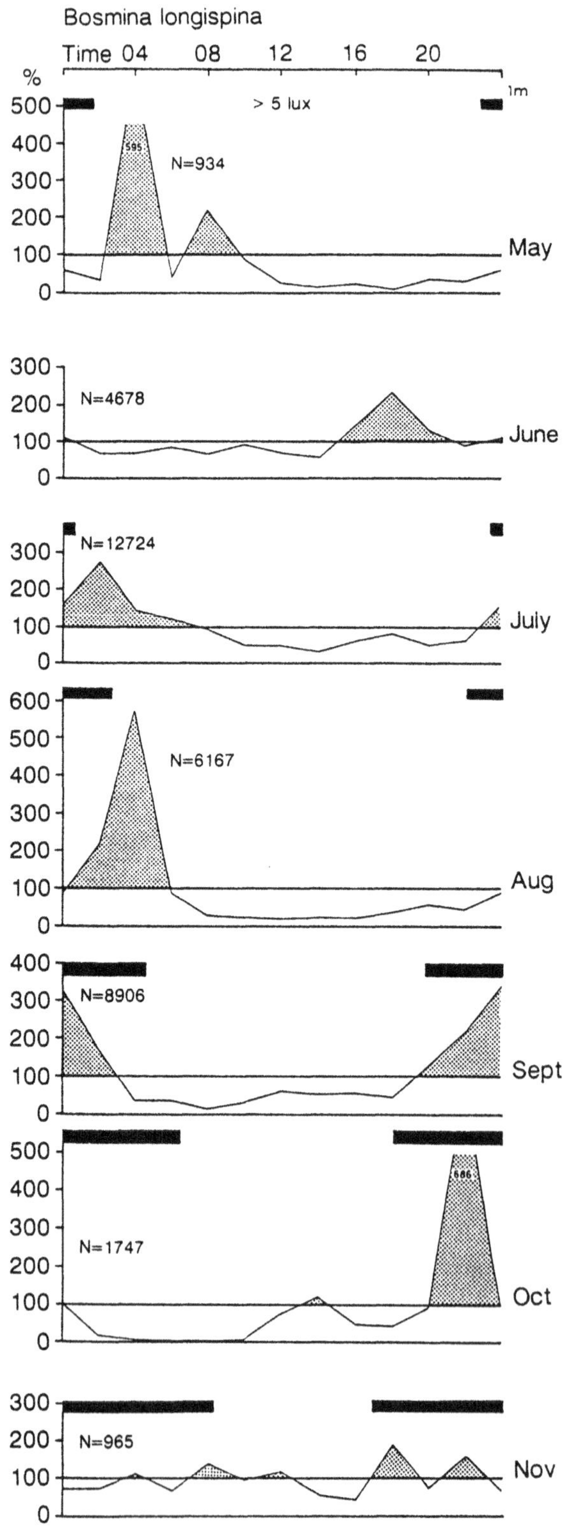

Fig. 6. Diel activity in *Bosmina longispina* at 1 m depth. Explanations in Fig. 3.

Discussion

The traps caught only swimming or creeping Cladocera. The method is therefore not suitable for direct quantitative studies, as there will be both inter- and intraspecific, as well as temporal variations in activity. It is believed, however, that the method allows one to look at general patterns of variation in abundance within a certain species throughout the year.

Most species were found between May and November and had active periods ranging from four to seven months. In northern Europe, it is common for littoral species of Cladocera to spend the winter as resting eggs (Järnefelt, 1956; Smyly, 1957; Flössner, 1964; Green, 1966), whereas adults are most abundant some time during spring, summer or autumn (Weigold, 1910; Straskraba, 1965; Lehtovaara & Sarvala, 1984). The Cladocera in Målsjøn fit well into this pattern.

To check whether more species were present than those recorded by the activity traps, and were perhaps inactive in winter, additional samples were taken with dip nets in January. These dip net samples did not yield any additional species.

None of the species in this study were found to have more than one sexual period per year. Ekman (1904) and Wesenberg-Lund (1894) also found only one period of gamogenesis in northern Sweden and Greenland, respectively. Berg (1929) and Wesenberg-Lund (1937) found that some species had two sexual periods in Denmark. According to Flössner (1972), it is common that Daphnidae, except *Daphnia*, have two sexual periods in Germany. Other families are normally monocyclic, but some dicyclic species may occur, especially in smaller water bodies. Frey (1982) described an intensive period of gamogenesis in the autumn as normal for the northern USA, while the same northern species tended to have a less coordinated and less intense sexual period in parts of the southern USA.

For most species, the most intense period of gamogenesis was in September–October in Målsjøn. Lilljeborg (1900) and Järnefelt (1956) found males of many of the same species at the same time in Sweden and Finland, respectively. In Denmark (Berg, 1929) and parts of northern Germany (Flössner, 1964), the sexual period for the same species seems to be most pronounced somewhat later, in October and November.

Gamogenesis is believed to be induced by unfavourable environmental conditions. The exact environmental stimuli needed for sexual reproduction are still uncertain (Kleiven *et al.*, 1992). However, there is considerable evidence for sexual periods being induced by photoperiod, food availability and chemical signals (crowding stimulus) (e.g. Stross & Hill, 1968; Shan, 1974; Hobæk & Larsson, 1990; Kleiven *et al.*, 1992). In Målsjøn, the photoperiod was shortened by four hours from mid-August to mid-September. Changes in food availability were not studied, and chemical stimuli could not be verified. Changes in water temperature have also been considered as a possible cue for gamogenesis, but less attention has apparently been paid to it in earlier investigations. The water temperature in the Målsjœn littoral decreased from 20 °C to 14 °C in August, and from 14 °C to 9 °C in September.

Under natural conditions, biological rhythms are normally synchronized over a 24-h period by external stimuli, any of which can be referred to as a 'Zeitgeber' (Aschoff, 1958; Cloudsley-Thompson, 1961). Light is considered the most important Zeitgeber, followed by temperature (Bünning, 1963; Hoffmann, 1969). Most species in this study, which exhibited a stable activity pattern over a certain period of time, were day-active. The synchronization with light indicates that light was the most important Zeitgeber.

In summer, when water temperature varied through the day, several species (*A. harpae*, *P. truncata*, *P. pediculus* and *D. brachyurum*) showed a tendency for an activity peak in the afternoon when the water temperature was highest. This indicates that their locomotor activity was modified by temperature variations, although these were usually less than 2 °C.

In other studies on the diurnal activity of littoral Cladocera, the highest numbers of individuals and greatest species abundances have been recorded at night (Szlauer, 1963; Whiteside, 1974; Whiteside *et al.*, 1978; Fairchild, 1981; Walls *et al.*, 1990). These studies were based on sampling techniques that measured vertical migrations, or the result of such movement patterns. The traps used in the present study measured horizontal movements on, or closely above the sediments. Since chydorids are adapted to creeping along submerged surfaces and are poor swimmers (Fryer, 1968), this method was considered adequate for studying the locomotor activity of the majority of littoral Cladocera. The inconsistent results from the different sampling techniques are interesting and should be considered complementary to each other in providing a more complete picture of diurnal activity. Most littoral cladocerans, especially the chydorids, are known to feed on periphyton and detritus. The horizontal movements during the day are probably connected with such activity along the bottom, as is the vertical crawling in the vegetation at night. There is also evidence of another feeding behaviour: filter feeding on suspended particulate matter directly from the water column (Fryer, 1968; Meyers, 1980). This could explain the vertical swimming activity observed at night in some species. An activity confined to the bottom during daytime may be an adaptation to avoid predation, especially from fish.

The sexual period seems to have influenced the activity pattern in some species. This was particularly pronounced in *A. elongatus*, which lost its stable day activity during the sexual period. One potential explanation is that activity connected with mating differs from feeding activity, which appears to be synchronized with light. It is well known that freshwater invertebrates may exhibit different types of activity during different periods of the day (Götz, 1949; Müller & Österdahl, 1970; Solem, 1973).

The size differences in *B. longispina* in the catches from depths of 0.2 and 1 m indicate that young animals lived closer to the shore than adults, and that the activity pattern differed for the two age groups. Since the traps were placed on the bottom, I have no records of activity higher up in the water column at the 1 m depth, whereas the traps at the 0.2 m depth covered the entire range from the bottom to the surface. I cannot exclude the possibility that *B. longispina*, which is also planktonic, is active closer to the surface during the day at 1 m depth. *Bosmina swarms*, which have been reported previously by others (Klemetsen, 1970; Gliwicz & Rybak, 1978; Fairchild, 1981), were observed on two occasions close to the surface during the day. The normal situation is, however, that the species when living in the pelagial, has a negative heliotactic behaviour, and migrates to deeper water during the day (Welch, 1952; Hutchinson, 1967). A possible explanation for the nocturnal activity at 1 m depth is that the vertical migration from deeper water in the pelagial also supplies parts of the littoral with animals. Welch (1952) found that diel migrations are not strictly vertical, but take place in all directions. Walls *et al.* (1990) found higher numbers and significantly larger individuals of *B. longispina* in night samples taken 0.5–5.0 m from the shoreline in Lake Pyhäjärvi, Finland. This also indicates an immigration from the pelagial at night.

Acknowledgments

I would like to thank John O. Solem for collaboration in the field work and valuable advice during this study. I am also grateful to Petter Larsson and two anonymous reviewers for constructive criticisms of an earlier version of the manuscript, and to Robert Binns for language improvement.

References

Aschoff, J., 1958. Tierische Periodik unter dem Einfluss von Zeitgebern. Z. Tierpsychol. 15: 1–30.

Berg, K., 1929. A faunistic and biological study of Danish Cladocera. Vidensk. Meddr dansk naturh. Foren. 88: 31–111.

Breder, C. M., 1960. Design for a fry trap. Zoologica. (N.Y.) 45: 155–159.

Bünning, E., 1963. Die physiologische Uhr. 2. utg. Berlin, Springer Verlag. 153 pp.

Cloudsley-Thompson, J. L., 1961. Rhythmic activity in animal physiology and behaviour. Academic Press, New York, 236 pp.

Edmondson, W. T. & G. G. Winberg, 1971. A manual on methods for the assessment of secondary productivity in fresh waters. IBP Handbook No. 17. Blackwell, Oxford, 358 pp.

Ekman, S., 1904. Die Phyllopoden, Cladoceren und freilebenden Copepoden der nordschwedischen Hochgebirge. Zool. Ib. Abt. f. Syst. 21: 1–170.

Fairchild, G. F., 1981. Movement and microdistribution of Sida crystallina and other littoral microcrustacea. Ecology 62: 1341–1352.

Flössner, D., 1964. Zur Cladocerenfauna des Stechlinsee– gebietes II. Ökologische Untersuchungen über die litoralen Arten. Limnologica (Berlin) 2: 35–103.

Flössner, D., 1972. Krebstiere, Crustacea. Kiemen- und Blattfüsser, Branchiopoda. Fischläuse, Branchiura. Die Tierwelt Deutschlands 60: 1–501.

Frey, D. G., 1982. Contrasting strategies of gamogenesis in northern and southern populations of Cladocera. Ecology 63: 223–241.

Fryer, G., 1968. Evolution and adaptive radiation in the Chydoridae (Crustacea, Cladocera): A study in comparative functional morphology and ecology. Phil. Trans. r. Soc., London, Ser. B. 254: 221–385.

Gliwicz, Z. M. & J. I. Rybak, 1976. Zooplankton. Chapter 5 in Pieczynska, E. (ed.), Selected problems of lake littoral ecology. University of Warsaw, Warsaw, Poland.

Green, J., 1966. Seasonal variation in egg production by Cladocera. J. anim. Ecol. 35: 77–104.

Götz, B., 1949. Tageszeit und Insektenaktivität. Naturwissenschaftliche Rundschau 2: 257–261.

Hoffmann, K., 1969. Zum Einfluss der Zeitgeberstärke auf die Phasenlage der synchronisierten circadianen Periodik. Z. vergl. Physiol. 62: 93–110.

Hobæk, A. & P. Larsson, 1990. Sex determination in Daphnia magna. Ecology 71: 2255–2268.

Hutchinson, G. E., 1967. A treatise on limnology. II. Introduction to lake biology and the limnoplankton. New York, John Wiley & Sons, Inc., 1115 pp.

Järnefelt, H., 1956. Materialen zur Hydrobiologie des Sees Tuusulanjärvi. Acta Soc. Fauna Flora Fenn. 71: 1–38.

Kleiven, O. T., P. Larsson & A. Hobæk, 1992. Sexual reproduction in Daphnia magna requires three stimuli. Oikos 65: 197–206.

Klemetsen, A., 1970. Plankton swarms in Lake Gjøkvatn, East Finnmark. Astarte 3: 83–85.

Lehtovaara, A. & J. Sarvala, 1984. Seasonal dynamics of total biomass and species composition of zooplankton in the littoral of an oligotrophic lake. Verh. int. Ver. Limnol. 22: 805–810.

Lilljeborg, W., 1900. Cladocera Sueciae. Nova Acta R. Soc. Scient. upsal. Ser. 3, 19: 1–701.

Meyers, D. G., 1980. Diurnal vertical migration in aquatic microcrustacea: Light and oxygen responses of littoral zooplankton. Pp. 80–90 in Kerfoot, W. C. (ed.), Evolution and ecology of zooplankton communities. Hanover, New Hampshire, Univ. Press of New England, 793 pp.

Müller, K. & L. Österdahl, 1970. Beobachtungen über das Laichen der Quappe Lota lota L. Oikos Suppl. 13: 130–133.

Pennak, R. W., 1962. Quantitative zooplankton sampling in littoral vegetation areas. Limnol. Oceanogr. 7: 487–489.

Ricker, W. E., 1968. Methods for assessment of fish production in fresh water. IBP handbook No 3. Oxford, Blackwell, 313 pp.

Sars, G. O., 1993. On the freshwater crustaceans occurring in the vicinity of Christiania. University of Bergen, Bergen, 197 pp.

Shan, R. K., 1974. Reproduction in laboratory stocks of Pleuroxus (Chydoridae, Cladocera) under the influence of photoperiod and light intensity. Int. Revue Ges. Hydrobiol. 59: 643–666.

Smyly, W. J. P., 1957. Distribution and seasonal abundance of Entomostraca in moorland ponds near Windermere. Hydrobiologia (Haag) 11: 59–72.

Solem, J. O., 1973. Diel rhythm of fry and young of Lota lota (Pisces). Oikos 24: 325–327.

Straskraba, M., 1965. Contributions to the productivity of the littoral region of pools and ponds. I. Quantitative study of the littoral zooplankton of the rich vegetation of the backwater Labicko. Hydrobiologia (Haag) 26: 421–443.

Stross, R. G. & J. C. Hill, 1968. Photoperiod control of winter diapause in the fresh-water crustacean, Daphnia. Biol. Bull. 134: 176–198.

Szlauer, L., 1963. Diurnal migrations of minute invertebrates inhabiting the zone of submerged hydrophytes in a lake. Schweiz. Z. Hydrol. 25: 56–64.

Walls, M., M. Rajasilta, J. Sarvala & J. Salo, 1990. Diel changes in horizontal microdistribution of littoral Cladocera. Limnologica 20: 253–258.

Weigold, H., 1910. Biologische Studien an Lyncodaphniden und Chydoriden. Int. Revue ges. Hydrobiol. Hydrogr. 3 (Suppl.): 1–118.

Welch, P. S., 1952. Limnology. New York, McGraw-Hill, 538 pp.

Wesenberg-Lund, C., 1894. Grønlands Ferskvandsentomostraca. I. Phyllopoda branchiopoda et Cladocera. Videns. Meddr dansk naturh. Foren. 56: 82–173.

Wesenberg-Lund, C., 1937. Ferskvandsfaunaen biologisk belyst. Invertebrata. Bd. 2. Gyldendalske Boghandel – Nordisk Forlag, København: 415–837.

Whiteside, M. C., 1974. Chydorid (Cladocera) ecology: Seasonal patterns and abundance of populations in Elk Lake, Minnesota. Ecology 55: 538–550.

Whiteside, M. C., J. B. Williams & C. P. White, 1978. Seasonal abundance and pattern of chydorid, Cladocera in mud and vegetative habitats. Ecology 59: 1177–1188.

Hydrobiologia **307**: 203–206, 1995.
P. Larsson & L.J. Weider (eds), Cladocera as Model Organisms in Biology.

Effects of predator-released chemicals on some life history parameters of *Daphnia pulex*

Andrzej Engelmayer
Department of Biology, 4700 Keele Street, Toronto, Ontario, M3J 1P3, Canada
Permanent address: Department of Hydrobiology, University of Warsaw, Nowy Swiat 67, 00-046 Warsaw, Poland

Key words: Chaoborus, Daphnia, Lepomis, life-history traits, predator-induction

Abstract

The effects of chemicals released by fish and *Chaoborus* larvae on some life history traits in *Daphnia pulex* were studied in an *in situ* enclosure experiment. The mean size of *Daphnia* individuals was smaller in the presence of fish-released cues. Also the minimal size of an egg bearing female in the presence of fish exudates was smaller than in the population exposed to the chemicals released by *Chaoborus* larvae as well as in the control population. Fish-released chemicals caused the increase in clutch size in *Daphnia*. There were no statistically significant differences between the studied life history parameters of the control and *Chaoborus* treatments. The results are discussed in reference to recent laboratory research.

Introduction

Predation is recognized as a major factor affecting the life history traits of freshwater cladocerans (Lynch, 1977; Dorazio & Lehman, 1983). It has been observed that in cladoceran populations/under the pressure of vertebrate predators like fish (which prefer larger prey), the mean body size, egg size, and size at first reproduction tend to decrease, with the simultaneous increase of clutch size (DeMott & Kerfoot, 1982; Vanni, 1987). On the other hand, invertebrate predators (which prefer smaller prey) cause a shift towards late reproduction and fewer but larger offspring (Spitze, 1991; Brett, 1992). There is increasing evidence that not only actual predator selectivity and size specific mortality of prey can cause the shifts in life histories of cladoceran populations. But the chemical cues of predators can induce changes in the behavior and morphology of prey (Dodson, 1988, 989), as well as can induce shifts in life history traits. It is known that chemicals released by fish cause an earlier reproduction, production of more but smaller eggs and overall body size reduction in *Daphnia* (Machaček, 1991; Stibor, 1992). In the presence of *Chaoborus*, individuals of some *Daphnia* clones delay their maturation and produce fewer but larger offspring (Lüning, 1992).

The aim of this study was to investigate and compare in an *in situ* experiment, the effect of chemicals released by vertebrate and invertebrate predators on some life-history parameters of *Daphnia*.

Material and methods

The experiment was conducted in the oligotrophic Ranger Lake (Ontario, Canada). The major group of vertebrate predators consists of juvenile fish (pumpkinseed, largemouth bass, smallmouth bass, yellow perch, common shiner). The impact of their predation pressure varies both seasonally and annually. The most important invertebrate predators are *Chaoborus* larvae. In this lake, there are three coexisting *Chaoborus* species (*C. flavicans*, *C. punctipennis*, *C. trivitatus*). The coexistence of three *Chaoborus* species with different behaviors and life cycles makes the predation pressure exerted by *Chaoborus*, to be almost constant throughout the season.

The experimental setup consisted of nine enclosures made of clear polyethylene sleeves sealed at the bottom and attached to a floating dock, which was anchored in the pelagic zone of the lake. The enclosures (approximately 0.7 m^3 volume and 260 cm deep)

were filled with filtered (40 μm) lake water. After filling the zooplankton community was established in all enclosures by emptying the contents of 10 vertical hauls (30 cm diameter 12 m depth, 40 μm mesh size) into the enclosures. Before adding zooplankton to the enclosures, all *Chaoborus* larvae were removed and the remaining zooplankton was equally divided into nine enclosures. Then, floating mesh-cages (80 μm mesh size) for predators were placed in each enclosure so that the water with dissolved chemicals could disperse but there were no actual prey-predator encounters. Three replicates of each of three treatments were established: 1. Fish – where the mesh-cage contained three individuals of 1 + pumpkinseed (*Lepomis gibbosus*), 2. *Chaoborus*, where the mesh-cage contained 300 *Chaoborus* larvae, and 3. Control – mesh-cage contained no predators. Fish and *Chaoborus* larvae were caught in Ranger Lake. The enclosures were covered with a net (mesh size of about 2 mm) in order to make it impossible for adult *Chaoborus* to lay eggs in the enclosures. The predators were fed with lake zooplankton every second day. To the Control mesh--cages, the same amounts of heat-killed zooplankton were added in order to minimize the differences in nutrient inputs.

During the experiment, 100 Chaoborus larvae were added twice to each mesh-cage of the *Chaoborus* treatment to compensate for the loss of animals due to death and pupation. Two fish died during the experiment, but were replaced by new ones within one day. The experiment was conducted for 35 days (25 May to 30 June 1992). Zooplankton samples were collected once weekly from each enclosure. Five samples in a vertical series at 0.5 m intervals were taken with a 0.5 m long, 5 dm^3 of volume Schindler-Patalas trap, filtered through a plankton net of 80 μm mesh size and fixed in 4samples were collected before adding the predators. Chlorophyll-*a* concentration was based on the samples collected on the same sampling days and analyzed by the laboratories of the Ontario Ministry of Environment.

Individuals of *Daphnia pulex* found in the samples were measured. A total number of 2784 individuals were analyzed. The number of eggs (clutch size) were determined for each individual. The minimal size of egg bearing females (considered approximately as size at first reproduction) was determined for each sample. Two-way ANOVA was performed on these data to test for the effect of treatment (Fish *vs Chaoborus vs* Control) and time (consecutive sampling days) on *Daphnia* body size and minimal size of an egg bearing female. The means were compared by the Tukey HSD test for pairwise comparison of means. The slopes and adjusted means of linear regressions of clutch size *vs* individual length were compared by ANCOVA using the BIOM package of Rohlf (1984).

Table 1. Summary of two-way ANOVA of the influence of treatment (Fish vs. *Chaoborus* vs. Control) and time on: (A) *Daphnia* mean body length and (B) size at first reproduction.

	Source	df	MS	F	P
A	Treatment	2	0.050	5.840	0.001
	Day	4	0.207	24.350	0.000
	Interaction	8	0.015	1.730	0.132
	Error	30	0.009		
B	Treatment	2	0.034	5.430	0.013
	Day	3	0.101	15.900	0.000
	Interaction	6	0.025	3.930	0.009
	Error	21	0.006		

Results

Mean *Daphnia* length increased in all enclosures during the experiment. However mean *Daphnia* body length was significantly smaller ($p = 0.05$) in the Fish treatment *vs Chaoborus* and Control treatments (Fig. 1, Table 1).

No gravid female was found in any sample during the first sampling day and in any fish treatment sample on the last sampling day. Therefore, ANOVA for the minimal size of egg bearing female was calculated for four sampling days (Table 1). The size at first reproduction was significantly lower ($p = 0.05$) in the Fish treatment *vs Chaoborus* and Control treatments (Fig. 1, Table 1).

There was no significant increase of egg number with *Daphnia* body size. No significant differences among the regression slopes were found (Table 2). However, the adjusted regression mean was significantly higher in the Fish treatment *vs Chaoborus* and Control treatments (Table 2) which means that in the presence of fish exudates *Daphnia* individuals had bigger clutch size relative to individuals of the same body size in the *Chaoborus* and Fish treatments. The *Chaoborus* and Control treatments did not show significant differences among adjusted means (Table 2).

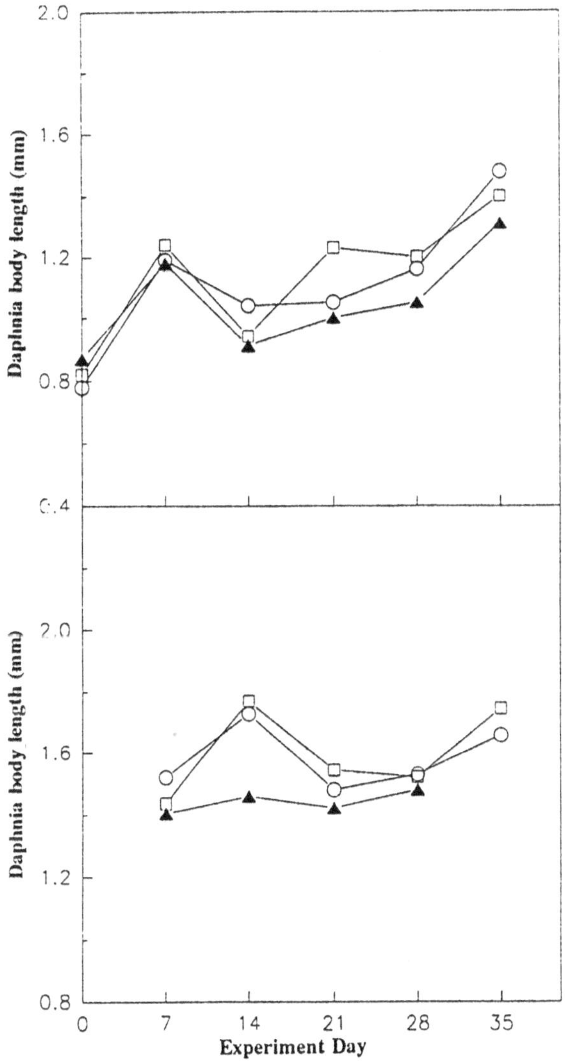

Fig. 1. Changes in mean *Daphnia* body length, each point represents mean of average body length in each replicate of a treatment – upper panel. Changes in minimal size of egg-bearing females, each point represents the mean of minimal lengths of egg-bearing individuals in each replicate of a treatment – lower panel. Filled triangles represent the Fish treatment, empty squares represent the *Chaoborus* treatment and empty circles represent the Control treatment.

During the experiment the Chlorophyll-*a* concentration varied between 0.2 to 1.4 μg dm^{-3} but there were no significant differences in Chlorophyll-*a* concentration between the treatments (at $p = 0.05$, ANOVA).

Table 2. Summary of ANCOVA testing for differences between treatments. F stands for Fish, CH for *Chaoborus* and C for Control. Calculation of adjusted means was based on pooled regression coefficient (b_{within}) for each pair of compared treatments.

	Treatments	df	F	*P*
Differences	F vs. C	1.94	0.385	0.536
among	F vs. CH	1.103	0.169	0.682
slopes	CH vs. C	1.126	0.172	0.679
Differences	F vs. C	1.95	6.736	0.010
among	F vs. CH	1.104	17.497	<0.001
adjusted means	CH vs. C	1.127	0.781	0.379

Discussion

There is increasing evidence that individuals of a single clone of *Daphnia* are able to distinguish between vertebrate and invertebrate chemical signals and undertake the proper defensive reaction (Dodson, 1988, 1989; Weider & Pijanowska, 1993). Laboratory studies have shown a very strong effect of fish or invertebrate released chemicals on various defensive responses like diel vertical migration (Dodson, 1988; Ringelberg, 1991a, b), on the production of defensive morphological structures like spines, helmets or neckteeth in many *Daphnia* species (Dodson, 1989; Hanazato, 1990: Lüning, 1992). Also life history traits like size and age at first reproduction, clutch size, eggs size can be modified by the presence of predator released chemicals (Machaček, 1991; Lüning, 1992; Stibor, 1992; Weider & Pijanowska, 1993).

Different clones of *Daphnia* differ in their sensitivity to chemical cues released by different predators (Lüning, 1992; Weider & Pijanowska, 1993). In natural conditions. the high predator pressure could lead to the elimination of clones which are not able to undertake the proper defensive response (Pace *et al.*, 1984; Spitze, 1991; Pijanowska *et al.*, 1993). In this experiment a natural lake population of *Daphnia* was used and the effect of chemicals might be obscured by the occurrence of a clonal mixture in the enclosures. Those clones, which do not respond to predator cues should have the highest reproductive success because animals which do not undertake costly defenses may optimize their reproduction. This is despite the observed differences between treatments. *Daphnia* individuals from the fish enclosures were smaller and the minimal size of egg-bearing females was smaller, which is consistent with recent laboratory studies (Machaček, 1991; Stibor, 1992; Weider & Pijanowska, 1993). Also, the comparison of relations between individual size and clutch size shows an effect of fish released cues similar to the results presented by Machaček (1991). However, in contrast to the results obtained by Lüning (1992), no significant effect of *Chaoborus* exudates on observed life history parameters in *Daphnia pulex* was found in this experiment.

The lack of reaction to *Chaoborus* exudates and the induction of life history shifts by fish might be explained by the different benefits of fixed and inducible defensive reactions under different predation regimes (Pijanowska, 1993). In Ranger Lake, fish density and the strength of fish predation may vary seasonally and annually, and this makes it very difficult to predict the timing of peak of fish predation. Therefore, under these conditions, the inducible reaction to fish exudates would be favored over the fixed one since it enables *Daphnia* to react more quickly to changes in fish predation pressure. On the other hand, under constant, strong predation pressure from *Chaoborus*, the risk of being eaten is always very high and the fixed reaction might be favored over the inducible one. Thus, removing *Chaoborus* larvae from the control enclosures did not cause a shift in the life history traits of *Daphnia pulex*. This might explain the lack of a significant difference between the control the *Chaoborus* treatments.

Acknowledgments

I thank Don McQueen for help and assistance. This study was supported by a grant from The Natural Sciences and Engineering Research Council of Canada.

Financial support from the University of Bergen and Grant No. 128 from the Polish Committee for Scientific Research to Joanna Pijanowska enabled me to take part in the Third Symposium on Cladocera, where the first version of this paper was presented. I would like to thank Julia Lüning, Petter Larsson and an anonymous reviewer for valuable comments and Larry Weider for linguistic help.

References

Brett, M. T., 1992. *Chaoborus* and fish-mediated influences on *Daphnia longispina* population structure, dynamics and life history strategies. Oecologia 89: 69–77.

DeMott, W. R. & W. C. Kerfoot, 1982. Competition among cladocerans: nature of the interaction between *Bosmina* and *Daphnia*. Ecology 63: 1949–1966.

Dodson, S. I., 1988. The ecological role of chemical stimuli for the zooplankton: predator-avoidance behavior in *Daphnia*. Limnol. Oceanogr. 33: 1431–1439.

Dodson, S. I., 1989. The ecological role of chemical stimuli for the zooplankton: predator-induced morphology in *Daphnia*. Oecologia 78: 361–367.

Dorazio, R. M. & J. T. Lehman, 1983. Optimal reproductive strategies in age-structured populations of zooplankton. Freshwat. Biol. 13: 157–175.

Hanazato, T., 1990. Induction of helmet development by a *Chaoborus* factor in *Daphnia ambigua* during juveniles stages. J. Plankton Res. 12: 1287–1294.

Lüning, J., 1992. Phenotypic plasticity of *Daphnia pulex* in the presence of invertebrate predators: morphological and life history responses. Oecologia 92: 383–390.

Lynch, M., 1977. Fitness and optima body size in zooplankton populations. Ecology 58: 763–774.

Machaček, J., 1991. Indirect effect of planktivorous fish on the growth and reproduction of *Daphnia galeata*. Hydrobiologia 225 (Dev. Hydrobiol. 71): 193–197.

Pace, M. L., K. Porter & Y. S. Feig, 1984. Life history variation within a parthenogenetic population of *Daphnia parvula* (Crustacea: Cladocera). Oecologia 63: 43–51.

Pijanowska, J., 1993. Diel vertical migration in zooplankton: fixed or inducible behavior? Arch. Hydrobiol. Beih. 39: 89–97.

Pijanowska, J., L. J. Weider & W. Lampert, 1993. Predator-mediated genotypic shifts in a prey population: experimental evidence. Oecologia 96: 40–42.

Ringelberg, J., 1991a. Enhancement of phototactic reaction in *Daphnia hyalina* by juvenile perch (*Perca fluviatilis*). J. Plankton Res. 13: 17–25.

Ringelberg, J., 1991b. A mechanism of predator-mediated induction of diel vertical migration in *Daphnia hyalina*. J. Plankton Res. 13: 83–89.

Rohlf, F. J., 1984. Biom. A package of statistical problems to accompany the text Biometry. Exeter Software, Setanket, New York.

Spitze. K., 1991. *Chaoborus* predation and life-history evolution in *Daphnia pulex*: temporal pattern of population diversity, fitness and mean life history. Evolution 45: 82–92.

Stibor, H., 1992. Predator induced life-history shifts in a freshwater cladoceran. Oecologia 92: 162–165.

Vanni, M. J., 1987. Effects of food availability and fish-predation on a zooplankton community. Ecol Monogr. 57: 61–88.

Weider, L. J. & J. Pijanowska (1993). Plasticity of *Daphnia* life histories in response to chemical cues from predators Oikos 67: 385–392.

Hydrobiologia **307**: 207–212, 1995.
P. Larsson & L.J. Weider (eds), Cladocera as Model Organisms in Biology.

The influence of a fish exudate on two clones of the hybrid *Daphnia galeata* × *hyalina*

Tineke Reede & Joop Ringelberg
Department of Aquatic Ecology, University of Amsterdam, Kruislaan 320, 1098 SM Amsterdam, The Netherlands

Key words: Daphnia, fish-exudate, intrinsic rate of increase, life-history

Abstract

Two *Daphnia* clones were isolated from different day depths during the period of diel vertical migration and were tested for their life-history responses to a fish exudate released by juvenile perch. Animals were exposed to fish exudate every 24 or 48 h. Both clones responded to the exudate by exhibiting earlier maturation and larger sizes of first clutches, which resulted in higher rates of population increase. Also, neonates were smaller in the presence of the exudate. It was found that the clone isolated from a deeper day depth ('migrating clone') was less sensitive to the exudate than the clone isolated from the surface waters, which was presumed to be non-migrating. The non-migrating clone responded by having smaller neonate sizes and smaller sizes at maturity than the migrating clone. The non-migrating clone responded to the fish chemical when it was exposed to it every 24 or 48 h, whereas the migrating clone only responded to the exudate if exposed to it every 24 h.

Introduction

The adaptive value of diel vertical migration (DVM) is generally believed to be a decrease in mortality due to visual predators (see Lampert, 1989, 1993 for reviews). In Lake Maarsseveen, *Daphnia galeata* × *hyalina* performs a DVM for some weeks when 0+ perch (*Perca fluviatilis*) are present in the open water, and are preying on daphnids. During the first few weeks, average population day depth increases. The *Daphnia* population consists of many genetically distinct clones (Spaak & Hoekstra, 1993). One possible explanation for the increasing day depth is that clones exhibit differential migration, and combined with predation, some clones react less to relative changes in light intensity and/or fish exudates, which enhance reactivity (Ringelberg, 1991 *et al.* and submitted). As a result, at the height of the migration period, daphnids are no longer found in the epilimnion during daytime.

Migrating and non-migrating clones, the two extreme types, are subject to different environmental conditions during the period of migration and therefore their life history traits should be different (Stich & Lampert, 1984) if we deal with the two extreme strategies. Some of these characteristics will be genetically determined and others phenotypically induced. Several examples of the latter can be found in the literature. For instance, Machacek (1991) reported larger clutch sizes and smaller eggs in *D. galeata* treated with fish exudates and Stibor (1992) found similar effects in *D. hyalina*. A chemical, released by *Chaoborus*, has been shown to cause delayed maturation and fewer but larger offspring in *D. pulex* (Luning, 1992). The effect of fish exudates on behavioural traits varies in different clones of *D. magna* (De Meester, 1993) and Weider & Pijanowska (1993) have shown that clones, isolated from different ponds, do have different life history responses to chemical cues released by predators such as age and size at maturation.

Whether it is advantageous to migrate or not, depends on the extent of predation pressure, food concentration and on the vertical temperature gradient. Also of importance is to what extent phenotypic changes and genotypical differences occur in pertinent life history traits? Therefore, we argue that exudates of juvenile perch might have a different effect on these traits in migrating and non-migrating clones from Lake

Table 1. Summary of experimental conditions, all of which were performed in the absence and presence of fish exudate

Treatment	1	2	3	4
Temperature	20 °C	20 °C	20 °C	20 °C
Average algal concentration (mg C/l)	0.57	0.57	1.14	1.14
Media refreshed every	24 h	48 h	24 h	48 h
Experiments carried out with clone	M3/O2	M3/O2	M3	M3
Number of animals	10/10	7/7	10	10

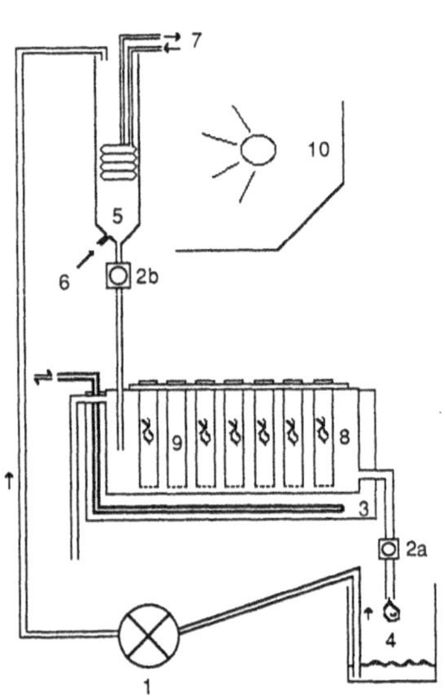

Fig. 1. Schematic representation of one unit of the flow through system.

Maarsseveen. Here, we report on the results of experiments to test this hypothesis.

Material and methods

Several individuals of the hybrid *Daphnia hyalina* × *galeata* were collected from Lake Maarsseveen I, at the beginning of the period of diel vertical migration in June 1992. Some were caught at a day depth of 0–7 m (non-migrating animals) and others at a day depth of 15–25 m (migrating animals). From these animals, clones were cultured and maintained in the laboratory under controlled conditions for several generations before the life-history experiments were started. Clones were kept in transparent glass jars filled with 1.0 l of filtered lake water. They were fed a surplus of log-phase cells of *Scenedesmus acuminatus* three times a week and the water in the vessels was renewed weekly. *Scenedesmus* was grown in Woods Hole MBL in continuous culture (Guillard, 1975).

In the experiments, two genetically distinct clones (K. Schwenk, pers. comm.) were used, a migrating one (hereafter referred to as clone M3) and a non-migrating one (hereafter referred to as clone O2). Neonates (< 12 hours old) were used in the experiments as shown in Table 1.

In Fig. 1, one unit of the culturing set-up is shown.

Individual animals were grown in tubes (9) in which the bottom consisted of a gauze (mesh-size 150 μm) through which algae can pass freely, but neonates cannot. Seven tubes were placed in a cuvette (8) and the latter was placed in a controlled-temperature waterbath (3). Algae (*Scenedesmus acuminatus*) were grown in reservoir bottles (7) which were illuminated by fluorescent lamps (10). The Skinner valves (2a and b) and the peristaltic pump (1) were connected to a timer. At certain times, Skinner valve 2a would open and the cuvette would empty into the beaker (4) (a small amount of medium remains behind for the animals to swim in). Then valve 2b would open and the cuvette (8) (and thus the tubes – 9) would be refilled with a fresh amount of algal suspension from the reservoir bottle. Lastly, the algal suspension in the beaker (4) would be pumped back into the reservoir bottle. In this way, the tubes were refreshed every six hours with medium from the bottles and the food concentration was kept as constant as possible throughout the experiments. The amount of algal suspension fed to the daphnids was checked regularly. Culture medium in the reservoir bottles was refreshed each day or every other day (Table 1) with log-phase cells of *Scenedesmus acuminatus* grown in continuous culture. The light/dark cycle was 16/8 hours.

The experiments were conducted with water from the hypolimnion (no young perch present) of Lake Maarsseveen, which was continuously filtered through a sandfilter at 20 °C. Water containing fish-exudate was obtained from an aquarium containing about 20 liters of lake-water and 10 individuals of perch (*Perca fluviatilis*) of approximately 3 cm length. Half the water in the aquarium was refreshed every day and feces were removed simultaneously. The 'fish-water' used in the experiments was filtered through a 0.2 μm mesh filter.

Neonates in the tubes were counted and removed daily. Also size and age at maturity (i.e. the day on

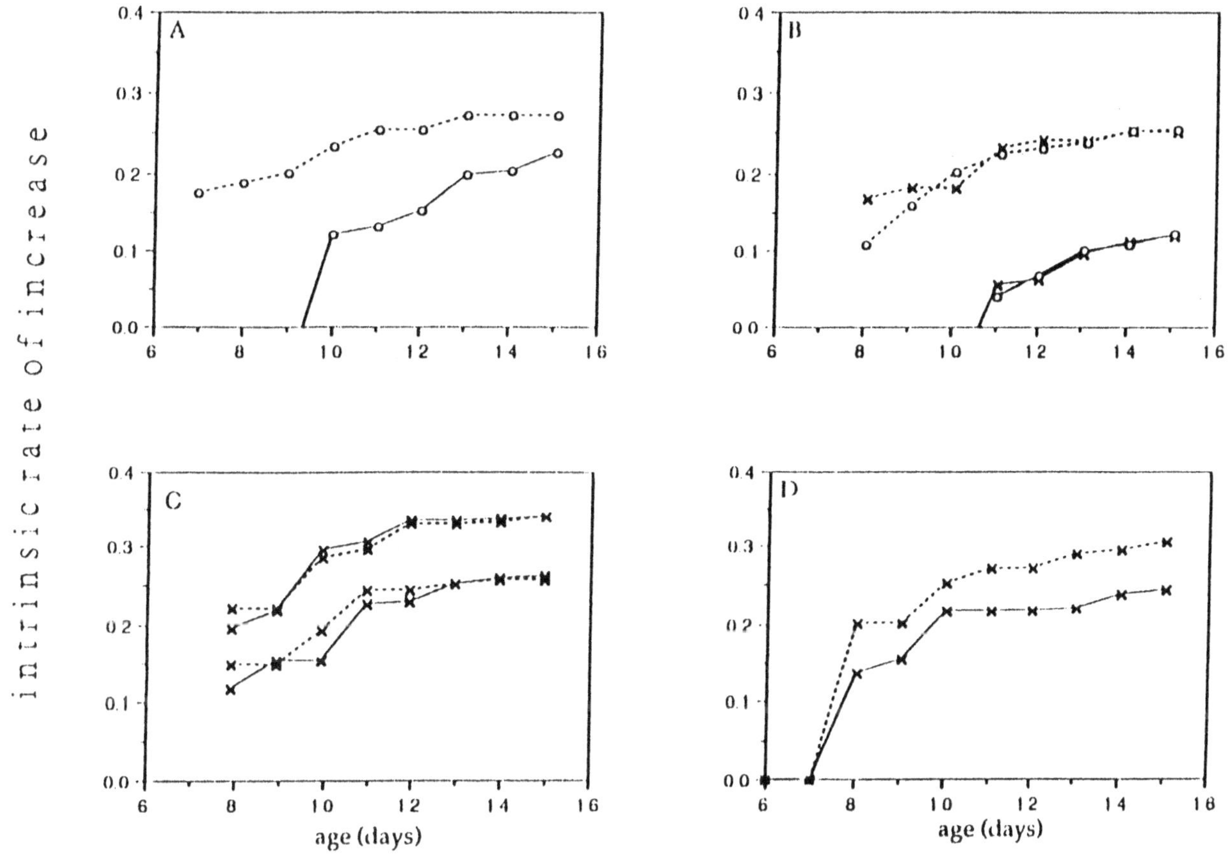

Fig. 2. Intrinsic rate of increase calculated for each age class in different clones of D. galeata × hyalina. Dotted lines represent fish treatments, solid lines represent the controls. Confidence limits are presented in Table 2. a. Clone O2. Food level: 0.57 mg C/l; fish exudate added every 48 hours; b. Clone O2 (open circles) and clone M3 (crosses). Food level: 0.57 mg C/l; fish exudate added every 24 hours; c. Clone M3. Two food levels: Top lines 1.14 mg C/l, bottom lines 0.57 mg C/l; fish exudate added every 48 hours; d. Clone M3. Foodlevel: 1.14 mg C/l; fish exudate added every 24 hours.

which the first clutch of eggs was deposited in the brood pouch) were determined. The experiments were continued until the calculated intrinsic rate of increase (r) reached an approximately constant value. This usually took 14–18 days.

The daphnids treated with fish exudate, were kept in 1/3 fish-water and 2/3 hypolimnion water. Control animals were kept in 100% hypolimnion water.

Population growth rates, ages and sizes at maturity and offspring sizes were determined. A Students' t-test was used to calculate differences between treatments and clones.

Population growth rate calculations

Population growth rates were calculated according to the Euler equation:

Table 2. Intrinsic rates of increase with 95%-confidence limits calculated on the fifteenth day of the experiments, as depicted in Fig. 2

Inset	Clone	Treatment	Food	r-value
A	O2	Control	0.57	0.225 ± 0.017
A	O2	Fish	0.57	0.272 ± 0.046
B	O2	Control	0.57	0.121 ± 0.047
B	O2	Fish	0.57	0.253 ± 0.025
B	M3	Control	0.57	0.120 ± 0.048
B	M3	Fish	0.57	0.251 ± 0.020
C	M3	Control	0.57	0.262 ± 0.059
C	M3	Fish	0.57	0.259 ± 0.047
C	M3	Control	1.14	0.344 ± 0.011
C	M3	Fish	1.14	0.341 ± 0.009
D	M3	Control	1.14	0.239 ± 0.077
D	M3	Fish	1.14	0.301 ± 0.041

$$1 = \sum e^{-rx} l_x m_x, \quad (1)$$

where r = intrinsic rate of increase for the population, x = age class in days, l_x = probability of surviving to age x, and m_x = age specific fecundity.

Since r cannot be isolated on one side of the equation, iterative calculations must be performed. However, this will leave us with only one value for r per data set from which the variance cannot be calculated. A variance was generated by applying the Jackknife method as described by Meyer *et al.* (1986).

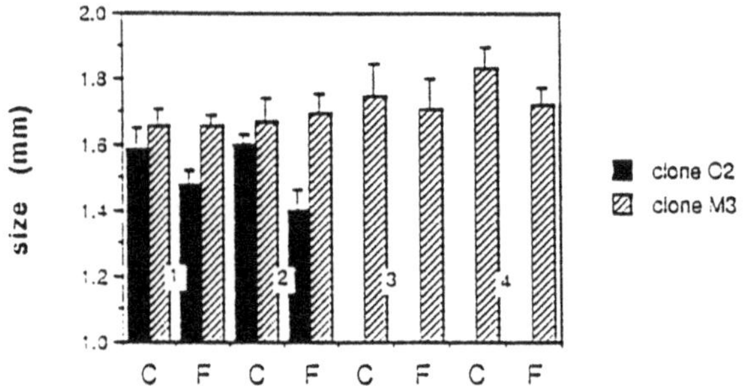

Fig. 3. Size at maturity for experimental conditions 1–4, as described in Table 1. C – control, F – fish treatment. Vertical bars represent 95% confidence limits.

Results

Population growth rates

Figure 2a shows that population growth rates of clone O2 were higher if fish exudate was added to the culture medium. The difference was significant up to an age of about 13 days if the medium was refreshed every 48 h (t = 12: $p < 0.03$, df = 12 / t = 13: $p = 0.34$, df = 12). If the medium was refreshed every 24 h, the difference was more pronounced and remained significant throughout the experiment. This was the same for both clones (Fig. 2b. clone O2: $p < 0.001$, df = 15. Clone M3: $p < 0.001$, df = 18). If the medium was refreshed every 48 h, no differences in population growth rates were found for clone M3 (Fig. 2c). This held for both food concentrations. Differences in growth rates due to different food levels however, were significant ($p < 0.01$, df = 18). Figure 2b also shows that with fish exudate administered every 24 h, no difference in population growth rate patterns between the two clones, was present.

If the medium was refreshed every 24 h, clone M3 also showed differences in growth rates although this difference was not significant at the highest food level (Fig. 2d. $p = 0.12$, df = 18). At the lowest food concentration, the difference was significant during the course of the entire experiment (Fig. 2b. $p < 0.001$, df = 18).

Age and size at maturity

Across all treatments, animals from clone M3 did not differ significantly in size (Fig. 3). Daphnids from clone O2 were smaller if fish exudate was present. The difference was significant if the media were refreshed every other day (Fig. 3. $p < 0.001$, df = 12). However, animals from clone O2 were always significantly smaller than those from clone M3 ($p < 0.001$, df = 12).

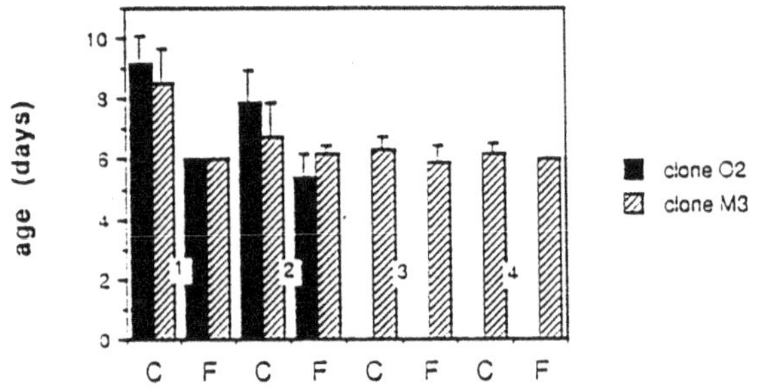

Fig. 4. Age at maturity for experimental conditions 1–4, as described in Table 1. C – control, F – fish treatment. Vertical bars represent 95% confidence limits.

Clone O2 reproduced significantly earlier if fish exudate was added every 24 or every 48 hours ($p < 0.001$, df = 12). With a daily administered dose of fish exudate, this also held true for clone M3 ($p < 0.001$, df = 18) and no significant clonal differences were observed (Fig. 4).

Offspring size

In Figs 5a and b, neonates from first clutches for both clones were significantly smaller in the presence of fish exudate ($p < 0.001$, df = 48 for both clones). Also, O2 neonates were significantly smaller than those of M3 under similar conditions ($p < 0.001$, df = 48). Although juveniles tended to become larger with increasing age of the mothers, neonates of O2 fish-treated animals were always significantly smaller than those of the control group ($p < 0.001$, df = 48). This difference disappeared in clone M3 as the mothers grew older.

Discussion

The population of *Daphnia* in Lake Maarsseveen in summer consists of *D. galeata* and the hybrid *D. galeata* × *hyalina*. Allozyme analysis has revealed that both taxa are composed of different clonal groups, in which the relative abundances change from year to year and within a single season (Spaak & Hoekstra, 1993). Of the specimens collected in the lake during the diel vertical migration period of 1992, clone O2 probably did not migrate since it was taken from the upper water layers during daytime. Clone M3 probably migrated because the parent animal was collected at a greater depth during daytime. Our experimental cultures revealed that differences in life histories between these two clones exist and that several traits changed differentially in the presence of exudates from juvenile perch. These perch are visual predators and larger *Daphnia* will be more vulnerable to predation than smaller ones. Therefore, a smaller size and an additional reduction in size in the presence of fish exudate, as exhibited by clone O2, will diminish predation risk. Animals from the migrating clone were larger and no corresponding decrease in size was observed.

The non-migrating daphnids from clone O2 produce more offspring in the early period of reproduction if exudates of perch are present. Thus, this clone increases its chances of successful reproduction before it is eaten. In addition, clone O2 produces smaller juveniles, which stand a better chance of evading visual predators. It is possible that subsequent generations, i.e. granddaughters of the experimental mothers, are even more reduced in size than the first generation, as was found for *Daphnia hyalina* by Stibor (1992).

The migrating clone M3 showed no reduction in body length, but clutch size increased in the presence of perch exudates. Since DVM decreases the risk of predation by fish, a reduction in size could be advantageous because invertebrate predation is reduced (Lynch, 1980). However, the risk of predation by fish is not reduced completely in migrating animals since at the height of the migration period, *Daphnia* are still found in the gut of young perch. Therefore, the production of large clutches as early as possible must be advantageous for the migrating animals as well. Since clone M3 does not reduce body length, while clutches increase in size in the presence of fish exudate, no trade-off between somatic growth and number of offspring seems to be present, at least not under the experimental conditions used in this study.

Another difference between the clones seems to be a higher sensitivity to fish exudate for clone O2. This clone responded to the less frequent addition of the exudate, while clone M3 did not. At low predator densities in the lake and probably at low concentrations of the kairomone, animals from clone O2 might already be able to shift their life history traits, while the migrating clone would not respond. Kairomone concentration effects exist as was demonstrated by Loose (1993) with respect to the extent of migration in large experimental plankton towers.

Both clones O2 and M3 have higher population growth rates if perch exudate is present. Differences between the fish and the control groups were significant at the beginning of the reproductive period, but such differences were not always significant later in the life cycle. This is the result of a larger clutch size earlier in life. Neonates hatching from eggs from smaller clutches are larger and have higher survival rates (Tessier & Consolatti, 1989). Therefore it is not always advantageous to have larger clutches in the absence of predatory fish. At the higher food level, clone M3 no longer showed any difference in population growth rate between the fish and control treatments. This might indicate a food-fish exudate interaction. Or, perhaps the population growth rate had reached a maximum and could not increase further. Another possibility is that if a higher amount of algae was added, this also means that the amount of bacteria introduced in the media was also higher. Since the fish exudate is probably degraded by bacteria, this could cause faster decomposition and hence a weaker response. This is a point for further study.

Two strategies of population survival, migration or superior life history traits, were described by Stich & Lampert (1984) for *D. hyalina* and *D. galeata*, respectively, in Lake Constance. The present study indicates that both strategies can be present within one hybrid population complex. Moreover, it is not simply an additional strategy, since in the migrating clone, the expression of some life-history traits also changed in the presence of fish exudates. The *Daphnia* population in Lake Maarsseveen may consist of several genotypes with different potentials to migrate in addition to different potentials for shifting life-history characters (Weider, 1984, 1985). At one end of the spectrum, strongly migrating animals were found with no life-history adaptations, while the other end may be represented by non-migrating animals with phenotypic plasticity in life-history traits. When juvenile fish predators appear in the lake, this would lead to partial and dif-

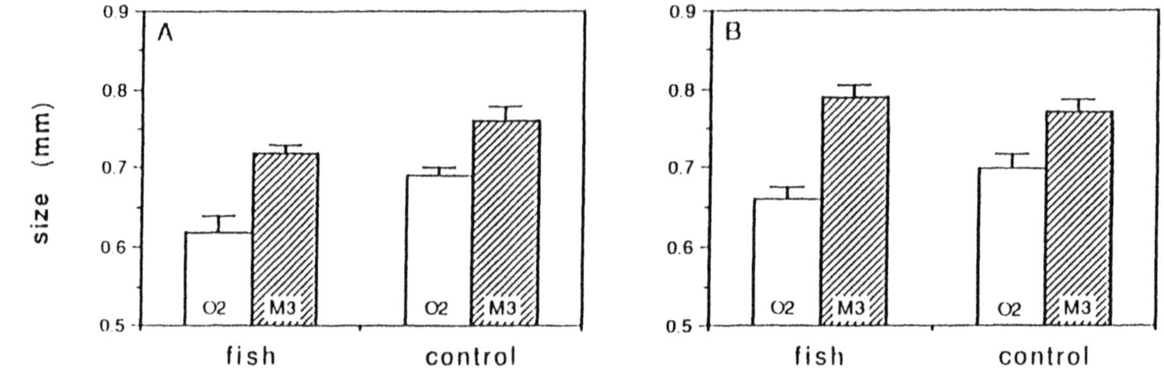

Fig. 5. Offspring sizes for both clones. Food concentration: 0.57 mg C/l, fish exudate added every 24 hours. a. means for the first two clutches; b. means for the last two clutches of the culture period. Vertical bars represent 95% confidence limits.

ferential migrations and shifts in relative clonal dominance, the extent of such clonal shifts would depend on food conditions and predation pressure. Changes both in amplitude of diel vertical migration (Ringelberg & Flik, submitted) and in clonal composition (Spaak & Hoekstra, 1993) were observed for different years in Lake Maarsseveen. However, more research has to be done before it is clear whether there is a dynamic genotypic shift related to vertical migration strategies.

Acknowledgement

We would like to thank K. Royackers who tested an earlier version of the culture apparatus.

References

De Meester, L., 1993. Genotype, fish-mediated chemicals, and phototactic behaviour in *Daphnia magna*. Ecology 74: 1467–1474.

Guillard, R. R. L., 1975. Culture of phytoplankton for feeding marine invertebrates. In Smith, W. L. & M. H. Chanley (eds), Culture of marine invertebrate animals. Plenum. New York: 29–60.

Lampert, W., 1989. The adaptive significance of diel vertical migration of zooplankton. Funct. Ecol. 3: 21–27.

Lampert, W., 1993. Ultimate causes of diel vertical migration of zooplankton: New evidence for the predator-avoidance hypothesis. Arch. Hydrobiol. Beih. 39: 37–44.

Loose, C. J., 1993. Daphnia diel verical migration behavior: Response to vertebrate predator abundance. Arch. Hydrobiol. Beih. Ergebn. Limnol. 39: 29–36.

Lüning, J., 1992. Phenotypic plasticity of *Daphnia pulex* in the presence of invertebrate predators: morphological and life history responses. Oecologia 92: 383–390.

Macháček, J., 1991. Indirect effect of planktivorous fish on the growth and reproduction of *Daphnia galeata*. Hydrobiologia 225 (Dev. Hydrobiol. 71): 193–197.

Meyer, J. S., C. G. Ingersoll, L. L. McDonald & M. S. Boyce, 1986. Estimating uncertainty in population growth rates: jackknife vs. bootstrap techniques. Ecology 67: 1156–1166.

Ringelberg, J., B. J. G. Flik, D. Lindenaar & K. Royackers, 1991. Diel vertical migration of *Daphnia hyalina* (sensu latiori) in Lake Maarsseveen: Part 2. Aspects of population dynamics. Arch. Hydrobiol. 122: 385–401.

Spaak, P. & J. R. Hoekstra, 1993. Clonal structure of the *Daphnia* population in Lake Maarsseveen: its implications for diel vertical migration. Arch. Hydrobiol. Beih. 39: 157–165.

Stich, H. B. & W. Lampert, 1984. Growth and reproduction of migrating and non-migrating *Daphnia* species under simulated food and temperature conditions of diurnal vertical migration. Oecologia 61: 192–196.

Tessier, A. J. & N. J. Consolatti, 1989. Variation in offspring size in Daphnia and consequences for individual fitness. Oikos 56: 269–276.

Weider, L. J., 1984. Spatial heterogeneity of *Daphnia* genotypes. Vertical migration and habitat partitioning. Limnol. Oceanogr. 29: 225–235.

Weider, L. J., 1985. Spatial and temporal genetic heterogeneity in a natural *Daphnia* population. J. Plankt. Res. 7: 101–123.

Weider, L. J. & J. Pijanowska, 1993. Plasticity of *Daphnia* life histories in response to chemical cues from predators. Oikos 67: 385–392.

Hydrobiologia **307**: 213–219, 1995.
P. Larsson & L.J. Weider (eds), Cladocera as Model Organisms in Biology.

Changes in epibiotic burden during the intermolt of *Daphnia* determined by hypodermal retraction stages

Ruth L. Willey[1] & Stephen T. Threlkeld[2]
[1]*Department of Biological Sciences (066), University of Illinois at Chicago, Chicago, IL 60607-7060, USA*
[2]*Department of Biology, University of Mississippi, University, MS 38677, USA*

Key words: Daphnia, eggstage, epibiosis, hypodermis, molt

Abstract

We calibrated four stages of hypodermal retraction and cuticle regeneration with five stages of parthenogenetic egg development in *Daphnia*. Using the hypodermal retraction stages, we found that epibiotic burden increased with elapsed intermolt time for juvenile, male, and female *Daphnia* bearing parthenogenetic or ephippial eggs. The rate of increase of burden was similar for adult females of two *Daphnia* species and for males and females of *D. pulex*. Rate of increase of burden was similar between juvenile and adult females of *D. galeata mendotae* and *D. pulex*.

Introduction

The burden of epibionts attached to the cladoceran carapace varies with cladoceran body size, species, and planktivore abundance (Threlkeld *et al.*, 1993). Because epibionts must recolonize the cladoceran carapace after each molt, and reproduction of epibionts occurs while attached to the carapace, epibiotic burden should increase with the amount of time that has elapsed since the previous molt. Green (1974), Xu (1992) and Threlkeld & Willey (1993) used variation in the stages of parthenogenetic eggs carried by adult females to determine how epibiotic burden increased during the cladoceran intermolt. These studies necessarily used only adult females, leaving open the question of whether developmental stage or sex might influence epibiotic burden development.

In much the same way that the stage of development of parthenogenetic eggs carried by females has been used to determine elapsed intermolt time (Threlkeld, 1979), a diverse literature on crustacean molting, particularly that of decapod crustacean molting, suggests that patterns of hypodermal retraction and cuticle regeneration could be used to determine the relative amount of time elapsed since molting (Richards, 1951; Passano, 1960; Skinner, 1985; Roer & Dillaman, 1993). We calibrated the observed variation in hypodermal-cuticle structure at the edge of the female carapace with the known sequence of development of parthenogenetic eggs in *Daphnia*. Presuming that the sequence of hypodermal retraction is similar in all *Daphnia* morphs, we compared the development of epibiotic burden on juvenile and adult, parthenogenetic and ephippial, and male versus female *Daphnia*.

Materials and methods

We calibrated four stages of hypodermal retraction (Table 1), visible at the ventro-posterior edge of the carapace, against the known sequence of five stages of development of parthenogenetic eggs (Threlkeld, 1979). We used samples of *Daphnia pulex*, collected on 3 September 1992, from a beaver pond in the Deer Creek drainage, Hinsdale County, Colorado, USA, and samples of *D. galeata mendotae*, collected on 15 November 1992, from the lagoon at Kresge Gardens in the Chicago Botanical Garden, Glencoe, Illinois, USA. The *D. pulex* were infested with the ciliate peritrich, *Vorticella campanula*. The epibiont attached to the surface of *D. galeata mendotae* was the euglenoid flagellate, *Colacium calvum*. The *Daphnia*, with their epibionts, were preserved in sucrose-formalin.

For scanning electron microscopic (SEM) observations, specimens were dehydrated in ethanol, critical point-dried from ethanol, coated with 60 nm Au,

Table 1. Morphological criteria for determining hypodermal stages in *Daphnia*.

Hypodermal stage	Criteria
1	Hypodermal cells abut against cuticle with smooth interface.
2	Interface between hypodermal cells and cuticle is refractile; in old samples in which shrinkage is present, a small space may occur between cells and overlying cuticle (apolysis; D1 stage of Drach, 1939).
3	Full hypodermal retraction; new spines are everted, show clearly beneath old cuticle, and extend freely into exuvial space, often without touching old cuticle overhead.
4	Exuvial space is reduced and new spines lie closely under old cuticle.

Table 2. Intercalibration of hypodermal retraction stages with the stages of development of parthenogenetic aggs (with calculated midpoints of elapsed intermolt time); number of *Daphnia pulex* for each set of stages.

Hypodermal retraction stages (midpoints)					
4 (0.84)			1	8	10
3 (0.67)		1	9	2	
2 (0.46)		5			
1 (0.17)	10	4			
	1 (0.15)	2 (0.46)	3 (0.67)	4 (0.77)	5 (0.92)
	Parthenogenetic Egg Stages (midpoints)				

and examined at 5–10 kV with the SEM (ISI DS-130). For bright-field, light microscopic observations, specimens were examined in a custom-made, 0.5 ml Sedgewick-Rafter cell which allowed the use of the 40 × objective.

The relative amount of elapsed intermolt time (EIT), based on a scale of 0 to 1, was established for 50 female *Daphnia* carrying parthenogenetic eggs whose developmental stages were determined using Threlkeld's (1979) eggstage method. An additional sample of *Daphnia* (N = 78) was examined comparing five stages of ephippial egg development (modified from Mellors, 1975) with the hypodermal retraction stages, and a second intercalibration of ephippial egg development and hypodermal retraction stages was achieved. Using these hypodermal retraction stages, we compared development of epibiotic burden during the intermolt in parthenogenetic and ephippial female *Daphnia*, in juvenile and adult female *Daphnia*, and in male and female *Daphnia*. The distinction between juvenile and adult females was determined by the minimum length of females bearing eggs (1.5 mm for both *Daphnia* species). Differences between males and females were calculated for all individuals of both sexes because of the difficulty in distinguishing between juvenile and adult males. Relative burden was based, for both epibiotic species, on a scale of: 0 (no epibionts); 1 (1–10 epibionts); 2 (11–20 epibionts); 3 (21–30 epibionts); and 4 (above 30).

The data were analyzed by regression and analysis of covariance. Statistical analyses were conducted with SYSTAT (Wilkinson, 1988).

Results

Changes in the interface between the hypodermis (epithelial cell layer) and the cuticle can be observed at the ventro-posterior margin of the *Daphnia* carapace (Fig. 1). Best results are achieved using a 40 × objective (approximately × 400 magnification) to examine the narrow strip of the inner surface of the carapace between the outer, spinulose margin and the submarginal row of setose setae (Fig. 2). Four distinct stages of hypodermal retraction are recognizable (Figs 3–7; Table 1). Table 2 shows the intercalibration of stages of parthenogenetic and ephippial egg development, and EIT, during each hypodermal retraction stage. The criteria for identifying ephippial egg stages are modified from Mellors (1975) (Table 3).

Samples from both localities, Colorado and Illinois, contained adult *Daphnia* bearing parthenogenetic eggs allowing the relationship of burden to EIT to be compared (Fig. 8). For the two respective epibiotic taxa, *Colacium calvum* and *Vorticella campanula*, burden increased linearly over time ($P<0.001$) with homogeneous slopes ($P=0.604$), but with significantly different Y-intercepts ($P<0.001$). An epibiotic burden index of 0.8 (approximately 8 cells per *Daphnia*) is predicted, immediately postmolt, for the peritrichs on *D. pulex*

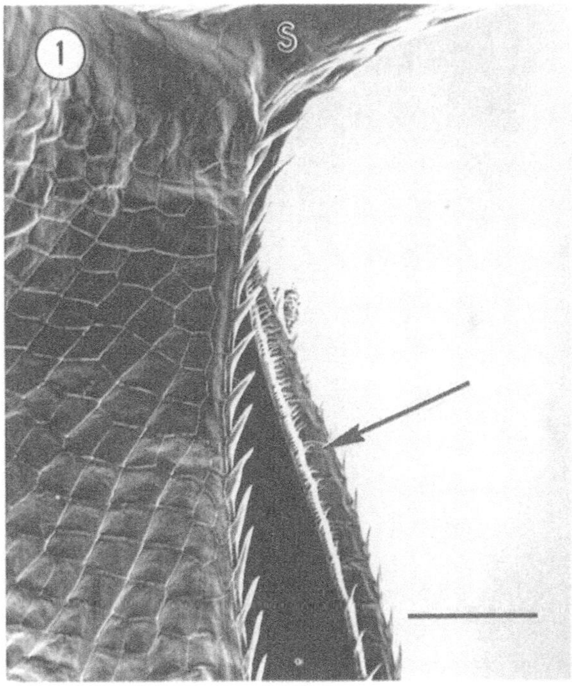

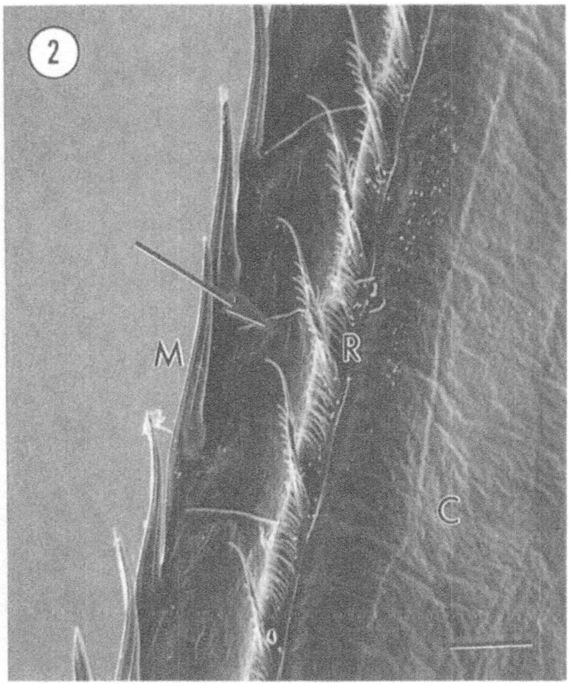

Figs 1–2. Scanning electron microscopic (SEM) view of the ventro-posterior margins of the carapace of *D. galeata mendotae* with the base of the posterior spine (S). The stages of hypodermal retraction are most clearly seen along the inner surface, between the outer margin and the sub-marginal row of setae, marked by the arrow. Scale line = 50 μm. *Fig. 2.* SEM of the inner surface of the carapace edge in the region, marked by the arrow in Fig. 1, showing the inner wall of the carapace (C), and the region for hypodermal examination (arrow) between the outer margin with spinules (M), and the submarginal rim with setose setae (R). Scale line = 10 μm.

Table 3. Description of stages of ephippial egg development during the intermolt of *Daphnia* with calibrated hypodermal retraction stages.

Modified ephippial eggstages	Mellors (1975) stages	Hypodermal retraction stages	Criteria
1	–	1.0	Eggs = 0; ephippial carapace milky white
2	1	1.0	Eggs = 2 white; ephippial carapace white
3	2	1.3	Eggs ≐ 2 brown; carapace light brown directly over eggs
4	3	3.3	Eggs = 2 brown; carapace brown over eggs and light brown just around egg area
5	4	4.0	Eggs= 2 black; carapace dark brown

whereas that for *C. calvum*, on *D. galeata mendotae*, is 0.

Comparison of burden development between females bearing parthenogenetic and ephippial eggs was conducted with the *D. pulex* samples from Colorado. No difference ($P = 0.49$) exists, for the rate of development of epibiotic burden of *Vorticella campanula*, between adult parthenogenetic and ephippial females of *Daphnia pulex* (Fig. 9).

Differences appear in the epibiotic burden on juvenile and adult female *D. pulex* from Colorado (Fig. 10). In both stages, burden of *Vorticella* increased throughout the EIT ($P<0.001$), but, on the adults, increased more rapidly (slope = 3.4) than on the juveniles (slope = 2.4) resulting in a weak interaction between the slopes ($P = 0.07$). Development of epibiotic burden of *Colacium calvum* on the juvenile and adult females of *D. galeata mendotae*, from Illinois, is

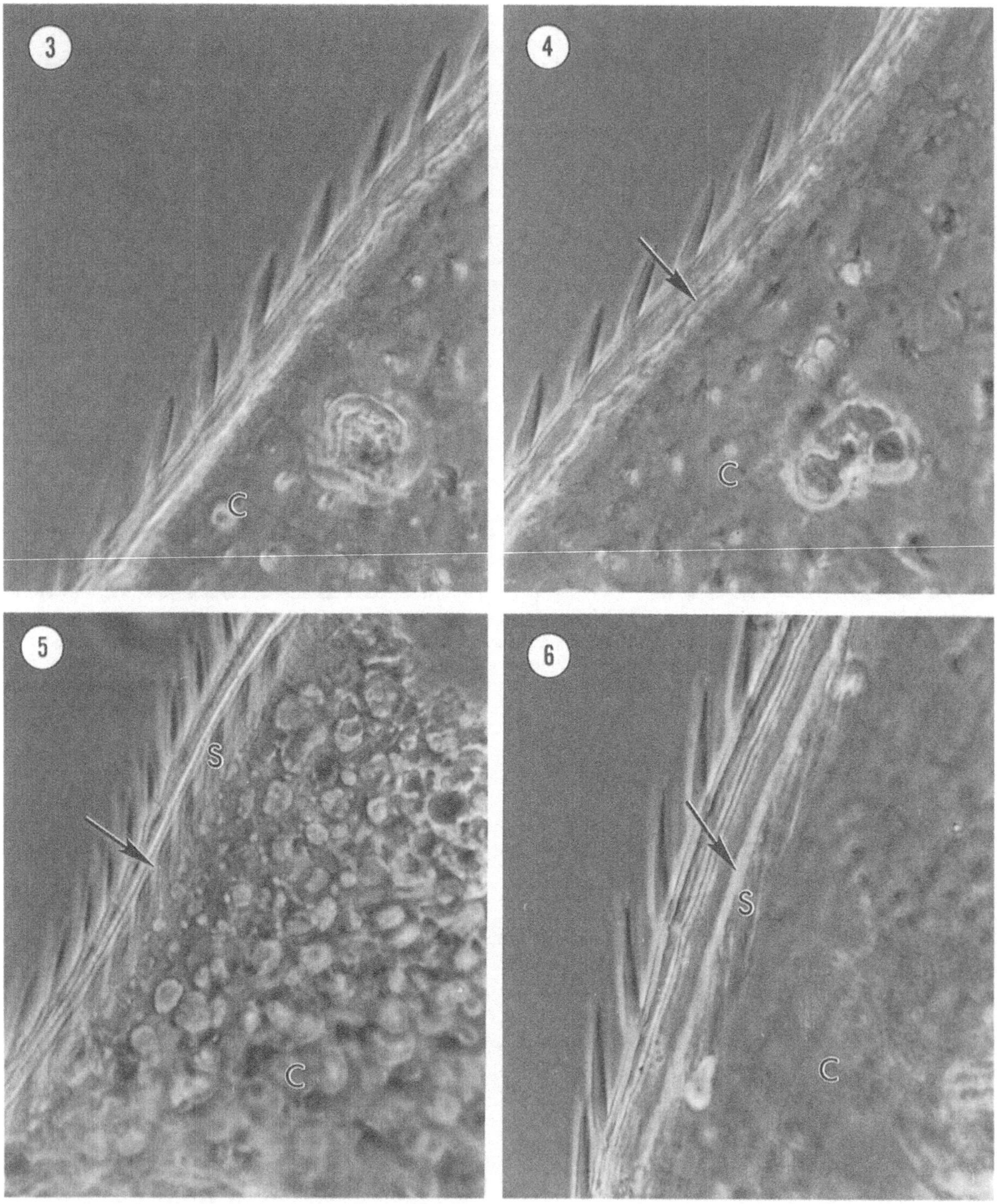

Figs 3–6. Four stages of hypodermal retraction during the intermolt of *Daphnia*. Fig. 3 = stage 1, Fig. 4 = stage 2, Fig. 5 = stage 3, Fig. 6 = stage 4. Cuticle (C); exuvial space (arrow); new spinules in exuvial space (S). Magnification, × 1000.

similar (Fig. 11). Elapsed intermolt time has a significant effect on development of burden ($P<0.001$) of either epibiotic taxon. The slope of burden increase for *Vorticella* on *D. galeata mendotae* adults (slope = 3.5) is essentially the same as that for *C. calvum* on adult *D. pulex* (slope = 3.4), but the juveniles of *D. galeata*

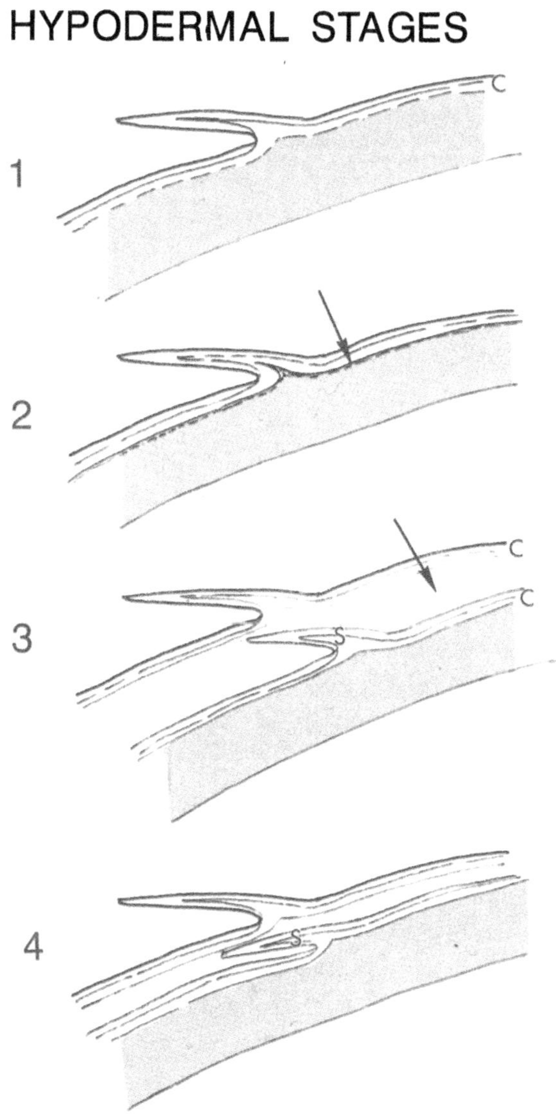

Figs 7. Diagram of hypodermal retraction stages. Hypodermal epithelium, stippled. Letters have same meaning as in Fig. 3.

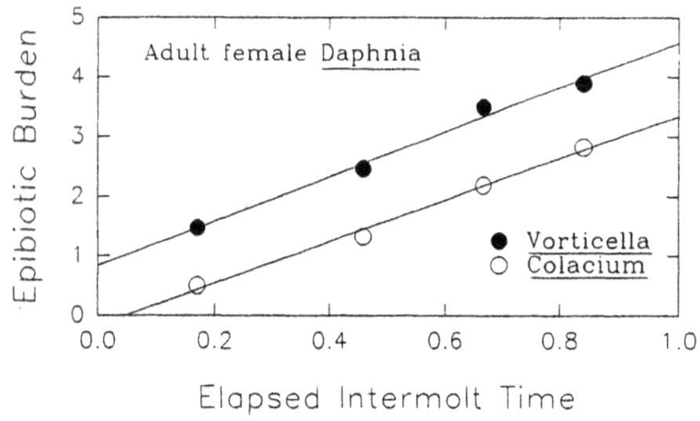

Figs 8. Epibiotic burden development on adult female *Daphnia* carrying parthenogenetic eggs; comparison of *D. pulex*: *Vorticella campanula* ($r^2 = 0.78$) and *D. galeata mendotae*: *Colacium calvum* ($r^2 = 0.76$).

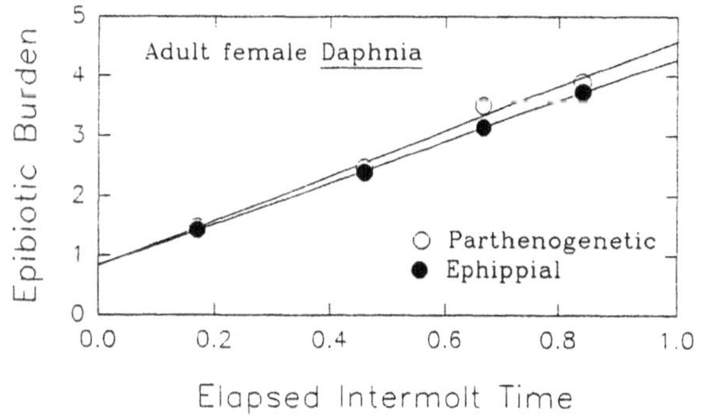

Figs 9. Development of epibiotic burden of *Vorticella campanula* on adult female *Daphnia pulex* carrying parthenogenetic or ephippial eggs.

mendotae have a lower slope (slope = 1.4) than that of *D. pulex* (slope = 2.4). As a result, a significant interaction occurs between developmental stages (juvenile or adult) and EIT ($P<0.001$) for the association with burden increase of *Colacium calvum*.

The average intermolt time of juvenile *Daphnia* females, of various species, has been estimated at approximately one-half that of adults (Ingle *et al.*, 1937; Anderson & Jenkins, 1942; Lei & Armitage, 1980; Lynch, 1989). If these estimates are used to express EIT of juveniles on the same scale as that for the adults (lower panels of Figs 10 and 11), no difference occurs between the rate of accumulation of epibionts by juvenile and adult female *D. galeata mendotae* ($P = 0.306$) or between juvenile and adult *D. pulex* ($P = 0.110$).

In the Colorado beaver pond, epibiotic *Vorticella* accumulation on *D. pulex* males and females was significantly affected by EIT (Fig. 12). The rates of increase were not significantly different ($P = 0.658$). For both sexes of *Daphnia*, the initial burden index was not different from 0. Evidence for variation in instar duration between males and females is equivocal (Munro & White, 1975), and intermolt (instar) duration of *Daphnia ambigua* has been shown by Lei & Armitage (1980) to be the same for males and females.

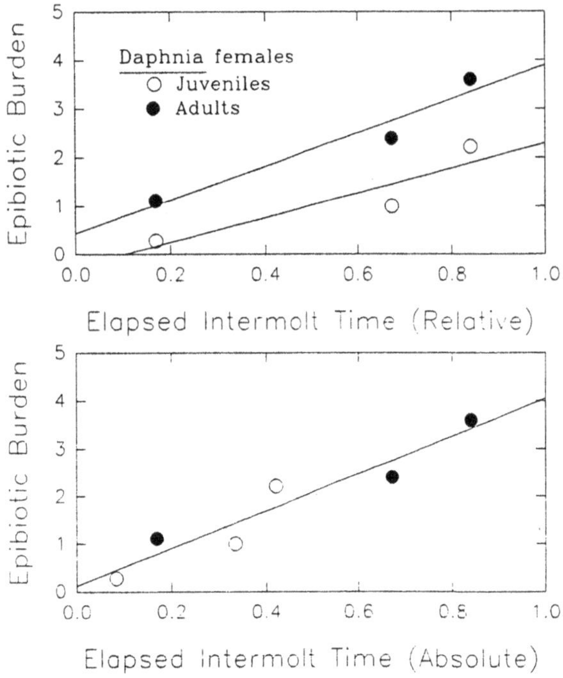

Figs 10. Comparison of epibiotic burden response, by *V. campanula*, on juvenile and adult female *D. pulex*. Upper panel is based on the same scale for both stages (juvenile: $r^2 = 0.53$; adult: $r^2 = 0.54$). Lower panel is based on a scale adjusted for a juvenile intermolt that is 50% that of the adult ($r^2 = 0.61$).

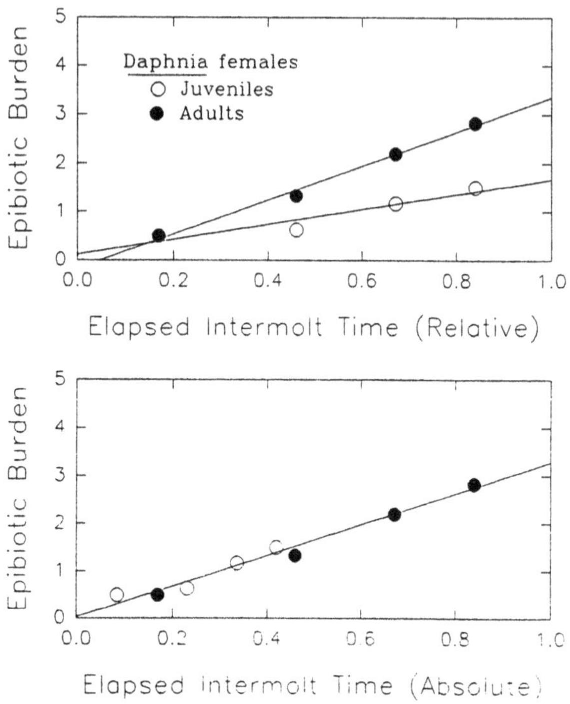

Figs 11. Comparison of epibiotic burden response, by *C. calvum*, on juvenile and adult female *D. galeata mendotae*. Upper panel is based on the same scale for both stages (juvenile: $r^2 = 0.30$; adult: $r^2 = 0.76$). Lower panel is based on a scale adjusted for a juvenile intermolt that is 50% that of the adult ($r^2 = 0.71$).

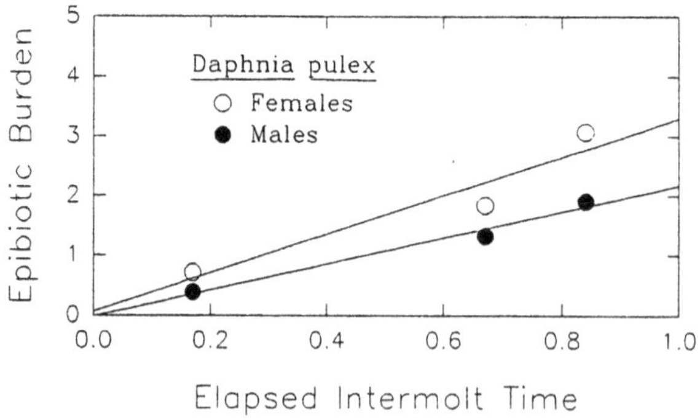

Figs 12. Development of epibiotic burden of *Vorticella campanula* on male and female *Daphnia pulex*. Males: $r^2 = 0.64$; Females: $r^2 = 0.47$.

Discussion

Daphnia has a typical crustacean integument (cuticle and hypodermis) (Dennell, 1960; Schultz & Kennedy, 1977). Growth, or increase in body length, occurs stepwise immediately after each molt while the cuticle is still soft (Edlen, 1937; Green, 1956). Successive stages during the intermolt period, based on changes in the integument, have been well worked out for decapod crustaceans (see Passano, 1960; Skinner, 1985). Similar changes, particularly those of proecdysis, are visible at the posterior edge of the translucent carapace of *Daphnia* at magnifications suitable for routine examination of zooplankton samples. Observations are best observed on the inner surface of the narrow strip of carapace between the outer spinules, and the submarginal setae (Fryer, 1991).

Epibionts colonize after each *Daphnia* molt and reproduce while attached. Their numbers should be low immediately postmolt, and increase during the intermolt. It is not known whether burden increase is a function of epibiont reproduction, subsequent epibiont attachment, or both. For all sexes, and stages, burden increased throughout the elapsed intermolt time. Rates of increase were similar for the adult females of the two species of *Daphnia*, for juvenile and adult females, and for males and females. Our results confirm that elapsed intermolt time is essential to the development of epibiotic burden.

Initial colonization by epibionts (burden index at the Y-intercept) varied between the two epibiotic systems examined. *Vorticella* appeared to attain a positive burden immediately postmolt, whereas *C. calvum* appeared to start from zero. Green (1974) showed that the peritrich, *Epistylis helenae*, could anticipate the molt of *D. pulex*. If *Vorticella* has the same ability to detect molt and detach, then free-living cells (telotrochs; Vacchiano, *et al.*, 1992) should be available for immediate recolonization when the molt is complete. If *C. calvum* cannot anticipate the *Daphnia* molt, then re-colonization may be delayed by the time necessary for the attached cells to regenerate their motile flagella, detach from the castskin, and locate a new substrate organism (see Ward & Willey, 1981).

Use of the hypodermal retraction criteria to determine intermolt stages has greatly expanded our capabilities to follow processes involving elapsed intermolt time during the growth of *Daphnia* species. Our studies have (1) helped determine differences in response of epibiotic cells to different taxa, sexes, and developmental stages of substrate organisms, and (2) provided initial insight into the role of elapsed intermolt time, of the substrate organisms, as well as possible variation in epibiotic colonization behavior in the regulation of epibiotic populations attached to freshwater, crustacean zooplankton.

Acknowledgments

We wish to thank Dr R. Taylor, Chicago Botanical Garden, Glencoe, IL; the Cebolla District Ranger office, USDA Forest Service, Gunnison, CO; and the Colorado Division of Wildlife, Glenwood Springs, CO, USA, for permission to collect zooplankton. Mr R. Wibel provided support for the electron micrographs.

References

Anderson, B. G. & J. C. Jenkins, 1942. A time study of events in the life span of *Daphnia magna*. Biol. Bull. 83: 260–272.

Dennell, R., 1960. Integument and exoskeleton. In T. H. Waterman, (ed.), The Physiology of Crustacea. Academic Press, NY: 449–472.

Edlen, A., 1937. Experimentelle Wachstumsstudien an *Daphnia magna*. Kungl. fysio. Sallsk. 49: 1–29.

Fryer, G., 1991. Functional morphology and the adaptive radiation of the Daphniidae (Branchiopoda: Anomopoda). Phil. Trans. r. Soc. Lond. (B) 331: 1–99.

Green, J., 1956. Growth, size and reproduction in *Daphnia* (Crustacea: Cladocera). Proc. zool. Soc. Lond. 126: 173–204.

Green, J., 1974. Parasites and epibionts of Cladocera. Trans. zool. Soc. Lond. 32: 417–515.

Ingle, L., T. R. Wood & A. M. Banta, 1937. A study of the longevity, growth, reproduction and heart rate of *Daphnia longispina* as influenced by limitations in quantity of food. J. exp. Zool. 76: 325–352.

Lei, C.-H. & K. B. Armitage, 1980. Growth, development and body size of field and laboratory populations of *Daphnia ambigua*. Oikos 35: 31–48.

Lynch, M., 1989. The life history consequences of resource depression in *Daphnia pulex*. Ecology 70: 246–256.

Mellors, W. K., 1975. Selective predation of ephippial *Daphnia* and the resistance of ephippial eggs to digestion. Ecology 56: 974–980.

Munro, I. G. & R. W. G. White, 1975. Comparison of the influence of temperature on the egg development and growth of *Daphnia longispina* O. F. Muller (Crustacea: Cladocera) from two habitats in southern England. Oecologia 20: 157–165.

Passano, L. M., 1960. Molting and its control. In T. H. Waterman (ed.), The Physiology of Crustacea. Academic Press, NY: 473–536.

Richards, A. G., 1951. The Integument of Arthropods. U. Minnesota Press, Minneapolis, 411 pp.

Roer, R. D. & R. M. Dillaman, 1993. Molt-related change in integumental structure and function. In M. N. Horst & J. A. Freeman (eds), The Crustacean Integument. CRC Press, Boca Raton, FL.: 1–37.

Schultz, T. W. & J. R. Kennedy, 1977. Analyses of the integument and muscle attachment in *Daphnia pulex* (Cladocera, Crustacea). J. Submicrosc. Cytol. 9: 37–51.

Skinner, D. M., 1985. Molting and regeneration. In D. E. Bliss & L. H. Mantel (eds), The Biology of Custacea, vol. 9. Academic Press, NY: 43–146.

Threlkeld, S. T., 1979. Estimating cladoceran birth rates: The importance of egg mortality and the egg age distribution. Limnol. Oceanogr. 24: 601–612.

Threlkeld, S. T., D. A. Chiavelli & R. L. Willey, 1993. The organization of zooplankton epibiont communities. Trends Ecol. Evol. 8: 317–321.

Threlkeld, S. T. & R. L. Willey, 1993. Colonization, interaction, and organization of cladoceran epibiont communities. Limnol. Oceanogr. 38: 584–591.

Vacchiano, E., A. Dreisbach, D. Locascio, L. Castaneda, T. Vivian & H. E. Buhse, Jr., 1992. Morphogenetic transitions and cytoskeletal elements of the stalked zooid and the telotroch stages in the peritrich ciliate *Vorticella convallaria*. J. Protozool. 39: 101–106.

Ward, K. A. & R. L. Willey, 1981. The development of a cell-substrate attachment system in a euglenoid flagellate. J. Ultrastr. Res. 74: 165–174.

Wilkinson, L., 1988. SYSTAT: The system for statistics. SYSTAT, Inc.

Xu, Z., 1992. The abundance of epizoic ciliate *Epistylis daphniae* related to their host *Moina macrocarpa* in an urban stream. J. Invert. Pathol. 60: 197–200.

Hydrobiologia **307**: 221–230, 1995.
P. Larsson & L.J. Weider (eds), Cladocera as Model Organisms in Biology.

An in situ test of the effects of food quality on *Daphnia* population growth

Bart T. De Stasio, Jr.[1,3], Lars G. Rudstam[2], Adrienne Haning[3], Pat Soranno[3] & Yvonne C. Allen[3*]
[1]*Biology Department, Lawrence University, Appleton, WI 54912, USA*
[2]*Cornell Biological Field Station and Department of Natural Resources, Cornell University, 900 Shackleton Point Road, Bridgeport, NY 13030-9750, USA*
[3]*Center for Limnology, University of Wisconsin, 680 North Park Street, Madison, WI 53706, USA*
**Current address: Community Dynamics and Habitat Unit, Fisheries and Wildlife Research Section, Ministry of Natural Resources, P.O. Box 5000, Maple, Ontario, L6A 1S9 Canada*

Key words: Daphnia, food limitation, Lake Mendota, cyanobacteria

Abstract

In years of low planktivory by fish, midsummer declines and low population abundances of *Daphnia pulicaria* in Lake Mendota, WI, USA have been attributed to poor food quality and low food abundance. This hypothesis has been proposed because of midsummer blooms of blue-green algae in this eutrophic urban lake. We tested the hypothesis by performing in situ food manipulations during the midsummer decline of the *D. pulicaria* population, and during the early autumn low population period. In July, animals held in clear plastic vials containing GF/F filtered lake water exhibited poor survivorship, ceased producing offspring after three days, and had low lipid reserves. Daphnids fed whole-lake water or filtered water enriched with *Chlamydomonas* survived at a significantly higher rate and produced abundant offspring. In September, there were no significant differences in survivorship among treatments, but animals fed water enriched with *Chlamydomonas* produced significantly more offspring than animals in the other treatments. The hypothesis of poor food quality causing the midsummer decline of the *Daphnia* population is not supported by our experiments. Because predation by juvenile fish does not appear to be important in regulating the midsummer *Daphnia* population during these years, alternate hypotheses for the midsummer decline are suggested. These include increased predation by invertebrate predators such as *Leptodora kindtii*, a life history shift by daphnids during midsummer to production of resting eggs, and a combination of low levels of planktivory and deteriorating feeding conditions acting together to cause the population decline.

Introduction

Understanding *Daphnia* population dynamics is a major topic in freshwater ecology because of the important role that daphnids play in lake ecosystems (Peters & de Bernardi, 1987; Lammens *et al.*, 1990). Numerous studies have focused on whether food availability or predation is the dominant mechanism determining the seasonal and interannual patterns of *Daphnia* abundances observed in lakes (Hall, 1964; Gliwicz, 1985; Larsson *et al.*, 1985; Threlkeld, 1979b, 1985; Benndorf & Horn, 1985; Carpenter & Kitchell, 1988; McQueen *et al.*, 1986; Sommer, 1989). Recently it has become clear that a combination of these factors and the timing of events interact to regulate *Daphnia* dynamics (Cryer *et al.*, 1986; McQueen *et al.*, 1989, Luecke *et al.*, 1990; Leibold, 1989; Rudstam *et al.*, 1993). The testing of multiple hypotheses at different times of the year is therefore required for a more complete understanding of *Daphnia* population dynamics.

The interannual and seasonal dynamics of *D. pulicaria* and *D. galeata mendotae* in Lake Mendota, Wisconsin, USA is strongly affected by the presence of an abundant population of cisco (*Coregonus artedii*) a cold water planktivore (Vanni *et al.*, 1990; Luecke *et al.*, 1990, 1992; Rudstam *et al.*, 1993). In years with high cisco abundance (i.e. before a major die off as in the summer of 1987), the large bodied *D. pulicaria* are prevented from increasing in abundance in spring because of predation by cisco. The smaller daphnid *D. galeata* increases in abundance and causes a relatively short clearwater phase. Because very few eggs are produced during the clearwater phase, poor feeding conditions have been assumed to result in the decline of daphnids during midsummer.

Following a dramatic die-off of cisco in 1987, the cisco population has remained at low levels (Fig. 1). Low cisco abundance, and therefore low springtime planktivory, have allowed *D. pulicaria* to increase in early spring and persist into early summer. The spring increase in *D. pulicaria* leads to a prolonged clearwater phase, and apparently allows *D. pulicaria* to prevent *D. galeata* from establishing a substantial population. During these years, *D. galeata* only occurs in very low numbers throughout the spring, summer and fall, but can increase following a late fall diatom bloom. Intensive studies of the behavior and feeding of larval and older planktivorous fishes in Lake Mendota during low cisco years, conducted throughout the ice-free period, indicate that only extremely low predation by fish on daphnids occurs during the summer period (Luecke *et al.*, 1992; Post *et al.*, 1992; T. Johnson, pers. com.) Overall, estimates of daphnid mortality due to the combined predation by fish and invertebrates such as *Leptodora kindtii* and *Mesocyclops edax* are lower than that required to cause the observed midsummer declines of *D. pulicaria* (Luecke *et al.*, 1992).

In contrast to most other eutrophic lakes, the midsummer zooplankton community of Lake Mendota contains very few cladocerans other than daphnids. Potential competitors with daphnids for food, such as *Bosmina* and *Chydorus*, typically occur at densities of less than 0.2 l^{-1} throughout the summer, while *Diaphanosoma* may reach pelagic densities of 1.0 ind. l^{-1} (Luecke *et al.*, 1992). However, during the summer egg ratios of daphnids often decline, while the phytoplankton community shifts to one mostly dominated by cyanobacteria species. Although *D. pulicaria* should be fairly resistant to starvation conditions (Threlkeld, 1976), we have always observed a midsummer decline concurrent with the shifts in phytoplankton composition. This led to the hypothesis that low food availability or poor food quality causes the drastic *Daphnia* population declines in midsummer.

We have tested the hypothesis that food quantity and quality are important factors in the midsummer

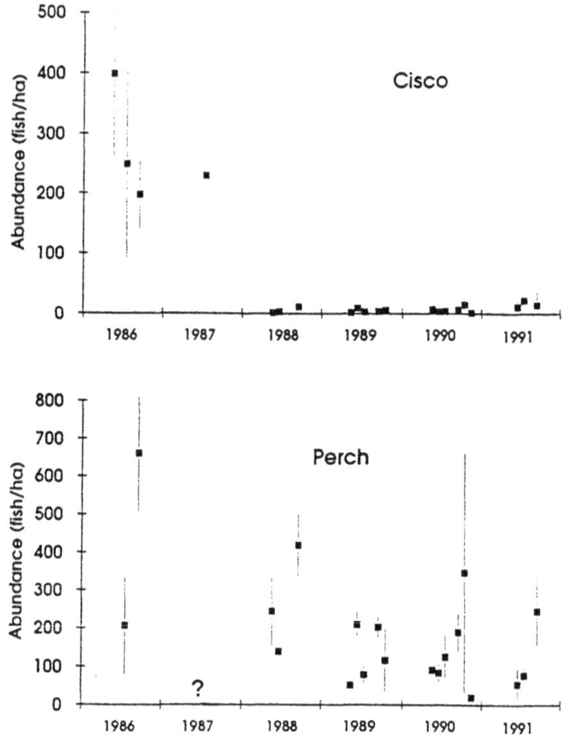

Fig. 1. Abundance (mean±2 s.e.m.) of cisco and yellow perch in Lake Mendota from 1986 through 1991. Abundance of fish was measured with hydroacoustics (single beam, 70 kHz) and divided into two size groups: targets −46 to −40 dB (approximate fish lengths 12 to 25 cm, primarily age-1 and older yellow perch as well as younger cisco, white bass, freshwater drum), and larger than −40 dB (fish larger than 25 cm, primarily age-3 and older cisco, as well as larger perch, white bass and freshwater drum). We used the species composition within these two size groups from gill net catches to estimate the abundance of the two species. These numbers include age-1 and older fish for both perch and cisco and age-0 cisco in the fall. A small number of age-0 cisco were present in the lake in 1989. No age-0 cisco were caught in the lake in 1991. Only cisco abundance was measured in 1987 because surveys were only conducted at night that year and a major proportion of the perch population moves inshore at night. Perch abundance increases in the fall of most years due to increased use of the open water by yearling perch. Perch larger than about 7 cm are vulnerable to these gill nets (age-0 perch in the fall and age-1 perch in the spring and summer should be caught in the net if they used the open water). Details of methods given in Rudstam & Johnson (1992) and also discussed in Rudstam *et al.* (1993).

decline of *D. pulicaria* and also that poor food conditions maintain low abundances during late summer and fall in Lake Mendota. Survivorship and offspring production by individual *D. pulicaria* were measured during in situ enclosure experiments where food supply was manipulated. Experiments performed during the midsummer decline and during the fall period of low abundances in 1991 indicate that poor food quality and quantity do not cause the population crash during July, but that the low population size during September may be due to poor feeding conditions.

Methods

Zooplankton and phytoplankton dynamics

Data on population dynamics and seasonal shifts in community composition of zooplankton and phytoplankton were collected at the deepest portion of the lake (deep hole station, maximum depth = 24 m). Comparisons of zooplankton collections made along onshore-offshore transects indicate that the *Daphnia* population dynamics can be assessed accurately using samples from this deep hole station (B. Byom & P. Soranno, pers. com.). Zooplankton samples were collected weekly throughout the ice-free season in 1991. A Wisconsin net (80 μm-mesh, 0.3 m diameter) was hauled from 20 m to the surface on each date to collect zooplankton. Based on calibrations of this net against a tube sampler and a Clarke-Bumpus sampler, a 70% net efficiency was employed in determining the volume of water sampled. Animals were preserved immediately in a 10% buffered sucrose-Formalin solution.

Zooplankton were identified and enumerated at 12 ×–50 × magnification. Egg ratios were calculated by dividing the number of eggs observed (including ephippia) by the number of adult daphnid females counted. Seasonal changes in daphnid size were determined by measuring the lengths of individuals on each date using an ocular micrometer at 50 × magnification. Animals were chosen haphazardly from samples, with between 16 and 65 specimens measured per sample.

Integrated phytoplankton samples were obtained by lowering a weighted 8 m length of Tygon tubing through the water column (inner diameter 2.5 cm), capping the upper end, and retrieving & preserving the sample in Lugol's solution. Phytoplankton analyses were performed by Phycotech, Inc. of Indiana. Samples were permanently mounted, identified and enumerated using inverted microscopy. Counts were converted to biovolume estimates following measurement of linear dimensions of each taxa and employing geometric formulae.

Food manipulation experiments

In situ food manipulations were conducted in clear plastic medicine vials (70 ml volume). All vials were incubated during the experiments in a large mesh nylon bag suspended at a depth of 2 m in University Bay, Lake Mendota. *Daphnia pulicaria* were collected from the epilimnion (1–2 m depth) and randomly assigned to vials in each food treatment. Fifty daphnids were tested in each treatment with a single daphnid placed in each vial. We selected animals of the approximate size at first reproduction in Lake Mendota for all experiments (i.e. 1.3 mm).

Animals in the starvation treatment were placed in vials containing epilimnetic water which had been filtered through a Millipore GF/F filter (0.45 μm pore size). The enriched treatment subjected animals to GF/F water which had been enriched by the addition of laboratory cultured *Chlamydomonas vulgaris* (a minimal amount of nutrient medium was also typically included as a result of our procedure). Final concentration of the food in this treatment was 5×10^4 cells ml^{-1}. The third treatment consisted of whole-lake water collected from the epilimnion also and passed through a 153 μm-mesh nylon net to remove zooplankton. Examination of the water used in each treatment indicated that the GF/F water contained no algae and that filtering water through the 153 μm-mesh net did not significantly alter the food quality or quantity compared to that available in the lake. Large species such as *Ceratium hirundinella*, *Closterium spp.*, abundant filaments and bundles of *Aphanizomenon flos-aquae*, as well as colonies of *Microcystis aeruginosa* were present in the whole-lake water.

Vials were filled with fresh treatment water at every monitoring time, and were completely sealed by plastic caps in between these times. Preliminary experiments on changes in dissolved oxygen concentrations in the vials over time indicated that oxygen dropped by less than 10% per day for each treatment. It took more than four days during these preliminary summer trials before oxygen levels dropped enough to cause significant mortality of individuals. We therefore assessed the status of all individuals every 24 hours in our experiments (except for the final two times in September when animals were checked every 48 hours). Temperatures during the July experiment were between 24 °C and 25 °C, and temperatures in September were 21 °C at the start and declined to 17 °C by the end of the study.

On each date, we determined if the daphnid was alive and the number of offspring released during the prior 24 hour period. If alive, the animal was transferred to a vial containing freshly prepared treatment water (offspring were not transferred). On the day an animal was found dead (or on the final day of the experiment) we determined the number of offspring (neonates) produced in the previous 24 hour period, the number of eggs held in the brood chamber, and the lipid content of the adult. We used a lipid index scale similar to that developed by Tessier & Goulden (1982) with zero representing no visible lipids and four indicating high lipid content. Lipid index data are presented only for the July experiment due to loss of data from the September experiment.

Differences in survivorship data among treatments were tested with Peto & Peto's log rank test (Pyke & Thompson, 1986). Wilcoxon's signed rank and the Kruskal-Wallis test were used to determine the significance of differences among the number of offspring produced in treatments. Statistical analyses were performed with the SYSTAT statistical software package (Wilkinson *et al.*, 1992).

Results

Daphnia and phytoplankton development during 1991

Daphnia pulicaria was the dominant daphnid in Lake Mendota during 1991. Following a decline in abundance of overwintering females in early spring, the population increased to a maximum of approximately 25 animals l^{-1} in early May (Fig. 2a). This abundance peak was preceded by a peak in egg production, and coincided with the spring clearwater phase (Fig. 3). A secondary peak in abundance occurred in mid-June, and was followed by a rapid population decline during the first two weeks of July (Fig. 2a). Egg-ratios declined almost one month prior to the population decrease. *Daphnia* abundances remained low during the summer and into the fall, with increases beginning in early October. Egg-ratios remained relatively low throughout the summer and fall, with only minor increases during the population increase in October (Fig. 2a).

The average length of *D. pulicaria* in 1991 fluctuated over the course of the seasons. Just prior to and during the rapid population increase in April animal size also increased (Fig. 2b). Following a decrease in length during late April, daphnid size remained fairly constant until early July when the average length decreased again. Animal length fluctuated around smaller sizes throughout the late summer and fall until an increase occurred in November when the population again increased in abundance (Fig. 2a). The ratios of adult to immature daphnid densities ranged between 0.35 and 0.75 during the midsummer decline, with no clear relationship to the changes in abundance.

The phytoplankton community underwent seasonal shifts in 1991 similar to those in 1988 and 1989. Following a spring increase in smaller edible algae (cryptophytes and diatoms), a spring clearwater phase occurred with drastic increases in water clarity during May (Fig. 3). After this clearwater phase, cyanobacteria species such as *Aphanizomenon*, *Microcystis* and the dinoflagellate *Ceratium* increased in importance and dominated the plankton through most of the summer. The shift to dominance by cyanobacteria occurred during the decline phase of the *Daphnia* population in early July (Fig. 2). Large bundles of *Aphanizomenon* and colonies of *Microcystis* were abundant during this period. In October, a bloom of diatom species occurred at the time of fall mixis in the lake.

Experiments during summer decline

There was a significant effect of food manipulation on the survivorship of individual *D. pulicaria* held at ambient temperature and photoperiod (Peto & Peto's log rank test: $X^2=12.76$; $P<0.005$). Animals fed a high concentration of *Chlamydomonas* exhibited the highest survivorship, with more than 60% of the animals still alive at the end of the five day experiment (Fig. 4a top). There was also a significant effect on number of offspring produced (Wilcoxon signed ranks test, $P<0.05$). Following a one or two day lag, the number of offspring produced by animals in the enriched treatment tripled (Fig. 4a bottom). Additional evidence of the good feeding conditions for these animals is seen in the high lipid content of animals in the enriched treatment (Table 1).

The survivorship of animals fed whole-lake water was not significantly different from that of animals fed the *Chlamydomonas* enriched water ($X^2=0.005$; $P>0.90$). Approximately 60% of the animals remained alive at the end of the experiment (Fig. 4a top). Offspring production increased over the course of the experiment for animals fed whole-lake water. Reproductive output for the whole-lake water fed animals was significantly lower than that observed for the

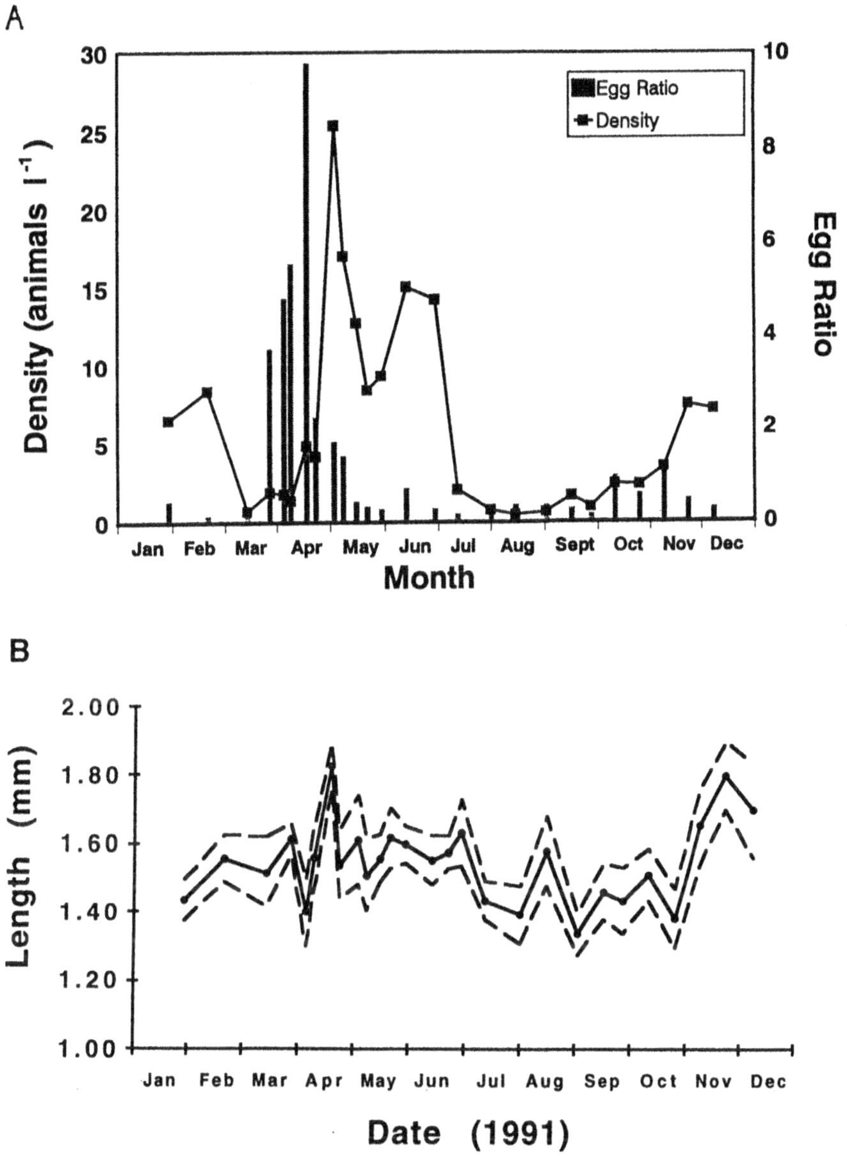

Fig. 2. Daphnia pulicaria population dynamics in Lake Mendota during 1991. (A) Mean abundances (line) and egg ratios calculated as eggs per adult female (bars graph). (B) Mean (solid line) and 1 s.e.m. interval (upper and lower dashed lines) of adult and immature lengths during 1991.

Chlamydomonas fed animals (Wilcoxon test: $P = 0.01$; Fig. 4a bottom) but followed the same general pattern as animals in the enriched treatment. Lipid content was somewhat lower than that of animals in the enriched food treatment (Table 1). Animals placed in GF/F filtered water died at a significantly faster rate than animals in either of the other two treatments (filtered vs. enriched: $X^2 = 8.81$, $P<0.005$; filtered vs. whole-lake water: $X^2 = 8.79$, $P<0.005$). Fewer than 30% of the animals survived the starvation conditions for five days (Fig. 4a top). Following an initial lag similar to that observed in the other treatments, animals held in filtered water failed to produce offspring after only three days (Fig. 4a bottom). These animals also exhibited a lower lipid index than animals in either of the other two treatments (Table 1).

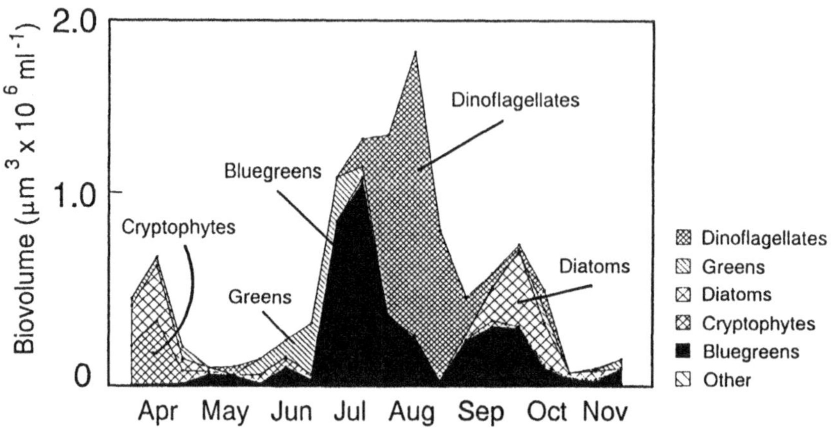

Fig. 3. Abundance and community composition of phytoplankton in Lake Mendota during 1991.

Table 1. Lipid index values for *Daphnia pulicaria* held in enclosures exposed to different food treatments in Lake Mendota, WI during July 1991. The lipid index can range from 0 (no lipids visible) to a maximum of 4. Lipid index values were determined at the end of the experiment. The number of surviving individuals is listed for each treatment.

Treatment	*n*	Mean lipid index	1 Standard error
Enriched GF/F	28	2.8	0.17
Whole-lake Water	22	1.7	0.37
GF/F Filtered	10	1.0	0.28

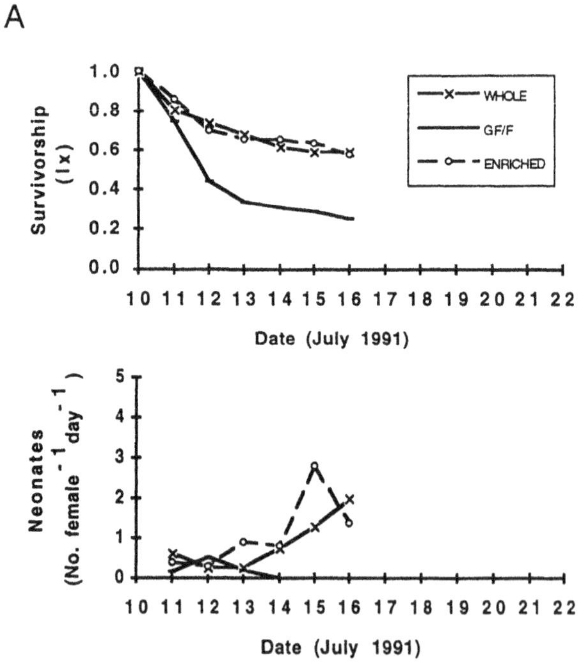

Fig. 4 A. Survivorship (top) and neonate production (bottom) of *Daphnia pulicaria* held in enclosures during (a) July experiments and (b) September experiments in Lake Mendota during 1991. Survivorship (l_x) is presented as the proportion of the original number in each treatment alive on each day. Treatments were GF/F filtered lake water (GF/F), GF/F filtered water enriched with *Chlamydomonas* (enriched), and lake water passed through a 153 μm mesh net (whole).

Experiments during fall period

There was no significant effect of the food manipulation on survivorship of *D. pulicaria* held at ambient conditions in mid-September 1991 ($X^2 = 3.59$; $P > 0.10$). Survival of animals in the *Chlamydomonas* fed treatment was again similar to that of animals in whole-lake water (Fig. 4b top). Animals held in filtered water exhibited a lower overall survivorship. However, this difference was not statistically significant.

Production of offspring was highest in the treatment with *Daphnia* fed *Chlamydomonas*, with a maximum of five neonates produced per adult daphnid per day (Fig. 4b bottom). Both the whole-lake water and the GF/F filtered water treatment animals produced relatively low, but non-zero numbers of offspring. Egg-ratios in the whole-lake water treatment did not differ significantly from those estimated from the plankton samples during this experiment (whole-lake water treatment mean = 0.9, plankton mean = 0.4, *t*-test: $P > 0.05$).

There was a seasonal affect on survivorship times in these experiments. Survival was better during the fall

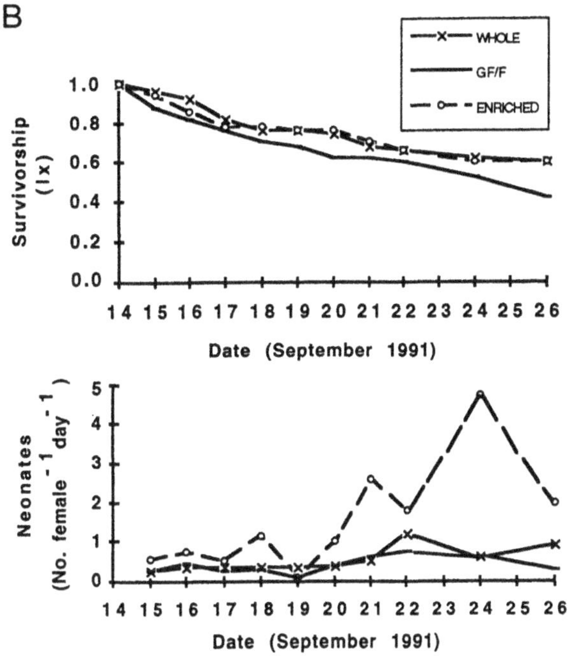

Fig. 4 B.

experiments (Fig. 4a & 4b – top). Comparisons of survivorship within a treatment among seasons indicate that mortality rates were significantly lower in September than in June (Peto & Peto's log rank tests: $X^2>9.0$; $P<0.005$ for all treatments). In the GF/F filtered water treatment, there was an approximate six-fold increase in starvation times, with 42% of the animals still alive after 12 days (Fig. 4b top).

Discussion

Population dynamics of Daphnia in Lake Mendota

Daphnia population dynamics in Lake Mendota have been shown to be determined by seasonal shifts in intensity of planktivory during years of high fish abundance (Luecke *et al.*, 1992; Vanni *et al.*, 1990; Rudstam *et al.*, 1993). In years with low fish abundance, previous studies indicate that planktivory by fish and invertebrate predators can not explain the midsummer decline of *D. pulicaria* (Luecke *et al.*, 1992; Rudstam *et al.*, 1993). These studies have proposed the hypothesis that the decline of *D. pulicaria* is caused by decreases in available food concentrations following the clearwater phase and perhaps increases of inedible cyanobacteria species during the summer (Vanni *et al.*, 1992, Rudstam *et al.*, 1993). This same explanation is employed in the PEG model of seasonal succession in lakes (Sommer *et al.*, 1986), and is frequently used to explain midsummer declines of daphnids (e.g. Gliwicz, 1985; Sommer, 1989).

Our in situ experimental analysis of this hypothesis for Lake Mendota indicates that individuals fed whole-lake water during the summer period when *D. pulicaria* is drastically declining survive and reproduce as well as animals fed an enriched diet, and do significantly better than starved animals. Daphnids held in enclosures and fed whole-lake water exhibited adequate survivorship and sufficient production of offspring to allow for the maintenance of population abundances in the absence of other mortality factors (Fig. 4a). This finding does not support the contention that low food abundance or poor food quality alone causes the midsummer decline of *D. pulicaria* in Lake Mendota.

During late summer and early fall, low food availability or poor food quality may explain the continued low daphnid abundances (Fig. 2a). *Daphnia* fed whole-lake water in September produced significantly fewer offspring than animals fed *Chlamydomonas* (Fig. 4b). This indicates that food resources in the lake at that time may not have been of sufficient quantity or quality to allow for population growth (Fig. 4b bottom). This conclusion is also supported by the fact that *D. pulicaria* can produce offspring at ambient lake temperatures when well fed, suggesting that low autumn water temperatures probably do not prevent population development during that time period. This explanation also is consistent with the observed increase in *D. pulicaria* abundance following the shift in phytoplankton composition in October from a cyanobacteria and *Ceratium* dominated community to one dominated by diatoms (Fig. 3).

A number of concerns can be raised concerning these types of enclosure experiments. First, overall survivorship in our enclosures was lower than normally expected in these types of experiments. This may have been due to soluble toxins produced by cyanobacteria in the lake, such as *Microcystis*, which are known to produce substances that can affect daphnid survivorship (e.g. DeMott *et al.*, 1991). Our filtering methods would not remove these types of substances from any of the treatments. In addition, although we attempted to select animals of similar sizes for starting the experiments, we did not know the feeding or reproductive history of the animals actually employed. Also, because there is a correlation between female size and number of neonates per brood, our estimate of the effect of food manipulation on offspring production in enclosures (Fig. 4 bottom) may be biased by growth during the experiment. However, even restricting our analysis to the early part of each experiment (when growth differences among treatments should be minimal) clearly indicates that starved animals cease producing eggs, while those fed either whole-lake water or enriched water continued to produce eggs (Fig. 4). Nevertheless, even considering these potential variables, comparisons against our food enriched and starvation (GF/F filtered) treatments indicate that daphnids fed whole-lake water were surviving, storing lipids, and producing offspring at rates higher than expected during a population decline due to poor feeding conditions.

Midsummer Daphnia population declines

Midsummer declines of daphnids have been studied extensively in other lakes, but no single cause has been found to explain this phenomenon in general. Variation in mortality due to predation by fish and invertebrates has been shown to be important in many studies (e.g. Hall, 1971; Sprules, 1972; DeMott, 1983). In Lake Mendota, however, predation pressure from age 1 and older fish was extremely low in 1991 because cisco did not recover after the 1987 die-off. Also, only weak year classes of other planktivores were produced from 1988 to 1990, possibly as a result of the on-going biomanipulation of piscivore populations (Kitchell, 1992; Luecke *et al.*, 1992; Rudstam *et al.*, 1993).

As a result of the whole-lake biomanipulation in Lake Mendota, there are very extensive data on the role of planktivores in controlling the population of *D. pulicaria*. Estimates of larval fish abundances at nine stations collected on a weekly basis with a purse seine indicate that both yellow perch and white bass had above average year classes in 1991 (T. Johnson, pers. com.). However, predation by these and other young-of-the-year fish is low during June and July. Larval fish are either too small to feed on *Daphnia* (e.g. black crappie, white bass, sunfishes), or have already grown and moved inshore to the littoral zone (e.g. yellow perch; Post *et al.*, 1992; Schael *et al.*, 1991). Previous studies of the movement of yellow perch have shown that these smaller individuals spend both day and night in the nearshore regions (McCarty, 1990). This is supported by the very low catch of fish typically caught in early July in vertical gillnets fished at the deep hole station (Luecke *et al.*, 1992; T. Johnson, pers. com.). Predation by white bass young-of-the-year could have been significant in September (J. Gala and T. Johnson, in prep.), although white bass biomass was an order of magnitude lower than cisco biomass prior to the 1987 die off (L. Rudstam, unpublished data) and estimates of consumption based on bioenergetics analyses indicate that peak predation by white bass occurred in early June, one month prior to the daphnid population decline in early July (T. Johnson, pers com.).

Although abundances of the invertebrate predator *Leptodora kindtii* typically increase concurrent with the decrease of *D. pulicaria* in early July, previous studies in Lake Mendota have found that estimates of consumption by *L. kindtii*, *Mesocyclops edax*, and fish on *D. pulicaria* are usually lower than the calculated death rates of daphnids during midsummer (Luecke *et al.*, 1992). However, it is difficult to obtain good estimates of *L. kindtii* density because these animals are not captured well using traditional zooplankton collection methods (Lunte & Luecke, 1990; Lathrop & Carpenter, 1992). Other estimates from the literature on feeding rates by *L. kindtii* are 2–5 times greater than those obtained in Lake Mendota, so it is possible that invertebrate predation may be more important than we have thought (e.g. Browman *et al.*, 1989). Observed increases in average daphnid size during the population increase in spring, and the rather large size of animals during the summer when population density was low are consistent with the idea that invertebrate predation on smaller animals may have been occurring (Fig. 2b). In addition, the potential for error in the calculation of death rates using measures of population growth (r) and egg ratios (b) is quite high unless rather restrictive conditions apply in the field population (Taylor, 1988). It would clearly be helpful to have more complete information on population dynamics and feeding rates by invertebrate planktivores such as *L. kindtii*.

A third hypothesis to explain the midsummer decline is that *D. pulicaria* shift to production of resting eggs (ephippia) during June and July each year. Although we do not have quantitative estimates on the number of ephippia produced, we have observed an increase in male *D. pulicaria* and ephippial egg production during June and early July in years of both high (1987) and low planktivore abundances (1988, 1991). This may represent a demographic response to the poor conditions that typically occur in Lake Mendota in late summer (either high planktivory, poor feeding conditions or a combination of both). If the deterioration of environmental conditions for daphnids in late summer is rather predictable (as appears to be the case), a strategy of shifting to ephippia production would be advantageous. There is a need for further investigation of cues for onset of ephippia production and the role of dormancy in daphnid midsummer dynamics.

Any hypothesis to explain the summer decline of *D. pulicaria* (as observed in most years of low planktivory following the clear-water phase; Vanni *et al.* 1990; Lathrop & Carpenter, 1992; Luecke *et al.*, 1992) must account for the observed changes in egg ratios during the period of decrease, in addition to the actual decrease in daphnid abundance. Food limitation is often the explanation given for the observed change in egg ratios, however predation by visual planktivores on egg-bearing females can also lead to decreases in egg ratios (Threlkeld, 1979a, 1985). There is a slight decrease in size of daphnids during the summer of 1991 which may indicate increased size-selective planktivory (Fig. 2b), but this is based on measurements of only 16 individuals. A number of other lines of evidence do not support the explanation that predation by fish caused the midsummer decline in 1991. As mentioned above, estimates of fish abundances indicate that predation intensity was no greater in 1991 than in the previous years in which predation was too low to explain observed mortality rates (i.e. 1988, 1989; Luecke *et al.*, 1992). Also, we have observed similar midsummer declines in all years of low planktivory regardless of the timing of larval fish populations. Therefore, temporal variation in mortality due to predation by juvenile and larval fish is not a likely explanation. Predation by *L. kindtii* may explain the midsummer dynamics of *D. pulicaria* if our estimates of abundance or feeding rate are extremely low. However, size-selective predation by both larval fish and invertebrate predators would remove smaller individuals, perhaps causing a shift to larger size daphnids.

In summary, it appears most likely that an explanation that includes the combined effects of deteriorating food conditions, low levels of planktivory by both fish and invertebrate predators, and life history responses to the more or less predictably poor conditions in late summer would fit best with the extensive data available from Lake Mendota. Manipulative experiments to test specific hypotheses, like those presented here, are a necessary part of our investigations of field populations, but multiple factors almost certainly determine the dynamics of natural systems. The challenge now is to design new methods of experimentally testing multiple hypotheses concerning the ecology and evolution of these relatively simple, yet dynamic, ecosystems.

Acknowledgments

This research was supported by the Sigma Xi Grants-in-Aid of Research Program, the American Museum of Natural History (Theodore Roosevelt Memorial Fund), the Federal Aid in Sport Fish Restoration Act under Project F-95-P, the Wisconsin Department of Natural Resources, and the Long-Term Ecological Research Project-Northern Temperate Lakes (NSF grant BSR 9011660). Additional funds were supplied by Lawrence University. We thank R. Gardetto, and C. Ramcharan for technical assistance, S. Dodson for the loan of algal cultures, and the Lake Mendota Biomanipulation Project crew for general aid and discussions. Helpful criticisms were kindly offered by E. De Stasio, T. Frost, P. Larsson, P. Olsen and two anonymous reviewers. This paper is contribution No. 169 of the Cornell Biological Field Station.

References

Benndorf, J. & W. Horn, 1985. Theoretical considerations on the relative importance of food limitation and predation in structuring zooplankton communities. Arch. Hydrobiol., Ergeb. Limnol. 21: 383–396.

Browman, H. I., S. Kruse & W. J. O'Brien, 1989. Foraging behavior of the predaceous cladoceran, *Leptodora kindtii*, and escape responses of their prey. J. Plankton Res. 11: 1075–1088.

Carpenter, S. R. & J. F. Kitchell, 1988. Consumer control of lake productivity. Bioscience 38: 764–769.

Cryer, M., G. Peirson & C. R. Townsend, 1986. Reciprocal interactions between roach, *Rutilus rutilus*, and zooplankton in a small lake: prey dynamics and fish growth and recruitment. Limnol. Oceanogr. 31: 1022–1038.

DeMott, W. R., 1983. Seasonal succession in a natural *Daphnia* assemblage. Ecol. Monogr. 53: 321–340.

DeMott, W. R., Q.-X. Zhang & W. W. Carmichael, 1991. Effects of toxic cyanobacteria and purified toxins on the survival and feeding of a copepod and three species of *Daphnia*. Limnol. Oceanogr. 36: 1346–1357.

Gliwicz, Z. M., 1985. Predation or food limitation: an ultimate reason for extinction of planktonic cladoceran species. Arch. Hydrobiol., Ergeb. Limnol. 21: 419–430.

Hall, D. J., 1964. An experimental approach to the dynamics of a natural population of *Daphnia galeata mendotae*. Ecology 45: 94–110.

Hall, D. J., 1971. Predator-prey relationships between yellow perch and *Daphnia* in a large temperate lake. Trans. am. Micros. Soc. 90: 106–107.

Kitchell, J. F. (ed.), 1992. Food web management: A case study of Lake Mendota. Springer-Verlag, Berlin, 553 pp.

Lammens, E. H. R. R., R. D. Gulati, M.-L. Meijer & E. van Donk, 1990. The first biomanipulation conference: a synthesis. Hydrobiologia 200/201 (Dev. Hydrobiol. 61): 619–627.

Larsson, P., S. Andersen, Y. Borsheim, P. Jakobsen & G. Johnsen, 1985. Individual growth of *Daphnia longispina* in the summer decline phase of the population. Arch. Hydrobiol., Ergeb. Limnol. 21: 341–350.

Lathrop, R. C. & S. R. Carpenter, 1992. Zooplankton and their relationship to phytoplankton. In J. F. Kitchell (ed.), Food web management: A case study of Lake Mendota. Springer-Verlag, Berlin: 127–150.

Leibold, M., 1989. Resource edibility and the effects of predators and productivity on the outcome of trophic interactions. Am. Nat. 134: 922–949.

Luecke, C., M. J. Vanni, J. J. Magnuson, J. F. Kitchell & P. T. Jacobson, 1990. Seasonal regulation of *Daphnia* populations by planktivorous fish: implications for the spring clearwater phase. Limnol. Oceanogr. 35: 1718–1733.

Luecke, C., L. G. Rudstam & Y. Allen, 1992. Interannual patterns of planktivory 1987–89: An analysis of vertebrate and invertebrate planktivores. In J. F. Kitchell (ed.), Food web management: A case study of Lake Mendota. Springer-Verlag, Berlin: 275–301.

Lunte, C. C. & C. Luecke, 1990. Trophic interactions of *Leptodora* in Lake Mendota. Limnol. Oceanogr. 35: 1091–1100.

McCarty, J. P., 1990. Diel periodicity of movement and feeding of yellow perch (*Perca flavescens*) in Lake Mendota, Wisconsin. Trans. Wis. Acad. Sci. Arts Lett. 76: 65–76.

McQueen, D. J., J. R. Post & E. L. Mills, 1986. Trophic relationships in freshwater pelagic ecosystems. Can. J. Fish. aquat. Sci. 43: 1571–1581.

McQueen, D. J., M. R. S. Johannes, J. R. Post, T. J. Stewart & D. R. S. Lean, 1989. Bottom-up and top-down impacts on freshwater pelagic community structure. Ecol. Monogr. 59: 289–309.

Peters, R. H. & R. de Bernardi (eds), 1987. 'Daphnia'. Mem. Ist. ital. Idrobiol. 45.

Post, J. R., L. G. Rudstam, D. M. Schael & C. Luecke, 1992. Pelagic planktivory by larval fishes in Lake Mendota. In J. F. Kitchell (ed.), Food web management: A case study of Lake Mendota. Springer-Verlag, Berlin: 303–317.

Pyke, D. A. & J. N. Thompson, 1986. Statistical analysis of survival and removal rate experiments. Ecology 67: 240–245.

Rudstam, L. G. & B. M. Johnson, 1992. Development, evaluation, and transfer of new technology. In J. F. Kitchell (ed.), Food web management: A case study of Lake Mendota. Springer–Verlag, Berlin: 507–524.

Rudstam, L. G., R. C. Lathrop & S. R. Carpenter, 1993. The rise and fall of a dominant planktivore: Direct and indirect effects on zooplankton. Ecology 74: 303–319.

Schael, D. M., L. G. Rudstam & J. R. Post, 1991. Gape limitation and prey selection in larval yellow perch (*Perca flavescens*), freshwater drum (*Aplodinotus grunniens*) and black crappie (*Pomoxis nigromaculatus*). Can. J. Fish. aquat. Sci. 48: 1919–1925.

Sommer, U., Z. M. Gliwicz, W. Lampert & A. Duncan, 1986. The PEG model of seasonal succession of planktonic events in fresh waters. Arch. Hydrobiol. 106: 433–471.

Sommer, U. (ed.), 1989. Plankton ecology – succession in plankton communities. Springer-Verlag, Berlin.

Sprules, W. G., 1972. Effects of size-selective predation and food competition on high altitude zooplankton communities. Ecology 53: 375–386.

Taylor, B. E., 1988. Analyzing population dynamics of zooplankton. Limnol. Oceanogr. 33: 1266–1273.

Tessier, A. J. & C. E. Goulden, 1982. Estimating food limitation in cladoceran populations. Limnol. Oceanogr. 27: 707–717.

Threlkeld, S. T., 1976. Starvation and the size structure of zooplankton communities. Freshwat. Biol. 6: 489–496.

Threlkeld, S. T., 1979a. Estimating cladoceran birth rates: the importance of egg mortality and the egg age distribution. Limnol. Oceanogr. 24: 601–612.

Threlkeld, S. T., 1979b. The midsummer dynamics of two *Daphnia* species in Wintergreen Lake, Michigan. Ecology 60: 165–179.

Threlkeld, S. T., 1985. Resource variation and the initiation of midsummer declines of cladoceran populations. Arch. Hydrobiol., Ergeb. Limnol. 21: 333–340.

Vanni, M. J., C. Luecke, J. F. Kitchell, Y. Allen, J. Temte & J. J. Magnuson, 1990. Effects on lower trophic levels of massive fish mortality. Nature 344: 333–335.

Vanni, M. J., J. Temte, Y. Allen, R. Dodds, P. J. Howard, P. R. Leavitt & C. Luecke, 1992. Herbivory, nutrients, and phytoplankton dynamics in Lake Mendota, 1987–1989. In J. F. Kitchell (ed.), Food web management: A case study of Lake Mendota. Springer-Verlag, Berlin: 243–273.

Wilkinson, L., M. Hill & E. Vang, 1992. SYSTAT: Statistics, Version 5.2 edition. SYSTAT, Inc., Evanston, IL, 822 pp.

Hydrobiologia **307**: 231–238, 1995.
P. Larsson & L.J. Weider (eds), Cladocera as Model Organisms in Biology.

Interrelationships between crustacean zooplankton and phytoplankton:

Results from 15 years of field observations at the mesotrophic Saidenbach Reservoir (Germany)

Wolfgang Horn & Heidemarie Horn
Sächsische Akademie der Wissenschaften zu Leipzig, Arbeitsgruppe 'Limnologie von Talsperren', Neunzehnhainer Str. 14, D-09514 Wünschendorf, Germany

Key words: phytoplankton, crustacean plankton, long-term observation, interrelationships, Secchi depth

Abstract

Fifteen years of data (1975–1990) on the phyto- and crustacean plankton in the meso/eutrophic Saidenbach Reservoir were analysed to reveal correlations between these groups or members of them. The weekly or fortnightly samples were collected from different depths and times, and were integrated to form seasonal averages. For the dominant organisms, summer means of abundance were plotted against one another. Among the Crustacea, *Daphnia galeata* exerts a strong influence on nanoplankton resulting in an increased Secchi depth in years with a high standing stock of Daphnia. No such correlations were found for Eudiaptomus and the Cyclopoida, which are not able to remove such small particles. The abundance of Daphnia showed no obvious effects on Cyanophyta, colony forming Chlorophyta and *Asterionella formosa*. This implies that in the Saidenbach Reservoir, factors other than crustacean grazing are of crucial importance for the population dynamics of the latter algal groups. It is shown, that the influence of the Crustacea on the phytoplankton is evident not only for short periods, i.e. in clear-water stages, or in biomanipulated lakes, but also can be observed under 'non-manipulated' conditions for longer time periods.

Introduction

Biomass and species composition in any body of water depend on a number of factors. Most important are the resources, e.g. nutrients, light, on the growth side and the losses, e.g. grazing, predation, on the other side. All these factors exist together in any ecosystem, and are commonly referred to as bottom-up and top-down effects, respectively. In laboratory experiments, e.g. for the determination of individual filtration and ingestion rates, in situ investigations, e.g. calculation of individual or population performance as community grazing rate, enclosures and experiments with parts or whole ecosystems, e.g. experimental lakes with different or no fish stock, the influence of the zooplankton on the phytoplankton and the cycling of minerals and energy have been studied.

Numerous investigations on the interactions between phyto- and zooplankton have been performed. Mostly, they have been based on short-term experiments in enclosures, laboratory or in situ experiments. Long-term observations of these relationships are scarce, particularly in waters where, in contrast to biomanipulated lakes, the fish stock is subjected to little direct anthropogenic influence.

Description of study site

Investigations were carried out in the drinking-water Saidenbach Reservoir (Erzgebirge, Sachsen, Germany). At its maximum water-level, it holds a volume of 22.4×10^6 m^3, the surface area is approximately 146 ha and the maximum depth is 45 m. The mean annual air temperature is about 7°C. The water is soft, relatively nutrient rich and characterized by meso/eutrophic conditions with increasing nutrient concentrations in the tributaries and in the water of the main basin. There was a doubling of soluble reactive phosphorus (SRP) during the study period. During the growing-season, phytoplankton development is mostly P-limited. Winter concentrations of SRP range from 10 to 15 $\mu g\ l^{-1}$ P, while epilimnetic summer concentrations range from 0 to 4 $\mu g\ l^{-1}$. Occassionally, Si-limitation also occurs for short periods (see Horn & Horn, 1990); nitrate concentrations are permanently above 25 mg l^{-1} nitrate, thus indicating no N-limitation.

Methods

Phytoplankton samples were taken (during the time of ice formation and break-up longer sampling intervals were needed) from about 10–14 different depths every week or every fortnight throughout the year, with a Ruttner-sampler, while crustacean plankton were collected with a closing net (mesh size 175 μm) from five different depth ranges, extending throughout the whole water column. The values for the different depths under investigation were depth and time integrated to form seasonal averages, so that every presented summer value is the result of nearly 100 zooplankton and 300 phytoplankton samples. The summer average includes all data from the beginning of the stagnation period (when water temperatures finally rise above 4°C) to its end, if vertical temperature differences no longer exist within the water column. For the present study, the summer average for the whole water column was used and not that of the epilimnion, because the vertical distributions of the Crustacea sometimes show higher concentrations in waters below the very shallow epilimnion in early summer. The biomass or biovolume of the zooplankton species was calculated from length-weight ratios on the basis of length measurements taken on each sampling date (see Bottrell *et al.*, 1976, and our own formulas). In the case of phytoplankton, self-ascertained standard volumes were used. The present study considers all years from 1975 to 1990 except for 1982. This year was excluded, because only monthly data from zooplankton were available.

For the well-known and expected causal connections among planktonic organisms (e.g. in food chains), and for the most important species/groups, the summer means of abundance were plotted against one another. The results were only examined for linear correlations and their significance (Grimm & Recknagel, 1985). No further statistical analyses were carried out, because the goal of this paper was to detect and to describe the prevailing processes and relationships between phyto- and zooplankton over entire growing seasons. In order to avoid the danger of deriving spurious connections from correlations, only those relationships for which there exists a large body of evidence from experimental and field observations were investigated statistically. To avoid the impression that between the tested variables, an ultimately valid quantitative relationship exists, the calculated linear regression line was not included in the figures. The essential statistical parameters for each analysis of the non-log transformed data are presented (Table 1 or in the Figures).

Results

In Fig. 1 the annual averages of the biovolume of the dominant phyto- and crustacean plankton groups over the time period 1975–1990 are presented. There are no obvious trends concerning the total biovolume of both, but a shift in the crustacean community after 1986 may be seen: the cladocerans – dominated by *Daphnia galeata* – decreased. For the phytoplankton, the increasing dominance of blue-greens in the late 1980's is most noteworthy.

Figure 2 shows the relationships between nanoplankton, that means all species smaller than 20 μm in length, and the total crustacean biovolume, and specific constituent groups. No diminution in grazing influence on the nanoplankton can be deduced from the correlation to the pooled biovolume of the whole crustacean community (see Table 1 also). In the following diagrams, the main compartments of the Crustacea were tested. *Daphnia galeata* seems to exert a strong influence on the nanoplankton. Low nanoplankton biovolumes were found in years with high densities of Daphnia. When looking at Eudiaptomus and the Cyclopoida, quite the reverse relationship to the nanoplankton becomes apparent, as the biovolumes of both crustacean groups are positively correlated with nanoplankton biovolume.

Two species among the nanoplankton were dominant in the Saidenbach Reservoir: *Rhodomonas pusilla* and *Chrysochromulina parva*. These organisms show a clear relationship to *Daphnia galeata* (see Fig. 3) as shown previously (Fig. 2) for the whole group of nanoplankton, which again indicates the influence of the filter-feeding Daphnia on the small phytoplankton species. The relationships between these two species and the copepods were absent or showed – as in the case of the nanoplankton as whole group – the reverse tendency, i.e. a positive correlation.

However, there were no obvious interrelationships at all between the two tested net plankton groups,

Table 1. Statistical parameters from correlation analyses. Due to multiple testing the critical probability value is p = 0.0026 (Rice, 1989). The number of data pairs in the correlations are mostly 15, but 1) indicates that 1977 has been omitted, and 2) that data only for the last 10 years are considered. + and − indicate positive or negative slopes.

Variable 1	Variable 2	r^2	p	Figure
Crustacea	Nanoplankton	0.34	0.0225	2
Daphnia galeata	Nanoplankton	0.54-	0.0018	2
Eudiaptomus gracilis	Nanoplankton	0.59+	0.0008	2
Cyclopoida	Nanoplankton	0.65+	0.0003	2
Crustacea	*Rhodomonas pusilla*	0.10	0.2509	
Daphnia galeata	*Rhodomonas pusilla*	0.60-	0.0007	3
Eudiaptomus gracilis	*Rhodomonas pusilla*	0.21	0.0858	
Cyclopoida	*Rhodomonas pusilla*	0.58+	0.0010	
Crustacea	*Chrysochromulina parva* 2)	0.07	0.4569	
Daphnia galeata	*Chrysochromulina parva* 2)	0.70-	0.0025	3
Eudiaptomus gracilis	*Chrysochromulina parva* 2)	0.33	0.0824	
Cyclopoida	*Chrysochromulina parva* 2)	0.20	0.1950	
Daphnia galeata	Green algae (colonies)	0.19	0.1043	4
Daphnia galeata	Blue-greens	0.15	0.1538	4
Crustacea	*Asterionella formosa*	0.15	0.1538	
Daphnia galeata	*Asterionella formosa*	0.10	0.2509	4
Eudiaptomus gracilis	*Asterionella formosa*	0.00	1.0000	
Cyclopoida	*Asterionella formosa*	0.01	0.7229	
Crustacea	Secchi depth 1)	0.03	0.5538	5
Daphnia galeata	Secchi depth 1)	0.77+	0.0000	5
Eudiaptomus gracilis	Secchi depth 1)	0.11	0.2466	5
Cyclopoida	Secchi depth 1)	0.45	0.0086	5
Cyclopoida	*Daphnia galeata*	0.40	0.0114	6
Cyclopoida	*Eudiaptomus gracilis*	0.14	0.1694	
Daphnia galeata	*Eudiaptomus gracilis*	0.10	0.2509	

or *Asterionella formosa* with any zooplankton group (Fig. 4, Table 1). Neither in the case of colony forming green algae, mainly *Coenochloris polycocca, Pandorina morum*, nor in the group of blue-greens, which were dominated by *Aphanothece clathrata, Gomphosphaeria lacustris, Aphanizomenon flos-aquae*, was any interrelationship apparent.

The high filter-feeding performance of *Daphnia galeata* results in an enhanced Secchi depth in years when this species has high standing stocks (Fig. 5). The asterisk marks a year with unusually high flushing rates and therefore especially high mineral turbidity. No such correlations however can be presented for Eudiaptomus and the Cyclopoida. These organisms are not able to remove the small algae, which affect the Secchi depth to a great extent.

Discussion

It is widely known and accepted, that nanoplankton is a good ingestable food source and, therefore, is intensively grazed upon by filter-feeding crustacean plankton. In particular, cladocerans like Daphnia can use it and as a consequence, suppress nanoplankton standing stocks to a great extent in situations where there is a high stock of Daphnia. The observed strong correlation between the two groups supports the hypoth-

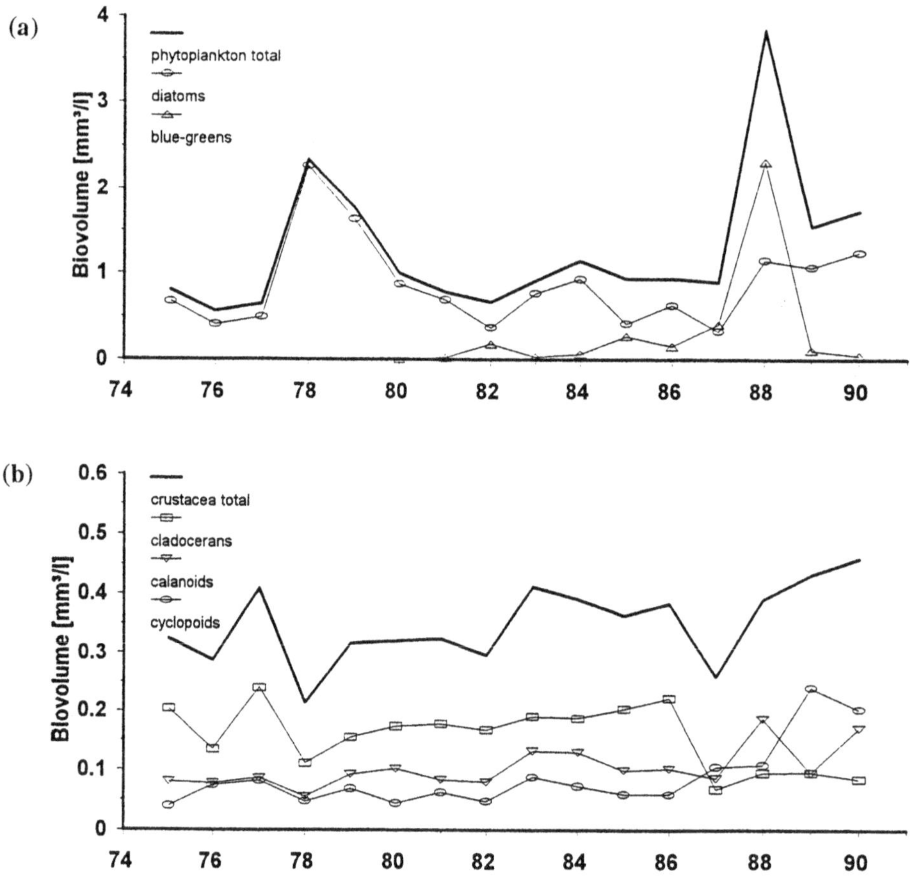

Fig. 1. The annual average of the biovolumes of the dominant groups of (a) phyto- and (b) crustacean plankton during the study period (naupliar stages were not considered in the averages).

esis that *Daphnia galeata* suppresses the nanoplankton as a whole (and certain nanoplanktonic species like *Rhodomonas pusilla* and *Chrysochromulina parva*) by grazing. Furthermore, our results support the notion that the copepods are not primarily 'micro-filter feeders', and they are not able to graze down particles as small as those filtered by Daphnia. The occassionally significant positive relation between Eudiaptomus or Cyclopoida and nanoplankton, which was caused by the simultaneously high values of these non-cladocerans and nanoplankton, can be attributed to the low density of the Cladocera in these years. This suggests the existence of an indirect effect of the Cyclopoida on the nanoplankton.

From the other groups of algae, it is also known that their population dynamics can be affected directly, e.g. by grazing or non-grazing, or indirectly, for instance by elimination of competitors, and changes in nutrient conditions by the zooplankton. The colony forming green algae can take advantage of the zooplankton suppressing other competing algae and of nutrient uptake via gut passage; algae with gelatinous sheaths are protected from a herbivore's digestive enzymes and are therefore hardly digestable, but they are permeable to certain nutrient ions (Porter, 1976). This proposed scenario is not likely to persist during the whole growing season, but to exist only for some weeks, more or less exclusively during the time of crustacean population development. Also blue-greens can profit from the Crustacea, which are suppressing their competitors. Certain blue-green species may also inhibit the zooplankton mechanically or by toxins. None of these proposed effects however were observed in Fig. 4 or could be derived from it. In the Saidenbach Reservoir, further effects on blue-green algae development should not be disregarded, e.g. the increasing nutrient conditions, Si-depletion or weather conditions influencing thermal stability (Horn & Horn, 1990). This also

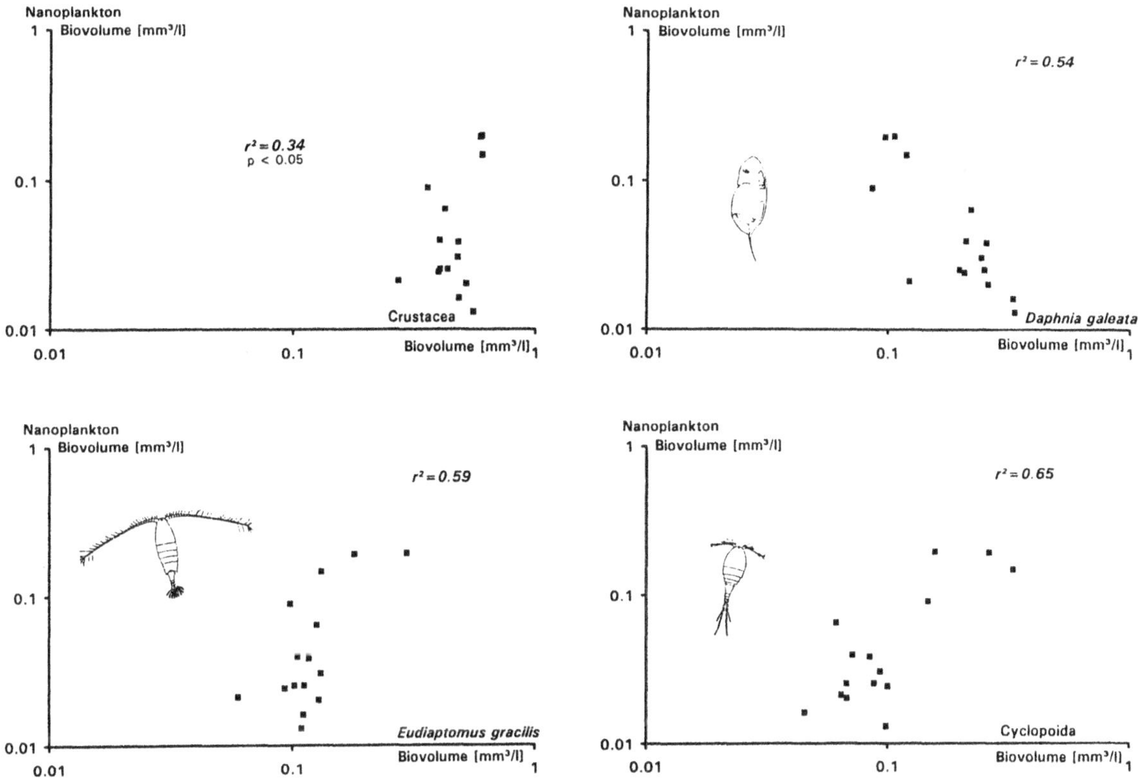

Fig. 2. The interrelationships between the summer means of the nanoplankton and the planktonic Crustacea (total), *Daphnia galeata*, *Eudiaptomus gracilis* and the Cyclopoida.

holds for the diatom *Asterionella formosa*, which was examined for possible interrelationships too. Although from our own (Horn, 1981) and other investigations (Infante, 1973; Geller, 1975), it is known that this alga can be ingested by Daphnia and in spring, the gut of the dominant Cyclops is obviously to a great extent filled with diatoms, no apparent correlation between this alga and the entire crustacean population, Daphnia and the Cyclopoida respectively, were evident (Fig. 4, Table 1). Our analysis reveals that the dynamics of this diatom species is essentially influenced by other factors. In particular, stratification conditions and sedimentary losses are very important in influencing the population development of *Asterionella formosa*. A further important but not evident fact for the diatoms is the elimination of nanoplanktonic competitors by Daphnia. This can improve the nutrient conditions for the diatoms by reducing P-competition and thus stimulates their growth. Because the diatoms are important for phosphorus export to the sediment and therefore for self-purification of the water column (Grim, 1967), this possible alga compositional shift (as a result of Daphnia grazing) can be a desired effect.

The filter-feeding activity of the *Daphnia galeata* population exerts a great influence on the Secchi transparency values, because these filter-feeders are able to eliminate the nanoplankton to a great extent. Small particles (i.e. nanoplankton) affect the Secchi depth values to a greater extent than comparable concentrations of larger particles. These two effects result in enhanced Secchi depths in years when Daphnia occur at high densities. Because Eudiaptomus and the Cyclopoida are apparently not able to remove such small particles, no 'enhancement effect' on the Secchi depth was observed, when copepods were abundant.

Reasons for the different values of Daphnia biovolume during the study period are assumed to be caused by changes in the fish population. Sport fishermen have reported a high standing stock of zooplanktivorous fishes, mostly larvae, fingerling, small fishes – e.g. stickleback, perch, roach, for the last few years. Unfortunately, no exact data are available as far as this ecosystem compartment is concerned. But it seems that

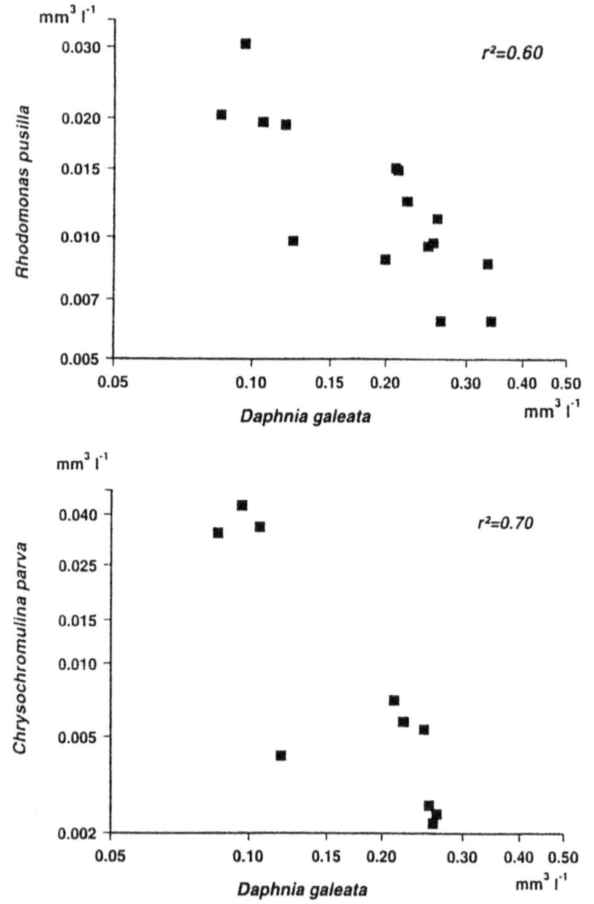

Fig. 3. The summer means of *Daphnia galeata vs* a) *Rhodomonas pusilla*, and b) *Chrysochromulina parva.*

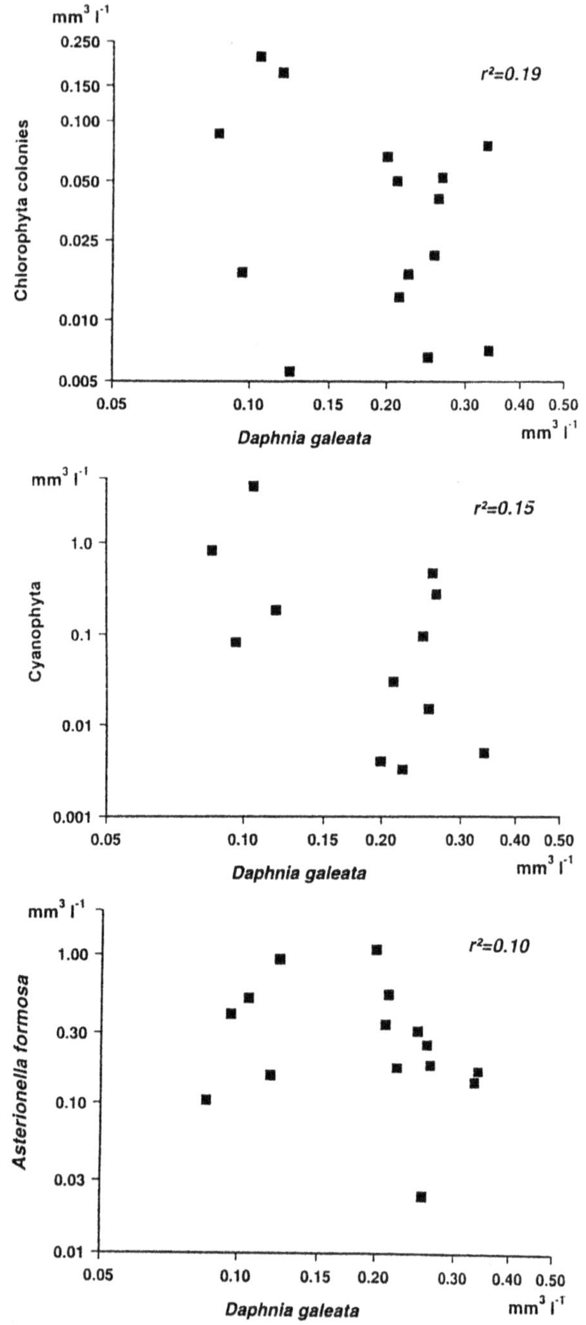

Fig. 4. The interrelationships between the summer means of *Daphnia galeata vs* colony forming a) Chlorophyta, b) Cyanophyta and c) *Asterionella formosa*.

under 'natural conditions', i.e. without anthropogenic heavy piscivorous fish stocking as in biomanipulation experiments, large fluctuations in biovolume and composition of the crustacean plankton may be normal; especially in reservoirs being subjected to large water level fluctuations with its effects on fish reproduction. Among invertebrate predators, only the cyclopoids may have an influence on Daphnia. *Leptodora kindtii* is also present, but at low abundance. However, we did not find a significant correlation between *Daphnia galeata* and cyclopoid abundances (Table 1, Fig. 6).

Conclusions

The influence of the crustacean plankton on the phytoplankton and vice versa can be observed not only in short-term experiments with extremely high standing stocks of Daphnia or in biomanipulated lakes. The well known and proven effects for short periods, e.g. clear water stages in ponds and lakes, also exist under 'non-manipulated' conditions for longer time periods. They can be confirmed using the summer means of long-term observations.

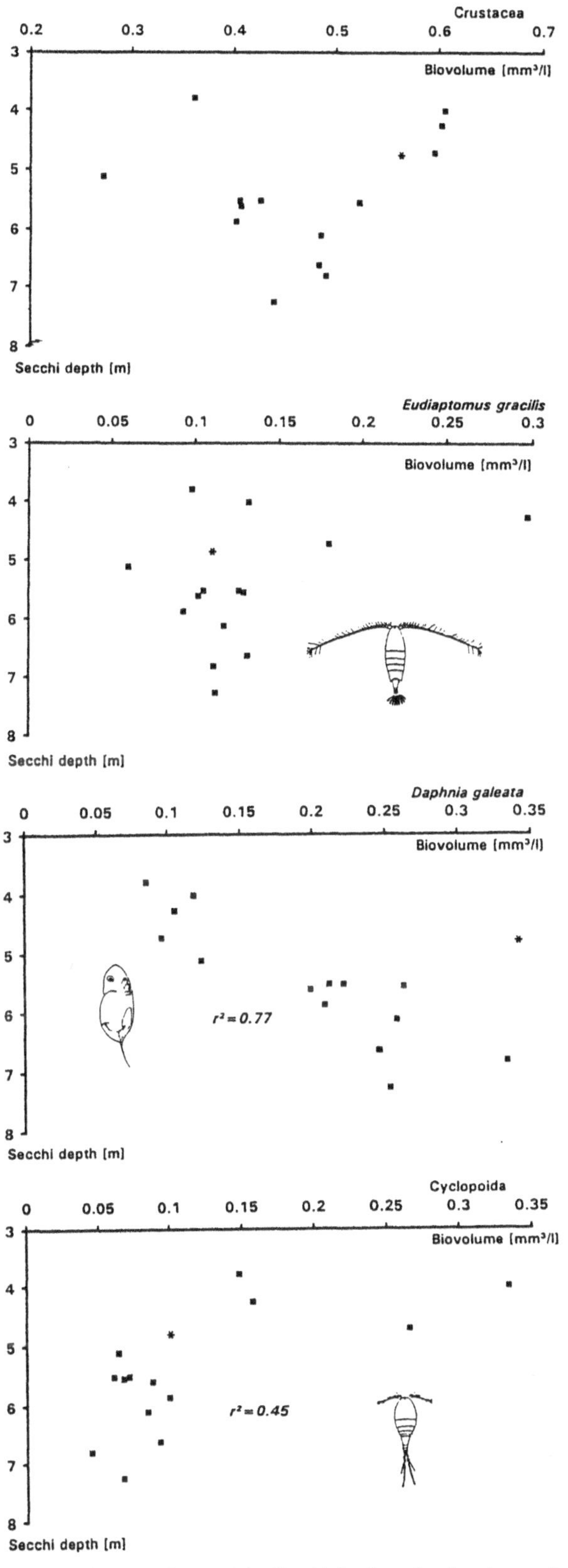

Fig. 5. The dependence of the Secchi depth on the abundance of the planktonic Crustacea (total, without naupliar stages), *Daphnia galeata, Eudiaptomus gracilis* and the Cyclopoida (summer means; the asterisk marks a year with high flushing rates and therefore high mineral turbidity; this year is not included in the analysis).

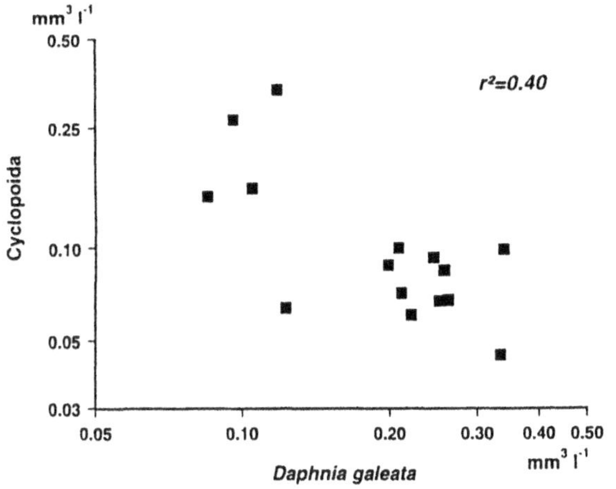

Fig. 6. The interrelationship between *Daphnia galeata* and the Cyclopoida (summer means of biovolume).

The influence of the crustacean plankton on the phytoplankton in the Saidenbach Reservoir can be demonstrated for selected groups only. *Daphnia galeata* appears to be the most important grazer. This species exerts obviously a strong influence on *Rhodomonas pusilla, Chrysochromulina parva* and the nanoplankton in general, and results in enhanced Secchi depths in years with a high standing stock of Daphnia. No such correlations were detected for Eudiaptomus and the Cyclopoida, which are apparently not able to graze small algae.

On the other hand, no direct influence of any Crustacea on the net phytoplankton could be derived with this method. Obviously, other factors, e.g. sedimentation, competition, seem to be more essential for the population dynamics of *Asterionella formosa*, colony forming green algae and blue-greens.

Acknowledgement

We thank the staff of the Hydrobiological Laboratory, Neunzehnhain for assistance in collecting samples, P. Larsson and two anonymous reviewers for their valuable comments and their linguistic corrections and J. Arnscheidt for improving the English text.

References

Bottrell, H. H., A. Duncan, Z. M. Gliwicz, E. Grygierek, A. Herzig, A. Hillbricht-Ilkowska, H. Kurasawa, P. Larsson & T. Weglenska, 1976. A review of some problems in zooplankton production studies. Norw. J. Zool. 24: 419–456.

Geller, W., 1975. Die Nahrungsaufnahme von *Daphnia pulex* in Abhängigkeit von der Futterkonzentration, der Temperatur, der Körpergrösse und dem Hungerzustand der Tiere. Arch. Hydrobiol. Suppl. 48: 47–107.

Grim, J., 1967. Der Phosphor und die pflanzliche Produktion im Bodensee. Gas- und Wasserfach (Wasser) 108: 1261–1271.

Grimm, H. & R.-D. Recknagel, 1985. Grundkurs Biostatistik. VEB Gustav Fischer Verlag, Jena, 156 pp.

Horn, W., 1981. Phytoplankton Losses due to Zooplankton Grazing in a Drinking Water Reservoir. Int. Revue ges. Hydrobiol. 66: 787–810.

Horn, H. & W. Horn, 1990. Growth of Blue-Green Algae in the Saidenbach Reservoir and its Relationship to the Silicon Budget. Int. Revue ges. Hydrobiol. 75: 461–474.

Infante, A., 1973. Untersuchungen über die Ausnutzbarkeit verschiedener Algen durch das Zooplankton. Arch. Hydrobiol. Suppl. 42: 340–405.

Porter, K. G., 1976. Enhancement of algal growth and productivity by grazing zooplankton. Science 192, No. 4246: 1332–1334.

Rice, W. R., 1989. Analyzing tables and statistical tests. Evolution 43: 223–225.

Hydrobiologia **307**: 239–251, 1995.
P. Larsson & L.J. Weider (eds), Cladocera as Model Organisms in Biology.

Life history characteristics and distribution of *Bythotrephes longimanus* Leydig (Crustacea, Onychopoda) in the Biesbosch reservoirs

Henk A. M. Ketelaars, Arco J. Wagenvoort, Reglindis F. Herbst*,
Patricia A. W. van der Salm & Ger-An J. de Jonge-Pinkster
Water Storage Corporation Brabantse Biesbosch, P.O. Box 61, 4250 DB Werkendam, The Netherlands
**Institute of Hydrobiology, Technical University Dresden, Mommsenstr. 13, D-01062 Dresden, Germany*

Key words: Bythotrephes longimanus, Daphniidae, diel distribution, horizontal distribution, invertebrate predators, *Leptodora kindti*, life history

Abstract

The diel and horizontal distributions of *Bythotrephes longimanus* Leydig, *Leptodora kindti* (Focke) and Anomopoda (cladocerans), as well as several life history parameters of each developmental stage of *B. longimanus* were studied in the non-stratified Biesbosch reservoirs. Anomopoda avoided the surface layer during the day only in the pelagic zone. In the surface layer near the shore, however, they were very abundant. Because densities in the pelagic zone during the day were very low, and high during the night, the population undoubtedly moved from the littoral zone or from the sediments to the pelagic. Accumulation of zooplankton at the littoral site is either the result of random dispersal or wind-induced movements. *Bythotrephes*' diel distribution pattern is possibly related to that of their prey, the anomopods. *Leptodora* mostly exhibited a 'normal' diel vertical migration with a small amplitude. Wind most likely influenced the horizontal distribution of the zooplankton. *Bythotrephes* body length increased with higher temperatures, whereas spine length was constant throughout the year. *Bythotrephes* can already be fertile in the first developmental stage, indicating that a rapid adaptation to a favourable environment is possible. The number of parthenogenetically produced eggs per ovigerous female was higher at the start of the growing season and constant throughout the rest of the year. Relatively few resting eggs and males were found in autumn.

Introduction

In 1987 *Bythotrephes longimanus* Leydig 1860 (Crustacea, Onychopoda), a predatory cladoceran previously unrecorded in the Netherlands, was discovered in one of the Biesbosch reservoirs (Visser, 1988). In the subsequent years, *B. longimanus* has become a permanent member of the zooplankton community in the three Biesbosch reservoirs. In general, this invasion did not clearly influence the abundance of another invertebrate predator, *Leptodora kindti* (Focke, 1844) (Crustacea, Haplopoda), which has been present in the reservoirs for many years, and it did not have a distinct effect on mean zooplankton densities and algal biomass. However, when *Bythotrephes* densities were high an important short-term impact on herbivorous daphnid densities was observed (Ketelaars & Van Breemen, 1993). In order to study the potential effect of *Bythotrephes* on the structure and functioning of the aquatic community in the Biesbosch reservoirs it is essential to have information on life-history parameters (Rosenberg, 1979).

The ecology of *B. longimanus* has not been studied extensively. Apart from extensive morphological data, Ischreyt (1930, 1939) also presented data on life-history traits of *B. longimanus*. Herbert (1954) studied tolerance of *B. longimanus* to low oxygen concentrations. Mordukhai-Boltovskaya (1956) reported on several aspects of the ecology of *B. longimanus*: distribution, seasonal dynamics and predation, the cycle of parthenogenetic reproduction (Mordukhai-Boltovskaya, 1957), predation rates at several temperatures (Mordukhai-Boltovskaya, 1958) and the influence of temperature on parthenogenetic reproduc-

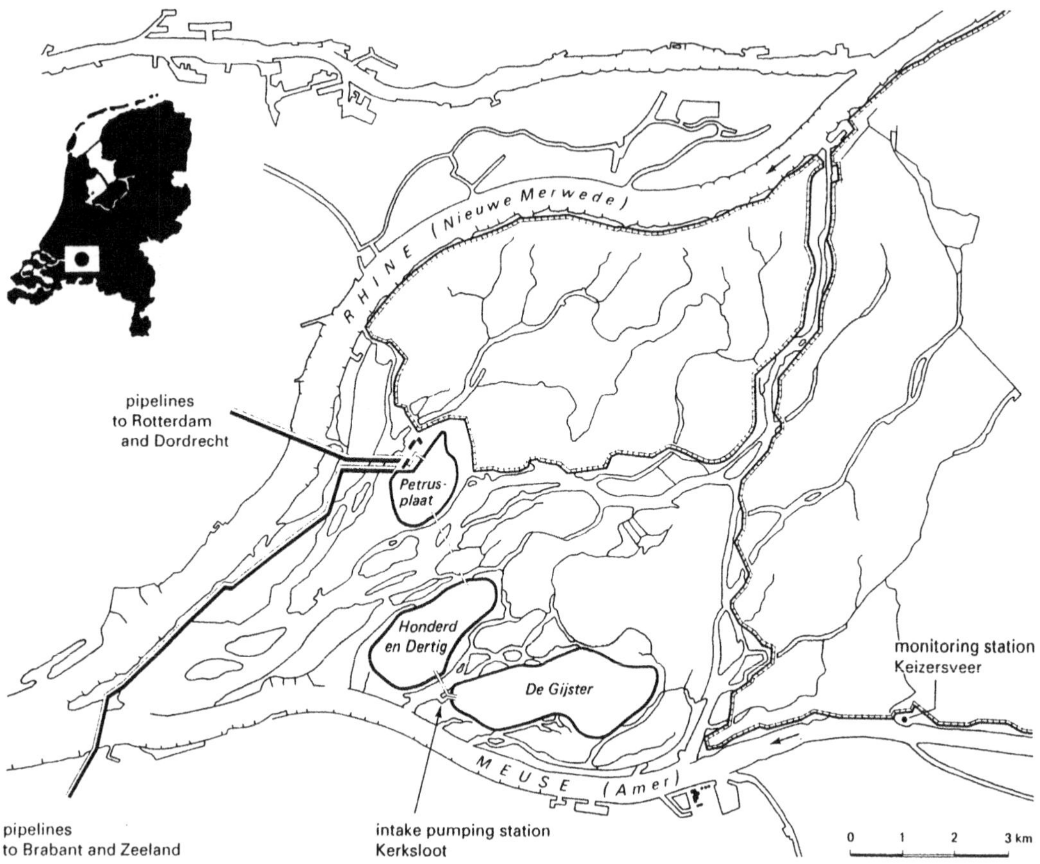

Fig. 1. Location of the storage reservoirs in the Biesbosch, the Netherlands.

tion (Mordukhai-Boltovskaya, 1959). Zozulya (1975) investigated the diel vertical migration and horizontal distribution in several reservoirs in Russia and described laboratory observations on the movement of *B. longimanus* in relation to the structure of its locomotory organs (Zozulya, 1978). Zozulya & Mordukhai-Boltovskoi (1977) described seasonal morphological variability of *B. longimanus* in several Russian reservoirs. Dzyuban (1979) gave some information on the vertical distribution of *Bythotrephes* in reservoirs along the Volga. Field and laboratory work on the resting eggs of *B. longimanus* has been carried out by Andrew & Herzig (1984) and Herzig (1985). In addition to these specific studies, some information on the ecology of *B. longimanus* can also be obtained from zooplankton (e.g. Southern & Gardiner, 1932; De Bernardi, 1981; Dokulil *et al.*, 1990) and fish predation studies (e.g. Guma'a, 1978; Jachner, 1991). The studies cited above contributed to our knowledge of the ecology of *B. longimanus*. Many aspects, however, remain to be studied.

The recent successful invasion of *B. cederstroemi* in the North American Great Lakes (Lehman, 1987) has stimulated ecological research on this species. Valuable information has been obtained on the impact of *B. cederstroemi* on phyto- and zooplankton communities (Sandgren & Lehman, 1990; Sprules *et al.*, 1990; Lehman, 1991), its embryonic and post-embryonic development (Yurista, 1992), the predator-defence mechanism of the spine (Barnhisel, 1991a, b), prey preferences and predation rates (Vanderploeg *et al.*, 1993) and seasonal size variation (Berg & Garton, 1988; Burkhardt, 1994).

A controversy exists as to whether *B. longimanus* and *B. cederstroemi* are two separate species or whether they can be considered as intraspecific forms of a single polymorphic species (Benisch, 1930; Ischreyt, 1934; Flößner, 1972; Mordukhai-Boltovskoi, 1968; Zozulya & Mordukhai-Boltovskoi, 1977). This fact should be taken into account when inferring ecological information from *B. cederstroemi* to *B. longimanus*.

Table 1. Some characteristics of the De Gijster and Petrusplaat water storage reservoirs in the Biesbosch, the Netherlands

		De Gijster	Petrusplaat
Water surface	ha	305	100
Volume	10^6 m^3	40	13
Maximum depth	m	27	15
Mean depth	m	13	13
Retention period	weeks	11	4
Chlorophyll-a	$\mu g\ l^{-1}$	14	5
t-Phosphorus	$\mu g\ l^{-1}$	180	70

*Annual mean (1992).

This article describes some life-history parameters (seasonal variation in body size and spine length and number of parthenogenetic and sexual eggs) for each developmental stage of *B. longimanus* as well as horizontal and vertical distributions of *B. longimanus* and *Leptodora kindti* in the Biesbosch reservoirs.

Description of the study area

Near the confluence of the Rhine and Meuse Rivers in the Netherlands, three interconnected reservoirs were built in the 1970's to supply drinking water production plants in the south-western part of the Netherlands with raw water (Fig. 1). The asphalt-concrete dikes surrounding the reservoirs rise 8 m above the surrounding area. Water level is 6.5 m higher than in the surrounding area. The reservoirs function as storage for periods of low river flow or inferior water quality. Storage also leads to considerable water quality improvements. About one eighth of the people in the Netherlands depend on the surface water stored in these reservoirs as their source of drinking water. The reservoirs are fed with moderately polluted, but nutrient-rich water from the Meuse River. Control of algal biomass in the Biesbosch reservoirs is achieved by light limitation through artificial mixing over sufficient depth and by zooplankton grazing (Oskam, 1978). Table 1 gives information on the two reservoirs used in this study.

Materials and methods

Vertical distribution

Vertical distribution of the cladocerans was determined at noon (16, 17 and 18 June 1993) and midnight (17 and 18 June 1993) in the Petrusplaat Reservoir. During each sampling bout, 10 samples were taken with a 2-l Ruttner sampler (Hydrobios, Kiel) at 1-m intervals from the surface to the bottom at a pelagic site (depth: 14 m). In addition five samples were taken with a 2-l Ruttner sampler at a littoral site (0 m depth). On 17 June five additional samples were taken at 2 m depth, because densities were low at the surface. The samples from each depth were sieved through a net (100 μm mesh size) and preserved in a sugar-formalin solution (Haney & Hall, 1973) with an end-concentration of 4% formalin. Cladocerans were identified to genus, and *B. longimanus* and *L. kindti* were always counted in the entire sample. When total cladoceran densities were high, the samples were split using a Folsom Plankton Splitter (Hydrobios, Kiel), the aliquot was counted, and numbers were calculated for the entire sample.

At the pelagic site, temperature and oxygen concentrations were measured at 4 m depth intervals at noon with a combined electrode (WTW-OXI 196). One liter samples for the determination of chlorophyll *a* were taken at 4 m depth intervals from the surface to the bottom on 14 and 18 June (11.00 a.m.). Chlorophyll *a* concentrations were determined according to the Dutch Standard Method (warm ethanol-extraction) (NNI, 1981). Transparency was measured at noon using a Secchi-disk.

Horizontal distribution

Horizontal distributions of cladocerans were studied in the De Gijster Reservoir on 14 and 17 September 1992 between 10.00 a.m. and 12.00 a.m. at nine locations with varying depths (see Fig. 6). Integrated samples from the surface to the bottom were taken with a plankton-net (mesh size 500 μm; diameter 40 cm; length 100 cm) with a conical top (diameter opening 17 cm; length 20 cm) (Hydrobios, Kiel). Samples were preserved as described above, and *B. longimanus* and *L. kindti* densities were determined in the entire sample. When total daphnid densities were high, the samples were split using a Folsom Plankton Splitter, the aliquot was counted and numbers were calculated for the entire sample. Wind velocity and direction were measured continuously during the day.

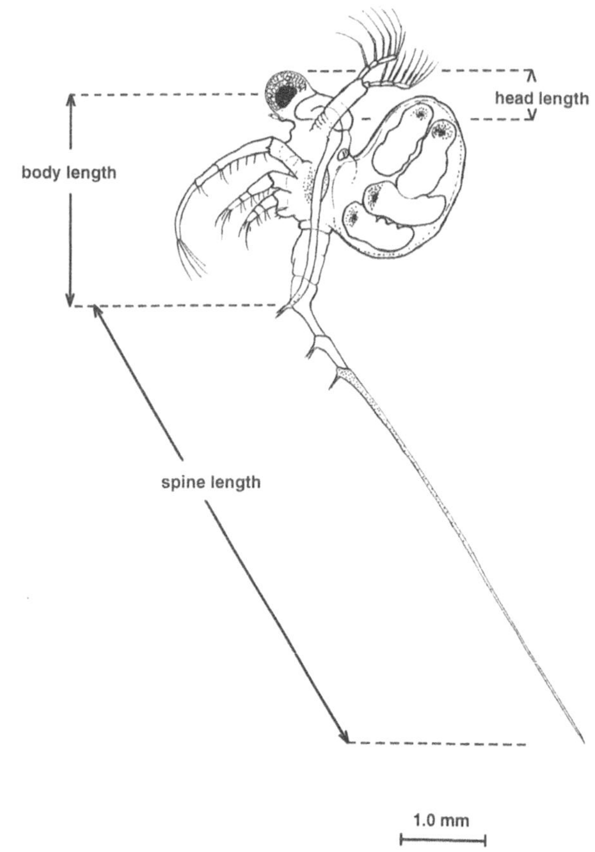

Fig. 2. Female *Bythotrephes longimanus* (third developmental stage) with length parameters measured.

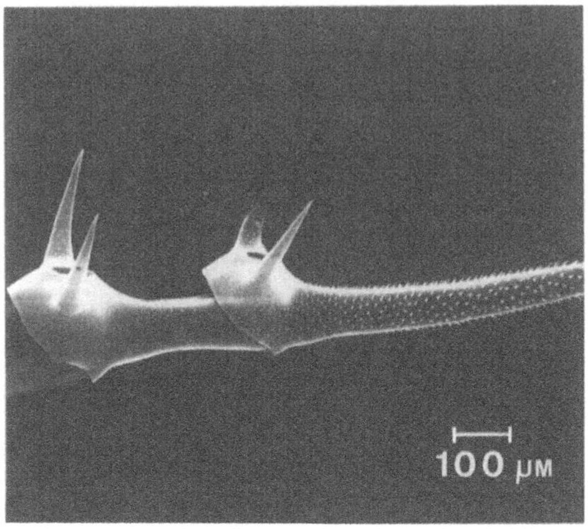

Fig. 3. Anterior part of the caudal appendage of *Bythotrephes longimanus* showing two pairs of lateral spines (Photo: Dr H. Kersken, Gelsenkirchen). Bar length. 100 μm.

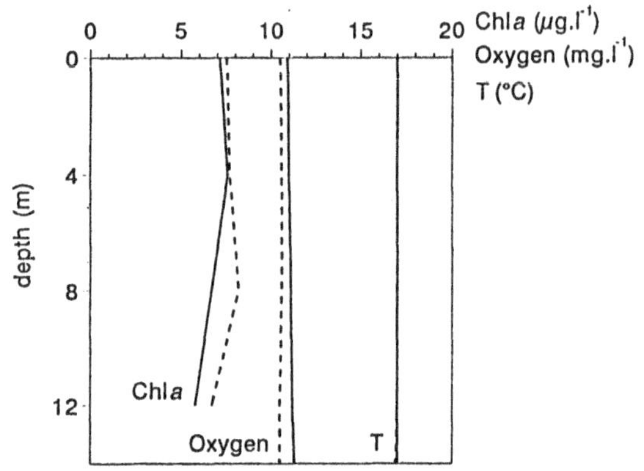

Fig. 4. Representative temperature-, oxygen- and chlorophyll *a* (chl *a*)-gradients in Petrusplaat Reservoir in June 1993. Chl *a*: —: 14 June; - - -: 18 June. Oxygen, temperature: —: 16 June; - - -: 18 June.

Life history characteristics

Samples were collected at regular intervals (weekly during and bi-weekly outside the growing season (June–Oct.)) in 1991 at three locations in the pelagic zone of De Gijster Reservoir (depths: 14, 15 and 26 m). Samples were taken between 10.00 a.m. and 12.00 a.m. with a 2-l Ruttner sampler at 2 m depth intervals from the surface to the bottom. In 1992, samples for life history characteristics were taken twice a week in September and weekly between October and December. Samples were sieved through a net (100 μm mesh size); the resulting composite sample was preserved as described above. Temperature was measured weekly in the outlet of the reservoir using a calibrated thermometer.

All *B. longimanus* were examined under a dissecting microscope (Wild M8, magnification 25×) and length parameters (Fig. 2) were measured for each developmental stage. Total body length was defined as half the measured head length plus measured body length (distance from the centre of the eye to the base of the caudal appendage). Only animals which possessed a complete tail spine were used for length measurements. The developmental stages of *B. longimanus* are characterized by the number of pairs of lateral spines (paired barbs) on the caudal appendage (tail spine). Neonates possess one pair of barbs and typically, a new pair is added at each moult, because moulting is restricted to the body and the exoskeleton is shed at the base of the lateral spines, leaving the old tail spine attached to the new spine (Ischreyt, 1930; Fig. 3). A difference in the number of instars exists between parthenogenetically (3 stages) and sexually (4 stages) produced *B. cederstroemi* (Yurista, 1992). *Bythotrephes* reproduces parthenogetically during summer and sexually later in the year. Parthenogenetic eggs are deposited into the brood pouch where they develop into embryos. Parthenogenetic offspring therefore can be counted as eggs or embryos, depending on the developmental state. Resting eggs are produced sexually and differ from parthenogenetic eggs by their colour (yellow to brown) and thickness of the shell (Herzig, 1985). The number of parthenogenetically produced offspring (eggs, embryos) and sexually produced offspring (resting eggs) in the brood pouches were counted. Males were distinguished by the possession of a pair of penises (Flößner, 1972).

Results

Vertical distribution

Temperature, oxygen and chlorophyll *a* measurements for Petrusplaat Reservoir are given (Fig. 4). No thermal or oxygen gradient was observed and only minor differences in chlorophyll *a* content along the depth gradient were observed. Secchi-transparency decreased from 4.5 m on 16 June to 4.1 m on 18 June.

The depth distributions of *B. longimanus*, *L. kindti* and anomopods (consisting almost entirely of the family Daphniidae) at both sampling sites in Petrusplaat Reservoir are given (Fig. 5). During daytime, *B. longimanus* and *L. kindti* were not present in the surface layer at the pelagic site. Daphnids were present in very low densities. On the three daytime sampling dates, *B. longimanus* densities at most depths were low. In the one shallow near-shore sampling site, *B. longimanus* densities were always high (on 16 June no samples were taken near shore). In contrast, *L. kindti* was present throughout the water column in low densities and in a slightly higher density near the shore (on one sampling date only, Fig. 5C). On the other two daytime sampling dates, *L. kindti* was present in high densities below 1–3 m, while the highest densities were found below 9 m. At the shallow near-shore site, however, its density was very low (Fig. 5E; on 16 June no samples were taken at the littoral site). At night, *B. longimanus* and *L. kindti* were present throughout the water column. On 17 June the highest *Bythotrephes* densities were found at 6–13 m, and on 18 June from 10–12 m, while the highest *Leptodora* densities were found at 3–9 m and 9–12 m.

At the pelagic site the highest anomopod densities during daytime were found below or at 6 m. At the shallow site, densities were much higher during the day. At night anomopod numbers in the surface layer of the pelagic site were low on 17 June and high on 18 June. Highest numbers were found at the deeper sites. At night, numbers varied at the littoral site: on 17 June density was as high as in the deeper part, whereas on 18 June density was rather low.

Horizontal distributions

Horizontal variations in spatial distributions were studied in De Gijster Reservoir on two occasions when a moderately to fairly strong wind was blowing from opposite directions (Fig. 6). On 14 September, *B. longimanus* and *L. kindti* mainly occurred at the eastern and south-eastern sampling sites, but *L. kindti* was present at more sites. Highest daphnid densities were found at the central site. At the other eight locations, daphnids were present in rather low numbers. On 17 September, when the wind was blowing out of the north-east, the horizontal distribution differed considerably. *B. longimanus* occurred predominately at the northern sites, the central site and the south-western site. *L. kindti* was present at all sites. Daphnids were present at all sites in varying numbers with the highest density at the central site.

Life history characteristics

Life-history parameters for *B. longimanus* are given (Figs 7 and 8). *B. longimanus* was detected for the first time in 1991 when the temperature was 13.9 °C and in 1992 when the temperature was 6.5 °C. The first female found carried seven parthenogenetic eggs, the highest number per female recorded in 1991. Neonates were found between June and mid-October and were found to carry eggs between June and the beginning of September (Fig. 7A). Second and third stage animals occurred between the end of June and the beginning of November. Second stage animals carried eggs between the end of June and mid-October. Third stage *Bythotrephes* carried eggs until the beginning of November. Whereas no first stage males were found, second stage males were present between the end of September, when temperatures started to decline, and the end of November. Average clutch size per ovigerous female for first, second and third stage animals was 4.5, 3.2 and 3.5 respectively in 1991 (sample sizes were 10, 20 and 21 animals respectively) and 3.2, 2.7

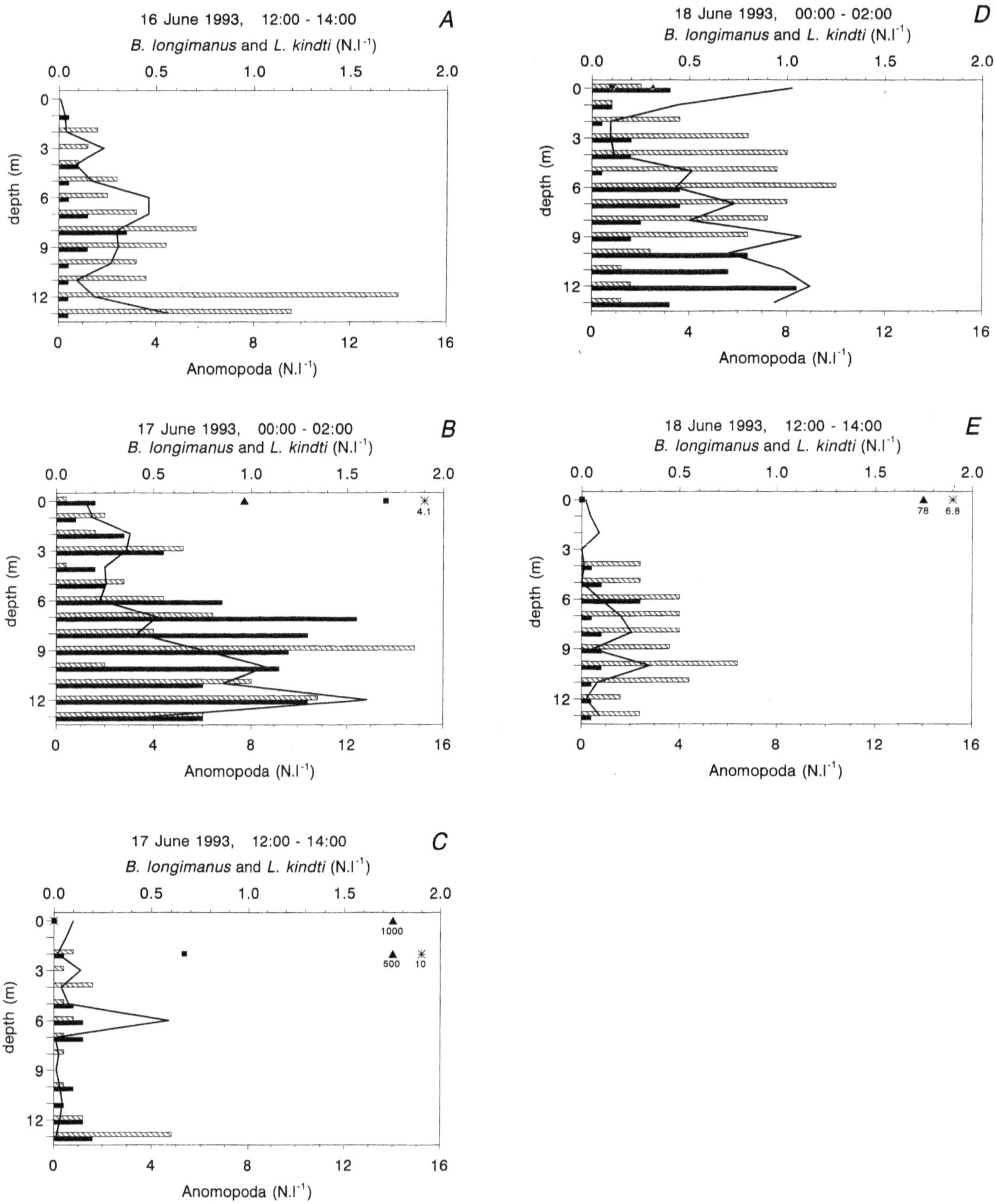

Fig. 5. Day and night depth distributions of *Bythotrephes longimanus*, *Leptodora kindti* and Anomopoda in Petrusplaat Reservoir on 16 and 17 June 1993. Pelagic site: ■: * *B. longimanus*; ▨ *L. kindti*; —: Anomopoda. Littoral site: : *B. longimanus*; ■: *L. kindti*; ▲: Anomopoda. If densities were too high to plot, values are indicated. On 16 June no samples were taken at the littoral site. On 17 June (12:00–14:00) the littoral site was sampled at two depths (0 and 2 m).

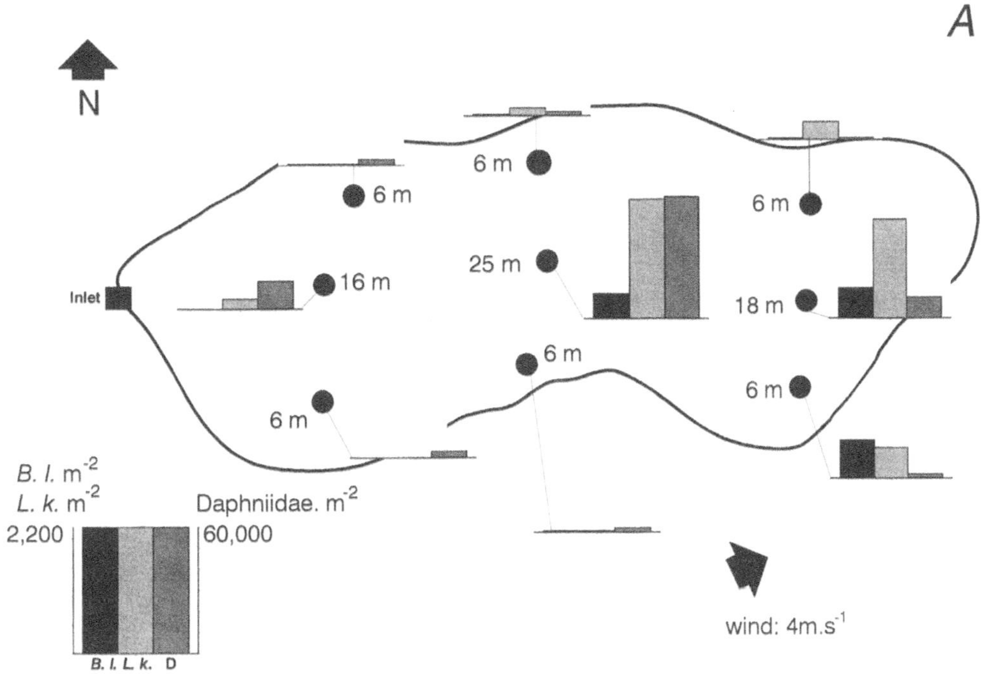

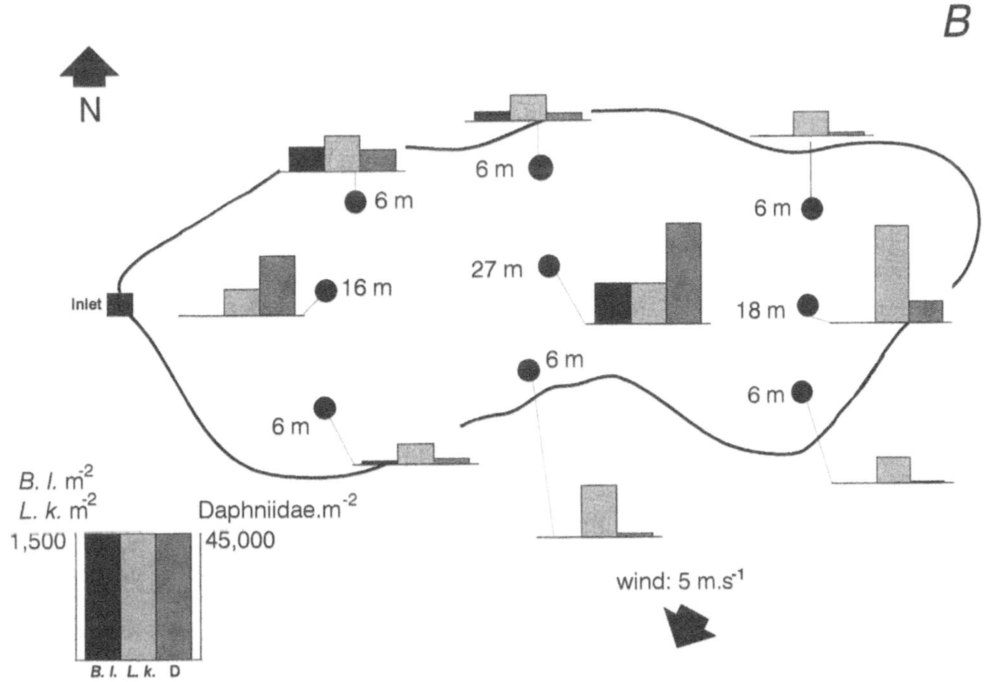

Fig. 6. Horizontal distribution of *Bythotrephes longimanus* (*B. l.*), *Leptodora kindti* (*L. k.*) and Daphniidae (D) in De Gijster Reservoir on 14 September (A) and 17 September (B) 1992. •: sampling site, with depth indicated. Wind direction and -force are also indicated.

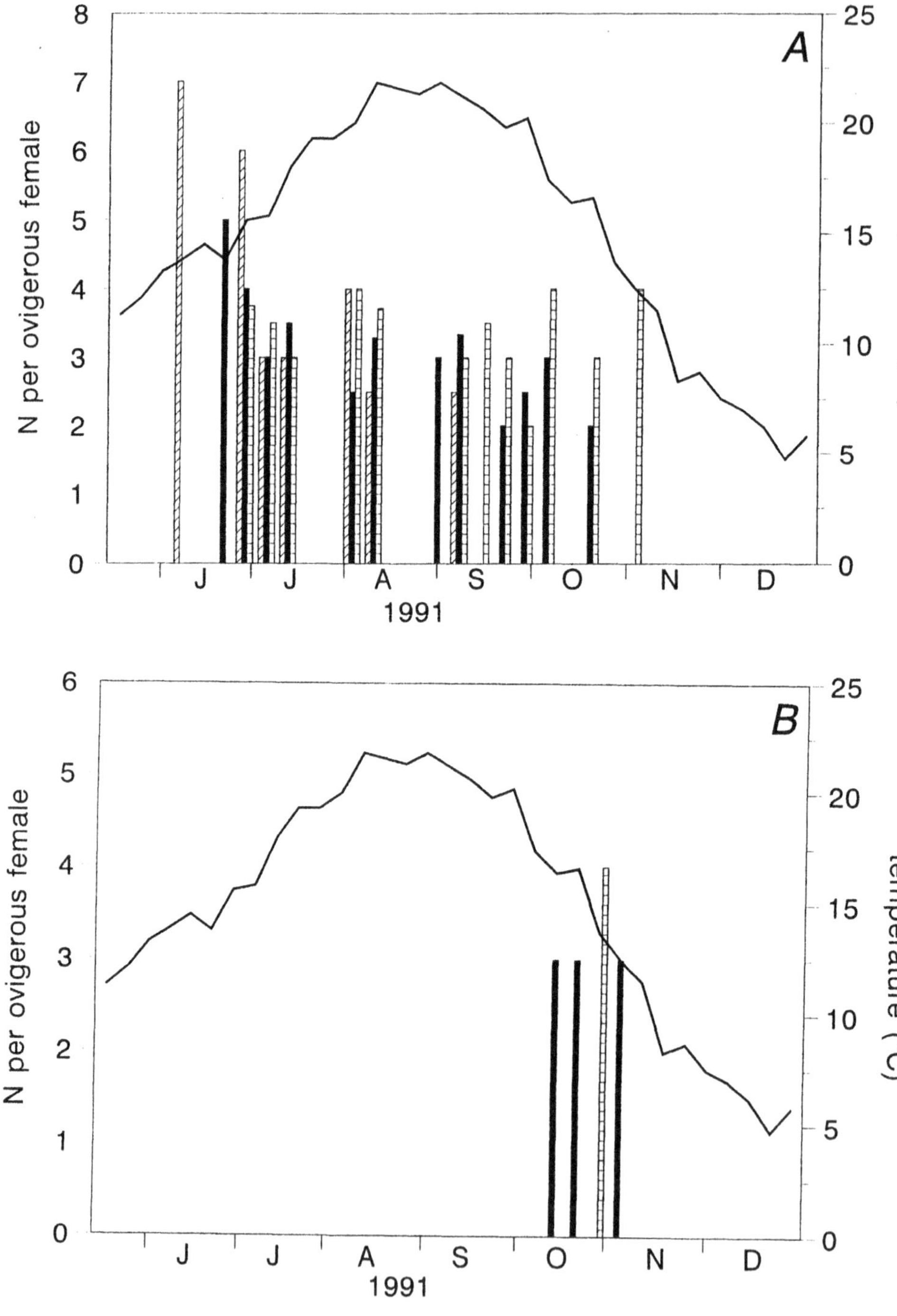

Fig. 7. Bythotrephes longimanus. Mean number of parthenogenetic eggs/embryos (A) and sexually produced resting eggs (B) per ovigerous female of each developmental stage and water temperatures in De Gijster Reservoir in 1991. ▨ : first stage; ■: second stage; ▤ : third stage; —: water temperature.

and 3.6, respectively in 1992 (sample sizes were 18, 33 and 51 animals respectively) (Data not shown in graph). In the beginning of the growing season (June), clutch sizes were higher as compared to the rest of the year. Sexually produced resting eggs were detected only occasionally (N = 11), and were observed in

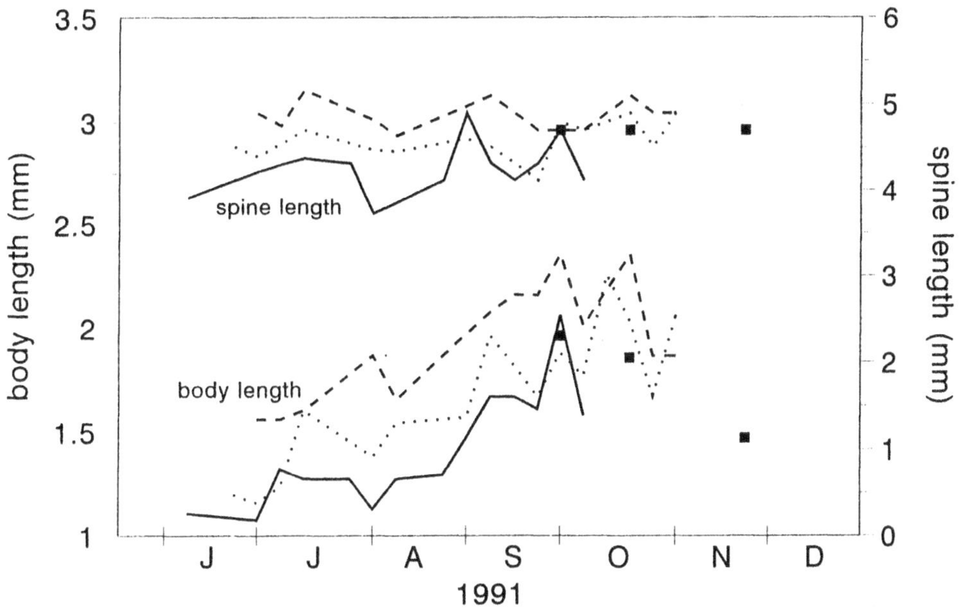

Fig. 8. Mean body length and mean spine length of *Bythotrephes longimanus* in De Gijster Reservoir in 1991. —: first stage; ······: second stage; ——: third stage; ■: males.

second as well as in third stage animals in October and at the beginning of November in 1991 (Fig. 7B) and between September and the beginning of December in 1992 (Data not shown). The number varied from 1–4 resting eggs per ovigerous female.

Body length increased gradually throughout the year, reaching a maximum in September-October at water temperatures of 18–20 °C (Fig. 8). First, second and third stage females increased their body size by 92%, 95% and 51% respectively (sample sizes were 23, 31 and 23 animals respectively). Body size of males decreased, but because only three males were found the decrease was not significant Spine length stayed more or less constant for all developmental stages of females and males throughout the year.

Discussion

Vertical distribution

From previous studies of the vertical migration of *Bythotrephes,* we know that *Bythotrephes* exhibits diel vertical migration, albeit sometimes only weakly. In general, *Bythotrephes* exhibits 'normal' DVM: it stays in the deeper parts of the lake during the day and just before sunset it migrates to the upper water or to the surface (Southern & Gardiner, 1932; Dzyuban, 1979; Lehman, 1987; Dokulil *et al.*, 1990; Nauwerck, 1993). According to Sebestyén (1933), Dzyuban (1979), Dokulil *et al.* (1990), Vijverberg (1991) and Nauwerck (1993), *Leptodora* also exhibits a 'normal' DVM. 'Normal' DVM seems to be a response to predation by visually orienting predators ('mortality avoidance hypothesis', Lampert, 1993). Predation pressure can be so strong that it is advantageous to leave the nutritionally and thermally favourable surface waters and seek refuge in the unfavourable pelagic, lower waters (Lampert, 1993). In the Biesbosch reservoirs, the deep pelagic is not an unfavourable environment because the reservoirs are well mixed and no thermal or nutritional (phyto- and bacterioplankton) gradient exists (see Fig. 4).

Our results show that the anomopods avoided the surface layer during the day only in the pelagic zone. In the near-shore, surface layer, however, they were very abundant. Because abundance in the pelagic zone during the day was very low and high during the night, the population must have moved to the pelagic zone during the evening. Active crustacean migration to the shore does not seem plausible because no physiological mechanism exists to explain this behaviour (J. Ringelberg, pers. comm.). Migration to the shore has been found for *Chaoborus flavicans* (Diptera) (Franke, 1987). Distance from the littoral site to the pelagic site is much too far for individuals to migrate the entire distance. The night-time pelagic population therefore must originate from the sediments. Migration to and from lake sediments has been found by Hart (1975). Accumulation of zooplankton at the littoral site could have been the result of non-directed movement, induced by wind. However, this is not the only factor because it can not explain the high densities of *Leptodora* during daytime on 18 June, at least if both anomopods and *Leptodora* are affected by wind in the same manner. Therefore random dispersal also must be involved.

The diel distribution pattern of *Bythotrephes* may be explained by the hypothesis that it pursued the anomopods, which are preferred prey items. At sufficiently high daphnid densities, *Bythotrephes* is less preyed upon by fish than daphnids, because it is protected by its long tail spine, which acts as a defense against fish predation (Barnhisel, 1991b; pers. observ.). No consistent distribution pattern was found for *Leptodora*. On two out of the three daytime sampling dates we did not find fewer animals in the water column during the day than during the night (i.e. 'diurnal deficit', De Stasio, 1993), as was the case for *Bythotrephes* and the Anomopoda. This suggests that *Leptodora* exhibited a 'normal' DVM, albeit with a small amplitude. This was also found by Stich (1989) and Dokulil *et al.* (1990). Low densities during daytime on 17 June cannot be explained. Perhaps wind induced a temporary movement away from the pelagic site. In fact, density at the littoral site was much higher than during daytime the next day. The difference in distribution patterns between *Bythotrephes* and *Leptodora* is a remarkable phenomenom. Both invertebrate predators feed on crustaceans and rotifers of the same size-class (300–700 μm) (Mordukhai-Boltovskaya, 1958). Possibly, the invader *Bythotrephes* is more opportunistic than *Leptodora*.

It has to be stressed that migration to deeper parts of the pelagic does not imply a demographic disadvantage to the zooplankton because no thermal or distinct nutritional gradient exists in the reservoirs. As a next step, it would be interesting to investigate the vertical distribution pattern of the zooplankton in spring when no invertebrate predators are present and predation pressure by juvenile fish is low.

Horizontal distribution

Horizontal variations in the spatial distribution of zooplankton depend on several factors. In addition to active avoidance of the littoral zone by Anomopoda and Copepoda, wind-induced surface currents and geostrophic effects of the Coriolis force due to the earth's rotation cause a nonrandom dispersion of zooplankton. Whenever a fairly strong wind persists for an appreciable period of time, the zooplankton tend to accumulate in the leeward epilimnion (Langford & Jermolajev, 1966; Wetzel, 1983). Since the average depth of the Biesbosch reservoirs is 15 m, the deflection angle caused by the Coriolis force is insignificant (Wetzel, 1983).

Wind influences on the Biesbosch reservoirs are high due to their wind-exposed location high above the surrounding area. The results from the horizontal distribution studies indicate that wind direction can have a direct influence on zooplankton distribution. On 14 September when the wind was blowing along the length-axis of the reservoir, highest zooplankton abundance was found at the leeward side, indicating transport by wind-induced water movements.

On 17 September, when a fairly strong wind was blowing out of the north-east, i.e. across the length-axis of the reservoir, no clear distribution pattern was observed. It is possible that near-shore populations in the northern part were protected by the dikes surrounding the reservoir and populations, which occurred at the deeper sites, were less affected by the wind. But because zooplankton distributions are patchy in time and space, systematic errors in density determinations are high, and it is difficult to detect a clear distributional pattern. Our results demonstrate that zooplankton in De Gijster Reservoir were heterogeneously distributed. This phenomenon is of key importance for obtaining representative samples.

Life history characteristics

Figure 7 shows that *B. longimanus* can be fertile in the first developmental stage. First stage animals carrying parthenogenetic eggs have also been reported by Ischreyt (1930) and Mordukhai-Boltovskaya (1957). Yurista (1992) states that females of *B. cederstroemi* with one or two pairs of barbs are typically immature. According to Ischreyt (1930), the eggs are retained in the brood pouch when the animal moults. The development of the eggs is therefore independent of moulting and it cannot be stated that only third stage animals are mature. Embryos in second stage animals were noticed frequently. Also, resting eggs were observed in second stage animals. Ischreyt (1939) found full-grown embryos in second stage *Bythotrephes* in some subalpine lakes, whereas in other lakes, only third stage animals carried full-grown embryos. Mordukhai-Boltovskaya (1957) also mentioned second stage *Bythotrephes* carrying embryos. Yurista (1992), who occasionally found second stage *B. cederstroemi* able to reproduce, explains this apparent discrepancy by the explanation that animals may occasionally moult without adding barbs. He does not explain how this moulting takes place. Under laboratory conditions *Bythotrephes* is difficult to culture (Mordukhai-Boltovskaya, 1957; Berg & Garton, 1988; Yurista, 1992; pers. observ.), probably because favourable environmental conditions are difficult to achieve. This could explain the fact that Yurista (1992) did not find fertile first stage *B. cederstroemi* and only occasionally noticed reproductive second stage animals.

The first resting eggs were found just after a period when water temperature dropped 4 °C over two weeks time. This agrees well with Herzig (1985) who concluded that sexual reproduction is restricted to the time of the year that is characterized by the greatest heat loss from the water and by a short-day photoperiod.

Whereas body size increased with increasing temperature, the number of eggs per ovigerous female was constant. The fact that neonate size increased throughout the year indicates that *Bythotrephes*, instead of producing more offspring, produced larger-sized offspring. This may provide *Bythotrephes* with a better survival potential because it can prey upon a greater size-range.

Until 1993 neither first stage males nor the four-barbed gametogenetically produced form, as observed by Yurista (1992) and Vekhov (1981) for *B. cederstroemi*, were observed here. This was probably due to the fact that both occur at low densities during only a short period of the year and because relatively small water quantities were sampled. Much larger water volumes, sampled in spring 1994, yielded several individuals of four-barbed *B. longimanus*.

Body size increased throughout the year, as was found by Burkhardt (1994). Factors that determine zooplankton body size are food concentration (e.g. Vijverberg, 1976), temperature (e.g. Culver, 1980) and fish predation (Brooks & Dodson, 1965). However, it is difficult to separate direct factors (e.g. a direct temperature effect due to optimum growth of *Bythotrephes* at temperatures of approximately 20 °C) from indirect factors (e.g. abundance of prey and predation rates of *Bythotrephes* and fish), because water temperature, abundance of prey and prey consumption rates are not independent of each other. Mordukhai-Boltovskaya (1958) determined that the maximum predation rate of *Bythotrephes* increased from 9 prey.d^{-1} at 10–15 °C to 25 prey·d^{-1} at 20–25 °C. Herzig and Auer (1990) reported that the predation rate of *Leptodora kindti* was influenced by temperature and prey density. The highest predation rate of *L. kindti* was observed between 19 °C and 20 °C; above 20 °C predation rates decreased. Higher predation rates at higher temperatures in addition to a direct effect of temperature on growth, can explain the greater body size of *Bythotrephes* in the warm summer months. Size-selective fish predation is the third factor that can determine size distribution. Burkhardt (1994), however, reported that temporal variation in body size of *B. cederstroemi* could not be accounted for by known gradients of planktivory. It should also be noted that body length only makes a small contribution to total length and thus will not likely affect size-selective fish predation.

The most characteristic feature of *Bythotrephes* is its long tail spine. Its basic function is as a stabilizer for forward motion and as a steering organ (Ischreyt, 1930; Zozulya & Mordukhai-Boltovskoi, 1977). It also acts as an anti-predator defense device (Barnhisel, 1991a, b). Although spine length can be more accurately measured than other length parameters (Berg & Garton, 1988) it can not be used as a measure of body size because it is not necessarily representative of body size (Ischreyt, 1934; Burkhardt, 1994). For example our data demonstrated that spine length stayed more or less constant for each developmental stage, while body size increased during the year. Barnhisel (1991b) offered to juvenile yellow perch (*Perca flavescens*), *B. cederstroemi* with varying portions of the spine removed. She found that perch showed greater preference to attack no-spined and half spined *Bythotrephes* and were less likely to reject, and more likely to ingest *Bythotrephes* with the spine removed. If *Bythotrephes* is able to adjust its spine length as a response to fish predation, then the small size variation in spine length that we found could therefore indicate that fish predation in our reservoirs is relatively low. In fact, regular fish stock assessments in the reservoirs show that fish abundance is low (Visser, 1993).

In 1991 and 1992 *Bythotrephes* was recorded for the first time at quite different water temperatures. Mordukhai-Boltovskaya (1956) found *Bythotrephes* when temperatures ranged from 10–12 °C, whereas Garton *et al.* (1990) found *B. cederstroemi* in Lake Erie at approximately 4 °C. According to Herzig (1985), resting eggs of *B. longimanus* hatch at ambient water temperatures of 4–5 °C above the sediment. He also states that resting eggs hatch irregularly over a long period of time during the year. Temperature has a large effect on the developmental rate of embryos (Yurista, 1992). Mordukhai-Boltovskaya (1956) found that generation time at lower temperatures is longer (23 days at 10 °C; 7 days at 25 °C). These findings indicate that the population does not grow rapidly in spring and early summer. Therefore, the probability that *Bythotrephes* is caught in the rather small-sized routine samples is low in spring and early summer. Larger volumes have to be sampled (Yurista, 1992), to determine when *Bythotrephes* really appears in the spring and early-summer plankton.

Acknowledgements

The authors are indebted to Francien Lambregts-van de Clundert, Hans van Loon, Jeanne van den Beemd-Gommers, Len van Halm and Meral Westerink for their assistance in the field and laboratory. We would also like to thank Joop Ringelberg for helpful discussions. Critical remarks on the manuscript by Jurgen Volz, Alois Herzig and an anonymous reviewer are greatly appreciated.

References

Andrew, T. E. & A. Herzig, 1984. The respiration rate of the resting eggs of *Leptodora kindti* (Focke 1844) and *Bythotrephes longimanus* Leydig 1860 (Crustacea, Cladocera) at environmentally encountered temperatures. Oecologia 64: 241–244.

Barnhisel, D. R., 1991a. The caudal appendage of the cladoceran *Bythotrephes cederstroemi* as defense against young fish. J. Plankton Res. 31: 529–537.

Barnhisel, D. R., 1991b. Zooplankton spine induces aversion in small fish predators. Oecologia 88: 444–450.

Benisch, J., 1930. Über ein Vorkommen von *Bythotrephes cederströmii* Schödler in Deutschland. Zool. Anz. 86: 333–335.

Berg, D. J. & D. W. Garton, 1988. Seasonal abundance of the exotic predatory cladoceran, *Bythotrephes cederstroemi*, in western Lake Erie. J. Great Lakes Res. 14: 479–488.

Brooks, J. L. & S. I. Dodson, 1965. Predation, body size, and composition of plankton. Science 150: 28–35.

Burkhardt, S., 1994. Seasonal size variation in the predatory cladoceran *Bythotrephes cederstroemi* in Lake Michigan. Freshwat. Biol. 31:97–108.

Culver, D., 1980. Seasonal variation in the sizes at birth and at first reproduction in Cladocera. In Kerfoot, W. C. (ed.), Evolution and Ecology of Zooplankton Communities. The University Press of New England, Hanover (NH); London: 358–366.

De Bernardi, R., 1981. Biotic interactions in freshwater and effects on community structure. Boll. zool. ital. 48: 353–371.

De Stasio Jr., B. T., 1993. Diel vertical migration by zooplankton: Population budgets and the diurnal deficit. Bull. mar. Sci. 53: 44–64.

Dokulil, M., A. Herzig & A. Jagsch, 1990. Trophic relationships in the pelagic zone of Mondsee, Austria. Hydrobiologia 191 (Dev. Hydrobiol. 53): 199–212.

Dzyuban, N. A., 1979. The invertebrates: A. The zooplankton (metazoic) of the Volga. In Mordukhai-Boltovskoi, Ph. D. (ed.), The River Volga and its life. Dr W. Junk Publishers, The Hague: 195–231.

Flößner, D., 1972. Krebstiere, Crustacea. Kiemen- und Blattfüßer, Branchiopoda, Fischläuse, Branchiura. Gustav Fischer Verlag, Jena, 501 pp.

Franke, C., 1987. Detection of transversal migration of larvae of *Chaoborus flavicans* (Diptera, Chaoboridae) by the use of a sonar system. Arch. Hydrobiol. 109: 355–366.

Garton, D. W., D. J. Berg & R. J. Fletcher, 1990. Thermal tolerances of the predatory cladocerans *Bythotrephes cederstroemi* and *Leptodora kindti*: relationship to seasonal abundance in western Lake Erie. Can. J. Fish. aquat. Sci. 47: 731–738.

Guma'a, S. A., 1978. The food and feeding habits of young perch, *Perca fluviatilis*, in Windermere. Freshwat. Biol. 8: 177–187.

Haney, J. F. & D. J. Hall, 1973. Sugar-coated *Daphnia*: a preservation technique for Cladocera. Limnol. Oceanogr. 18: 331–333.

Hart, R. C., 1975. The substrate-bin – a new sampling device for studying diel vertic(al migratory movements on to and off lake sediments. Freshwat. Biol. 6: 155–159.

Herbert, M. R., 1954. The tolerance of oxygen deficiency in the water by certain Cladocera. Mem. Ist. ital. Idrobiol. 8: 97–107.

Herzig, A., 1985. Resting eggs – a significant stage in the life cycle of crustaceans *Leptodora kindti* and *Bythotrephes longimanus*. Verh. int. Ver.. Limnol. 22: 3088–3098.

Herzig, A. & B. Auer, 1990. The feeding behaviour of *Leptodora kindti* and its impact on the zooplankton community of Neusiedler See (Austria). Hydrobiologia 198 (Dev. Hydrobiol. 60): 107–117.

Ischreyt, G., 1930. Über Körperbau und Lebensweise des *Bythotrephes longimanus* Leydig. Arch. Hydrobiol. 21: 241–324.

Ischreyt, G., 1934. Über *Bythotrephes cederstroemii* Schödler. Int. Revue ges. Hydrobiol. Hydrogr. 31: 181–202.

Ischreyt, G., 1939. Über den *Bythotrephes* subalpiner Seen. Arch. Hydrobiol. 34: 105–129.

Jachner, A., 1991. Food and habitat partitioning among juveniles of three fish species in the pelagial of a mesotrophic lake. Hydrobiologia 226: 81–89.

Ketelaars, H. A. M. & L. W. C. A. Van Breemen, 1993. The invasion of the predatory cladoceran *Bythotrephes longimanus* Leydig and its influence on the plankton communities in the Biesbosch Reservoirs. Verh. int. Ver. Limnol. 25: 1168-1175.

Langford, R. R. & E. G. Jermolajev, 1966. Direct effect of wind on plankton distribution. Verh. int. Ver. Limnol. 16: 188–193.

Lampert, W., 1993. Ultimate causes of diel vertical migration of zooplankton: New evidence for the predator-avoidance hypothesis. Arch. Hydrobiol./Beih. Ergebn. Limnol. 39:79–88.

Lehman, J. T., 1987. Palearctic predator invades North American Great Lakes. Oecologia 74: 478–480.

Lehman, J. T., 1991. Causes and consequences of cladoceran dynamics in Lake Michigan: implications of species invasion by *Bythotrephes*. J. Great Lakes Res. 17: 437–445.

Mordukhai-Boltovskaya, E. D., 1956. Some notes on the biology of *Leptodora kindtii* Focke and *Bythotrephes* Leydig in Rybinsk reservoir. Dokl. Akad. Nauk SSSR 110: 688–691. (In Russian).

Mordukhai-Boltovskaya, E. D., 1957. Parthenogenetic breeding of *Leptodora kindtii* (Focke) and *Bythotrephes* Leydig. Dokl. Akad. Nauk SSSR 112: 1133–1135.

Mordukhai-Boltovskaya, E. D., 1958. Preliminary notes on the feeding of the carnivorous cladocerans *Leptodora kindtii* and *Bythotrephes*. Dokl. Akad. Nauk SSSR 122: 828–830.

Mordukhai-Boltovskaya, E. D., 1959. About influence of temperature on parthenogenetic reproduction of *Bythotrephes*. Bull. Inst. Biol. Vodokhr. 5: 21–22. (In Russian).

Mordukhai-Boltovskoi, F. D., 1968. On the taxonomy of the Polyphemidae. Crustaceana 14: 197–209.

Nauwerck, A., 1993. Migration strategies of crustacean zooplankton in Lake Mondsee. Arch. Hydrobiol./Beih. Ergebn. Limnol. 39: 109–122.

NNI, 1981. Water. Spectrofotometrische bepaling van het gehalte aan chlorofyl *a*. Nederlandse Norm NEN 6520. NNI, Delft. (In Dutch).

Oskam, G., 1978. Light and zooplankton as algae regulating factors in eutrophic Biesbosch reservoirs. Verh. int. Ver. Limnol. 20: 1612–1618.

Rosenberg, D. M., 1979. Freshwater benthic invertebrate life histories: current research and future needs. J. Fish. Res. Bd Can. 36: 289–345.

Sandgren, C. G. & J. T. Lehman, 1990. Response of chlorophyll-*a*, phytoplankton, and microzooplankton to the invasion of Lake Michigan by *Bythotrephes*. Verh. int. Ver. Limnol. 24: 386–392.

Sebestyén, O., 1933. The daily vertical migration of *Leptodora kindtii* Focke (Crustacea, Cladocera). Arb. Ung. Biol. Forsch. Inst. 6: 112.

Southern, R. & A. C. Gardiner, 1932. Reports from the Limnological Laboratory. II. The diurnal migrations of the Crustacea of the plankton in Lough Derg. Proc. R. ir. Acad. 40: 121–159.

Sprules, G. W., H. P. Riessen & E. H. Jin, 1990. Dynamics of the *Bythotrephes* invasion of the St. Lawrence Great Lakes. J. Great Lakes Res. 16: 346–351.

Stich, H. B., 1989. Seasonal changes of diel vertical migrations of crustacean plankton in Lake Constance. Arch. Hydrobiol. Suppl. 83: 355–405.

Vanderploeg, H. A., J. R. Liebig & M. Omair, 1993. *Bythotrephes* predation on Great Lakes' zooplankton measured by an in situ method: implications for zooplankton community structure. Arch. Hydrobiol. 127: 1–8.

Vekhov, N. V., 1981. Some features of *Bythotrephes longimanus* (Cladocera, Cercopagidae) biology in tundra lakes. Vestn. Zoo. (USSR) 5:72–75. (In Russian).

Vijverberg, J., 1976. The effect of food quantity and quality on the growth, birth rate and longevity of *Daphnia hyalina* Leydig. Hydrobiologia 51: 99–108.

Vijverberg, J., 1991. Variability and possible adaptive significance of day-time vertical distribution of *Leptodora kindtii* (Focke) (Cladocera) in a shallow eutrophic lake. Hydrobiol. Bull. 25: 85–91.

Visser, P., 1988. *Bythotrephes longimanus* Leydig in Nederland. Hydrobiol. Bull. 22: 211. (In Dutch).

Visser, P., 1993. De ontwikkeling van de visstand in de spaarbekkens van het Waterwinningbedrijf Brabantse Biesbosch in de periode 1974–1987. WBB, Werkendam, 59 pp. (In Dutch).

Wetzel, R. G., 1983. Limnology. Saunders College Publishing, Philadelphia, 767 pp.

Yurista, P. M., 1992. Embryonic and postembryonic development in *Bythotrephes cederstroemii*. Can. J. Fish. aquat. Sci. 49: 1118–1125.

Zozulya, S. S., 1975. Some aspects of the behaviour of *Bythotrephes longimanus* (Leydig). Povedeniye Vodnykh Bespozvonochnykh 30–32. (In Russian).

Zozulya, S. S., 1978. On movement of *Bythotrephes longimanus* Leydig. Gidrobiol. Zh. 14: 14–17. (In Russian).

Zozulya, S. S. & F. D. Mordukhai-Boltovskoi, 1977. Seasonal variability of *Bythotrephes longimanus* (Crustacea, Cladocera). Dokl. Akad. Nauk SSSR 232: 493–495.

Hydrobiologia **307**: 253–261, 1995.
P. Larsson & L.J. Weider (eds), Cladocera as Model Organisms in Biology.

Competition or niche segregation between *Holopedium* and *Daphnia*; empirical light on abiotic key parameters

Dag O. Hessen[1], Bjørn A. Faafeng[2] & Tom Andersen [1]
[1] *University of Oslo, Biological Institute, P.O. Box 1027, Blindern, N-0315 Oslo, Norway*
[2] *Norwegian Institute for Water Research., P.O. Box 173, Kjelsås, N-0411 Oslo, Norway*

Key words: Cladocera, competition, abiotic parameters, calcium

Abstract

A survey of 346 Norwegian lakes revealed some consistent patterns with regard to the occurrence of *Holopedium gibberum* versus *Daphnia* spp. Both biomass and frequency of *Holopedium* were negatively correlated with pH and Ca-concentration, and this species rarely occurred at pH > 8, or Ca-concentrations > 10 mg l^{-1}. On the contrary, the various *Daphnia* species, although showing different susceptibilities to fish predation and different preferences for lake productivity, were all positively correlated with Ca. Statistical analysis as well as laboratory experiments and biochemical considerations, suggest that *Holopedium* may be a superior competitor at very low Ca-concentrations, but is replaced by *Daphnia* at higher pH and increasing Ca-content. High pH in itself is possibly disadvantageous for *Holopedium* by preventing formation of the muco-polysaccaride mantle.

Introduction

The cladocerans *Holopedium gibberum* and *Daphnia* spp. are common inhabitants of lakes in the northern hemisphere. While *Daphnia* species have colonized almost every type of lake, though with different species found in lakes of different productivity, *Holopedium* is largely confined to oligo- and mesotrophic softwater lakes (Hamilton, 1958; Pejler, 1965). Their distribution pattern will obviously differ along gradients of both abiotic and biotic parameters, which will define their multidimensional niches. When extensive niche overlap occurs, competitive interactions are expected. Intraspecific competition would be a major factor in plankton communities, since members of the same species, or clone mates, necessarily have nearly complete niche overlap in many respects. Interspecific competition will, in contrast, be confined to narrower zones of niche overlap. However, since species (or higher taxa) rather than individuals are the conventional unit in ecological studies, most emphasis has been devoted to interspecific interactions. Regarding *Holopedium* and *Daphnia*, quite pronounced competitive interactions might be expected to occur, since they are both filter-feeders largely competing for the same range of food particles (Hessen, 1985; Hessen *et al.*, 1989), and co-occurrence frequently occurs in soft-water lakes which are poor in nutrients with low food availability and low fish predation pressure.

Evidently, coexistence and pronounced niche overlap of potential competitors is commonly observed (Jacobs, 1977; Seitz, 1980; Matveev, 1983, 1985; Ghilarov, 1984; Geller, 1985), and such stable coexistence may persist for years (Hessen *et al.*, 1989). Yet species with close spatial and nutritional niches would be expected to influence each other negatively; the competitive outcome is unpredictable, and competitive exclusion often is not observed (Allen, 1973; Neill, 1975; Lynch, 1978; Goulden *et al.*, 1982). The mechanisms allowing coexistence of zooplankton competitors are numerous, and it might be questioned whether coexistence of competitors should still be considered a 'paradox' (Hutchinson, 1961; Ghilarov, 1984).

Among biotic interactions promoting coexistence, modifying predation is perhaps the most powerful, and frequently reported in zooplankton communities (Allen, 1973; Jacobs, 1977; Tessier, 1986a, b). Fish predation is well recognized as a major determinant for

Table 1. Major parameters for the 342 lakes. Phytoplankton, zooplankton, chlorophyll a, total phosphorus and total nitrogen from July and August samples, calcium and pH from August samples. Phytoplankton and zooplankton as μg DW/l, other parameters as μg/l.

	Average	Median	Max.	Min.
Phytoplankton	115.8	28.2	3500	4.0
Zooplankton	80.4	51.8	1078	0.5
Chlorophyll a	5.6	2.2	131	0.4
Total P	12.4	7.0	246	1.8
Total N	377	288	2539	67
Ca	4.0	2.4	44	0.4
pH	6.5	6.5	8.7	4.7

size distribution and occurrence of *Daphnia* species, and this holds also for *Holopedium*. Stenson (1973) found fish predation to be the most important factor for *Holopedium* abundance in eight forest lakes. Also age structure, differential utilization of food at various life stages, stored energy and presence of other species may determine the competitive outcome between two species (Allen, 1973; Neill, 1975; Lynch, 1978; Goulden *et al.*, 1982; Tessier, 1986; Hoenicke & Goldman, 1987).

While all these biotic parameters are well recognized as modifiers and determinants for competitive outcome, the abiotic factors are less well-studied. In particular for *Holopedium*, the total absence of this species in nutrient-rich lakes is striking, and one may speculate whether this restricted distribution may be explained by physicochemical constraints rather than competitive inferiority. *Holopedium* is recognized as an indicator of oligotrophic lakes (Pejler, 1965), of low calcium (Ca), low pH (Hamilton, 1958; Yan *et al.*, 1988), and low humic content (Weiser, 1942). Hamilton (1958) states that *Holopedium* is restricted to lakes with pH less than 7.5 and Ca < 20 mg l^{-1}. While it frequently occurs in humic lakes, this is by no means a prerequisite (Yan *et al.*, 1988). Its avoidance of eutrophic, high pH and high Ca waters seem, however, to be a generic property of *Holopedium*. A regional survey of 346 Norwegian lakes, most of them extreme softwater lakes, offered a possibility to test some of these hypotheses. Here we examine and focus on the role of lake productivity and central chemical parameters like pH (both high and low) and calcium (Ca) on the relative success of *Holopedium* versus *Daphnia* species.

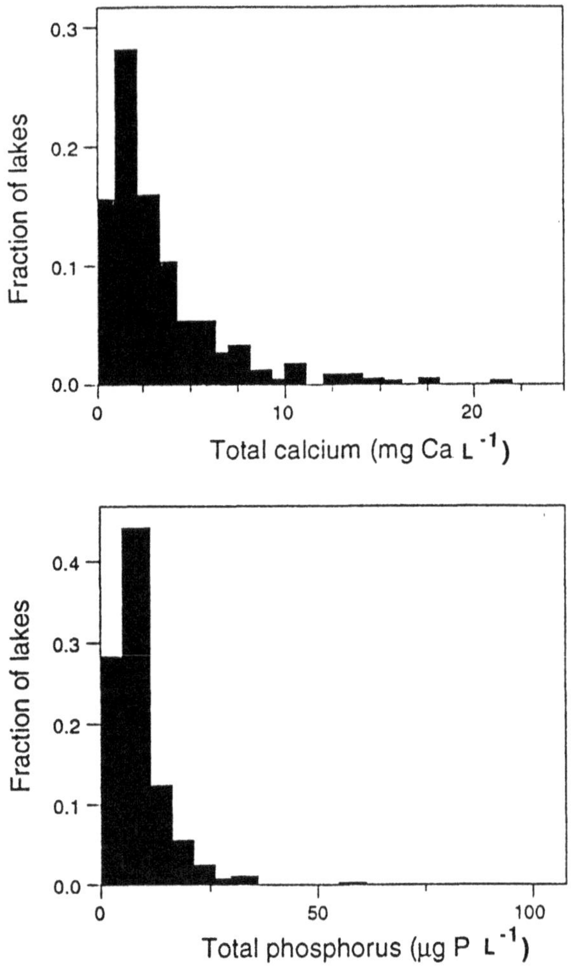

Fig. 1. Distribution of total phosphorus and calcium in the 346 lakes surveyed.

Materials and methods

During 1988, 346 lakes in Norway were sampled in May, June, July and August for the following parameters: total phosphorus (TP), total nitrogen, chlorophyll a and phytoplankton. Zooplankton were sampled in July and August from an integrated sample from surface to twice the secchi-depth. Except from smaller volumes in a few, turbid lakes, 10 l of water from a van Dorn sampler were mixed and screened through a 45 μm mesh. The samples were fixed with acid Lugol's solution, and biomasses were determined from established length-dry-weight regressions (Bottrell *et al.*, 1976). For the present analysis, we selected samples only from July and August for all parameters, to match the zooplankton data. pH and calcium concentrations were analyzed from the August samples. A description

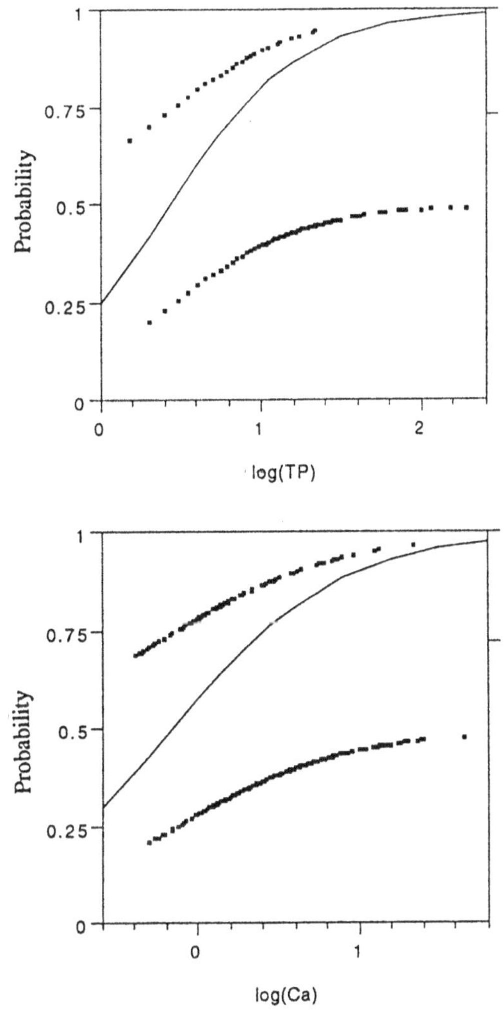

Fig. 2. Univariate logistic regressions (eq. 1) of *Daphnia* spp. occurrence on total phosphorus (TP; upper panel) or calcium (Ca; lower panel). Based on presence/absence data from 346 lakes for all *Daphnia* species pooled. Solid line: probability of occurrence as predicted by the model. Dots: ranked concentrations in individual lakes without *Daphnia* spp. (above solid line), and with *Daphnia* spp. (below solid line).

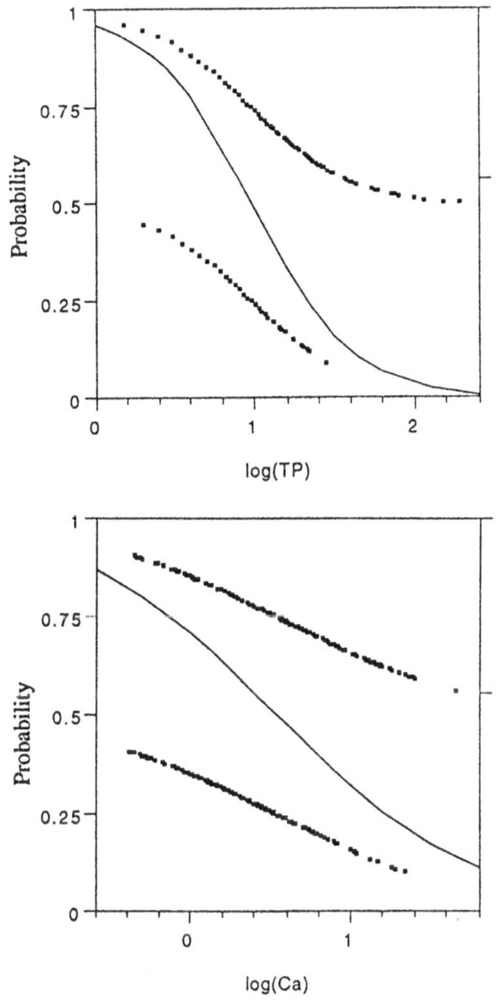

Fig. 3. Univariate logistic regressions (eq. 1) of *Holopedium gibberum* occurrence on total phosphorus (TP; upper panel) or calcium (Ca; lower panel). Based on presence/absence data from 346 lakes. Solid line: probability of occurrence as predicted by the model. Dots: ranked concentrations in individual lakes without *Holopedium* (above solid line), and with *Holopedium* (below solid line).

of important lake parameters are given in Table 1. The lakes were selected according to certain criteria; they should preferably be of the same size. Ninety-four percent of the lakes had a surface area > 1 km^2 (average 11.8 km^2), and all had a maximum depth > 15 m. The 20 largest Norwegian lakes were included among these. Alpine lakes and lakes with high flushing rates were avoided, as were strongly acidified lakes. Only 2% of the lakes had a pH < 5.0. The lakes covered latitudes between 65°N and 77°N, and represented a range from oligotrophy to eutrophy, with total P in the range 3–246 $\mu g\ l^{-1}$, but with the majority of lakes below 50 μg tot P l^{-1} (Fig. 1) and a median of 7.0 tot P l^{-1}. The same sampling and analytical procedures were followed at all localities. All parameters were analyzed from integrated samples over twice the secchi-depth transparency (max 0–10, min 0–2 m). P and N were analyzed by a Technicon Autoanalyzer after persulfate digestion. Chlorophyll *a* was measured spectrophotometrically after acetone extraction, and calcium was measured by atomic absorption spectrophotometery.

As a simple screening test of pH-tolerance in *Holopedium*, a series of incubations were performed in low-buffered lake-water over a range of pH. Populations of *Holopedium* and *Bosmina longispina* from a humic, acidic lake (pH 4.5) were incubated in the laboratory in untreated lake water and pH of 6.1, 6.5, 7.5 and 8.9. pH adjustments were performed by adding 0.1 N NaOH (from 3 drops to 1.2 ml in 250 ml beakers), which would change the ionic strength marginally. We added 10 individuals of *B. longispina* and *H. gibberum* to each beaker in non-replicated treatments. The animals were fed equal amounts of the green algae *Selenastrum capricornutum* (in excess) and survival, egg production and status of the gelatinous mantle were monitored for 14 days.

Zooplankton relationship to Ca and P was assessed by logistic regression. This is a general technique for describing ordinal or nominal dependent variables (for example, species presence/absence data) as functions of (continuous or discrete) dependent variables (see e.g. Trexler & Travis (1993) for a recent review of ecological applications). If we denote the probability that a given event occurs by p, the standard logistic regression model assumes that the probability ratio $\ln(p/1-p)$ is a linear function of the independent variable x:

$$\ln\frac{p}{1-p} = a_0 + a_1 x. \quad (1)$$

Logistic regression models can also be used to describe situations where a species is assumed to have a unimodal optimum in an environmental gradient, by fitting the probability ratio to a quadratic polynomial in the environmental variable with a polynomial logistic regression (ter Braak & Looman, 1986). By a slight extension of the method of ter Braak & Looman (1986), we can make a probabilistic model of a species' occurrence in a two-dimensional environmental gradient by fitting species presence/absence data to a logistic regression model with a bivariate quadratic function (a quadratic response surface) in the independent environmental variables x and y:

$$\ln\frac{p}{1-p} = a_0 + a_1 + a_2 x^2 + a_3 y + a_4 y^2 + a_5 xy. \quad (2)$$

We used JMP 2.04 (SAS Institute Inc. Cary, NC) for all statistical computations in this work, and MATLAB 4.0 (The MathWorks Inc. Natick, MA) for visualization of response surfaces.

Results

As would be expected, TP was a good predictor of algal biomass (log TP *vs* Log Chl. a: $r^2 = 0.79, p < 0.0001$), and this close correlation justifies the use of TP as a trophic state variable. Similarily, pH and Ca were also well correlated (pH vs. Log Ca: $r^2 = 0.62, p < 0.0001$). On the other hand, TP was weakly correlated with Ca ($r^2 = 0.12$), yet significant owing to the high sample size, and not significantly related to pH ($r^2 = 0.006$). This suggests that abiotic bicarbonate buffering rather than primary production was the main determinant of pH, at least during August when the abiotic samples were taken. An *a priori* constraint of the data would thus be that the role of Ca versus TP (or pH versus algal biomass) could be assessed, while the effects of Ca relative to pH, or TP relative to algal biomass could not be easily separated. Acidic lakes were deliberately avoided in this study, however, and lake pH *per se* would not be expected to be a major determinant to distribution patterns in this survey. The pH range was small, only seven lakes (2%) with pH < 5 and 10 lakes (2.9%) > 7.5. Nearly 80% of the lakes were within a pH-range of 6.0–7.5.

In general, cladocerans were by far the most dominant subgroup among the metazoan zooplankton, contributing nearly 50% of total biomass, followed by the calanoid copepods (*c.* 25% of total biomass), while cyclopoid copepods and rotifers contributed on average only slightly more than 10% each. Of a total of 346 lakes, some cladoceran species were represented in more than 90% of these. *Bosmina* spp. was by far the most abundant group, with *Bosmina longispina* as the most abundant species. *Daphnia* spp. were also present in the majority of the surveyed lakes, though with highly differing species compositions in the various lakes. *Holopedium gibberum* is a common inhabitant of oligotrophic, softwater lakes in Scandinavia (Pejler, 1965), and occurred in 136 of the surveyed localities.

Simple logistic regression models (eq. 1) for the occurrence of *Holopedium gibberum* and *Daphnia* spp. along gradients of Ca and TP (Fig. 2) revealed a strongly increasing probability that *Daphnia* spp. would be present with increasing concentrations of both Ca and TP ($p < 0.0001$ for both). The opposite pattern was found for *Holopedium* (Fig. 3), though this species was apparently more negatively influenced by TP than Ca. The probability of *Holopedium* occurring above a TP of 40 $\mu g\ l^{-1}$ was only 10%, while its occurrence probability was more than 80% for TP below 3 $\mu g\ l^{-1}$.

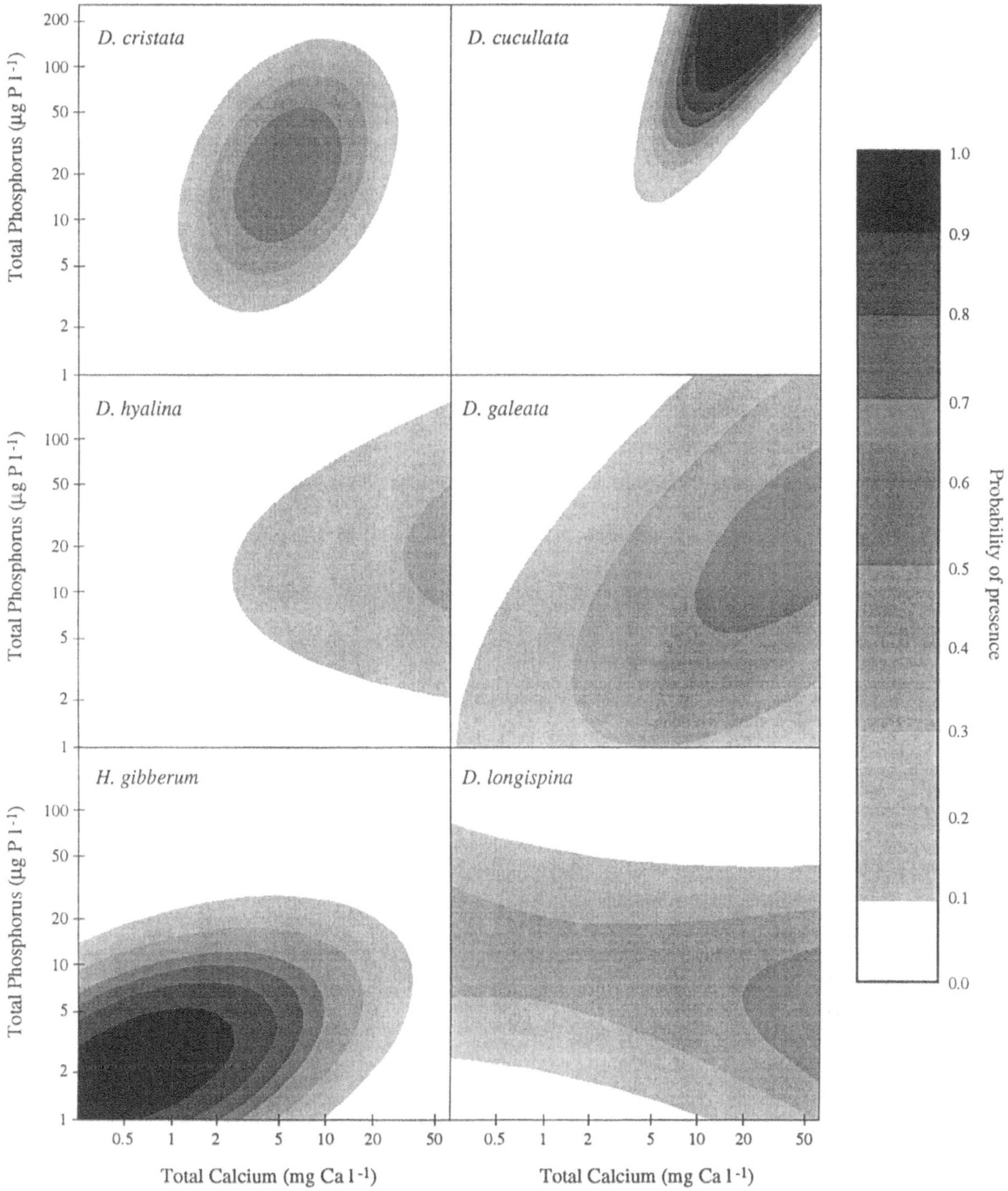

Fig. 4. Response surfaces for bivariate logistic regressions (eq. 2) of individual species' occurrences in a two-dimensional gradient of total phosphorus (TP) and calcium (Ca). Regressions based on presence/absence data from 346 lakes. Areas with same shading have same probability range for occurrence of a species (probability scale on the right side).

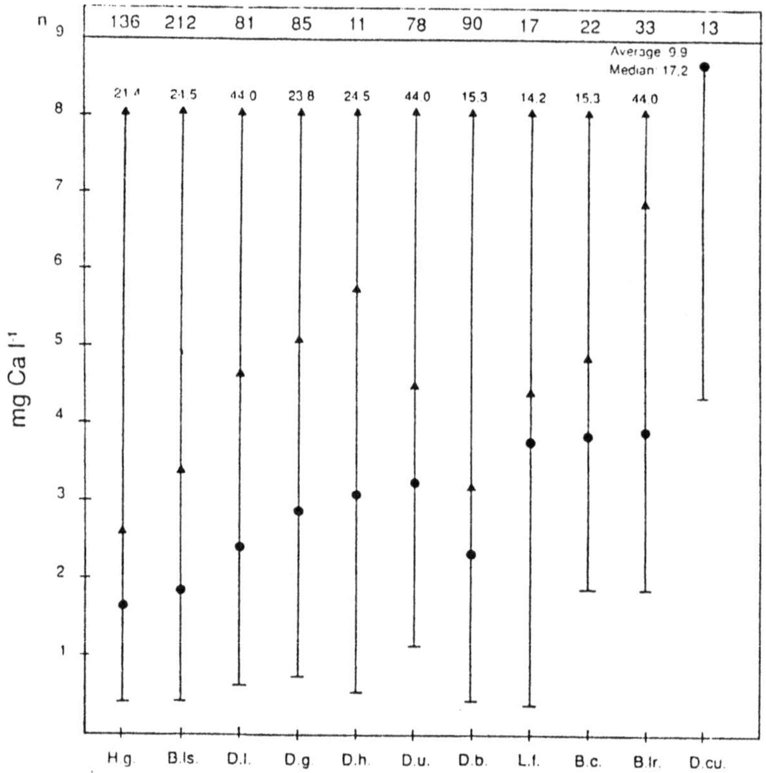

Fig. 5. Occurrence along Ca-concentrations for common cladoceran species included in the survey. Total range given as vertical line with maximum concentrations for recorded occurrence given above. Circles denote mean and triangles denote median concentrations of Ca in lakes where the species occur. N = number of localities. H.g.: *Holopedium gibberum*, B.ls.: *Bosmina longispina*, D.l.: *Daphnia longispina*, D.g.: *Daphnia galeata*, D.h.: *Daphnia hyalina*, D.u: *Daphnia cristata* D.b.: *Diaphanosoma brachyurum*, L.f.: *Limnosida frontosa*, B.lr.: *Bosmina longirostris*, D.cu: *Daphnia cucullata.*

The negative relationship was highly significant ($p < 0.0001$) for both parameters.

When decoupling *Daphnia* spp. at the species level, a more subtle pattern emerges. Figure 4 shows two-dimensional logistic regression models (eq. 2), which predict the presence of various *Daphnia* species in relation to total P and Ca. The justification for this statistical treatment is based on the recognition that Ca and TP are weakly correlated variables, while Ca and pH as well as TP and phytoplankton biomass were highly correlated. All *Daphnia* species (particularly *D. hyalina* and *D. cucullata*) showed preference for high Ca, and *D. cucullata* also showed preference for high TP. On the contrary, *Holopedium* was negatively related to both these parameters. This would indicate that Ca was relatively more important than TP for most daphnids, while *Holopedium* avoided both high TP and high Ca.

As illustrated in Fig. 5, most of the pelagic herbivorous cladocerans do, however, occur over most of the range of Ca covered in this study. The extremes at

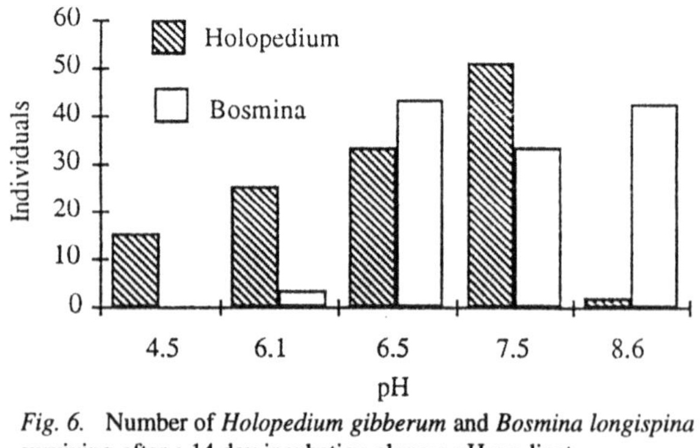

Fig. 6. Number of *Holopedium gibberum* and *Bosmina longispina* surviving after a 14 day incubation along a pH gradient.

the lower and upper ends of the scale are *Holopedium* and *D. cucullata*, respectively. *Holopedium* are found largely over the same range of geographical and lake type distribution as the most common *Daphnia* species, *D. longispina, D. galeata* and to some extent *D. cristata*, and quite frequently co-occur with these species.

The same holds for *Bosmina longispina*. *Holopedium* is, however, clearly negatively related to the presence of all *Daphnia* species, while it is nearly neutral to *Bosmina longispina* (Hessen *et al.*, 1995). This suggests some indication of competitive interactions, yet an apparent avoidance could just as well be attributed to physicochemical constraint and optima.

The simple test for pH tolerance in *Holopedium* indicated an increased success with pH increasing from 4.5 to 7.5, but reduced survival at pH 8.6 within 14 days (Fig. 6). Initial pH was in this case 8.9, but it was drifting towards 8.5 at the end of the experiment. After 14 days, pH of the control treatment was raised to 8.2, giving total mortality within 10 days (data not shown in Fig. 6). The animals in this treatment could be more susceptible than those taken directly from the lake, but nevertheless this suggests that a pH above 8 would be harmful to *Holopedium*. Individuals in this high pH treatment did not produce eggs. There was no indication of a pH-mediated erosion of the mucopolysaccharide mantle, prior to death. *Bosmina*, which were present in the same flasks, seemed to be unaffected at this pH.

Discussion

The 346 lakes surveyed covered quite a narrow range of lake productivity and calcium content, and with few exceptions, most species occurred over most of this range. There was however quite a distinct segregation between *Holopedium gibberum* and *Daphnia* spp., the former primarily occupying low-calcium and low productivity lakes. While there obviously are competitive interactions between the various *Daphnia* species as well, this discussion will focus on some of the major physiological and life-history strategy differences between *Daphnia* and *Holopedium*, and how this could affect their distribution and competitive abilities under different environmental conditions. Both groups are highly susceptible to fish predation, yet the observed distribution patterns for *Holopedium* relative to *Daphnia* could not be attributed to the fish communities of these lakes (Hessen *et al.*, 1995).

A most obvious difference between the two genera is in their external structure, with *Daphnia* possessing the ordinary calcium/chitin layer like most other crustaceans, while *Holopedium* is surrounded by a gelatinous muco-polysaccharide mantle. The relative role of a carapace compared to the mantle in energy balance and carbon metabolism of the animals is not straightforward. In *Daphnia*, exuviae constitute close to 10% of body carbon (Lynch *et al.*, 1986), and on average the animals pay a tax of almost 5% of body carbon daily in molting losses (Andersen, 1993). The metabolic costs of the mantle synthesis is less clear. Tessier (1986a) demonstrated that the sheath could make up more than 30% of dry weight (DW) in juveniles, but this declines to less than 8% in adults. Release of surplus carbon as extracellular polysaccharides is a common property of bacteria and certain algae, and it can be speculated that *Holopedium*'s mantle is made of 'waste' carbon allocated into a structural component. If so, *Daphnia* would allocate a greater proportion of ingested carbon into non-somatic tissue relative to *Holopedium*. This remains simply speculative so far, and the competitive success of *Holopedium vs Daphnia* at low food is not clear-cut. Tessier (1986b) indicated that the mantle would be a competitive drawback. *Holopedium* have, however, far higher specific filtering rates than *Daphnia* (Hessen, 1985; Hessen *et al.*, 1989) which certainly could be profitable under low food concentrations.

Niche separation in the preferred size spectrum of food would obviously counter-act competitive effects. While *Daphnia* feed on bacteria and algae with almost equal efficiency, *Holopedium* is regarded as a typical 'macrofiltrator' (Hessen, 1985; Persson, 1985) preferring particles in the size-range of 10–30 μm equivalent diameter. It is, however, still capable of including a large proportion of bacteria in its diet (Hessen *et al.*, 1989). Though even if they might be arbitrarily regarded as 'micro-' and 'macrofiltrators' in relative terms, they do indeed have a considerable overlap in food size spectra.

The differences between an ordinary carapace and a polysaccharide mantle implies also fundamental biochemical differences which might have consequences beyond energy or carbon budgets. There are at least three properties of the mantle, relative to the carapace, that might be relevant to the observed distributional patterns. First, the mantle is low in Ca. Second, it is low in phosphorus, and third, the acid mucopolysaccharides might not be easily synthesized at high pH.

Some aspects of relative calcium requirements may be obtained from analysis of elemental content. Yan & Mackie (1989) reported a Ca content of 2.9% of DW for *Daphnia pulex* (range 2.1–4.8), in contrast to only 0.4% (range 0.2–0.5) in *Holopedium*. The high Ca-content is apparently a generic property of *Daphnia* spp. (see Yan & Mackie, 1989). This seven-fold difference in relative Ca-content suggests a higher requirement for this mineral in *Daphnia*. This may also be because a major proportion of Ca would be associated with the carapace, and hence would need to be re-allocated after molting. While there is so far no direct evidence for the possibility of strict Ca-deficiency in zooplankton, the strong positive correlation between Ca and occurrence of *Daphnia* is supportive of such an hypothesis. Tessier & Horwitz (1990) proposed Ca-deficiency in crustacean zooplankton as a major reason for their observed transition from large-bodied to small-bodied species along with decreasing lake alkalinity. Since there also are different Ca-requirements among the crustaceans, and in softwater lakes like those of the present study where Ca-concentrations frequently are below 1 $\mu g\ l^{-1}$, we suggest that low Ca correspondingly may promote the competitive superiority of the low-calcium demanding *Holopedium*.

The high specific content of phosphorus in *Daphnia* relative to other planktonic crustaceans is well-recognized (Andersen & Hessen, 1991). While *Daphnia*, on average have a value of 1.5% of dry-weight as P, *Holopedium* has a significantly lower P-content (close to 1%). This difference in P-content is partly because of the presence of P in a Daphnid's exoskeleton. The difference is not trivial, since *Daphnia* thus have a higher P-requirement and there is recent evidence that *Daphnia* at times may be limited by a shortage of P (Hessen, 1992; Sterner & Hessen, 1994).

The statistical analysis gave evidence also for a strong effect of pH on the relative success of *Daphnia* spp. vs. *Holopedium* as shown previously (Yan *et al.*, 1988). *Holopedium* is in general far more tolerant to low pH than *Daphnia*, but in the present material where acid lakes were deliberately avoided, the observed pattern was attributed to *Holopedium*'s avoidance of lakes with high pH. The close coupling between Ca and pH does not allow a strict separation of the effects of these two parameters. A direct toxic effect of high Ca is unlikely, and thus the apparently negative effect of high Ca would most likely be an indirect effect of increased competitive ability of *Daphnia* (and possibly other zooplankton groups) under high Ca-availability. In nature, *Holopedium* do occur in lakes at pH above 8. The simple laboratory test did, however, indicate reduced survival for *Holopedium* at high pH, and we suspect high pH, in itself, to have the potential of affecting zooplankton community composition. This could be linked to the acid mucopolysaccharides of *Holopedium*'s mantle, which might be susceptible to high pH, though there was no visual evidence for a direct erosion of the mantle under high pH.

In conclusion, central physical and chemical factors like pH and Ca-concentration may have profound impacts on zooplankton community composition, and determine competitive abilities of dominant herbivores. The different species have different strategies to cope with these physicochemical challenges. Most remarkably, the family *Holopedidae*, in Europe represented by the single species *H. gibberum*, has chosen a structural strategy quite distinctly different from other Cladocera, by having replaced an ordinary carapace with a gelatinous mantle. While this probably is advantageous for reasons like protection against invertebrate predators and buoyancy control, it would probably also be an energetically favourable structure under low food conditions and in acidic environments. Also, the mantle diminishes requirements for minerals like phosphorus and calcium, which could be restricting factors in nutrient poor soft waters. Although the test here (Fig. 6) is only indicative, it nevertheless suggests that high pH is detrimental to *Holopedium*. In addition, *Holopedium* is apparently also a less efficient competitor under more eutrophic conditions. Extremes in pH or (low) Ca could thus place physiological constraints on the distributions of *Holopedium* and *Daphnia* which would operate in opposite directions, and thus affect their physicochemical niches. Changes in abiotic and biotic parameters within these limits will largely determine their competitive success. In evolutionary terms, one might speculate whether this may be regarded as a less successful strategy, since the family *Holopedidae* is represented by only two species in contrast to the numerous species in most other herbivorous cladoceran families.

References

Allen, J. D., 1973. Competition and the relative abundances of two cladocerans. Ecology 54: 485–498.

Andersen, T, 1993. Grazers as sources and sinks for nutrients. Dr. phil. thesis, University of Oslo, 263 pp.

Andersen, T. & D. O. Hessen, 1991. Carbon, nitrogen, and phosphorus content of freshwater zooplankton. Limnol. Oceanogr. 36: 807–814.

Botrell, H. H., A. Duncan, Z. M. Gliwicz, E. Grygierek, A. Herzig, A. Illbricht-Ilkowska, H. Kurasawa, P. Larsson & T. Weglenska, 1976. A review of some problems in zooplankton production studies. Norw. J. Zool. 24: 419–456.

Geller, W., 1985. Production, food utilization and losses of two coexisting, ecologically different *Daphnia* species. Arch. Hydrobiol. Beih. 21: 67–79.

Ghilarov, A. M., 1984. The paradox of the plankton reconsidered; or, why do species coexist? Oikos 43: 46–52.

Goulden, C. E., L. L. Henry & A. J. Tessier, 1982. Body size, energy reserves, and competitive ability in three species of cladocera. Ecology 63: 1780–1789.

Hamilton, J. D., 1958. On the biology of *Holopedium gibberum* Zaddach (Crustacea: Cladocera). Verh. int. Ver. Limnol. 13: 785–788.

Hessen, D. O., 1985. Filtering structures and particle size selection in coexisting Cladocera. Oecologia 66: 368–372.

Hessen, D. O., 1992. Nutrient element limitation of zooplankton production. Am. Nat. 140: 799–814.

Hessen, D. O., T. Andersen & A. Lyche, 1989. Differential grazing and resource utilization of zooplankton in a humic lake. Arch. Hydrobiol. 114: 321–347.

Hessen, D. O., B. Faafeng & T. Andersen, 1995. Replacement of herbivorous zooplankton species along gradients of ecosystem productivity and fish predation pressure. Can. J. Fish. aquat. Sci. (in press).

Hoenicke, R. & C. R. Goldman, 1987. Resource dynamics and seasonal changes in competitive interactions among three cladoceran species. J. Plankton Res. 9: 397–417.

Hutchinson, G. E., 1961. The paradox of the plankton. Am. Nat. 95: 137–145.

Jacobs, J., 1977. Coexistence of similar zooplankton species by differential adaptation to reproduction and escape in an environment with fluctuating food and enemy densities. II. Field data analysis of Daphnia. Oecologia (Berl.) 35: 35–54.

Lynch, M., 1978. Complex interactions between natural coexploiters – *Daphnia* and *Ceriodaphnia*. Ecology 59: 552–564.

Lynch, M., L. J. Weider & W. Lampert, 1986. Measurement of the carbon balance in *Daphnia*. Limnol. Oceanogr. 31: 17–33.

Matveev, V. F., 1983. Estimating competition in cladocerans using data on dynamics of clutch size and population density. Int. Rev. ges. Hydrobiol. 68: 785–798.

Matveev, V. F., 1985. Competition and population time lags in *Bosmina* (Cladocera, Crustacea). Int. Rev. ges. Hydrobiol. 70: 491–508.

Neill, W. E., 1975. Experimental studies on microcrustacean competition, community composition and efficiency of resource utilization. Ecology 56: 809–826.

Pejler, B., 1965. Regional-ecological studies of Swedish freshwater zooplankton. Zool. Bidr. Upps. 36: 407–515.

Persson, G., 1985. Community grazing and the regulation of in situ clearance and feeding rates of planktonic crustaceans in lakes in the Kuokkel area, northern Sweden. Arch. Hydrobiol. Suppl. 70: 197–238.

Seitz, A., 1980. The coexistence of three species of *Daphnia* in the Klostersee. I. Field studies on the dynamics of reproduction. Oecologia (Berl.).45: 117–130.

Stenson, J. A. E., 1973. On predation and *Holopedium gibberum* (Zaddach) distribution. Limnol. Oceanogr. 18: 1005–1010.

Sterner, R. W. & D. O. Hessen, 1994. Algal nutrient limitation and the nutrition of aquatic herbivores. Ann. Rev. Ecol. Syst. 25: 1–29.

ter Braak, C. J. F. & C. W. N. Looman, 1986. Weighted averaging, logistic regression and the Gaussian response model. Vegetatio 65: 3–11.

Tessier, A. J., 1986a. Life history and body size evolution in *Holopedium gibberum* Zaddach (Crustacea: Cladocera). Freshwat. Biol. 16: 279–286.

Tessier, A. J., 1986b. Comparative population regulation of two planktonic Cladocera (*Holopedium gibberum* and *Daphnia catawba*). Ecology 67: 285–302.

Tessier, A. J. & R. J. Horwitz, 1990. Influence of water chemistry on size structure of zooplankton assemblages. Can. J. Fish. aquat. Sci. 47: 1937–1943.

Trexler, J. C. & J. Travis, 1993. Nontraditional regression analyses. Ecology 74(6): 1629–1637.

Weiser, J., 1942. *Holopedium gibberum* auf der Böhmische-Märischen Anhöhe. Arch. Hydrobiol. 39: 334–352.

Yan, N. & G. L. Mackie, 1989. Seasonal patterns in metal levels of the net plankton of three Canadian Shield lakes. Sci. Total Env. 87/88: 439–461.

Yan, N. D., W. Keller, J. R. Pitblado & G. L. Macjie, 1988. *Daphnia-Holopedium* relationships in Canadian Shield lakes ranging in acidity. Verh. int. Ver. Limnol. 23: 252–257.

Hydrobiologia **307**: 263–272, 1995.
P. Larsson & L.J. Weider (eds), Cladocera as Model Organisms in Biology.

Interactions between Nile tilapia (*Oreochromis niloticus*) and cladocerans in ponds (Khartoum, Sudan)

Fatin Abdel Razig Elhigzi, Shawkat Ara Haider & Petter Larsson
Institute of Zoology, University of Bergen, N-5007, Norway

Key words: tropical cladocera, cladocera populations, size distribution, tilapia feeding, individual fish growth rate

Abstract

Juvenile Nile tilapia (*Oreochromis niloticus*) are omnivorous, and the question asked in this study is how they affect on their environment? Do they mainly act as predators on the cladoceran zooplankton or do they compete with the cladocerans for phytoplankton? This problem was studied in three ponds with and three ponds without small tilapia (3–5 cm). The fish growth rate, the succession of plankton species and the changes in abiotic conditions, were monitored over a period of 67 days. The fish biomass was kept low and the mean was approximately constant (12.6 g m^{-2}) during the experiment. Phosphate was added to avoid phytoplankton nutrient limitation. Although the diet of Nile tilapia contained both phytoplankton and zooplankton, the fish affected the ecosystem in a similar way as zooplanktivorous fish. The fish ponds got more phytoplankton due to increase of Chlorophyta. Effects on the other phytoplankton groups Euglenophyta, Bacillariophyta, Cryptophyta and Cyanophyta could not be registered. The ponds without fish had higher densities of *Daphnia lumholtzi* and *D. barbata*. The other Cladocerans seemed less influenced by fish presence. The relative fish growth rate was most positively correlated with the density of *Daphnia lumholtzi*, *Diaphanosmoa excisum* and *Bosmina longirostris*. Tilapia seemes to have two feeding modes: (1) preying on large zooplankton and (2) unselective filtration of small planktonic organisms such as phytoplankton. In our experiment the first feeding mode affected the ecosystem more than the second.

Introduction

Since the innovative papers of Hrbacek *et al.* (1961) and Brooks & Dodson (1965), it has been repeatedly demonstrated that visually feeding zooplanktivorous fish selectively remove large and more conspicous zooplankton, mainly the large cladocerans (reviewed in Zaret, 1980; Gulati *et al.*, 1990; Carpenter & Kitchell, 1993). Left back are small crustaceans and rotifers. The large cladocerans are efficient phytoplankton grazers (Shapiro & Wright, 1984; Lampert *et al.*, 1986), and if they are removed, algal biomass may increase.

Studies on how community structure is affected by fish, are almost exclusively done with carnivorous fish (i.e. zooplanktivores and piscivores). The effects of omnivorous fish on ecosystems have been examined only in a few cases (Drenner *et al.*, 1984; Drenner *et al.*, 1986; Threlkeld, 1987; Diana *et al.*, 1991). The question is whether fish with a combined diet (i.e. feeding both on phytoplankton and zooplankton) will affect the ecosystem in the same way as zooplanktivorous fish or not. Will they act as predators or competitors to the cladocerans.

If they act as competitors, one should expect that phytoplankton instead of increasing in biomass, should be reduced in the presence of fish. Instead of reducing large Cladocera, i.e. *Daphnia*, these forms should increase relative to other zooplankton groups, as they are regarded as the best competitors among the zooplankton (Gliwicz, 1990).

To investigate this question, we selected Nile tilapia (*Oreochromis niloticus*). As juveniles they are omnivorous, but they switch to a more pure phytoplankton diet when they become larger (Moriarty *et al.*, 1973; Bowen, 1982). However, if kept under aquaculture conditions, they still retain a flexible diet, and because of this, *O. niloticus* is of worldwide economic importance in aquaculture (Pullin, 1991). *Oreochromis niloticus* is also the only indigenous species cultivated in the Sudan (Mishrigi, 1974; George, 1975; Yousif, 1988).

In this study the development of biotic and abiotic factors in ponds with and without fish was studied at the Fisheries Research Centre (FRC), Agricultural Research Corporation, Ministry of Agriculture and Natural Resources, Khartoum, Sudan.

Material and methods

Experimental design

The experiment lasted for a period of 67 days, from the 24th of February to the 2nd of May 1990. The ponds were drained and the fish harvested, counted and measured on 9th May (day 74). Nutrients were only analysed for the first 28 days.

Six rectangular concrete ponds (10×4 m^2), arranged in a row, were filled to a depth of 1 m with water pumped from the White Nile River. Wire screens with 2 mm mesh size at the inlets of all ponds prevented the entrance of fish larvae. Water was added to maintain a constant level during the experiment. Three replicate ponds were randomly assigned for the fish treatment and three for controls without fish. The three fish ponds were each stocked with 200 young tilapia from the White Nile. They had sizes ranging between 3–5 cm in total length. The ponds were fertilized with triple superposphate (TSP) to avoid nutrient limitation. TSP was spread daily on all ponds at a dose of 8 g pond^{-1} d^{-1} (60 kg ha^{-1} month^{-1}) which is believed to be optimal dosing (Maar *et al.*, 1966; Shedadeh, 1976).

Sampling and analysis

All ponds were sampled every other day for physical, chemical and biological variables during the first month of the experiment. During the second month, ponds were sampled twice and three times weekly for biotic and abiotic variables, respectively.

Integrated water samples from five different stations per pond were collected randomly. A 5 l plexiglass tube (1 m in length) closed with a silicon stopper, was used to sample the whole water column. Samples were taken between 10.00 and 11.00 hrs. during each sampling bout. Surface and close to bottom dissolved oxygen, surface water temperature (oxygenmeter, YSI), conductivity, and pH (Radiometer) were measured directly in the field. Total phosphorus and total nitrogen were analyzed at Hordaland Fylkes Laboratorium in Bergen. Samples were filtered through glass fiber filters (GF/F), and were fixed with 1% concentrated H_2SO_4.

Phytoplankton samples (20 ml) were taken directly from the integrated sample and fixed with Lugol's solution. The algae were identified to genus and counted under an inverted microscope (Utermöhl, 1958). Biovolume was estimated from the numbers and mean volume of the algal cells. Zooplankton composition was estimated by filtering the reminder of the integrated water sample through a 45 μm net, and preserving it with 4% formalin.

Sizes (mm) of cladocerans were measured as body lengths (excluding spines and helmet peaks). Mean individual size for each species was determined by randomly selecting 50 individuals (if not so many all in the samples) per sampling date in each pond.

Fish were caught with a drawing net on six different dates during the course of the experiment. The length and weight of the fish were measured before they were released back into the pond. Individual growth rate, g, per day was calculated as

$$g = 2(w_2 - w_1)/\{(w_1 + w_2)(t_2 - t_1)\},$$

where w_1 and w_2 are the mean individual weights from each pond at time t_1 and t_2 respectively (Ricker, 1975). A substantial portion of each fish population was caught during each sampling bout, but the catches provided only rough approximations of the population size. Exact numbers of fish in each pond were known when stocking at the beginning of the experiment and at the end when the ponds were emptied. On each fish sampling date, stomach samples from two fishes per pond were collected to see if the diet contained both phytoplankton and zooplankton.

Statistics

A Mann-Whitney U-test was used to determine the differences between ponds with and without fish. The measured values from all dates and ponds were ranked and the difference in sum of ranks were tested. Zero values of densities (zooplankton) or biovolumes (phytoplankton) were omitted when registered in all ponds on the same date.

Correlation analyses were conducted to analyse the interrelationship between individual fish growth rate and environmental factors. The data were log $(x+1)$

Table 1. Number and weight of fish stocked, sampled/reintroduced and harvested in the three ponds with fish: F1, F2 and F3. Each sampling date is identified with the day number in the experiment. Mean individual weights of the fish and the fish biomass in the ponds, are given in grams. The value marked with * is extrapolated from a smoothed growth curve. Due to technical problems F3 was not sampled that date

Day number	Type	Individuals			Mean weights			Mean biomasses		
		F1	F2	F3	F1	F2	F3	F1	F2	F3
1	Stocking	200	200	200	1.59	2.82	2.82	318	564	564
11	Sampl./reintro.	38	48	43	1.95	3.08	3.32			
22	Sampl./reintro.	59	51	19	2.72	4.61	3.86			
33	Sampl./reintro.	30	7	–	4.16	5.99	6.6*			
44	Sampl./reintro.	29	28	28	7.72	9.81	10.28			
67	Sampl./reintro.	12	7	5	22.06	16.86	14.98			
74	Harvesting	31	25	12	27.32	21.45	15.83	847	536	190

transformed to achieve more normal distributed data and to get homogeneous variances.

Since both the Mann-Whitney U-test and the correlation tests were simultaneously done on 19 and 16 comparisons respectively, only $p<0.0025$ for the Mann-Whitney U-test and $p<0.003$ for the correlations, were considered as significant (Rice, 1989). Differences in individual sizes were tested with a single classification ANOVA. For this analysis $p<0.007$ was accepted as significant.

The data were analyzed using Statistic (Analytical Software, Tallahassee, Florida) and FigP for Windows (Biosoft, Cambridge, UK) computer programs.

Results

Fish

The mean biomass of fish (about 12 g m^{-2}), was approximately the same in the beginning and at the end of the experiment (Table 1). The 200 juveniles stocked in each pond at the start, were reduced during the course of the experiment. This was due to predation from a bird, the pied kingfisher, *Ceryle rudis*, which was observed frequently in the ponds. The growth of the fish, however, did to some extent compensate for the losses due to kingfisher predation. Pond F2 had approximately the same biomass at the start and at the end of the experiment, while in F1 the fish biomass increased by about 270%. In F3 the biomass decreased to about 34% of the original stocked biomass. The development of the three separate fish populations was quite similar until day 44, but from that day on, the three fish ponds diverged. Individual growth was lowest in pond F3, while the highest was in F1. Although there were variable fish densities in the fish ponds, they developed differently from those without fish.

The stomach content analysis detected mainly algae, but also rotifers and cladocerans of most species frequently occurred in the fish diet (Elhigzi, 1993). The most abundant animal in the stomachs was *Bosmina longirostris*, which was also by far the most abundant crustacean species in the ponds.

Zooplankton

The zooplankton species found in the ponds were all common in the White Nile (Brooks & Rzoska, 1954; Green, 1971; Hanna & Schiemer, 1993). The same species of zooplankton occurred in all treatments (Elhigzi, 1993). An estimated percentage of the density of the main zooplankton groups is given in (Table 2). Rotifers were most common in all ponds, while cladocerans were the second most common group among the metazoan zooplankton. In biomass the Cladocerans were dominating. Neither the copepods ($p=0.297$) the rotifers ($p=0.204$) nor the Cladocerans ($p=0.885$, Fig. 1) as groups, showed any significant difference in density between the ponds with fish and ponds without fish.

Cladocera

Among the seven species of Cladocera found, *Bosmina longirostris*, *Moina dubia*, and *Ceriodaphnia cornuta* (unhorned form) were the most abundant (Fig. 2a, b & c). *Diaphanosoma excisum*, *Daphnia lumholzi* and

Table 2. Mean % composition (based on numbers) of the main metazoan zooplankton in the ponds with and without fish

Zooplankton	Without fish	With fish
Rotifera	59.08	51.65
Cladocera	32.87	39.24
Cyclopoida (Copepoda)	5.22	7.23
Calanoida (Copepoda)	2.83	1.77

Table 3. Mean size (mm) and ANOVA probabilities for the size difference between cladocerans in ponds without fish (NF) and with fish (F) during the study period. Significant probabilities are in bold, and N is number of animals measured

Species	NF	F	P	N
M. dubia	0.57	0.57	0.689	4059
B. longirostris	0.33	0.32	0.035	4690
C. cornuta	0.42	0.40	**0.000**	4769
D. excisum	0.68	0.66	0.027	2514
D. lumholtzi (h)	0.79	0.79	0.629	1117
D. lumholtzi (m)	1.17	1.15	0.647	459
D. barbata	0.83	0.70	**0.001**	395

Daphnia barbata occurred at intermediate densities (Fig. 2d, 3a, b & c). Few specimens were found of an undetermined *Alona* species. *Daphnia lumholtzi* occurred as two morphs, a helmeted form and a form with a rounded head, the *monaca* form (Green, 1967). The helmeted form was more abundant (Fig. 3a).

Although all species of Cladocera were present in all ponds, their populations seemed to be affected differentially by the presence of fish. *Daphnia*, both as a group, and separated into species and morphs seemed to be more abundant in the ponds without fish than those with (Fig. 3). *Daphnia* as a group ($p = 0.000$), the morph *D. lumholzi (h)* ($p = 0.002$) and *D. barbata* ($p = 0.000$) were significantly less abundant in the fish ponds compared with those without. For the *monaca* form the difference were not found significant ($p = 0.049$).

The development of the other cladoceran populations showed no clear differences between ponds with and ponds without fish (Fig. 2). In the ranking tests no significant difference could be detected neither for *M. dubia* ($p = 0.122$), *B. longirostris* ($p = 0.316$), *C. cornuta* ($p = 0.006$) nor *D. excisum* ($p = 0.192$).

Table 4. Correlations between relative fish growth per day, and some biotic and abiotic parameters. The data were taken from the five periods in each pond when fish growth and environmental parameters were simultaneously measured ($N = 15$). The mean relative fish growth per day is calculated from the mean fish weight at the beginning and at the end of each growth period. The arithmetic means of the environmental parameters in the same periods are used for correlation. Phytoplankton was measured as biovolume, $\mu mm^3 \, l^{-1}$ and zooplankton as density, ind l^{-1}. All the data are $\log_e$ (x+1) transformed. Positive or negative correlations are marked + or −. Critical significance level is set to p=0.003.

Variable 1	Variable 2	r^2	p
Relative fish growth rate per day	Cladocera:	0.475	0.005
	M. dubia	0.014	0.068
	B. longirostris	0.651+	0.000
	C. cornuta	0.368	0.016
	D. excisum	0.686+	0.000
	Daphnia spp.	0.655+	0.000
	D. lumholzi (h)	0.705+	0.000
	D. lumholzi (m)	0.363	0.018
	D. barbata	0.027	0.555
	Phytoplankton	0.007	0.761
	Chlorophyta	0.027	0.559
	Euglenophyta	0.089	0.279
	Bacillariophyta	0.014	0.678
	Cyanophyta	0.322	0.028
	Abiotic parameters		
	Temperature (°C)	0.257	0.053
	Time (age increase)	0.288	0.039

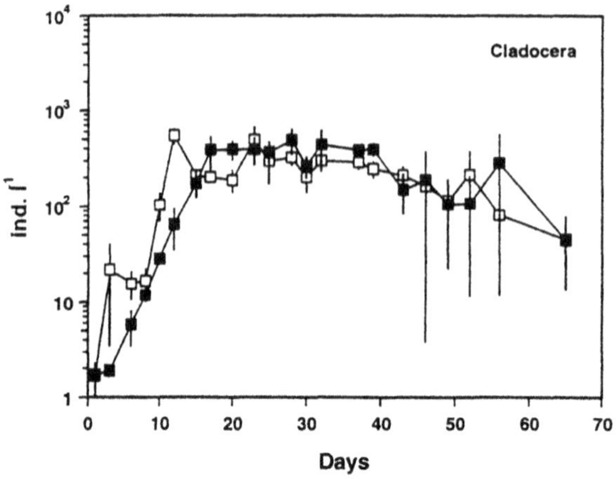

Fig. 1. Mean density of Cladocera in ponds with fish, filled squares, and without fish, open squares. Error bars are ±1 S.E.

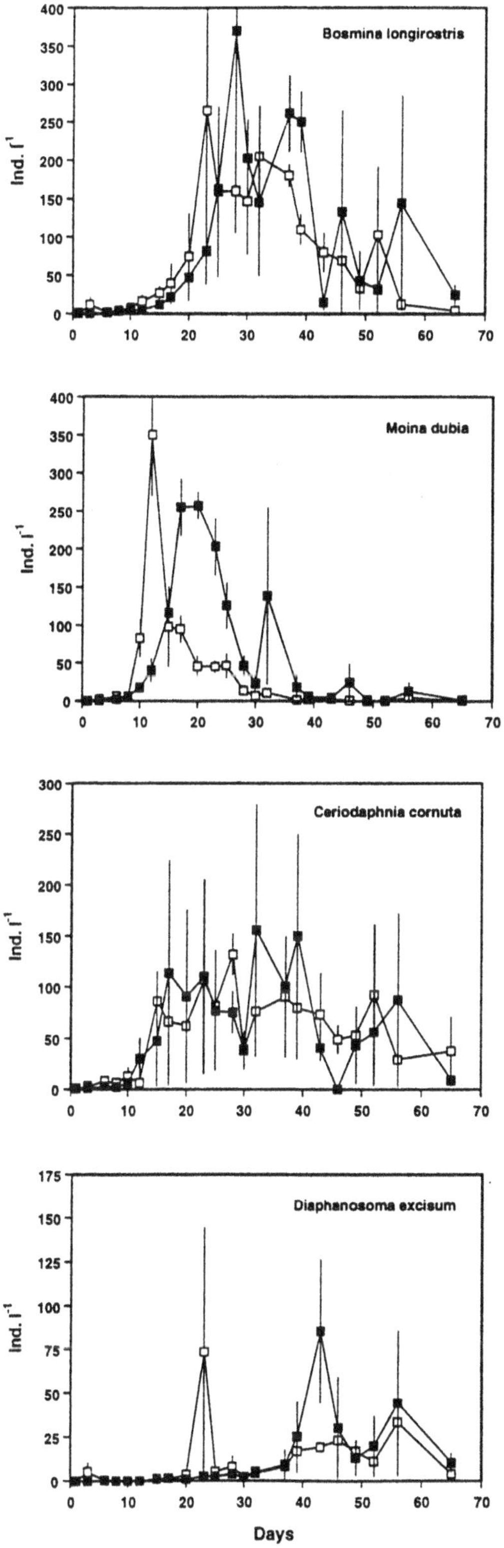

Fig. 2. Mean density of a) *Bosmina longirostris*, b) *Moina dubia*, c) *Cerio-daphnia cornuta* and d) *Diaphanosoma excisum*. Filled squares are densities in ponds with fish, and open squares ponds without fish. Error bars are ±1 S.E.

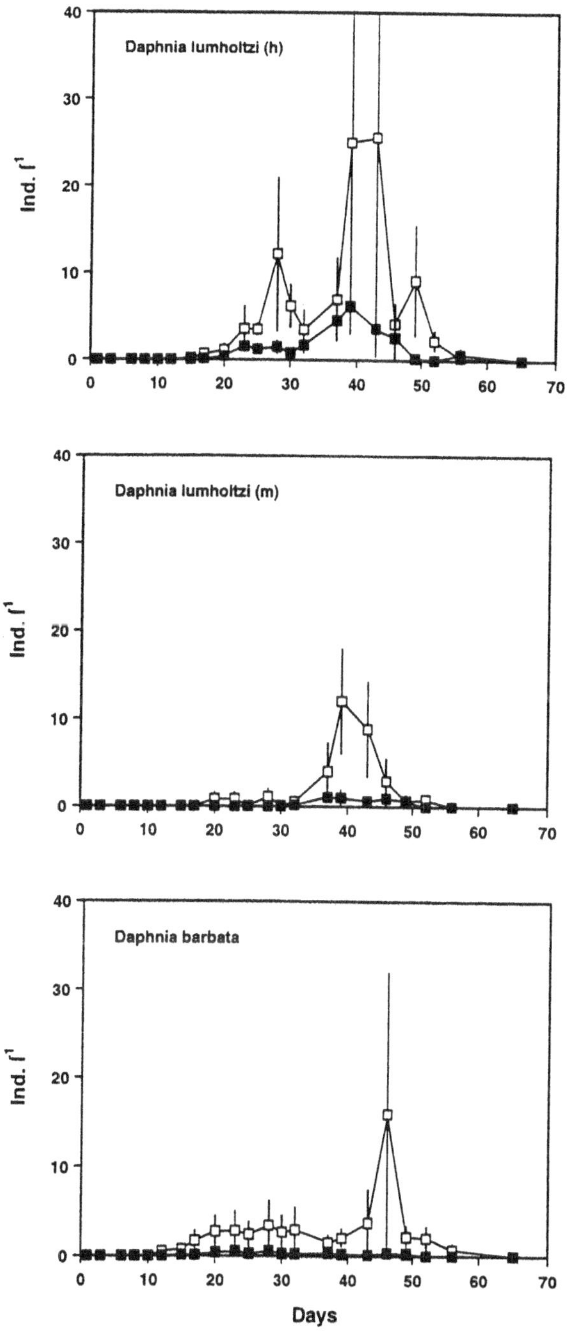

Fig. 3. Mean density of (a) *Daphnia lumholtzi* helmeted form, (b) *D. lumholtzi* f. *monacha* and (c) *D. barbata* in ponds with fish, filled squares, and without fish, open squares. Error bars are ±1 S.E.

The different species of Cladocera seemed to replace each other in the development of the ponds, both in those with and those without fish. The caotic picture from the development of the separate species seemed to be smoothed out in the total density of cladocerans (Fig. 1). A rapid increase in the first 15–20 days was followed by a slightly decreasing plateau.

Cladoceran size structure

The mean individual size (body length in mm) of the various cladoceran species in the ponds without fish, was found to be either larger or equal to the mean size in the ponds with fish (Table 3). Only *Ceriodaphnia cornuta* and *D. barbata* were found significantly larger in the ponds without fish.

Phytoplankton

Eighty-one genera of phytoplankton were recorded (Haider, 1993). In biovolume the Bacillariophyta dominated in the fishless ponds (*Melosira, Navicula, Synedra*). In the fish ponds the Bacillariophyta dominated the first month of the experiment, but then was replaced by Chlorophyta (*Chlorella, Chlamydomonas, Oocystis, Scenedesmus, Ankstrodesmus*). The other algal groups contributed with less than 20% in biovolume (Fig. 4).

The development of the total biovolume of the phytoplankton is shown in Fig. 5. It was found significantly more phytoplankton in the ponds with fish than in those without ($p = 0.000$). Separated into phytoplankton groups it was found that the difference was due to a much higher biovolume of Chlorophyta ($p = 0.000$). For the Euglenophyta (*Phacus, Trachelomonas*) ($p = 0.120$), Cyanophyta (*Anabaena, Microcystis, Oscillatoria*) ($p = 0.463$) and the Bacillariophyta ($p = 0.956$) no significant differences were found.

Water characteristics

Oxygen and pH, which both are affected by the level of primary production, increased during the experiment (Figs 6a & 6b). In spite of significant differences in algal biovolume, the differences in oxygen ($p = 0.005$) and pH ($p = 0.012$) were not found significant.

Nitrogen compounds were always above any limiting level (315–696 μgN l^{-1} with fish, 285–718 μgN l^{-1} without fish). Due to daily input of phosphate the concentration of total phosphate never became limiting (with fish: 94–459 μgP l^{-1}, without fish: 95–513 μgP l^{-1}). In the beginning of the experiment, the temperature in all ponds was 19 °C. It dropped in the second week to 15 °C, but increased again to 27 °C at the end of the experiment. The temperature in all ponds were almost identical. The conductivity in the water increased from the beginning (160–180 μS cm^{-1}) to the end of the experiment (260–300 μS cm^{-1}) due to evaporation.

Fish growth and food density

When individual fish growth rate was correlated with the amount of various food items in the ponds, Cladocera as a group was not sufficiently correlated with relative fish growth to be significant (Table 4). Divided in species and genera, however, *B. longirostris, D. excisum, D. lumholzi (h)* and *Daphnia* spp.(Fig. 7), were found highly correlated with relative fish growth. *Moina dubia, C. cornuta* and *D. barbata*, however, failed to be significantly correlated.

Individual fish growth rate was neither correlated to total phytoplankton nor to any phytoplankton group (Table 4). Among the abiotic factors there were no correlation with time (age increase during the experiment) or temperature. A possible difference between ponds was tested by ANOVA, but it failed to be significant.

Discussion

Although the densities of Nile tilapia were low, even compared to natural ecosystems (Backiel *et al.*, 1980), the zooplankton and phytoplankton communities in ponds with and without fish developed differently. The Nile tilapia appeared to affect the pond ecosystem in the same way as fish with a pure animal diet (Hall *et al.*, 1970; Stenson, 1972; Nilsson & Pejler, 1973; Andersson *et al.*, 1978; Stenson *et al.*, 1978; Zaret, 1980; Hurlbert & Mulla, 1981; Langeland, 1982; Lazzaro, 1987).

Planktivorous fish are known to be size-selective predators that prey selectively on largest zooplankton (Brooks & Dodson, 1965; Zaret, 1980). Our data are concurrent with this observation since the mean size of all the cladocerans in the fishless ponds was either equal to or larger than the mean size in the fish ponds.

Only few *Daphnia* were found in the fish stomachs, compared to for instance, *Bosmina longirostris* (Elhigzi, 1993). This was probably because of the much lower density of daphnids compared with the bosminids. *Daphnia* was probably the most preferred food item, and if they were reduced in density, the second or third preferred item would then be most abundant in the stomachs. There was also a strong cor-

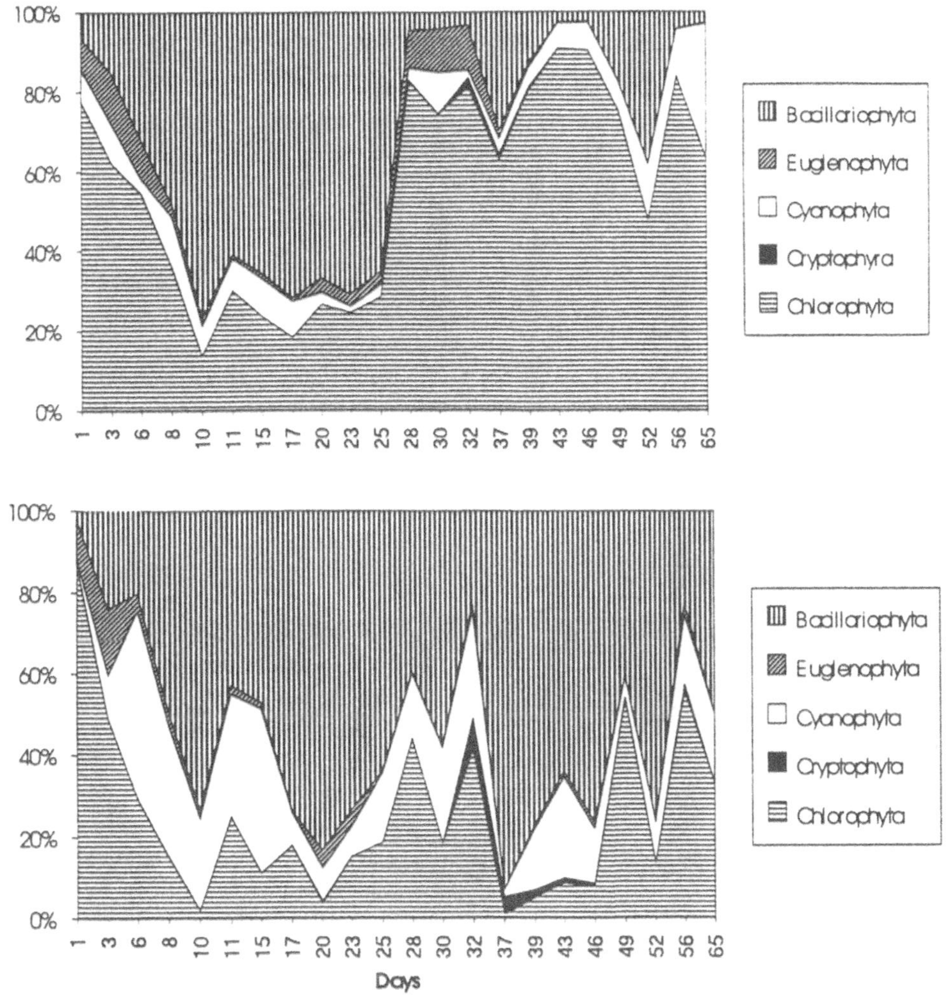

Fig. 4. Proportions of various phytoplankton groups (a) in ponds with and (b) ponds without fish. The figure is based on biovolumes.

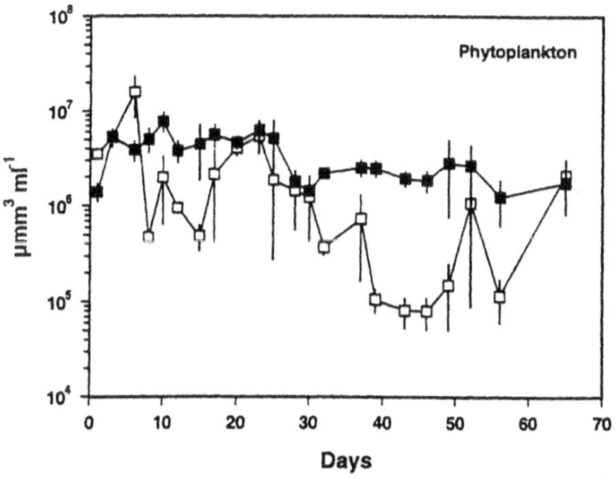

Fig. 5. Mean density of biomass (biovolume) of phytoplankton in ponds with fish, filled squares, and without fish, open squares. Error bars are ±1 S.E.

relation between the density of both *B. longirostris* and *D. excisum* and the fish growth rate. The fish obviously searched both species, but it could not be detected any difference in abundance in the fish ponds compared with the abundance in the fishless ponds. Our interpretation is that although both *B. longirostris* and *D. excisum* were preyed upon, they got approximately the same density as in the fishless ponds due to a weaker competition from the *Daphnia* species.

The phytoplankton biomass was higher in the ponds with fish than in those without. This was in spite of the fact that the two types of ponds had similar densities of total Cladocera, Copepoda and Rotifera. The only striking difference we could observe was the difference in *Daphnia* density. Our interpretation is thus that the difference in amount of algal is due to *Daphnia* grazing. This is not a great surprise and is in agreement with the

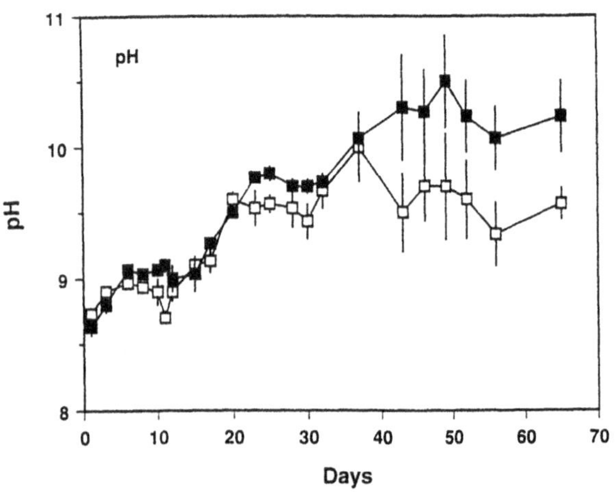

Fig. 6. Mean (a) oxygen concentrations and (b) pH in ponds with fish, filled squares, and ponds without fish, open squares. Error bars are ±1 S.E.

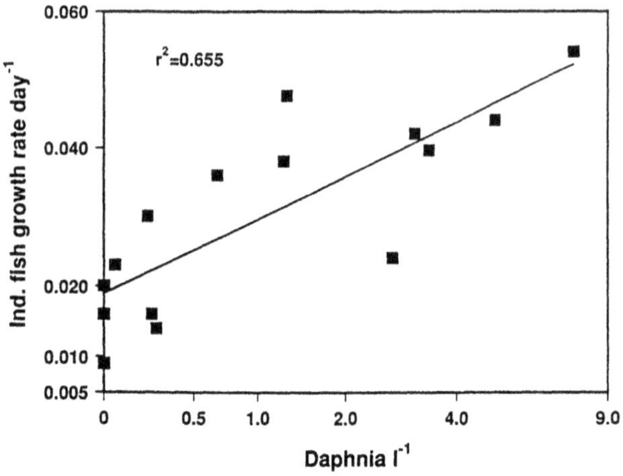

Fig. 7. Correlation between individual fish growth rate per day and density of *Daphnia* spp.. The data are $\log_e(x + 1)$ transformed.

size-efficiency hypothesis (Brooks & Dodson, 1965), recent findings by Gliwicz (1990) and many of the authors sited above. It underlines, however, that the main effect of Nile tilapia on the phytoplankton was indirect through decimation of the populations of large Cladocera.

Nile tilapia seemed not to affect the amount of phytoplankton directly by grazing. Although the stomach contents of the fish contained mainly phytoplankton, algal biomass was higher in the presence of fish than without. The phytoplankton group which increased, was Chlorophyta including edible forms for the Cladocera (Haider, 1993).

Increased invertebrate predation in the ponds without fish, could have been a factor affecting the densities of the various species (Hall, 1964; Richard *et al.*, 1975), but the only possible invertebrate predators found were *Thermocyclops neglectus* and *Asplanchna* sp., in very low densities (Elhigzi, 1993). Therefore, invertebrate predation was assumed to be negligible during the experiment.

The Nile tilapia, in the size range we studied, was growing better when some of the caldoceran species occurred in high densities. They probably had a strong preference for them. The Nile tilapia obviously searched these food items and were even able to reduce the populations of the largest forms. Although the stomachs of Nile tilapia were full of phytoplankton, phytoplankton densities seemed not affect their individual growth rate. We believe that the Nile tilapia has two different feeding modes: (1) size-selective predation, similar to most zooplanktivorous fish species and (2) a nonselective filter-feeding on smaller items like phytoplankton. In the fish size-group we studied the first feeding mode obviously was the most important.

Diana *et al.* (1991) conducted an experiment similar to ours. They tested the trophic cascade hypothesis (Carpenter *et al.*, 1985) in aquaculture ponds containing tilapia at different densities. In contrast to our results, they hardly detected any cascading effect, namely that fish predation should effect not only the prey, but also lower trophic levels. The only significant difference they found was reduced zooplankton densities, particularly small zooplankton, in ponds without fish. Large zooplankton or phytoplankton were not affected by the fish. In our experiment, cascading effects could be recognized in a change in Chlorophyta biovolume. The difference between the results from the two experiments was probably due to the fact that Diana *et al.* (1991) used much larger tilapia (initial mean weight 32 g) than we did (initial mean weight 2.4 g). The larger fishes had presumably already switched to almost exclusively filter-feeding, and thus might not have affected the larger zooplankton forms. Diana *et al.* (1991) had also had much higher fish biomass than we had (i.e. their values ranged initially from 32–96 g m^{-2} compared to our 12 g m^{-2}) This might explain why small zooplankton species were reduced in the presence of so much fish, simply by the presumed nonselective filter-feeding. Diana *et al.* (1991) did not go into great taxonomic detail. When they made their analyses, they discriminated between large zooplankton (cladocerans and copepods) and small zooplankton (protozoans, rotifers and nauplii). In our study we found that copepods as a group were not affected in the presence of fish, while daphnids were reduced. Since these two groups were lumped together by Diana *et al.* (1991), any changes in abundance might have counterbalanced each other.

For further research on the effects of Nile tilapia on it's environment, it will be important to go into more detail concerning their switch in diet. It is difficult to explain the advantage of switching from an omnivorous diet with preference for large zooplankton, to a diet dominated by smaller items like phytoplankton and small zooplankton forms. Such a diet switch does not appear to be in agreement with optimal foraging theory (Krebs & McCleery, 1984). However, the Nile tilapia, in their natural environment are suffering other constraints than in artificial ponds, i.e. competition from more specialized fishes, territorial behaviour from conspecifics and predation risks. In such an environment, a high degree of alertness to predators and competitors is necessary. Passive filter feeding, which do not need active selection of food items, might then be the most profitable feeding strategy.

Acknowledgments

We thank the director, Osman M. Saeed, and the staff at the Fisheries Research Center, Khartoum for the experimental work at Shaggara, Sudan, and NORAD (Norwegian Agency for Development Cooperation), NUFU (Norwegian Universities' Committee for Development of Research and Education) and NAVF (Norwegian Research Council for Sciences and Humanities) for financial support. We give special thanks to Kari Larsson, Mohamed Maglad and Asma A. R. Ahmed for their assistance in the field work, to the fishermen Elheik, Abdel Salam, Shoal and Yahia for their continuous assistance in the sampling procedure during the experiment, to Anders Hobæk and Torbjørn H. Ergon for help with the statistical programs and Glenn Bristow and Larry Weider for help with the English. Also, we thank two anonymous reviewers and Larry Weider for valuable comments and suggestions on an earlier version of this manuscript.

References

Andersson, G., H. Berggren, G. Cronberg & C. Gelin, 1978. Effects of planktivorous and benthivorous fish on organisms and water chemistry in eutrophic lakes. Hydrobiologia 59: 9–15.

Backiel, T., J. E. Thrope & J. F. Kitchell, 1980. Fish. In E. D. Cren, R. H. Lowe-McConnell (eds). The functioning of freshwater ecosystems. Cambridge University Press, London: 307–334.

Bowen, S. H., 1982. Feeding, digestion and growth-Quantitative consideration. In R. S. V. Pullin & R. H. Lowe-McConnell (eds), The Biology and culture of Tilapias. Proceedings of the International Conference on the biology culture of Tilapias, Rockefeller Foundation, Bellagio, Italy, ICLARM, Manila, Philippines: 141–156.

Brooks, A. J. & J. Rzoska, 1954. The influence of Gebel Aulia Dam on the development of Nile plankton. J. anim. Ecol. 23: 101–115.

Brooks, J. L. & S. I. Dodson, 1965. Predation, body size, and competition of plankton. Science 150: 26–35.

Carpenter, S. R., J. F. Kitchell & J. R. Hodgson, 1985. Cascading trophic interactions and lake productivity. BioScience 35: 634–639.

Carpenter, S. R. & Kitchell (eds), 1993. The trophic cascade in lakes. Cambridge studies in ecology. Cambridge University Press, 385 pp.

Diana, J. S., D. J. Dettweiler & C. K. Lin, 1991. Effect of Nile tilapia (*Oreochromis niloticus*) on the ecosystem of aquaculture ponds, and its significance to the trophic cascade hyphothesis. Can. J. Fish. aquat. Sci. 48: 183–190.

Drenner, R. W., S. B. Taylor, X. Lazzaro & D. Kettle, 1984. Partical grazing and plankton community impact of an omnivorous cichild. Trans. am. Fish. Soc. 113: 397–402.

Drenner, R. W., S. T. Threlkeld & M. D. McCracken, 1986. Experimental analysis of the direct and the indirect effects of an omnivorous filter-feeding clupeid on plankton community structure. Can. J. Fish. aquat. Sci. 43: 1935–1945.

Elhigzi, F. A. R., 1993. Interactions between Nile Tilapia (*Oreochromis niloticus*) and the pond community at various fish densities. Thesis, University of Bergen, Norway, 69 pp.

George, T. T., 1975b. Observations on the growth of *Tilapia nilotica* L. in tropical freshwater fish ponds treated with different fertilizers. In Report of the Symposium on Aquaculture in Africa. Accra, Ghana. CIFA Tech. Pap. 4, FAO Rome Suppl. 1: 625–641.

Gliwicz, Z. M., 1990. Food thresholds and body size in cladocerans. Nature 343: 638–640.

Green, J., 1967. The distribution and variation of *Daphnia lumholtzi* (Crustacea: Cladocera) in relation to fish predation in Lake Albert, East Africa. J. zool., Lond. 151: 181–197.

Green, J., 1971. Associations of Cladocera in the zooplankton of Lake sources of the White Nile. J. Zool., Lond. 165: 373–414.

Gulati, R. D., E. H. R. R. Lammens, M.-L. Meijer & E. van Donk (eds), 1990. Biomanipulation – Tool for Water Management. Developments in Hydrobiology 61. Kluwer Academic Publishers, Dordrecht, 628 pp. Reprinted from Hydrobiologia 200/201.

Haider, S. A., 1993. Pond experiments with omnivorous Fish Nile Tilapia (*Oreochromis niloticus*); Effects of Succession of Phytoplankton and Cladoceran Populations. Thesis, University of Bergen, Norway, 54 pp.

Hall, D. J., 1964. An experimental approach to the dynamics of a natural population of *Daphnia galeata mondotae*. Ecology 45: 94–112.

Hall, D. J., W. E. Cooper & E. E. Werner, 1970. An experimental approach to the production dynamics and structure of freshwater animal communities. Limnol. Oceanogr. 15: 839–928.

Hanna, N. S. & F. Schiemer, 1993. The seasonality of zooplanktivorous fish in an African reservoir (Gebel Aulia Reservoir, White Nile, Sudan). Part I: Limnological cycle and the fish community dynamics. Hydrobiologia 250: 173–185.

Hrbacek, J., M. Dvorakova, V. Korinek & L. Prochazkova, 1961. Demonstration of the effect of the fish stock on species composition of zooplankton and the intensity of metabolism of the whole plankton association. Verh. int. Ver. Limnol. 14: 192–195.

Hurlbert, S. H. & M. S. Mulla, 1981. Impacts of mosquito fish (*Gambusa affinis*) predation on plankton communities. Hydrobiologia 83: 125–151.

Krebs, J. R. & R. H. McCleery, 1984. Optimization in behavioural ecology. In Krebs, J. R. & N. B. Davies (eds), Behavioural ecology. 2nd edn. Blackwell, Oxford: 91–121.

Lampert, W., W. Fleckner, H. Rai & B. E. Taylor, 1986. Phytoplankton control by grazing zooplankton: a study on the spring clear-water phase. Limnol. Oceanogr. 31: 478–490.

Langeland, A., 1982. Interactions between zooplankton and fish in a fertilized lake. Holarct. Ecol. 5: 273–310.

Lazzaro, X., 1987. A review of planktivorous fishes: their evolution, feeding behaviour, selectivities, and impacts. Hydrobiologia 146: 97–167.

Maar, A., M. A. E. Moritimer & I. van der Lingen, 1966. Fish culture in Central East Africa. FAO Publ. 53608-66/E 158 p. In J. W. Miller (1975), Fertilization and Feeding Practices in Warm Water Pond Fish Culture in Africa. CIFA/T4 Suppl. 1: 513–541.

Mishrigi, S. Y., 1974. Observations on the effect of fertilizers in Sudanese fish ponds. Revue Zool. afr. 88: 735–740.

Moriarty, C. M. & D. J. V. Moriarty, 1973. Quantitative estimation of daily ingestion of phytoplankton by *Tilapia nilotica* and *Haplochromis nigripinnis* in Lake George, Uganda. J. Zool., London 171: 15–24.

Nilsson, N. A. & B. Pejler, 1973. On the relation between fish fauna and zooplankton composition in North Swedish lakes. Rep. Inst. Freshwat. Res. Dronttningholm 53: 51–77.

Pullin, R. S. V., 1991. Chichilds in aquaculture. In M. H. A. Keenleyside (ed.), Chichlid fishes behaviour, ecology and evolution. Chapman and Hall. Fish and Fisheries Series 2: 281–299.

Rice, W. R., 1989. Analyzing tables of statistical test. Evolution 43: 223–225.

Richard, R. C., C. R. Goldman, T. C. Frantz & R. Wickwire, 1975. Where have all the *Daphnia* gone? The decline of a major cladoceran in Lake Tahoe, California-Nevada. Verh. int. Ver. Limnol. 19: 835–842.

Ricker, W. E., 1975. Computation and interpretation of biological statistics of fish population. Fish. Res. Bd Can. Bull.: 191: 382 pp.

Shapiro, J. & D. I. Wright, 1984. Lake restoration by biomanipulation. Round Lake, Minnesota, the first two years. Freshwat. Biol. 14: 371–383.

Shehadah, Z. H., (ed.), 1976. Report of the FAO/CIFA Symposium on Aquaculture in Africa, Accra, Ghana. FAO/CIFA Tech. Pap. 4: 36 pp.

Stenson, A. J. E., 1972. Fish predation effects on the species composition of the zooplankton community in eight small forest lakes. Rep. Inst. Freshwat. Res. Drottningholm 52: 132–148.

Stenson, A. J. E., T. Bohlin, L. Henrikson, B. I. Nilsson, H. G. Nyman, H. G. Oscarson & P. Larsson, 1978. Effects of fish removal from a lake. Verh. int. Ver. Limnol. 20: 794–801.

Threlkeld, S. T., 1987. Experimental evaluation of trophic-cascade and nutrient mediated effects of planktivorous fish on plankton community structure. In W. C. Kerfoot & A. Sih (eds), Predation: direct and indirect impact of aquatic communities, New England.: 161–173.

Utermöhl, H., 1958. Zür Vervollkommung der quantitativen Erfassung des Planktons (mit besonderer Berücksichtigung des Ultraplanktons). Verh. int. Ver. limnol. 5: 567–596.

Yousif, O. M., 1988. Tilapia culture in Sudan. Naga, The ICLARM Quarterlt. 11: 13.

Zaret, T. M., 1980. Predation and freshwater communities. New Haven and London, Yale University Press, 187 pp.

Hydrobiologia **307**: 273–282, 1995.
P. Larsson & L.J. Weider (eds), Cladocera as Model Organisms in Biology.

Leptodora kindti: efficient predator and preferred prey item in Neusiedler See, Austria

Alois Herzig
Biologische Station Neusiedler See, A-7142 Illmitz, Austria

Key words: Leptodora kindti, population dynamics, *Diaphanosoma mongolianum*, *Leptodora* predation, *Leptodora* as prey

Abstract

In the Neusiedler See, *Leptodora kindti* appeared in the plankton in April/May; the growing season lasted until November. Its optimum conditions prevailed during the summer months, when densities of >500 ind. m^{-3} were recorded. First, males appeared by the mid/end of August, while maximum resting egg production occurred in September. At temperatures below 10 °C the population declined.

At high population levels, *L. kindti* caused suppression of its preferred prey item, the juveniles of *Diaphanosoma mongolianum*. Twenty-five to 45% of juvenile *Diaphanosoma* can be eliminated by *Leptodora*. This predation impact is a pronounced but short-term event. During the periods of high *Leptodora* densities, a significant relationship between *Leptodora* predation and the death rate of *Diaphanosoma* is found.

The older developmental instars of *L. kindti* are themselves endangered by the impact of planktivorous fish (i.e. *Pelecus cultratus*, *Alburnus alburnus*), which show a preference for *Leptodora* larger than 6 mm. Frequently, elimination by fish corresponds with high death rates of *Leptodora*. During the periods when adult *Leptodora* occurred in numbers >50 ind. m^{-3}, 55% of the variance connected with its death rate can be attributed to fish predation. The success of this population of *L. kindti* is discussed within the framework of predator-prey interactions.

Introduction

Leptodora kindti is a large, highly transparent zooplankter, which is a voracious predator. It feeds by grasping the prey and pushing it into its trapping basket formed by the thoracic limbs (Browman *et al.*, 1989; Herzig & Auer, 1990). It eats a wide variety of mainly cladoceran species, but it is also reported that copepods and rotifers are consumed (Cummins *et al.*, 1969; Edmondson & Litt, 1987; Lunte & Luecke, 1990) and occasionally it is cannibalistic (Mordukhai-Boltovskaya, 1958). Full-grown *Leptodora* most readily capture prey in the size range of 0.6–0.8 mm, which tend to be brightly coloured, bear a fragile carapace and are slow moving and not too vigorous. At high population densities (>500 ind. m^{-3}), its predation pressure can lead to a marked decline in prey populations; e.g. *Daphnia* spp. (Hall, 1964; Wright, 1965; Hillbricht-Illkowska & Karabin, 1970; Karabin, 1974; de Bernardi & Canali, 1975; de Bernardi & Guissani, 1975; Havel, 1985; Hovenkamp, 1990; Lunte & Luecke, 1990), *Bosmina* (Branstrator & Lehman, 1991), and *Diaphanosoma* (Auer, 1989; Herzig & Auer, 1990; Herzig, 1994).

On the other hand, *L. kindti* itself is a preferred prey item of planktivorous fish despite the fact that it is extremely transparent; the heavily pigmented area within the compound eye may be the most important component of conspicuousness. Examples of positive selectivity for *Leptodora* are found for *Coregonus* spp. (de Bernardi & Guissani, 1975; Luecke *et al.*, 1990; Akaret, 1993), *Perca* spp. (Craig, 1978; Luecke *et al.*, 1990), *Pelecus cultratus* and *Alburnus alburnus* (Herzig & Winkler, 1983; Herzig *et al.*, 1993).

In the Neusiedler See, a shallow alkaline lake with a mean depth of 1.1 m and a surface area of 321 km^2 (for more information see Herzig, 1994), *Leptodora* is the only invertebrate predator within the zooplankton community of the pelagic zone. In June 1977, it appeared for the first time; most probably its resting eggs were swallowed by waterfowl and introduced into lake through their feces. Since then, it has increased in numbers and has reached maximum densities of up to 2500 ind. m^{-3} in recent years (Herzig, 1994). As shown earlier (Herzig & Auer, 1990), *D. mongolianum* is the preferred prey in the Neusiedler See. *Arctodiaptomus spinosus* (Copepoda) is also captured but because of its vigour, it escapes and direct feeding on this copepod was never observed. Feeding on rotifers (e.g. *Hexarthra, Filinia*) was observed several times, especially for young stages of *Leptodora*, which sometimes feed on *Filinia longiseta* (Herzig & Auer, 1990; Auer, pers. com.). The size of *D. mongolianum* that are attacked by adult *Leptodora* varied between 0.5 and 0.9 mm; this means that mainly juvenile instars are selected.

In the years when *Leptodora* occurred in low numbers (maximum <150 ind. m^{-3}), the *Diaphanosoma* population was either not affected or only to a negligible extent by this invertebrate predator. More recently, when *Leptodora* occurred in peak densities of 400–2600 ind. m^{-3} and summer mean abundance varied between 200 and 300 ind. m^{-3}, clear coincidences of higher *Leptodora* densities and depressed numbers of *Diaphanosoma* became apparent (Herzig, 1994).

The aim of the present paper is to examine the role of *Leptodora* within the pelagic food web. Its impact as a predator on its major prey item — *Diaphanosoma mongolianum* — is shown for the years 1987–1991. Its population dynamics and the potential elimination of *Leptodora* by planktivorous fish is described and discussed for the years 1990 and 1991.

Materials and methods

Plankton

Samples were collected with vertical net hauls (mesh size: 40 μm and 250 μ). This resulted in integrated samples (50 – 120 l lake water filtered) without net clogging, since most of the suspended particles are smaller than 20 μm. Sampling was performed at four stations weekly and at 6–10 stations monthly. Data from all stations were averaged (geometric mean).

Collection of fish older than 0+

Quantitative and semi-quantitative data were gathered for fish older than 0+. Quantitative measurements were made with a BioSonics 105 dual beam echo sounder (frequency 420 kHz; log R TVG-function) in the years 1990 and 1991. Due to the lake's shallowness, a horizontal sounding was applied (for details see Bobek, 1990).

Since 1990 a semi-quantitative fishing program was conducted at a sampling interval of 4–14 days. Gill nets (30 m long, 1.5 m high) with a mesh size (knot to knot) of 10, 15, 20, 24 and 30 mm served as fishing gear. Because of the high fish densities, the gill nets were set for only 0.5 hours (for details see Herzig, 1994).

Calculation of birth and death rates of D. mongolianum and L. kindti

The population rate of change (*r*) was calculated using the equation

$$r = (\ln N_t - \ln N_0)/t,$$

where N_o equals the initial population size, N_t the final population size and *t* the time between the two observations.

The birth rate (*b*) was calculated using Paloheimo's model (Paloheimo, 1974)

$$b = \ln(E/N_{\mathrm{ad}} + 1)/D_e,$$

where *E* equals the number of eggs, N_{ad} the number of adult females and D_e the duration of embryonic development. The embryonic development times were calculated using the experimentally defined relationships between temperature and embryonic development time. For *D. mongolianum* the corresponding equation is

$$D_e = 24(t - 10.08)^{-1.04} \text{(Herzig, 1984)},$$

and for *L. kindti* the regression was calculated using my own data and the data of Kuzichkin (1975); the equation is

$$D_e = 3107(t + 1.8)^{-2.21} (n = 25; \text{temperature range}: 10 - 26°\text{C}).$$

The death rate (*d*) was calculated by subtraction:

$$d = b - r.$$

Determination of the predation rate of L. kindti

Experiments with zooplankton from the Neusiedler See showed that *D. mongolianum* is the preferred prey of *L. kindti*. Predation rates were determined for temperatures between 12 °C and 22 °C and for prey densities of 30 ind. l^{-1} up to 180 ind. l^{-1}. For further details see Herzig & Auer (1990). From this set of variables, a multiple regression was calculated:

$$Y = -5.005 + 0.352X_1 + 0.016X_2,$$

where Y is the predation rate (ind. *Leptodora*$^{-1}$ d^{-1}) and X_1 the temperature (°C) and X_2 the prey density (ind. l^{-1}) (Herzig, 1994). This regression was used for the calculation of a daily predation rate.

Determination of the consumption of Alburnus alburnus and Pelecus cultratus

The calculation of the amount of food consumed by planktivorous fish is based on the amount of food in the gut and the rate of gastric evacuation (details and results described in Herzig, 1994). The total gut contents of 450 *P. cultratus* and 600 *A. alburnus* were analyzed. The gastric evacuation rate (R) was estimated by applying an exponential function (Persson, 1982). The equation of Elliott & Persson (1978) was used for the calculation of the amount of food consumed:

$$C_t = S \times R \times 24,$$

where R is the rate of gastric evacuation and S is the amount of food (in numbers) in the gut. As the duration of the exposure of the nets was only 0.5 hours, it is assumed that the bias caused by gastric evacuation between being caught, killed and preserved is negligible.

Results

Leptodora abundance is positively correlated with lake temperature. The growing season lasted from April/May until November, while maximum densities were recorded during the summer months at temperatures of 18 °C–22 °C. Hatching of the resting eggs started at temperatures of 8 °C– 10 °C, and mass hatching occurred at the end of May/beginning of June when water temperature rapidly increased from 15 °C up to 20 °C. This period was followed by a very pronounced population increase, which is strongly correlated with the population development of *Diaphanosoma*.

A multiple regression analysis of the rate of population increase of *Leptodora* (Y), the rate of temperature increase (X_1) and the birth rate of *Diaphanosoma* (X_2) revealed that 20% of the variance of the population increase of *Leptodora* can be explained by these two variables ($R^2 = 0.2$; $F = 5.82$; $D.F. = 2{,}47$; $p < 0.0055$). A stepwise regression with the same set of variables showed that the birth rate of *Diaphanosoma* is already responsible for 14% of the variance of *Leptodora*'s rate of population increase.

Peak densities of *Leptodora* were reached by end of June/beginning of July and by mid/end of August and were in a range of about 500 to >2000 ind. m^{-3}. The duration of the period of high densities strongly depended on the intensity of the predation pressure of planktivorous fish.

The first males appeared by mid/end August and maximum resting egg production occurred in September. At temperatures below 10 °C, the population declined and went into diapause.

To evaluate the plausibility of the predator — prey interaction between *Leptodora* and *Diaphanosoma*, the daily elimination rates of *Diaphanosoma* by *Leptodora* were determined. The daily predation rate was calculated by using the multiple regression model defined in Herzig (1994). This predation rate was multiplied by the density of adult *Leptodora* and related to the density of juvenile *Diaphanosoma*. At higher predator densities (>300 ind. m^{-3}), a substantial elimination (25–45%; on one occasion >50%) can be attributed to *Leptodora* predation (Fig. 1).

If *Diaphanosoma* death rates and its daily elimination are superimposed on *Leptodora* density (Fig. 1), it becomes obvious that *Leptodora* does have the potential to control *Diaphanosoma*. Its impact is a pronounced but short term event. For a few days only, sometimes for a maximum of two weeks, predators and prey occurred in appropriate numbers. For the periods of maximum elimination of *Diaphanosoma* by *Leptodora*, a regression analysis of the death rate of *Diaphanosoma* (Y) and the percentage of juvenile *Diaphanosoma* standing stock eliminated (X) reveals a significant relationship (Fig. 2):

$$Y = 0.51X + 0.1$$

$$(n = 11; R^2 = 0.87; F = 58.22; p = 0.0001).$$

In Fig. 3 a comparison of birth and death rates of *Leptodora* reveals that high birth rates are accompanied by high death rates. The negative death rates at the time of onset of the population development indicate

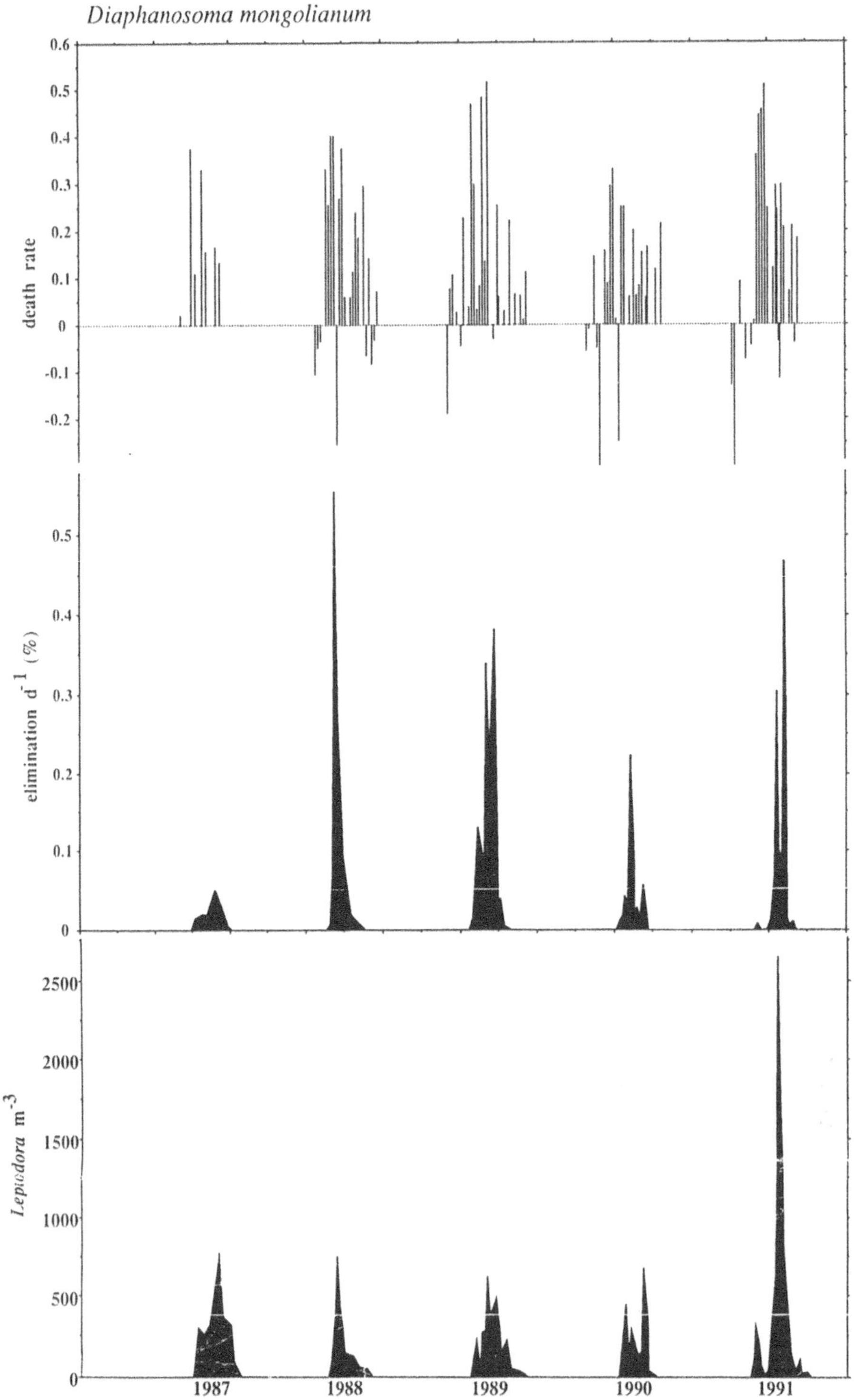

Fig. 1. Death rate of *Diaphanosoma mongolianum*, percentage of young *Diaphanosoma* eliminated by *Leptodora*, and numbers of *Leptodora* m^{-3}.

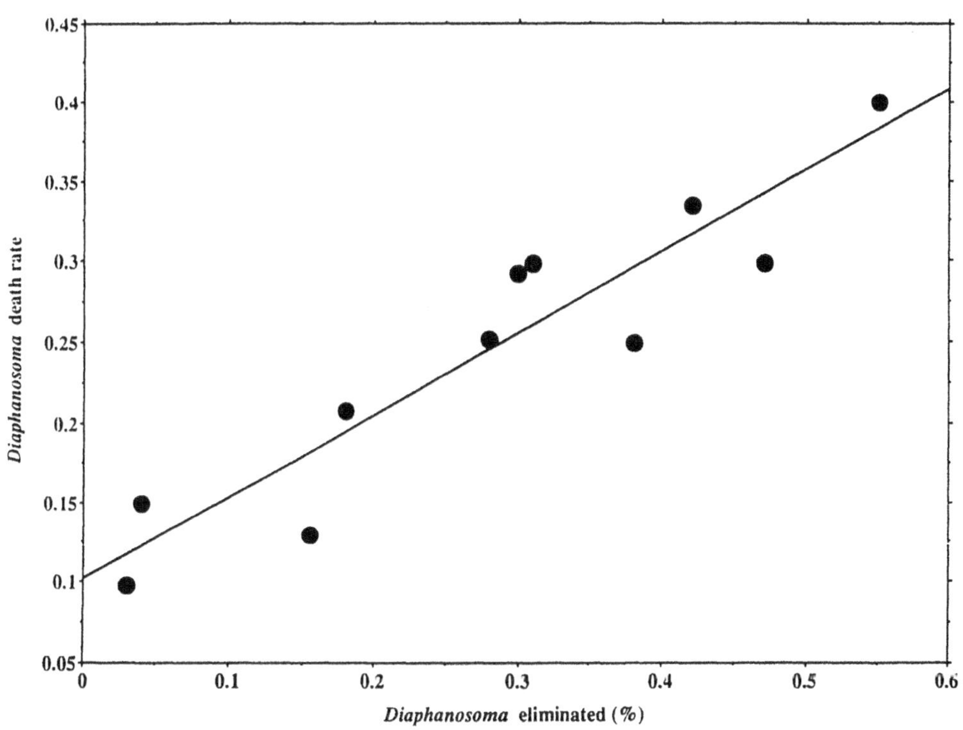

Fig. 2. *Diaphanosoma mongolianum*: Relationship between percentage of juvenile standing stock eliminated by *Leptodora* m^{-3} and the estimated death rate.

hatching of resting eggs. Later on in the year, negative death rates are the result on an inadequacy of the birth rate estimates, which were caused by extremely low egg numbers. According to earlier observations on *Diaphanosoma* (Leitner, 1990), wind generated turbulences may lead to losses of the eggs out of the brood pouch.

To estimate the potential influence of planktivorous fish, their daily elimination of *Leptodora* was estimated by multiplying the consumption by the corresponding fish density. Currently two cyprinid fishes — *Pelecus cultratus* and *Alburnus alburnus* — are the main planktivores. On the basis of hydro- acoustic measurements, numbers of *P. cultratus* were found to range between 0.06 to 1.29 ind. m^{-3} and numbers of *A. alburnus* were between 0.04 and 2.35 ind. m^{-3}. During summer, the highest total numbers of planktivores were in a range of 1.0 to 3.64 ind. m^{-3}, while the lowest densities varied between 0.56 and 1.81 ind. m^{-3}. *Pelecus* and *Alburnus* fed mainly on *Diaphanosoma* and *Leptodora* during the summer months. As soon as adult *Leptodora* appeared in numbers >50 ind. m^{-3}, according to gut analyses, both fishes started feeding on it, selecting the larger specimens (>6 mm) (Liu & Herzig, in prep.).

If the calculated elimination is superimposed on the death rates of *Leptodora*, it becomes obvious that in general a pronounced fish predation corresponds with high death rates. Between the number of *Leptodora* elimination by fish (X) and the death rates (Y), a significant positive relationship can be defined for the periods when the abundance of adult *Leptodora* exceeded 50 ind. m^{-3} (negative death rates are not included) (Fig. 4):

$$Y = 0.00247X + 0.07$$

$$(n = 29; R^2 = 0.55; F = 32.55; p = 0.0001).$$

Fifty-five percent of the variance connected with the death rate can be attributed to elimination by fish.

The quantitative relationships for the years 1990/1991 are summarized (Fig. 5). The elimination rates are expressed as percentages of the corresponding standing stock. *P. cultratus* and *A. alburnus* exert a very pronounced predation pressure on adult *Leptodora*, but a longer lasting and higher predation pressure is exerted on adults of *D. mongolianum*. Predation pressure by juvenile fish acts on the younger stages of the zooplankton, while larger fishes (>40 mm) try to feed on larger items. The impact of young fish is most pronounced on juvenile *Diaphanosoma*, but smaller

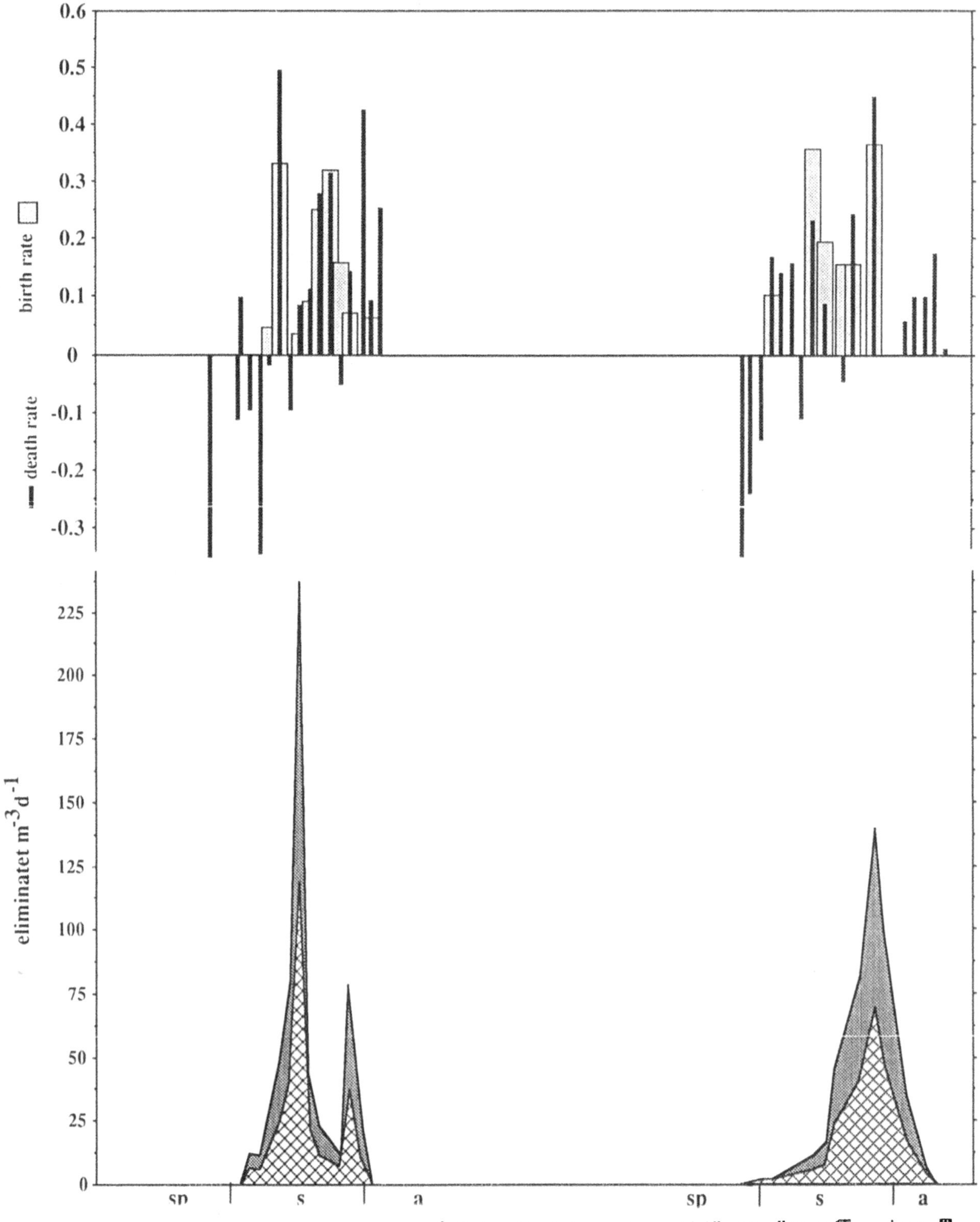

Fig. 3. *Leptodora kindti*: Birth and death rate; numbers m^{-3} eliminated by *Pelecus cultratus* and *Alburnus alburnus* (▩ maximum, ▩ minimum fish numbers).

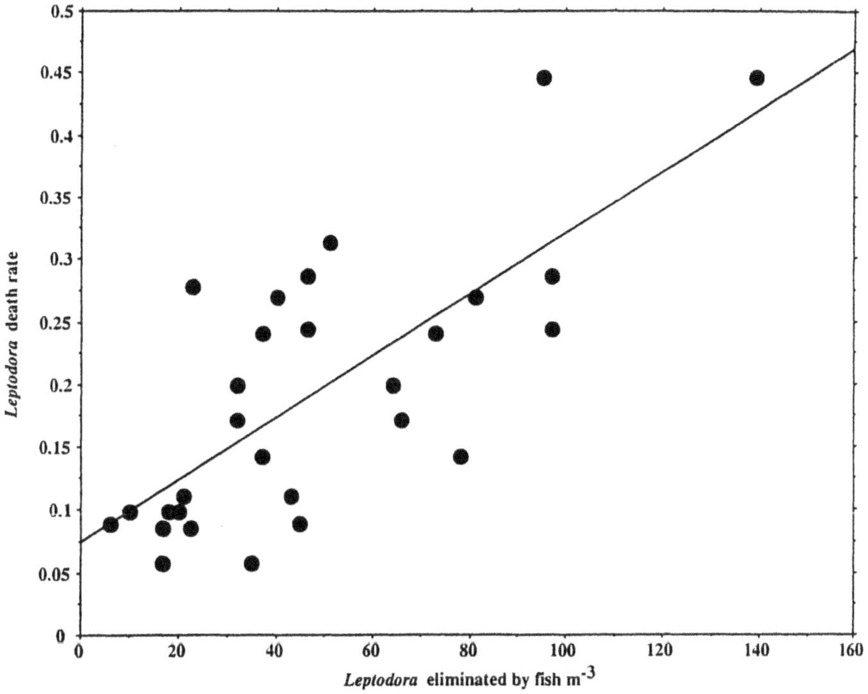

Fig. 4. Leptodora kindti: Relationship between numbers eliminated by fish m^{-3} and death rates.

Leptodora are also eaten. On the other hand, *Leptodora* prefers juvenile *Diaphanosoma* and thus impacts more heavily on the smaller size classes.

Discussion

Phenology

Leptodora is known as a summer form. In shallow lakes, it appears in April/May (e.g. Rybinsk Reservoir — Mordukhai — Boltovskaja, 1956; Balaton — Sebestyén, 1960; Sanctuary Lake/Pymatuning Reservoir — Cummins *et al.*, 1969; Lake Vechten — Hovenkamp, 1990). In deep lakes succesful development starts in June/July (e.g. Greifensee — Mittelholzer, 1970; Lauerzersee — Odermatt, 1970; Lago Maggiore — de Bernardi & Canali, 1975; Lake Erie — Garton *et al.*, 1990; Hallwiler See, Sempacher See — Akeret, 1993). Abundance is positively correlated with temperature (Garton *et al.*, 1990 found a correlation coefficient of 0.77), with peak densities in summer, which can reach maximum numbers of >2000 ind. m^{-3} in some lakes (Neusiedler See, Sanctuary Lake, Mikolajskie, Lake Vechten). In all lakes from which *Leptodora* has been described, temperatures below 10 °C induce diapause.

Leptodora — the predator

L. kindti does have the ability to regulate the population dynamics of its prey species. In the Neusiedler See at high densities (> 300 ind. m^{-3}), a substantial elimination (maximum >50%) of juvenile *Diaphanosoma* can be attributed to *Leptodora* feeding; during such periods a significant relationship between *Leptodora* predation and the death rate of *Diaphanosoma* can be established. As in the Neusiedler See, in Lago Maggiore the increase in the death rate of *Daphnia hyalina* is linked with the population growth of *Leptodora* and *Bythotrephes* (de Bernardi & Guissani, 1975). Hall (1964), Wright (1965), Cummins *et al.* (1969) and Dodson (1972) also report a relationship between death rates in *Daphnia* and the rapid increase of the population density of *Leptodora*. In Lake Vechten, nearly all juvenile mortality of *D. hyalina* and *D. cucullata* corresponded with the maximum occurrence of *Leptodora* at the end of July and in August (Hovenkamp, 1990). Similarly, in Lake Michigan *Leptodora* accounted for nearly 100% of *Bosmina* mortality (Branstrator & Lehmann, 1991). In Lake Mendota,

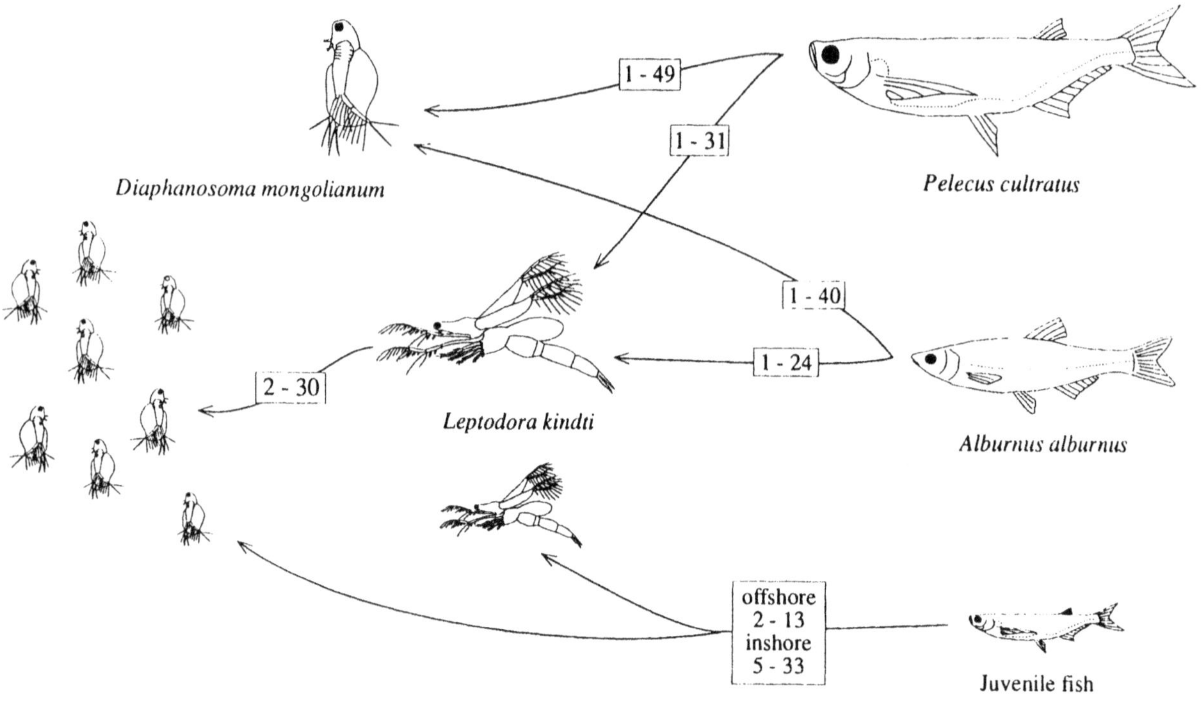

Fig. 5. Summary of the *Leptodora* centered predator — prey relationships for the summer months 1990 and 1991; elimination rates expressed as percentages of the corresponding standing stock.

Leptodora predation likely accounted for a substantial proportion of the mortality estimated for *D. galeata* during a few weeks in mid-summer. Even *Conochilus* colonies can be reduced by *Leptodora* by >40% (Lunte & Luecke, 1990), an elimination rate which equals that estimated for another rotifer, *Hexarthra mira* (Herzig & Auer, 1990).

In conclusion, the feeding of *Leptodora* pin — points a clear size selectivity, and the short duration of pronounced predatory impact and the intensity of this feeding can be recognized on the basis of an analysis of population dynamics of the prey species.

Leptodora — the prey

Planktivorous fish are known as important predators in aquatic systems. They do have the ability to regulate population dynamics of planktonic organisms. An enormous amount of research activity on selective predation has been undertaken, size and visibility of prey being the most frequent explanation for it (e.g. Zaret & Kerfoot, 1975; O'Brien *et al.*, 1976; Kerfoot, 1980; Langeland, 1982; Lazzaro, 1987; Gliwicz & Pijanowska, 1989). A large number of publications deal with the impact of planktivorous fish on cladoceran populations, some of them describe and discuss their impact on the large predatory organisms *Leptodora* and *Bythotrephes*.

In Lago Maggiore, *Leptodora*, *Bythotrephes* and *Daphnia* were selected by *Coregonus* sp. and electivity increased with size; a clear seasonality in prey preference did occur reflecting the population development of the large prey species (de Bernardi *et al.*, 1987). Similarly, Akaret (1993) described that *Coregonus* sp. from the Hallwilersee preferred *Bythotrephes*, while in summer the second most important food item was *Leptodora*. As in the Neusiedler See, as soon as *Leptodora* appeared in the water column of Lake Mendota it became the preferred prey in the diet of yellow perch (*Perca flavescens*) (23–89%) and together with *Daphnia* it formed the diet of cisco (*Coregonus artedii*) (Luecke *et al.*, 1990). In lake Windermere, *Leptodora* was important in the diet of perch (*Perca fluviatilis*), the average percentage in the gut varying between 30% (July/August) and 40% (September) (Craig, 1978). In all the examples given, the extreme seasonal variation in the degree to which planktivorous fish can regulate the various planktonic prey populations is stressed.

Cummins *et al.* (1969) found on the basis of preliminary results that fish predation on *Leptodora* was roughly of the same order of magnitude as the estimated death rate of *L. kindti*. In the Neusiedler See, a significant relationship between the number of *Leptodora* eliminated by fish and its estimated death rates can be established for the time when adult *Leptodora* occur in numbers >50 ind. m^{-3}.

The impact of juvenile fish was analyzed recently (Herzig *et al.*, 1993). Juveniles of *P. cultratus* fed on *L. kindti* especially at night; at this time *Leptodora* constituted 13%–42% of the gut content (expressed as weight). The contribution to elimination was on average 4.8% (range: 0.4–13%). Early summer mortalities of *Leptodora* might be the result of the feeding activity of juvenile fish (Herzig, 1994).

Since *L. kindti* first appeared in the plankton of the Neusiedler See, fish have preyed upon it. On the basis of gut analyses of *P. cultratus* caught in August 1977 and the relative abundance of it at that time, the daily elimination of *Leptodora* was estimated as 19%; a maximum of about 30% could have occurred (Herzig & Winkler, 1984). At this time, the estimated daily elimination by fish was high enough to control the population development of *Leptodora*. Despite the fact that the density of planktivorous fish has increased continuously since then (Herzig *et al.*, 1993), the population density of *Leptodora* has also increased within this period of time. It seems that features such as refuges, spatial heterogeneity, existence of alternative prey species combined with predator switching can contribute to stabilize this predator — prey interaction (cf. Murdoch & Oaten, 1975). In the Neusiedler See, as in other shallow lakes (Sebestyén, 1960; Cummins *et al.*, 1969) *Leptodora* tends to stay close to the lake bottom during daytime and move to the surface layers at night. This might reduce the predator pressure by fish at least during daytime.

Furthermore, *D. mongolianum* serves as 'good' food for *P. cultratus* and *A. alburnus* throughout May — September. In earlier years (without *Leptodora*) it was *the* preferred prey item. With the occurrence of *L. kindti* in the lake and being a preferred prey item of the two planktivores, the predatory impact on *D. mongolianum* should then be reduced. Actually, the long-term data indicate that as a result of rapid eutrophication (1968–1975) *D. mongolianum* has increased in numbers. As a consequence of the improved food conditions, numbers of planktivorous fish have also increased (especially *P. cultratus*) and this has resulted in a higher predation pressure on zooplankton and especially the *Diaphanosoma* population became depressed (Herzig, 1994). With *Leptodora* becoming more abundant during the summer months and being selectively eaten by planktivorous fish, predation pressure on *D. mongolianum* is partly reduced and since then it has shown a tendency towards an increased standing stock. On the other hand, *Leptodora* itself does impact on the *Diaphanosoma* population as well; but this effect exists only for a short period of time. The population development of *Diaphanosoma* is not really limited by this predator.

In conclusion, the presence of *Leptodora* as another species of prey seems to contribute to stabilizing the pelagic predator – prey relationships in the Neusiedler See.

Acknowledgments

The help in the field by R. Klein, and in the laboratory by F. Rauchwarter is acknowledged. I am grateful to G. Wolfram for his help with computer work. My sincere thanks are due to T. Andrew for correcting the English and casting a critical eye on the development of this paper. The criticisms and suggestions of anonymous reviewers were helpful and are acknowledged. This work was funded by the Ministry of Science and Research, the Ministry of Environment, Youth and Family and the Local Government of Burgenland.

References

Akeret, B., 1993. Zur Biologie von *Chaoborus flavicans*, *Leptodora kindti* und *Bythopheres longimanus* unter dem Einfluß interner Restaurierungsmaßnahmen in drei Schweizer Seen. Diss. ETH Zürich, 108 pp.

Auer, B., 1989. Vorkommen und Ernährungsbiologie von *Leptodora kindti* (Focke) im Neusiedler See. Diplomarbeit Univ. Wien, 60 pp.

Bobek, M., 1990. Applied hydroacoustics in cyprinid research. Proc. Symp. 'Fisheries in the year 2000', Royal Holloway and Bedford New College: 287–297.

Branstrator, D. K. & J. T. Lehman, 1991. Invertebrate predation in Lake Michigan: Regulation of *Bosmina longirostris* by *Leptodora kindti*. Limnol. Oceanogr. 36: 483–495.

Browman, H. I., S. Kruse & W. J. O'Brien, 1989. Foraging behavior of the predaceous cladoceran, *Leptodora kindti*, and escape responses of their prey. J. Plankton Res. 11: 1075–1088.

Craig, J. F., 1978. A study of the food and feeding of perch, *Perca fluviatilis* L., in Windermere. Freshwat. Biol. 8: 59–68.

Cummins, K. W., R. R. Costa, R. E. Rowe, G. A. Moshiri, R. M. Scanlon & R. K. Zajdel, 1969. Ecological energetics of a natural population of the predaceous zooplankter *Leptodora kindti* Focke (Cladocera). Oikos 20: 189–223.

de Bernardi, R. & S. Canali, 1975. Population dynamics of pelagic cladocerans in Lago Maggiore. Mem. Ist. ital. Idrobiol. 32: 365–392.

de Bernardi, R. & G. Guissani, 1975. Population dynamics of three cladocerans of Lago Maggiore related to predation pressure by a planktophagous fish. Verh. int. Limnol. 19: 2906– 2912.

de Bernardi, R., G. Guissani & M. Manca, 1987. Cladocera: Predators and prey. Hydrobiologia 145 (Dev. Hydrobiol. 35): 225–243.

Dodson, S. I., 1972. Mortality in a population of *Daphnia rosea*. Ecology 53: 1011–1013.

Edmondson, W. T. & A. H. Litt, 1987. *Conochilus* in Lake Washington. Hydrobiologia 147: 157–162.

Elliott, J. M. & L. Persson, 1978. The estimation of daily rates of food consumption for fish. J. anim. Ecol. 47: 977–991.

Garton, D. W., D. J. Berg & R. J. Fletcher, 1990. Thermal tolerances of the predatory cladocerans *Bythotrephes cederstroemi* and *Leptodora kindti*: Relationship to seasonal abundance in Western Lake Erie. Can. J. Fish. aquat. Sci. 47: 731- -738.

Gliwicz, Z. M. & J. Pijanowska, 1989. The role of predation in zooplankton succession. In U. Sommer (ed.), Plankton Ecology. Succession in Plankton Communities, Springer Verlag, New York, Berlin, Heidelberg: 253–296.

Hall, D. J., 1964. An experimental approach to the dynamics of a natural population of *Daphnia galeata mendotae*. Ecology 45: 94–112.

Havel, J. E., 1985. Predation of common invertebrate predators on long- and short-featured *Daphnia retrocurva*. Hydrobiologia 124: 141–149.

Herzig, A., 1984. Temperature and life cycle strategies of *Diaphanosoma brachyurum*: an experimental study on development, growth, and survival. Arch. Hydrobiol. 101: 143–178.

Herzig, A., 1994. Predator — prey relationships within the pelagic community of Neusiedler See. Hydrobiologia 275/276 (Dev. Hydrobiol. 94): 81– 96.

Herzig, A. & B. Auer, 1990. The feeding behaviour of *Leptodora kindti* and its impact on the zooplankton community of Neusiedler See (Austria). Hydrobiologia 198 (Dev. Hydrobiol. 60): 107–117.

Herzig, A. & H. Winkler, 1983. Beiträge zur Biologie des Sichlings — *Pelecus cultratus* (L.). Öster. Fischerei 36: 113–128.

Herzig, A., E. Mikschi, B. Auer, A. Hain, A. Wais & G. Wolfram, 1993. Untersuchungen zur Populationsökologie und Ernährungsbiologie der wirtschaftlich genutzen und der wichtigsten wirtschaftlich nicht genutzen Fische des Neusiedler Sees. Abschlußbericht, BM, f. Wissenschaft & Forschung, BM Umwelt, Jugend & Familie, Burgenländische Landesregierung, 125 pp.

Hillbricht-Illkowska, A. & A. Karabin, 1970. An attempt to estimate consumption, respiration and production of *Leptodora kindti* (Focke) in field and laboratory experiments. Pol. Arch. Hydrobiol. 17: 81–86.

Hovenkamp, W., 1990. Instar-specific mortalities of coexisting *Daphnia* species in relation to food and invertebrate predation. J. Plankton Res. 12: 483–495.

Karabin, A., 1974. Studies on the predatory role of the cladoceran *Leptodora kindti* (Focke) in secondary production of two lakes with different trophy. Ekol. pol. 22: 295–310.

Kerfoot, W. C., 1980. Commentary: transparency, body size and prey conspicuousness. In Kerfoot W. C. (ed.), Evolution and Ecology of Zooplankton Communities. The Univ. Press of New England, Hanover, New Hampshire: 609–617.

Kuzichkin, A. P., 1975. Effect of temperature on the seasonal population dynamics and fecundity of *Leptodora kindti* (Focke) (Cladocera) in the vicinity of point discharge of warm effluent of the Konakovo power plant. Hydrobiol. J. 11: 53– 56.

Langeland, A., 1982. Interactions between zooplankton and fish in a fertilized lake. Holarct. Ecol. 5: 273–310.

Lazzaro, X., 1987. A review of planktivorous fishes: Their evolution, feeding behaviours, selectivities, and impacts. Hydrobiologia 146: 97–167.

Leitner, C., 1990. Einfluß von Turbulenz und Trübe auf *Diaphanosoma brachyurum* (Lievin) (Cladocera, Crustacea) im Neusiedler See. Diplomarbeit, Univ. Wien, 53 pp.

Luecke, C., M. J. Vanni, J. J. Magnuson, J. F. Kitchell & P. T. Jacobson, 1990. Seasonal regulation of *Daphnia* populations by planktivorous fish: Implications for the spring clear-water phase. Limnol. Oceanogr. 35: 1718- 1733.

Lunte, C. C. & C. Luecke, 1990. Trophic interactions of *Leptodora* in Lake Mendota. Limnol. Oceanogr. 35: 1091–1100.

Mittelholzer, E., 1970. Populationsdynamik und Produktion des Zooplanktons im Greifen- und Vierwaldstättersee. Schweiz. Z. Hydrol. 32: 90–149.

Mordukhai-Boltovskaya, E. D., 1956. Beiträge zur Biologie von *Leptodora kindtii* Focke und *Bythotrephes* Leydig im Rybinsker Stausee. Dokl. Akad. Nauk. SSSR Biol. Sci. Sect. 110: 688–691.

Mordukhai-Boltovskaya, E. D., 1958. Preliminary notes on the feeding of the carnivorous cladocerans *Leptodora* and *Bythotrephes*. Dokl. Akad. Nauk. SSSR. Biol. Sci. Sect. 122: 828–830.

Murdoch, W. W. & A. Oaten, 1975. Predation and population stability. Adv. ecol. Res. 9: 1–131.

O'Brien, W. J., N. A. Slade & G. L. Vinyard, 1976. Apparent size as the determinant of prey selection by bluegill sunfish (*Lepomis macrochirus*). Ecology 57: 1304–1310.

Odermatt, D. M., 1970. Limnologische Charakterisierung des Lauerzersees mit besonderer Berücksichtigung des Planktons. Schweiz. Z. Hydrol. 32: 1–75.

Paloheimo, J. E., 1974. Calculation of instantaneous birth rate. Limnol. Oceanogr. 19: 692–694.

Persson, L., 1982. Rate of food evacuation in roach (*Rutilus rutilus*) in relation to temperature, and the application of evacuation rate estimates for studies on the rate of food consumption. Freshwat. Biol. 12: 203–210.

Sebestyén, O., 1960. On the food niche of *Leptodora kindtii* in the open water communities of Lake Balaton. Int. Revue ges. Hydrobiol. 45: 277–282.

Wright, J. C., 1965. The population dynamics and production of *Daphnia* in Canyon Ferry Reservoir. Limnol. Oceanogr. 10: 583–590.

Zaret, T. M. & W. C. Kerfoot, 1975. Fish predation on *Bosmina longirostris*: body size selection versus visibility selection. Ecology 56: 232–237.

Hydrobiologia **307**: 283–289, 1995.
P. Larsson & L.J. Weider (eds), Cladocera as Model Organisms in Biology.

Cyclomorphosis as a factor explaining success of a *Daphnia* hybrid in Tjeukemeer

Piet Spaak
Netherlands Institute of Ecology, Centre for Limnology, Rijksstraatweg 6, 3631 AC Nieuwersluis, The Netherlands
Present adress: Max-Planck-Institut für Limnologie, Abteilung Ökophysiologie, Postfach 165, D-24302 Plön, Germany (e-mail: spaak @ mpil-ploen.mpg.d400.de)

Key words: Cyclomorphosis, *daphnia*, interspecific hybridization, invertebrate predation, life histories

Abstract

Recent studies have shown that *Daphnia* hybrids are intermediate to their parental species for several life history traits, and that they combine advantageous traits of the parental species, thereby decreasing their risk of being preyed upon by planktivorous fish. In this study *D. galeata*, *D. cucullata* and their interspecific hybrid are compared with respect to cyclomorphosis and juvenile growth as a potential mechanism to avoid invertebrate predation. Helmets and tail spines are known to reduce predation risk, whereas fast juvenile growth reduces the period that *Daphnia* are vulnerable to predation. In laboratory cultures, I tested the hypothesis that, with regard to cyclomorphosis, juvenile hybrids are more comparable to the juveniles of *D. cucullata*, and with respect to growth rate, they are more comparable to *D. galeata*. The hybrid did not differ significantly from *D. galeata* in the time needed to reach 0.9 mm body length. In contrast, the relative spine and helmet length of juvenile hybrids were more comparable with those of *D. cucullata*, whereas in adult instars these characters were more similar to *D. galeata*. Thus during the first juvenile instars, the hybrids combine the relative long helmet and spine of *D. cucullata*, with the faster growth of *D. galeata*. This combination of characteristics might reduce hybrid vulnerability to invertebrate predation in field situations.

Introduction

Cyclomorphosis in *Daphnia* has received much attention during the last decades (for review see Jacobs, 1987). Both proximate (environmental) causes and ultimate (fitness) aspects of cyclomorphosis have been well-studied. Water temperature and turbulence contribute to the formation of helmets or head spines (Hrbáček, 1959; Jacobs, 1961). Recently, it has been found that also kairomones, released by potential predators are an important proximate cause for the production of helmets (Larsson & Dodson, 1993). Experimental work has demonstrated that helmets can be an effective defense mechanism against invertebrate predators (Sorenson & Sterner, 1992). In this paper, I will concentrate on the ultimate aspects of cyclomorphosis in relation to the coexistence of two *Daphnia* species and their hybrid.

For many *Daphnia* populations interspecific hybrids and hybrid taxa have been reported (Hebert, 1985; Wolf & Mort, 1986; Lieder, 1987; Hebert *et al.*, 1989, 1993; Taylor & Hebert, 1992). Especially in the *D. longispina* group, all possible hybrids between the three species (e.g. *D. cucullata*, *D. galeata* and *D. hyalina*) have been found (Wolf & Mort, 1986; Gießler, 1987), and at times, they coexist in the same lake (Wolf, 1987; Müller & Seitz, 1993). It is still not clear whether these hybrids are regularly formed by the parental species, although recent work suggests that certain *Daphnia* hybrids may be recent products of interspecific crosses (Spaak, 1994). It is not clear which characters of parental species and hybrids are important in maintaining their coexistence.

Daphnia species adapt differently to escape from predation. Four basic defensive strategies as summarized by Zaret (1975), include spatial/temporal refugia, reduction in body size, reduced visibility, and behavioural mechanisms (e.g. vertical migration). Recent investigations have shown that a combination of specific parental life history traits (size at maturity; intrinsic rate of increase, *r*) can enhance the relative fitness of hybrids during specific periods of the year, and can lead to temporal hybrid superiority (Spaak & Hoekstra, 1995). Because of the high abundance of *Daphnia* hybrids compared with their parental species in many lakes (Schwenk & Spaak, unpublished data), a similar or even more effective strategy used by the hybrid to circumvent invertebrate predation is expected.

Invertebrate predators, *Leptodora kindtii* and *Chaoborus* sp, are known to selectively eat daphnids $<$ 0.9 mm in length (Allan, 1973; Cooper, 1983; Herzig & Auer, 1990; Brandstrator & Lehman, 1991). It has been shown that daphnids with a high helmet or a long tail spine have a reduced risk of being preyed upon by *Chaoborus*, copepods and *Leptodora* (Mort, 1986, 1989; Pijanowska, 1990).

In this study I test: 1) if the *D. cucullata* × *galeata* hybrid is intermediate to the parental species for spine and helmet size, and juvenile growth, as has been noted for body size and clutch size (Weider, 1993; Spaak & Hoekstra, 1995); and 2) whether the hybrids combine specific advantageous traits of both parental species (i.e. the high helmet of *D. cucullata* and the higher growth rate of *D. galeata*). Such a strategy may help explain the success of *Daphnia* hybrids compared with their parental species.

I conducted a life table experiment under controlled laboratory conditions with different clones of *D. galeata*, *D. cucullata* and their hybrid, obtained from the same lake to study the effects of temperature and food level on juvenile growth, helmet length, and tail spine length in the different taxa. In particular I wanted to investigate how traits, like the relative size of the spine and helmet, change for the hybrid compared with the parental species during successive instars.

Materials and methods

The clones studied were isolated from the Tjeukemeer, a shallow eutrophic lake in the northern part of the Netherlands. In this lake, copepods prey upon *Daphnia* during spring and *Leptodora kindtii* preys upon *Daphnia* during summer and autumn (J. Vijverberg, pers. com.). Other invertebrate predators such as *Chaoborus*, *Neomysis* and water mites (*Unionicola aculeata*), occur usually in low densities. *Daphnia* clones were collected on 24 July 1990 using a tow-net with a mesh size of 335 μm. From a composite sample from five sites in the lake, females with parthenogenetic eggs were picked randomly and reared individually in the laboratory. They were assayed at four allozyme loci: phosphoglucomutase (*Pgm*, EC 5.4.2.2), phosphoglucose isomerase (*Pgi*, EC 5.3.1.9), aldehyde oxidase (*Ao*, EC 1.2.3.1) and glutamate oxaloacetate transferase (*Got*, EC 2.6.1.1). All electrophoresis was carried out on Titan III® cellulose acetate plates following standard methods (Hebert & Beaton, 1989). I obtained 309 clones with known genotype. Relative frequencies of the multilocus genotypes for *Got-Pgi-Pgm* in the lake in 1989 (Spaak & Hoekstra, unpublished data) were used to select clones of the most abundant genotypes for use in the life history experiment. *Got* was used to identify the taxa (Wolf & Mort, 1986). For *D. cucullata*, fourteen clones were selected and fifteen each for *D. galeata* and the hybrid. Clones were not replicated; the total design consisted of 264 individual cultures.

The results presented in this paper were obtained in the same experiment as described in Spaak & Hoekstra (1995), although from this data set only the 1990 data were used. This was done to circumvent an unbalanced design while no *D. cucullata* were present in the 1991 part of the experiment.

The experiment was conducted at three temperatures (12.5 °C, 17.5 °C and 22.5 °C ± 0.2) and two food levels (0.62 and 1.64 μg C ml^{-1}) consisting of a 1:1 mixture of *Chlamydomonas globosa* and *Scenedesmus obliquus* from continuous cultures. We used artificial *Daphnia* medium (Lynch *et al.*, 1986), to obviate effects of possible differences in water quality. The medium was modified by adding to it one-eighth volume of freshly-collected water from Lake Loosdrecht, which was filtered through a 15 μm mesh sieve. Buffering capacity of the medium was increased by adding 50 mg L^{-1} $Ca(HCO_3)_2$. This prevented a high egg abortion rate as observed by Spaak & Hoekstra (1995). For experiments at 22.5 °C, the medium was refreshed daily, at 17.5 °C on alternate days and for 12.5 °C every three days. The daphnids were cultured individually in 100-ml test tubes with 90 ml medium. The tubes were placed randomly in water baths at a 16:10h light/dark photoperiod, which were illuminated from the top with fluorescent light tubes (4 W/m^2). For further details on culturing see Spaak & Hoekstra (1995).

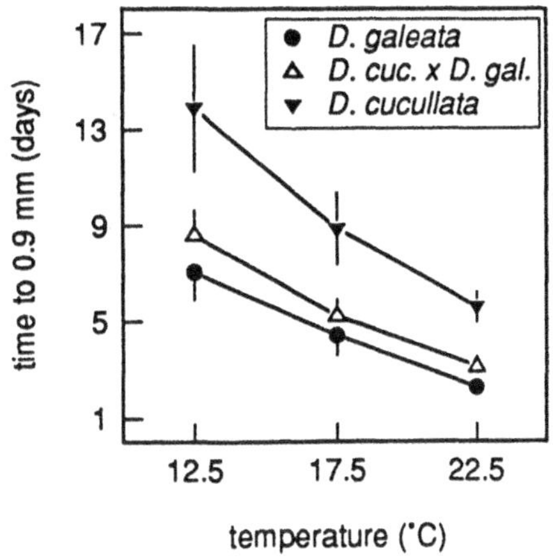

Fig. 1. Plots of means for the number of days necessary to reach the size of 0.9 mm for *Daphnia galeata*, *D. cucullata* and *D. cucullata* × *galeata*. DAY_09MM is plotted against three temperatures (12.5, 17.5 and 22.5 °C). Error bars indicate 95% confidence intervals.

For cultures, female neonates less than 12 h of age (17.5 and 22.5 °C) or less than 24 h old (12.5 °C) were used; four to six neonates were placed in each tube. These neonates were daughters of clones that were raised for at least one generation under the specific food level/temperature conditions. Since the age of the mothers differed between and within cultures, neonates were not uniform in length, and no special selections were made. While refreshing the medium, one animal was picked at random for measurements. In order to reduce the effect of handling damage, first juvenile instars were not returned to the tubes. As soon as eggs were observed in the brood pouch, the number of animals per tube was reduced to one. Lengths were measured under a microscope to the nearest 0.02 mm. Body length from the top of the compound eye to the base of the tail spine, helmet length from the top of the compound eye to the top of the helmet and tail spine length from the base of the the tail spine to end of the spine were measured. This was done from the first juvenile instar to the fourth adult instar. Egg stages (modified after Gulbrandsen & Johnsen, 1990) were recorded to enable a more accurate estimate of the age at maturity. All neonates were removed from the culture. From the 264 clones, 24 died before maturation and another six died before reaching the fourth adult instar.

The parameters used in the analysis are: body length, (LENGTH), relative tail spine length, SPINE, (length of tail spine/body length), relative helmet length, HELM (helmet length/body length) and the number of days necessary to reach a size of 0.9 mm.

Table 1. Results of three-way ANOVAs for the number of days necessary to reach the size of 0.9 mm. (DAY_09MM), log transformed

Source	DF	MS	F	P
Temperature	2	18.516	144.05	0.0000
Food level	1	0.359	2.79	0.0960
Taxon	2	10.471	81.47	0.0000
T × F	2	0.035	0.28	0.7596
T × Ta	4	0.126	0.98	0.4197
F × Ta	2	0.065	0.51	0.6021
T × F × Ta	4	0.043	0.33	0.8545
Error	222	0.129		

length, DAY-09MM. Tests for normality and correlations between mean and variance (Sokal & Rohlf, 1981) showed that transformation was required for DAY-09MM (logarithmic). The data on tail spine and helmet lengths and body size were analyzed with a four-way analysis of variance with repeated measurements (CSS, 1992). This was done because the seven measurements of spine and helmet length are not independent variables. In this analysis, these measurements were treated as a single dependent variable and instar became the fourth main effect, in addition to food level, temperature and taxon. Type III SS were used in all ANOVAs because of the unbalanced experimental design (unequal number of clones per taxa) (Shaw & Mitchell-Olds, 1993). If one to three data points were missing in a series of length measurements, cell means were used to replace the missing values. This was done in 87 cases (5.3%) for spine length and 82 cases (5.0%) for body length. The first seven instars were analysed. DAY_09MM was analyzed using a three-way analysis of variance. Significant differences between groups of means were tested with a Tukey multiple comparisons test with unequal sample sizes. Planned comparisons (contrast analysis) were used to test for differences between groups within an interaction term (Sokal & Rohlf, 1981).

Results

The time needed to reach a body length of 0.9 mm differed strongly between the taxa and temperatures, but no significant interactions were found (Table 1). *D. galeata* and the hybrid did not differ significantly at the three temperatures tested (Fig. 1a), except at

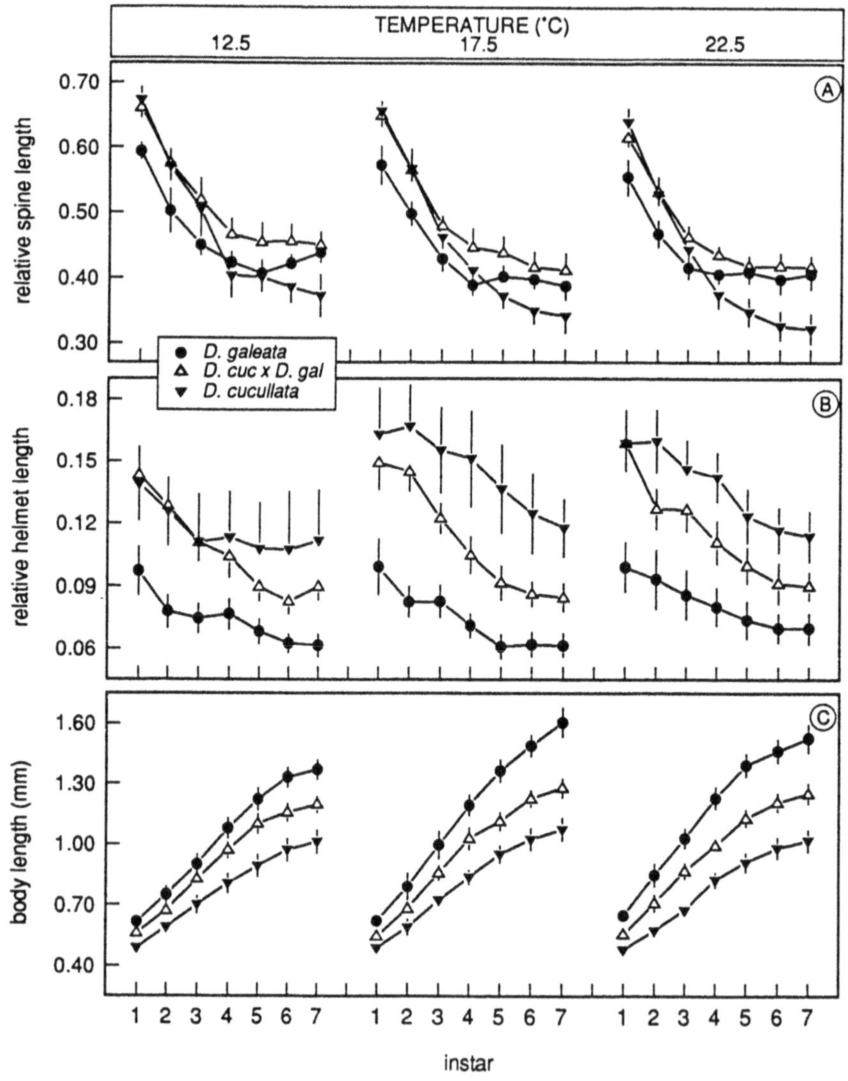

Fig. 2. Interaction plots of means and 95% confidence intervals for relative tail spine lengths (A), helmet lengths (B) and body size (C) for *Daphnia galeata*, *D. cucullata* and *D. cucullata* × *galeata* against instar number and temperature.

22.5 °C (HSD = 0.03, $p < 0.05$). *D. cucullata* needed approximately 50% more time to reach a body length of 0.9 mm, and was significantly different from the two other taxa in all cases. Body length similarity between *D. galeata* and the hybrid was temperature dependent (Fig. 2c). The higher the temperature, the fewer number of instars in which both taxa did not differ significantly in body size. This is confirmed by a significant temperature × taxon and temperature × instar interaction (Table 2).

The taxa differed significantly in their relative spine and helmet lengths. During the first juvenile instars, helmet and spine lengths of *D. cucullata* and the hybrid were relatively longer than in *D. galeata*. On the other hand, the relative spine length of *D. cucullata* during the last instars was shorter than in both other taxa. A four-way ANOVA revealed significant taxon and instar differences for all traits. There were significant temperature effects for HELM and SPINE, but no food level effects (Table 2). Besides the significant taxon and temperature effects, the analysis of relative helmet length revealed different relations between both helmet and spine length and instar, for the taxa reflected by a significant taxon × instar interaction. The interaction plots (Figs 2a, b) showed that the relative spine length of the hybrid was not significantly different from *D.*

Table 2. Results of four-way ANOVAs with repeated measurements for the characters: relative spine length (SPINE), relative helmet length (HELM) and body length (LENGTH)

Character:			HELM			SPINE			LENGTH		
Source	DF	DF Err	MS	F	MS Err	MS	F	MS Err	MS	F	MS Err
Temp	2	218	0.025	5.72**	0.004	0.196	25.08***	0.008	0.137	2.63	0.052
Food level	1	218	0.013	2.99	0.004	0.024	3.02	0.008	0.151	2.91	0.052
Taxon	2	218	0.428	99.24***	0.004	0.331	42.32***	0.008	9.531	183.56***	0.052
Instar	6	1308	0.070	141.05***	0.000	1.762	806.40***	0.002	10.078	2616.85***	0.004
T × F	2	218	0.003	0.72	0.004	0.008	1.09	0.008	0.047	0.90	0.052
T × Ta	4	218	0.010	2.25	0.004	0.007	0.96	0.008	0.254	4.90***	0.052
F × Ta	2	218	0.006	1.33	0.004	0.003	0.33	0.008	0.031	0.60	0.052
T × I	12	1308	0.001	2.39**	0.000	0.005	2.12*	0.002	0.009	2.38**	0.004
F × I	6	1308	0.000	0.65	0.000	0.002	0.94	0.002	0.004	1.12	0.004
Ta × I	12	1308	0.004	7.19***	0.000	0.065	29.94***	0.002	0.177	46.05***	0.004
T × F × Ta	4	218	0.009	2.07	0.004	0.006	0.76	0.008	0.038	0.73	0.052
T × F × I	12	1308	0.001	1.05	0.000	0.002	0.85	0.002	0.009	2.34**	0.004
T × Ta × I	24	1308	0.001	1.40	0.000	0.002	0.76	0.002	0.005	1.42	0.004
F × Ta × I	12	1308	0.001	1.12	0.000	0.004	1.80*	0.002	0.002	0.45	0.004
T × F × Ta ×I	24	1308	0.000	0.59	0.000	0.003	1.18	0.002	0.002	0.56	0.004

* $p < 0.05$, ** $p < 0.01$, *** $p < 0.001$.

cucullata for the first three instars, but it did differ significantly during the last two instars, but did not differ from *D. galeata*. This pattern was observed for all temperatures (Fig. 2a) although the slope of the individual plots differ. For example *D. cucullata*, had significantly shorter spine lengths for adult instars at 17.5 °C and 22.5 °C than at 12.5 °C. HELM for the first instars showed a pattern comparable with SPINE (Fig. 2b). During the first instar, *D. cucullata* and the hybrid did not differ significantly. During the seventh instar, the relative helmet length of the hybrid was intermediate to those of *D. galeata* and *D. cucullata*. The differences between the first and seventh instar were the largest for the hybrid in all cases.

Discussion

The *Daphnia* hybrid in this study differed significantly from both parental species for all investigated traits (DAY_09MM, HELM, SPINE, LENGTH). The patterns of the hybrid traits were not intermediate, as observed in earlier studies (Weider & Wolf, 1991; Weider, 1993; Spaak & Hoekstra, 1995), but in almost all situations, hybrids were more similar to one or the other parental species. Interestingly, the similarity of a given hybrid trait to either *D. galeata* or *D. cucullata* depends not only on the specific trait but is also age-dependent. The relative spine length of juvenile hybrids does not differ significantly from that of *D. cucullata*; on the other hand, adult hybrids have a relative spine length which does not differ significantly from that of *D. galeata*. The pattern observed for the hybrid with respect to SPINE was different to HELM. During the first juvenile instar, HELM did not differ significantly between *D. cucullata* and the hybrid at all temperatures, but at 17.5 °C and 22.5 °C during adult instars, HELM showed intermediate values between both parental species as was observed for body length (Fig. 2c).

Whereas the morphology of the hybrid juveniles was similar to *D. cucullata* juveniles, juvenile growth was more similar to *D. galeata* at all temperatures. These results are similar to earlier observations (Weider & Wolf, 1991; Spaak & Hoekstra, 1995) which showed that the intrinsic rate of increase (r) for *D. galeata* and *D. cucullata* × *galeata* were quite similar. Spaak & Hoekstra (1995) proposed that the ecological relevance of this observation was that the relatively high r combined with the smaller adult body size (1.1 mm, 1.3 mm and 0.9 mm for the hybrid, *D. galeata* and *D. cucullata*, respectively) of the hybrids gives them a selective advantage in lakes with strongly positive, size-selective predation by fish. It is remarkable that, although the hybrid and *D. galeata* differ in their size at maturity (1.1 mm and 1.3 mm respectively) (Weider & Wolf, 1991; Spaak & Hoekstra, 1995), the time needed to reach the size of 0.9 mm does not differ at two of the three temperatures (Fig. 1a). Only at 12.5 °C did *D. galeata* and the hybrid differ significantly.

Significant food level effects were not found throughout this study. For this reason, both food levels were pooled together in Figs 1 & 2. Although no food level effects on size-related traits were found, food level effects were detected for reproductive traits (Spaak & Hoekstra, 1995), which suggests that at least the low food level was below the incipient limiting level for the maintenance of reproduction.

With respect to invertebrate predation risk, first juvenile instars of hybrids blend features of both parental species such that their predation risk may be decreased, i.e. they incorporate the relative longer helmet and longer tail spine of *D. cucullata* with the higher growth rate of *D. galeata*. It is not clear if the strategy of the hybrid (i.e. investing in somatic growth and in morphological changes), requires more energy compared to the parental species. Studies on the feeding rate of hybrids compared with the parental species may shed more light on this subject.

Size-selective invertebrate predation is well known (Pastorok, 1981; Cooper, 1983; Krylov, 1992). For *Leptodora kindtii*, which prefers prey sizes < 0.9 mm (Mordukhai-Boltovskaya, 1958; Herzig & Auer, 1990), feeding rates of six *D. retrocurva* ind. d^{-1} (Havel, 1985) to 14 *D. pulex* ind. d^{-1} (Browman *et al.*, 1989) have been reported. This illustrates how important it is for a juvenile daphnid to grow as quickly as possible. Of course the absolute size of 0.9 mm is arbitrary. In this study, it is used to compare the juvenile growth of the three taxa; a size of 0.8 mm or 1 mm would have given comparable results.

Other authors have shown that *Daphnia* with longer tail spines and higher helmets have a reduced risk of being preyed upon (Havel & Dodson, 1984; Hebert & Grewe, 1985; Pijanowska, 1990). The induction of these longer spines and helmets is, at least partially, stimulated by a kairomone released by invertebrate predators (Krueger & Dodson, 1981; Havel, 1985). The one-eighth volume of lake water, which was added to the *Daphnia* medium to circumvent egg abortion, might have contained such a kairomone. Since the densities of possible invertebrate predators in Lake Loosdrecht, from which this water was taken, are not known, it is not clear how strong my results are influenced by chemical factors released by predators.

Daphnia helmets become smaller in culture (Jacobs, 1961; Tollrian, 1990; Hanazato & Ooi, 1992). This is a possible explanation why the 'summer' HELM values that I found in the laboratory for *D. cucullata* are comparable with HELM values found by Pijanowska (1990) outside the summer season under field conditions. Her field data for relative helmet and tail spine lengths for different size classes are in accordance with my data.

Whether the differences between the hybrid and its parental species, which I found in this study, effect the relative fitness of these taxa in the lake remains unclear. *D. cucullata* × *galeata* hybrids are found at high densities during summer in Tjeukemeer (Spaak, 1994). Since numerous factors (food availability, fish predation, turbulence/light, invertebrate predation) play a role in the population dynamics of these taxa, I can only state that the traits I studied might result in a higher fitness of the hybrid during periods of invertebrate predation. Whether this indeed is the case, will depend strongly on all the other factors that play a role, but were not taken into account in this study. In conclusion, this study demonstrates that *D. cucullata* × *galeata* shows a combination of traits in laboratory experiments which, if also present in the field, would most likely decrease their invertebrate predation risk compared to the parental species.

Acknowledgements

Rob Hoekstra did a large part of the practical work and Steven Visser collected the animals. I thank Jos van Damme, Ramesh Gulati, Rob Hoekstra, Jürgen Jacobs, Luc de Meester, Joop Ringelberg, Klaus Schwenk, Onno van Tongeren, Koos Vijverberg, Larry Weider and an anonymous reviewer for comments on earlier drafts and stimulating discussions.

References

Allan, J. D., 1973. Competition and the relative abundances of two cladocerans. Ecology 54: 484–498.

Brandstrator, D. K. & J. T. Lehman, 1991. Invertebrate predation in lake Michigan – Regulation of *Bosmina longirostris* by *Leptodora kindtii*. Limnol. Oceanogr. 36: 483–495.

Browman, H. I., S. Kruse & W. J. O'Brien, 1989. Foraging behavior of the predaceous cladoceran, *Leptodora kindtii*, and escape responses of their prey. J. Plankton Res. 11: 1075–1088.

Cooper, S. D., 1983. Selective predation on cladocerans by common pond insects. Can. J. Zool. 61: 879–886.

CSS: STATISTICA. Release 3.1., 1992. Statsoft, Inc. Tulsa Oklahoma, USA.

Gießler, S., 1987. Microevolution und Populationsgenetik im *Daphnia galeata/hyalina/cucullata*-Komplex (Crustacea: Cladocera) eine Freilandsanalyse. Ph.D thesis, University of Munich, Germany.

Gulbrandsen, J. & G. H. Johnsen, 1990. Temperature-dependent development of parthenogenetic embryos in *Daphnia pulex* de Geer. J. Plankton Res. 12: 443–453.

Hanazato, T. & T. Ooi, 1992. Morphological responses of *Daphnia ambigua* to different concentrations of a chemical extract from *Chaoborus flavicans*. Freshwat. Biol. 27: 379–385.

Havel, J. E., 1985. Predation of common invertebrate predators on long- and short-featured *Daphnia retrocurva*. Hydrobiologia 124: 141–149.

Havel, J. E. & S. I. Dodson, 1984. *Chaoborus* predation on typical and spined morphs of *Daphnia pulex*: Behavioral observations. Limnol. Oceanogr. 29: 487–494.

Hebert, P. D. N., 1985. Interspecific hybridization between cyclic parthenogens. Evolution 39: 216–220.

Hebert, P. D. N. & P. M. Grewe, 1985. *Chaoborus*-induced shifts in the morphology of *Daphnia ambigua*. Limnol. Oceanogr. 30: 1291–1297.

Hebert, P. D. N. & M. J. Beaton, 1989. Methodologies for allozyme analysis using cellulose acetate electrophoresis. Helena laboratories Beaumont, Texas.

Hebert, P. D. N., M. J. Beaton, S. S. Schwartz & D. J. Stanton, 1989. Polyphyletic origins of asexuality in *Daphnia pulex*. 1. Breeding-system variation and levels of clonal diversity. Evolution 43: 1004–1015.

Hebert, P. D. N., S. S. Schwartz, R. D. Ward & T. L. Finston, 1993. Macrogeographic patterns of breeding system diversity in the *Daphnia pulex* group. 1. breeding systems of Canadian populations. Heredity 70: 148–161.

Herzig, A. & B. Auer, 1990. The feeding behaviour of *Leptodora kindtii* and its impact on the zooplankton community of Neusiedler See (Austria). Hydrobiologia 198 (Dev. Hydrobiol. 60): 107–117.

Hrbáček, J., 1959. Circulation of water as a main factor influencing the development of helmets in *Daphnia cucullata* Sars. Hydrobiologia 13: 170–185.

Jacobs, J., 1961. Cyclomorphosis in *Daphnia galeata mendotae* Birge, a case of environmentally controlled allometry. Arch. Hydrobiol. 58: 7–71.

Jacobs, J., 1987. Cyclomorphosis in *Daphnia*. Mem. Ist. ital: Idrobiol. 45: 325–352.

Krueger, D. A. & S. I. Dodson, 1981. Embryological induction and predation ecology in *Daphnia pulex*. Limnol. Oceanogr. 26: 219–223.

Krylov, P. I., 1992. Density-dependent predation of *Chaoborus flavicans* on *Daphnia longispina* in a small lake – the effect of prey size. Hydrobiologia 239: 131–140.

Larsson, P. & S. I. Dodson, 1993. Invited review – chemical communication in planktonic animals. Arch. Hydrobiol. 129: 129–155.

Lieder, U., 1987. The possible origin of *Daphnia cucullata procurva* Poppe 1887 in the lakes of the Pomerian Lakeland by hybridization in the past. Hydrobiologia 145 (Dev. Hydrobiol. 35): 195–200.

Lynch, M., L. J. Weider & W. Lampert, 1986. Measurement of the carbon balance in *Daphnia*. Limnol. Oceanogr. 31: 17–33.

Mordukhai-Boltovskaia, E. D., 1958. Preliminary notes on the feeding of the carnivorous cladocerans *Leptodora kindtii*. Dokl. Biol. Sci. Sect. 122: 828–830.

Mort, M. A., 1986. *Chaoborus* predation and the function of phenotypic variation in *Daphnia*. Hydrobiologia 133: 39–44.

Mort, M. A., 1989. Cyclomorphosis in *Daphnia galeata mendotae*: Variation and stability in phenotypic cycles. Hydrobiologia 171: 159–170.

Müller, J. & A. Seitz, 1993. Habitat partitioning and differential vertical migration of some *Daphnia* genotypes in a lake. Arch. Hydrobiol. Beih. Ergeb. Limnol. 39: 167–174.

Pastorok, R. A., 1981. Prey vulnerability and size selection by *Chaoborus* larvae. Ecology 62: 1311–1324.

Pijanowska, J., 1990. Cyclomorphosis in *Daphnia*: an adaption to avoid invertebrate predation. Hydrobiologia 198 (Dev. Hydrobiol. 60): 41–50.

Shaw, R.G. & T. Mitchell-Olds, 1993. ANOVA for unbalanced data – an overview. Ecology 74: 1638–1645.

Sokal, R. R. & F. J. Rohlf, 1981. Biometry. W.H. Freeman and Co San Francisco, USA.

Sorensen, K. H. & R. W. Sterner, 1992. Extreme cyclomorphosis in *Daphnia lumholtzi*. Freshwat. Biol. 28: 257–262.

Spaak, P., 1994. Genetical ecology of a coexisting *Daphnia* hybrid species complex. Ph.D thesis, University of Utrecht, The Netherlands.

Spaak, P. & J. R. Hoekstra, 1995. Life history variation and the coexistence of a *Daphnia* hybrid with its parental species. Ecology (in press).

Taylor, D. J. & P. D. N. Hebert, 1992. *Daphnia galeata mendotae* as a cryptic species complex with interspecific hybrids. Limnol. Oceanogr. 37: 658–665.

Tollrian, R., 1990. Predator-induced helmet formation in *Daphnia cucullata* (Sars). Arch. Hydrobiol. 119: 191–196.

Weider, L. J., 1993. Niche breadth and life history variation in a hybrid *Daphnia* complex. Ecology 74: 935–943.

Weider, L. J. & H. G. Wolf, 1991. Life history variation in a hybrid species complex of *Daphnia*. Oecologia 87: 506–513.

Wolf, H. G., 1987. Interspecific hybridization between *Daphnia hyalina, D galeata* and *D. cucullata* and seasonal abundance of these species and their hybrids. Hydrobiologia 145 (Dev. Hydrobiol. 35): 213–217.

Wolf, H. G. & M. A. Mort, 1986. Interspecific hybridization underlies phenotypic variability in *Daphnia* populations. Oecologia 68: 507–511.

Zaret, T. M., 1975. Strategies for existence of zooplankton prey in homogeneous environments. Verh. int. Ver. Limnol. 19: 1484–1489.

Hydrobiologia **307**: 291–297, 1995.
P. Larsson & L.J. Weider (eds), Cladocera as Model Organisms in Biology.

Morphological variation and population dynamics of *Bosmina (E.) longicornis kessleri* Uljanin *(Crustacea: Anomopoda: Bosminidae)* in the Bukovec Reservoir (S.E. Slovakia)

Igor Hudec
Institute of Zoology and Ekosozology, Slov. Acad. Sci., Löfflerova 10, SK-040 01 Košice, Slovak Republic

Key words: Bosmina, Eubosmina, morphological variation, population dynamics, Carpathian region, reservoirs

Abstract

Bosmina (Eubosmina) longicornis kessleri and *B. (Bosmina) longirostris* are minor components among the crustacean zooplankton in the pelagial of Bukovec Reservoir. *B. (E.) l. kessleri* is a recent colonist in the Carpathian region and was first found in the Bukovec reservoir in 1986. The present study gives some information about the niche occupied by *B. (E.) l. kessleri* in the zooplankton community; the seasonal abundance of zooplankton and phenotypic variability (cyclomorphosis) of *B. (E.) l. kessleri* are also examined.

The most recent review and the last revision of the genus *Bosmina* was published by Lieder (1983a,b) which included morphological, phylogenetic and biogeographical distinctions. In this revision, *B. (E.) kessleri* Uljanin was recombined as a well-defined subspecies and placed in the *B. (E.) mixta* Lilljeborg, 1901 group. Later, Lieder (1991) recombined the *B. (E.) mixta* group in *B. (E.) longicornis* Schödler, 1865. An earlier revision of the same genus, based on lateral head pores, was presented by Kořínek (1971).

Bosmina (Eubosmina) longicornis kessleri Uljanin, 1874 is found in the circumbaltic region, especially in the east, but also in the lakes along the northern fringe of the Alps (Lieder, 1983a,b). Its detection in the Bukovec Reservoir is the first report from the Carpathian region.

In the Bukovec Reservoir, it coexists with *Bosmina (Bosmina) longirostris*, but both species prefer different water strata.

The *B. (E.) longicornis kessleri* population is evidently quite young, because Bukovec Reservoir was not finished until 1975. The reservoir serves as an additional source of drinking water for Košice. It was constructed on the upper part of the Ida Brook at an elevation of 416.7 m above sea level. The reservoir has a maximum depth of 46.2 m with a surface area of 1.02 km^2 and a length of 3.5 km. Most of the drainage area consists of mixed deciduous forest without any natural lakes or artificial ponds.

Methods

Quantitative zooplankton samples from different depths were collected monthly from December 1986 to December 1987 using a 2-l Friedinger bottle. A total of 20 l of water was taken during each sampling bout.

Water samples were filtered through a No. 20 plankton net, organisms were fixed in 4% formaldehyde, and counted in a Sedgwick-Rafter cell.

Representative plankton samples were taken monthly during 1987–1990 using a No. 20 plankton net with vertical hauls taken from 15 m in the pelagial.

The following parameters were analysed: body length, body height, antennules and mucrone length follow Black (1980).

Some characteristics of the pelagial in the Bukovec Reservoir with respect to eutrophication are given (Table 1)

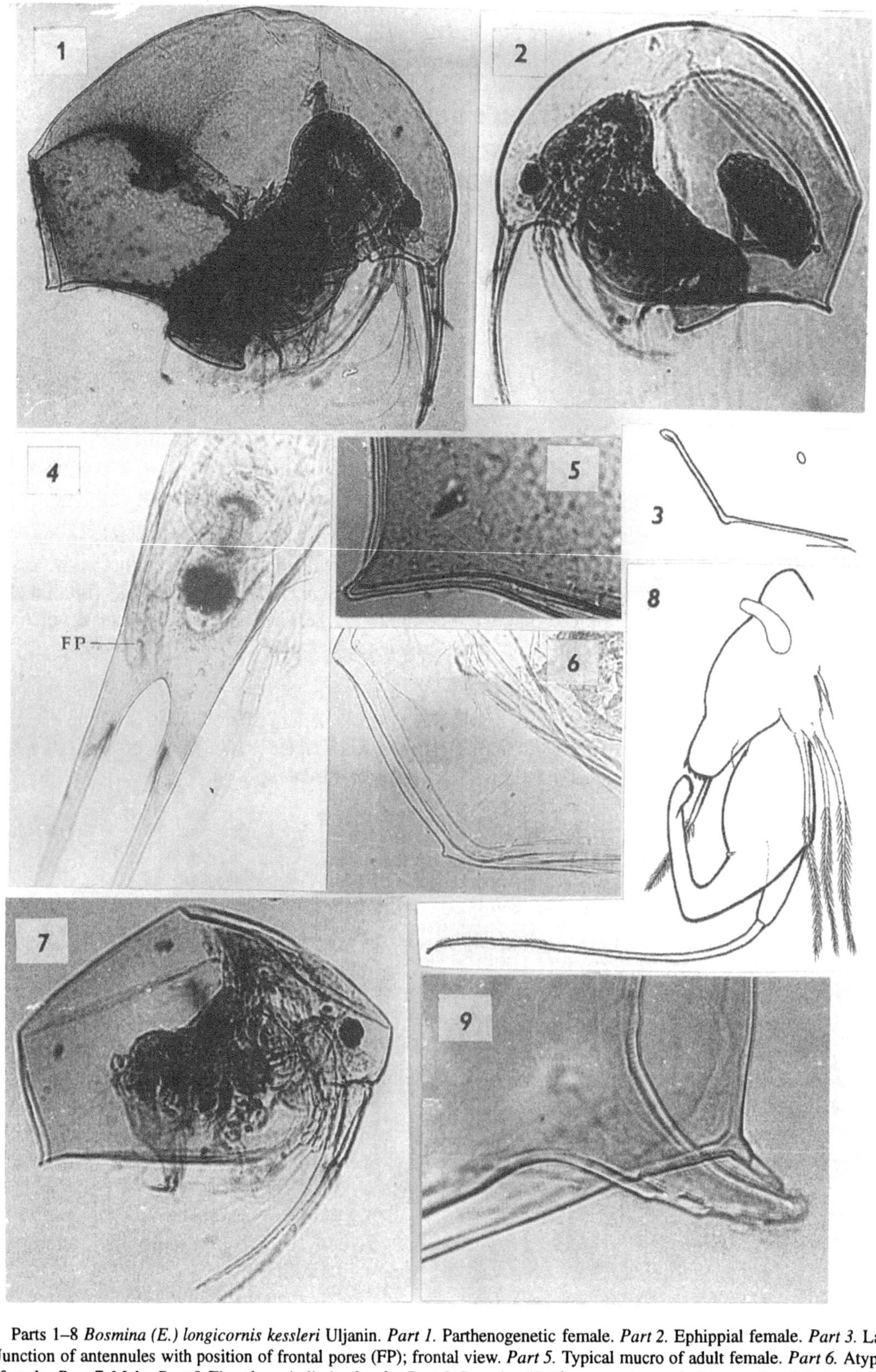

Plate 1. Parts 1–8 *Bosmina (E.) longicornis kessleri* Uljanin. *Part 1.* Parthenogenetic female. *Part 2.* Ephippial female. *Part 3.* Lateral pore. *Part 4.* Junction of antennules with position of frontal pores (FP); frontal view. *Part 5.* Typical mucro of adult female. *Part 6.* Atypical mucro in adult female. *Part 7.* Male. *Part 8.* First thoracic limb of male. *Part 9. Bosmina (B.) longirostris*: atypical mucro in adult female.

Table 1. Some characteristics of the Bukovec Reservoir from 1987 to 1990 (surface samples)

Year	Transparency (cm)		Chlorophyll-a (μg l^{-1})		P. tot. (mg l^{-1})		N : P	
	A[1]	B[2]	A[1]	B[2]	A[1]	B[2]	A[1]	B[2]
1987	470 ± 120	460	2.4 ± 1.3	2.5	0.04	0.05	43	35
1988	400 ± 100	370	4.3 ± 2.4	5.1	0.11	0.03	49	66
1989	440 ± 140	380	3.6 ± 2.5	3.8	0.17	0.22	13	7.5
1990	570 ± 110	550	2.9 ± 1.2	3.2	0.05	0.06	29	24

[1] Annual average ($\bar{X} \pm s_x$).
[2] Production period (April–September) ($\bar{X}$).

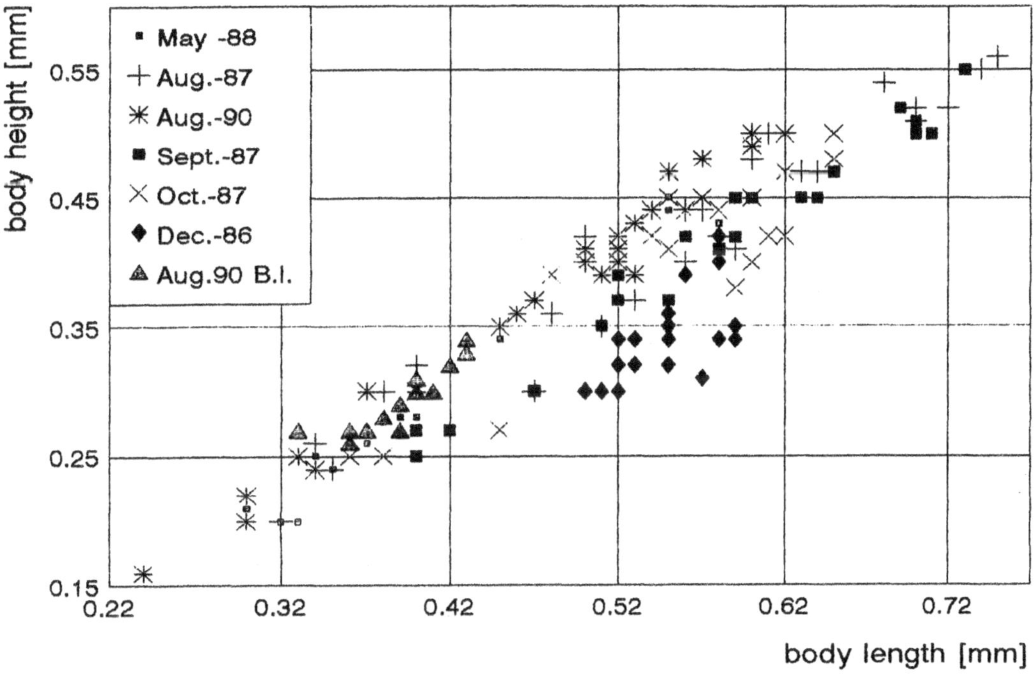

Fig. 1. Morphological measurements for *Bosmina (E.) longicornis kessleri* Uljanin during different parts of the year (for comparison *Bosmina (B.) longirostris* (B.l.) – relationship between body length and body height.

Results and discussion

Bosmina (E.) longicornis kessleri Uljanin, 1874

The general shape of females and males are given (Plate 1).

Body length to body height (Fig. 1). In spring the population was dominated by juveniles of different stages. Adults were found only sporadically.

The greatest morphological variability was observed in the summer (Table 2). During July and August the largest parthenogenetic females were detected with: body lengths up to 0.74 mm and body heights up to 0.56 mm (Fig. 1). Most animals carried 2–3 eggs, rarely four.

Sexual femals were 0.62 ± 0.02 mm in length, but parthenogenetic females were up to 0.74 mm long.

Animals from December (Table 2, Fig. 1) were slightly smaller (not longer than 0.60 mm). Adult parthenogenetic females usually carried either no eggs or a single egg.

Lateral pores (Plate 1, Part 3): Their position corresponds with *Eubosmina* type (Kořínek, 1971).

Antennules: they appeared disproportionately longer in juveniles, but in absolute measurements they

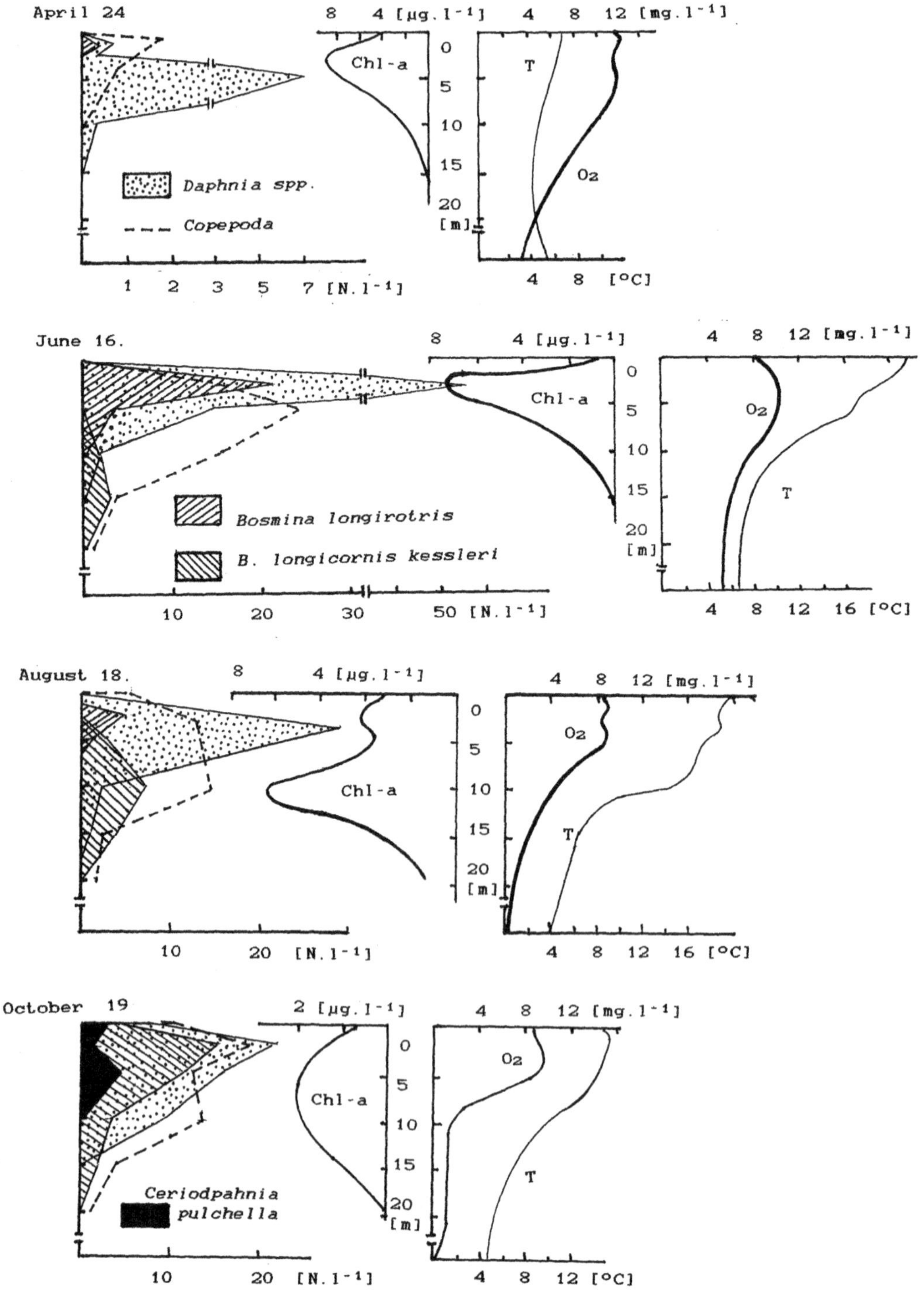

Fig. 2. Vertical distributions of Cladocera and *Copepoda* in Bukovec Reservoir during 1987. Chlorophyll-*a* (chl-*a*), temperature (T) and oxygen (O_2) vertical distributions are given.

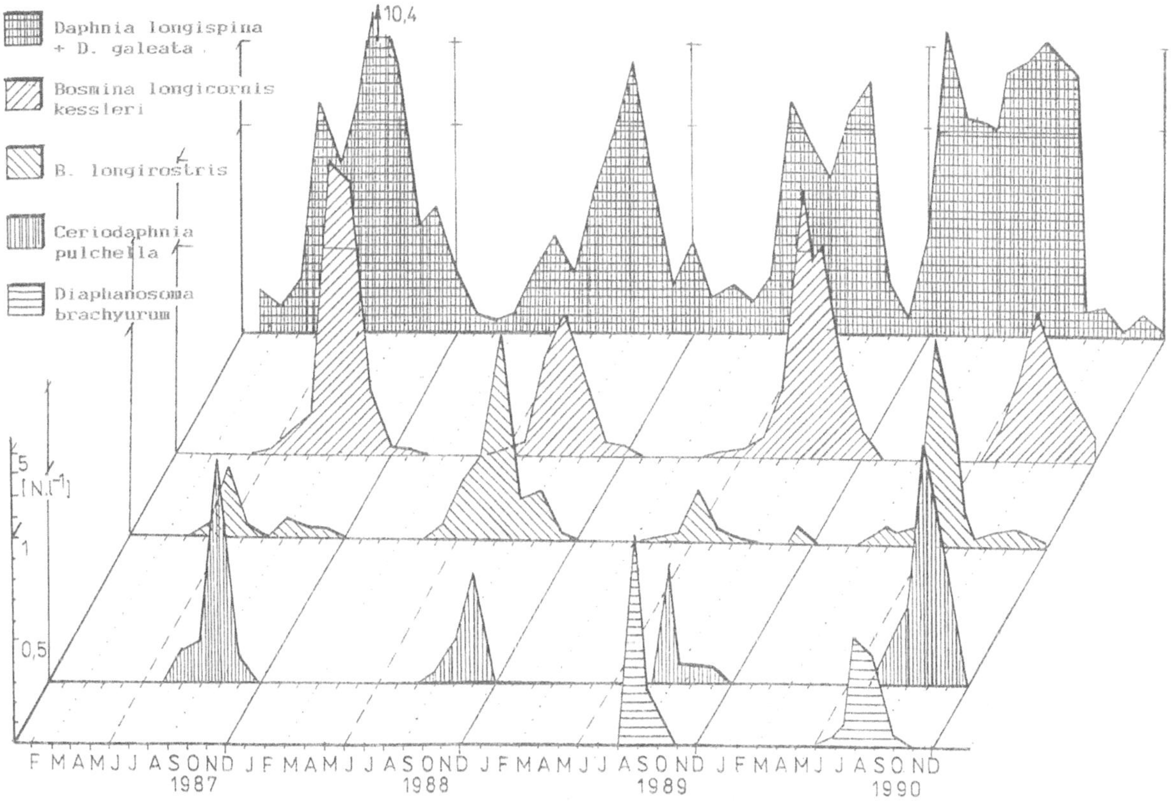

Fig. 3. Seasonal distribution and abundance of Cladocera in the Bukovec Reservoir during 1987–1990.

were nearly identical in length for all stages (Table 2). The antennules usually have 14 segments (90% of all populations), but not less than 12. They are mostly 0.21 mm long (from tip to base of the aesthetascs) and 0.04–0.05 mm between their base and the base of the aesthetascs. During July 1988 and August 1990, *B. (E.) l. kessleri* was preyed upon by *Eudiatomus gracilis* SARS[1] (as inferred from a reduction or absence of antennules). *E. gracilis* was most abundant in the same water strata as *B. (E.) l. kessleri*. We did not find *B. longirostris* with reduced or absent antennules (summer individuals were not longer than 0.43 mm, not higher than 0.34 mm and antennules were not longer than 0.15 mm). Were smaller individuals completely preyed upon by this copepod?

Similarly *B. longirostris* with body lengths greater than 0.45 mm were preyed upon by *E. gracilis* in Starina Reservoir (N. E. Slovakia).

[1] The following four copepods were mostly found in the epipelagial: *Cyclops strenuus* Fischer, 1851, *C. vicinus* Uljanin, 1875, *Thermocyclops crassus* (Fischer, 1853) and *Mesocyclops leuckarti* (Claus, 1857).

Antennules of *B. (E.) l. kessleri* (Plate 1, Part 4) are parallel in length from a dorsal view, while *B. (B.) longirostris* exhibit a 'V' formation of its antennules.

Mucro: (Plate 1, Part 5) it has a characteristic triangular corner, a blunt pointed end and a slightly protruded line along the ventral margin. The mucro is usually around 0.05 mm in length (Table 2). Its ventral margin in neonates, first stadium and also in males, follows the line of the ventral margin of the valves (the length of the mucro was measured from the tip of the submarginal seta).

In the summer, some individuals with slightly longer mucrones were detected. However, I did not find such elongated shapes (from German populations) as noted by Lieder (1983b). I found females with reduced mucrones which resembled *B. (E.) coregoni coregoni* (August 1990). Mucrones were different on each valve and on the other hand different mucrones were also found in late autumnal populations of *B. (B.) longisrostris*. Could this be a result of hybridisation between *B. (E.) l. kessleri* and *B. (B.) longirostris* (Plate 1, Parts 6, 9)?

Table 2. Morphological characters measured for *Bosmina (E.) longicornis kessleri* Uljanin from plankton samples (all measurements in mm)

Data collected			N	Carapace Length	Carapace Height	Antennules	Mucro
May	16,	1988	16	0.40 ± 0.10	0.30 ± 0.09	0.21 ± 0.02	0.048 ± 0.00
June	15,	1988	23	0.47 ± 0.10	0.36 ± 0.08	0.19 ± 0.05	0.048 ± 0.00
July	29,	1987	42	0.49 ± 0.08	0.39 ± 0.08	0.20 ± 0.007	0.049 ± 0.00
July	18,	1988	19	0.50 ± 0.09	0.40 ± 0.06	0.20 ± 0.04	0.050 ± 0.00
Aug.	20,	1987	29	0.57 ± 0.12	0.42 ± 0.10	0.20 ± 0.01	0.051 ± 0.00
Aug.	8,	1988	23	0.55 ± 0.10	0.41 ± 0.09	0.20 ± 0.05	0.049 ± 0.00
Aug.	9,	1990	18	0.45 ± 0.12	0.35 ± 0.10	0.20 ± 0.02[1]	0.045 ± 0.01
Sept.	9,	1987	23	0.57 ± 0.10	0.41 ± 0.08	0.21 ± 0.01	0.050 ± 0.00
Sept.	29,	1988	16	0.55 ± 0.08	0.40 ± 0.06	0.20 ± 0.03	0.050 ± 0.00
Oct.	19,	1987	22	0.56 ± 0.08	0.42 ± 0.07	0.20 ± 0.01	0.051 ± 0.00
Oct.	24,	1988	16	0.54 ± 0.06	0.39 ± 0.05	0.20 ± 0.02	0.050 ± 0.00
Oct.	8,	1989	23	0.53 ± 0.10	0.39 ± 0.08	0.20 ± 0.02	0.047 ± 0.00
Nov.	9,	1987	16	0.56 ± 0.04	0.42 ± 0.04	0.20 ± 0.01	0.051 ± 0.00
Nov.	23,	1988	15	0.54 ± 0.04	0.39 ± 0.03	0.20 ± 0.02	0.050 ± 0.00
Dec.	15,	1986	20	0.55 ± 0.03	0.34 ± 0.03	0.185± 0.005	0.049 ± 0.00
Dec.	7,	1987	14	0.56 ± 0.05	0.35 ± 0.04	0.190± 0.005	0.050 ± 0.00
				Bosmina longirotris			
Aug.	9,	1990	15	0.39 ± 0.03	0.29 ± 0.02	0.13 ± 0.01	0.045 ± 0.00

[1] N = 14.

Species composition and seasonal abundance

The species composition of all pelagial net zooplankton was poor. All Cladocera that were detected are given in Fig. 3. *Diaphanosoma brachyurum* was introduced (possibly by waterflow) in late summer 1988.

The reservoir is also on the annual migratory route of cranes *Grus grus*. This is a possible mechanism for the introduction *B. (E.) l. kessleri* into the reservoir.

From the species composition and apparent introduction of new species not only of zooplankton but of phytoplankton as well (*Ceratium hirudinella*: first record in 1988), it is evident that ecological niches of plankton in the Bukovec reservoir are vacant and also possibly in other water supply reservoirs of Slovakia (personal observation). It seems that the natural introduction of zooplankton into new artificial reservoirs is not as common as one might predict.

B. (E.) l. kessleri reached population maxima in August–September (Figs 2, 3), while *B. (B.) longirostris* exhibited population maxima in May–June following the spring phytoplankton maximum. During typical summer stagnation, *B. longirostris* nearly disappeared from the pelagial, while in autumn and winter the population reappeared.

B. (B.) longirostris had its maximal density in the upper part of the epilimnion (Fig. 2), while *B. (E.) l. kessleri* was nearly absent for most of the year. *B. l. kessleri* exhibited its maximal density in the lower part of epilimnion in the water temperature strata ot 8–14°C (which extended through the thermocline), while *B. (B.) longirostris* appeared less frequently or was absent. Around the end of September to mid-October, sexual females and males of *B. (E.) l. kessleri* were found near the surface (12–14°C water temperature).

A similar vertical habitat partitioning was also observed between *B. (B.) longirostris* and *B. (E.) coregoni* in Bohemia (Straškraba & Hrbáček, 1966), but not in Lake Constance (Müller, 1985).

Flössner (1972) characterized *B. longirostris* as a eurythermic species that occurs between 0.5 to around 25°C, with thermal optimum between 8–18°C and which prefers nutrient-rich smaller waterbodies (e.g. ponds). Bukovec is a mesotrophic reservoir that is oligotrophic after the spring maximum phytoplankton bloom. It seems that the oligo-mesotrophic conditions in this reservoir are the main reason why *B. longirostris* nearly disappears from the summer plankton, since it is common and abundant in other eutrophic Slovakian reservoirs (personal observations).

Based on the geographical distributions of these two bosminid species, I deduce that the temperature optimum of *B. (E.) longicornis kessleri* is slightly lower than for *B. longirostris*.

The question remains: can these two bosminid species inter-breed in the Bukovec Reservoir? Morphological analyses (Plate 1, Parts 6, 9) suggests that if a hybridization occurs, it is quite rare.

Summary

Bosmina (E.) longicornis kessleri Uljanin is evidently a recent colonist (post-1976) in the Bukovec Reservoir. It was first observed in 1986.

This species prefers deeper cooler water strate in the reservoir in contrast to *Bosmina (B.) longirostris* which is most abundant in the upper portion of the epilimnion. Only sexual females of *B. (E.) l. kessleri* were found near the surface in early autumn.

The body length/height ratio is almost constant with the exception of some values from December 1986.

Antennal length and length of the mucrones of adults was nearly constant throughout the year. Larger individuals were generally found in summer. Individuals of *B. (E.) l. kessleri* were partly preyed upon by *Eudiaptomus gracilis* (based on absent or damaged antennules) during the summers of 1989 and 1990.

Acknowledgements

I thank Dr L. J. Weider and two anonymous reviewers for critically reading and making helpful comments on the manuscript.

References

Black, R. W., 1980. The genetic component of cyclomorphosis in Bosmina. In W. C. Kerfoor (ed.), Evolution and ecology of zooplankton communities. Univ. Press. New England, Hannover, New Hampshire: 456–469.

Flössner, D., 1972. Krebstiere, *Crustacea* Kiemen- und Blattfüsser, *Branchiopoda* Fischläuse, *Branchiura*. Tierwelt Deutschlans 60. Fischer Verl., Jena, 501 pp.

Kořínek, V., 1971. Comparative study of head pores in the genus *Bosmina* Baird *(Crustacea, Cladocera)*. Věst. Československ. společ. zool. 35: 275–296.

Lieder, U., 1983a. Revision of the genus Bosmina Baird, 1845 *(Crustacea, Cladocera)*. Int. Revue ges. Hydrobiol. 68: 121–139.

Lieder, U., 1983b. Die Arten der Untergattung Eubosmina Seligo, 1900 (Crustacea, Cladocera, Bosminidae). Mitt. zool. Mus., Berl., 59: 195–292.

Lieder, U., 1991: The *Bosmina kessleri*-like morphotype of *Eubosmina* in Lake Muskoka, Ontario, Canada, as putative interspecific hybrids. Hydrobiologia 225 (Dev. Hydrobiol. 71): 71–80.

Müller, H., 1985. The niches of *Bosmina coregoni* and *Bosmina longirostris* in the ecosystem of Lake Constance. Verh. int. Ver. Limnol. 22: 3137–3143.

Straškraba, M. & J. Hrbáček, 1966: Net-plantok cycle in Slapy Reservoir during 1958–1960. In J. Hrbáček (ed.), Hydrobiological studies 1, Academia, Prague: 113–153.

Index

Page numbers are the first page of a paper in which the entry is discussed.

GPSR Compliance

The European Union's (EU) General Product Safety Regulation (GPSR) is a set of rules that requires consumer products to be safe and our obligations to ensure this.

If you have any concerns about our products, you can contact us on ProductSafety@springernature.com

In case Publisher is established outside the EU, the EU authorized representative is:

Springer Nature Customer Service Center GmbH
Europaplatz 3
69115 Heidelberg, Germany

Batch number: 10762579

Printed by Printforce, the Netherlands